The Physics ToolBox

A Survival Guide for Introductory Physics

Kirsten A. Hubbard
Debora M. Katz
United States Naval Academy

With a Forward by
Sheila Tobias

Australia • Canada • Mexico • Singapore • Spain
United Kingdom • United States

SURVIVE YOUR INTRODUCTORY PHYSICS CLASS!

If you must take an introductory physics class, then you *need* this book. It is full of important information designed to help you:

- Ace exams
- Understand lecture
- Sharpen your math skills
- Work homework problems
- Study more efficiently
- And get help when you need it!

The Physics ToolBox is a "paperback mentor," or guide, specifically created to supplement (not replace) other materials in your introductory physics course. It provides all of the critical material needed to succeed in a typical physics class—including material overlooked elsewhere—such as:

- An introduction to the nature of physics *and* science, so that you have a better idea of *why* you need to take physics;
- A look at "what to expect" and "how to succeed" in physics classes, including a step-by-step guide to solving problems;
- A 38 page "Study Success Guide" (in the Appendix) that will help you improve your study skills, both in and out of your physics class;
- A verbal overview of all the concepts you will learn in your course, written in a way that unites ideas and draws a "big picture;"
- An *extensive* review of all of the math you will *need* (and some you won't!) to solve the many, many problems encountered in your class;
- And, much, much more!

The Physics ToolBox is written so that you can skip around the text, reading only those sections that *you* need to succeed...*when* you need them!

Sheila Tobias, a nationally-recognized researcher and author on science education issues,writes in her Forward to *The Physics ToolBox*:

> "I welcome *The Physics ToolBox* [because it offers] multiple points of contact, missing links, problem-solving walk-throughs, highways and biways around and into the various topics of introductory physics."

Printed in the United States of America

1 2 3 4 5 6 7 05 04 03 02 01

ISBN 0-03-034652-5

For more information about our products, contact us at:
Thomson Learning Academic Resource Center
1-800-423-0563

For permission to use material from this text, contact us by:
Phone: 1-800-730-2214
Fax: 1-800-731-2215
Web: www.thomsonrights.com

Asia
Thomson Learning
60 Albert Complex, #15-01
Alpert Complex
Singapore 189969

Australia
Nelson Thomson Learning
102 Dodds Street
South Street
South Melbourne, Victoria 3205
Australia

Canada
Nelson Thomson Learning
1120 Birchmount Road
Toronto, Ontario M1K 5G4
Canada

Europe/Middle East/South Africa
Thomson Learning
Berkshire House
168-173 High Holborn
London WC1 V7AA
United Kingdom

Latin America
Thomson Learning
Seneca, 53
Colonia Polanco
11560 Mexico D.F.
Mexico

Spain
Paraninfo Thomson Learning
Calle/Magallanes, 25
28015 Madrid, Spain

CONTENTS AT A GLANCE

PART III: THE ESSENTIAL MATH

TABLE OF CONTENTS

Some sections (and portions of sections) contain advanced material and can most likely be skipped by non-physics majors. A distinction between regular and advanced material will be made clear in the text. See the Note to the Student for more information.

PART II: PHYSICS BLUEPRINT

ACKNOWLEDGMENTS

At the risk of forgetting some important people, the authors of *The Physics ToolBox* would like to thank certain individuals who have made this book possible.

First, we gratefully thank Sheila Tobias for submitting a Foreword to this book, and for her thorough and enlightening review of our manuscript. Her suggestions much improved the book.

Second, we are very thankful for the extraordinary reviews by John Jewett, which vastly enhanced the readability, accuracy, and tone of the work.

Third, we want to thank all of the people at the United States Naval Academy who also helped review the manuscript, including John Ertel, Larry Tankersly, and Alan Whiting.

Fourth, we would like to extend a special thank you to Eleanor Hubbard for her careful proofreading of the text.

Fifth, we must thank the people at Saunders College Publishing who made this book possible, including John Vondeling, Ed Dodd, Angus McDonald, and Jay Campbell.

And finally, we are deeply indebted Kirsten's husband, Brian Lewis, and Debora's friend, Adam Stone, for reading drafts, supplying feedback, and helping in many ways large and small over the last few years. Without the support they have provided, this book simply would not have been possible. Thank you.

Science must be understood as…a gutsy, human enterprise, not the work of robots programmed to collect pure information.
STEPHEN JAY GOULD
Science Writer, 1941 -

NOTE TO THE INSTRUCTOR

Over thirty percent of all students enrolled in college-level physics courses today—some three hundred thousand students every year—will not major in physics or engineering; rather, they are taking physics classes to fulfill a non-physics major requirement.

Yet, despite the excess of supplementary materials bundled with scientific textbooks today, there are no texts, ancillary guides, or helpbooks targeted to the special needs of this large and unique population. We, the authors of *The Physics ToolBox,* wanted to fill this gap.

For guidance, we looked to the research of progressive physics educators* such as Sheila Tobias, a nationally-recognized science educator and author. In the books *They're Not Dumb, They're Different* and *Breaking the Science Barrier* (written with physicist Carl T. Tomizuka), Tobias and Tomizuka examine, and try to provide new approaches to, a non-major student population they call the "second tier".

Second tier students are not second rate: they are intellectually curious, just as smart as physics majors, and perform well in other subjects. More importantly, they want to succeed in their physics class. However, they are not (yet) drawn to science or, by extension, scientific pedagogy.

This puts the second tier at a disadvantage: their unfamiliarity with typical modes of science instruction challenges the study skills and personal attitudes that succeeded for them in other courses. For example, according to Tobias' experimental population, second tier students find:

Science has to be made accessible to long-distance runners—slower to start, needing more rest-and-recovery time—and not just to sprinters.
DUDLEY HERSCHBACH
Nobel Laureate in Chemistry, 1932 -

* See the bibliography for a complete list of sources.

- scientific classroom atmospheres undesirably competitive and lacking in community;
- presentation of scientific material—including lectures, reading, and exam structure—too much focused on technique, and too little focused on concepts and context; and
- that their math skills are critical to success, but not well grasped.

Rightly or wrongly, the second tier's discomfort with traditional science education makes it harder for them to succeed in college-level physics classes.

Fortunately, this situation is ameliorated when science instruction takes some of the learning needs of the second tier into account. For instance, according to Tobias, second tier students:

- are verbal, not mathematical, learners, so they need wordy, concept-driven explanations;
- need to be presented with a "big picture," otherwise, they cannot construct understanding from details; and
- need to feel welcomed in the classroom, and be able to interact with their peers, in order to succeed.

Although physics instructors cannot hope—or even wish—to meet all of the learning needs of the second tier (after all, to name one example, we cannot eliminate mathematics from physics instruction), we can offer them tools to succeed in physics via a pedagogical format that speaks to their strengths.

Here is where *The Physics ToolBox* hopes to fill a void. This book endeavors to:

- provide a more verbal and "big picture" approach to introductory physics, to supplement (*not* replace) existing readings and lecture formats;
- show students how to create a learning community to help them help themselves when they feel overwhelmed or unable to cope;
- help students identify and change their expectations about scientific learning; and
- aide students in developing an alternative scientific-mathematical mindset instead.

To meet these goals, we have divided the book into three parts:

1. A "success" section (Part I), in which potential pedagogical problems for the second tier are identified, and common sense solutions are offered (including the very important solution of self-change);
2. A physics "blueprint," or verbal overview of important introductory physics topics (Part II); and
3. A math review (Part III) meant to remediate typical second tier student's deficiencies in mathematics.

Part I is intended to be read prior to the beginning of class; Part II is designed to be read piecemeal, as each topic appears in class; and Part III is intended to be read if and when mathematical difficulties present themselves. In addition, this book includes a 38-page Appendix, offering generalized study skills for overall academic success at the college level.

It should be plain, then, that *The Physics ToolBox* is *not* (nor is it meant to be) a traditional physics text. For example, we have liberally engaged in "skipping to the punch line," deleting proofs and equations and even entire topics (most notably optics and circuit theory) in order to draw connections between important concepts. Furthermore, information that does make it into the text (especially in the "blueprint" section) is heavy on explanations and light on equations.

To illustrate the different mindset of the second tier, consider this excerpt from Laura Fermi's memoir. Laura, the wife of Enrico, is a classic example of a second tier student. Notice her response to her husband's instruction, a classical physics (first tier) teaching style.

Enrico introduced me to the Maxwell equations. Patiently I learned the mathematical instruments needed to follow each passage...until I had digested my lesson and made it material of my own brain. Thus we arrived at the end of the long demonstration: the velocity of light and that of electromagnetic waves were expressed by the same number. "Therefore," Enrico said, "light is nothing but electromagnetic waves."

"How can you say so?"

"We have just demonstrated it."

"I don't think so. You proved only that through some mathematical abstractions you can obtain two equal numbers. But now you talk of equality of two things. You can't do that. Besides, two equal things need not be the same thing."

I would not be persuaded, and that was the end of my training in physics.

We have also reordered the traditional presentation of the physics canon somewhat, in the hope that students will better see a storyline there. And, perhaps most "disagreeably," we have written the entire math section as a series of tips—without context, background, or detail—so that students can quickly get the information they need to learn *physics*...without also needing to learn subtleties of math at the same time.

In short, we have omitted the kind of exposition that makes physics pleasing to professionals like you—such as reasoning from first principles, mathematical deduction, subtlety, and beauty—and replaced it with basic, "down and dirty" information that we think will help students get a good grasp of fundamental concepts. Although there are more elegant, and certainly more rigorous, ways of presenting the concepts covered in this book, we believe that certain students need to begin physics more simply.

Now, we understand—as physics instructors ourselves—that you may have deep-seated concerns about this approach. However, we would argue that some students cannot appreciate the deeper level of understanding that makes physics beautiful until they have constructed a skeletal framework of the *entire picture*. Our approach, therefore, does not completely eliminate mathematics, reasoning, or conceptual understanding—and it certainly does not replace the rigor of traditionally-taught classical physics—but it does attempt to reach the non-major student in other, less traditional ways.

In Chapter 2 we argue that a hallmark of "thinking like a physicist" is the ability to apply a limited arsenal of (appropriate) tools to a wide variety of problems. Majors students already have a well-stocked, though apprentice-level, physics toolbox containing the basic math skills, problem solving savvy, and scientific mindset needed for success in introductory college-level physics. Second tier students also have a well-stocked toolbox, but not one suitable for physics. This book strives to provide a Physics Toolbox to students that need one.

Although *The Physics ToolBox* was written with the needs of the second tier in mind, the final result, we believe, is a book that a *wide range* of students can enjoy and use. We have tested portions of it on all types of physics students—majors and non-majors alike—and have received very enthusiastic responses. After all, even majors need a math review now and then.

We therefore hope that you, and your students, will also find this book useful.

KIRSTEN HUBBARD
DEBORA KATZ, Ph.D.
June, 2001

FOREWORD
by SHEILA TOBIAS

When I advanced the notion of the "second tier" after engaging a small number of intellectually able but determinedly "nonscientific" learners in introductory physics and chemistry courses in the fall of 1989, I never intended the group that I had identified to be thought of as likely to respond to any one set of new or modified pedagogies or techniques.* Rather, I believed then as I do now that only *they*—and not the master teachers and textbook writers aiming to meet their needs—would be able to determine what "works" and what doesn't "work"; and that, therefore, it would be best to provide them with a *range* of materials for them to choose among.

Thus, I welcome *The Physics ToolBox*, not because it is necessarily the perfect "fit," but because, contrary to many of my reform colleagues' beliefs, there is no perfect fit. What second tier students of science need, in my experience, is redundancy, variety, alternative explanations, words and more words, diagrams, multiple approaches, and lots of back-and-fill; and *The Physics ToolBox* offers the newcomer to physics all of these and more.

Used in conjunction with a standard textbook, as an *optional addition* to lecture/labs/review sessions and even group work, this volume will provide students *who need it* with multiple points of contact, missing

---

* Tobias, Sheila. *They're Not Dumb, They're Different* (1990); and Sheila Tobias & Carl T. Tomizuka, *Breaking the Science Barrier* (1992).

links, problem-solving walk-throughs, highways and (as important) byways around and into the various topics of introductory physics.

Just as there is no "royal road" to physics—to play on a familiar anecdote told about Euclid—there may be no single pedestrian road as well. Instructors owe it to their students to let them find the footing they seek as they construct their own personal understanding of what is so "obvious" to others. Eventually, they should come to appreciate the rigor and parsimony of the classical approach. But only if they stay and succeed. And staying and succeeding is precisely what *The Physics ToolBox* hopes its student readers will be encouraged to do.

SHEILA TOBIAS
TUCSON, ARIZONA
May, 1999

NOTE TO THE STUDENT: WHY A TOOLBOX? HOW DO I USE IT?

Hello! And welcome to *The Physics ToolBox!*

This book is designed to help you *get the most* out of your introductory physics course. How so?

When you enrolled in your physics class—whether by choice or by school requirement—you broke ground on a new "physics home." Learning college-level physics is a lot like building a custom house: you start with a vision of your dream home, then you draw up a blueprint, and finally, you build it using tools and some elbow grease.

The Physics ToolBox helps you complete construction on your own "physics home." *Part I: Physics Is Fun* provides a view of the house to come; *Part II: Physics Blueprint* develops a blueprint or "big picture" overview to guide its assembly; and, *Part III: The Essential Math*, as well as the *Appendix: Studying for Success*, offers tools to help you build it.

Building Your Physics House

Now, like construction in real life, building a "physics home" from scratch can be intimidating (after all, "everyone" knows that physics is "hard"). However, this book is designed to help you *get the most* out of your introductory physics course. It provides you with study skills advice, math help, tips on dealing with your instructor and your classmates, and, best of all, a fresh look at the information that makes up a typical introductory physics course—including an entire section devoted to helping you develop your own "big picture" understanding of the material. So, if you want to succeed in your physics class, you *need* this book.

For example, some of the great stuff in *The Physics ToolBox* includes:

- Advice on how to succeed in a typical introductory physics class (Part I), including what "physics" is and how it is taught today;
- A recasting of physics material (Part II) so that you can make connections *between* topics—connections that you might not otherwise be able to make in the barrage of details typically encountered in an introductory physics course;
- Six chapters (Part III) of critical math review and tips (*Don't worry*, though! You don't have to *read* all of these chapters: just refer to them as math topics arise in your class.); and
- 38 pages of study skills information that will help you improve your study habits—both in and out of your physics class.

Furthermore, *The Physics ToolBox* is written so that *the very act of reading it is practice* for your introductory physics class. For example, the chapters and sections of this book are numbered like typical physics texts; the book's tone is midway between scientific exposition and light reading (probably, right where you are); and, just like your regular physics textbook, each chapter of *The Physics ToolBox* includes many examples (further inquiries into text material), exercises (worked-out problems), and other boxes (supplemental information) that need to be dealt with *as you read along* in order to fully understand the text.

This book contains so much information, in fact, that *The Physics ToolBox* features several directional devices to help you find the material that you need, when you need it. For example, the book:

Finger Pointer! ☞

- has hundreds of "Finger Pointers" (such as the one in the margin to the left of this paragraph) that direct you to related or referenced material. As you read through the book, you will soon get used to looking in the margin for directions to additional information;
- includes a top-notch index to direct you to just the right page for what you need. Just like the index in your regular textbook, our index is great for "skipping around," so that you don't always have to read the book straight through.

- alerts you to advanced-level material via cross-hatching in the margin (as shown to the left). If you are not a physics major, you can usually skip this material!

Other great features of *The Physics ToolBox* include big, empty margins for you to write in (please do!), as well as a collection of inspiring, physics-related quotes from scientists and non-scientists alike.

However, remember that how much you get out of *The Physics ToolBox* completely depends upon how much effort you put in: houses are not built by toolboxes alone. *You* need to work the exercises, read the examples and other boxes, try out the study skills, meet with your instructor, and form a physics team (discussed in Chapter 4), in order for this book to help you.

So, as you begin your physics class, we wish you all of the best in building your physics dream home. We sincerely wish that, with the help of this *ToolBox*, you will come to love physics as much as we do. At the very least, though, we hope that you will obtain a better understanding of the beautiful physics house that already exists all around you.

And now, *The Physics ToolBox!*

PART I: PHYSICS IS FUN

Welcome to physics!

No matter what you may have heard before, physics is actually tons of fun to learn and explore. In your introductory physics course, you will study nothing less than the *way the world works*, finally getting answers to such timeless questions as, Why is the sky blue? Why does the ocean have waves? Why does the earth orbit the sun (and not the other way around)?

All science is either Physics or stamp collecting.
ERNEST RUTHERFORD
American Physicist, 1871 - 1937

The problem with physics *classes*, however, is that often the interesting stuff gets lost in a strange, new learning environment. Everything from your instructor to your classmates; from your homework to your exams; from your laboratories to *all of that math*, is unfamiliar and (perhaps) intimidating.

The day I went into physics class it was death.
SYLVIA PLATH
American Poet, 1932 - 1963

Luckily, there is a solution to this frustrating situation: *getting the rules to the game.* If you knew what to expect from your physics course before it began, you could start off on the right foot and increase your chances for success. Therefore, the purpose of Part I of *The Physics ToolBox* is to introduce you to typical characteristics of college-level physics instruction—to help you *learn how to learn physics*—so that you are fully prepared for the new physics learning style.

We start (in *Chapter 1: What Is Physics?*) by examining why physics is taught to non-physicists in the first place. After that, we take a closer look at the nature of science (*Chapter 2: What Is Science?*), so that you can begin to see why physics is taught the way it is. Next, we present some challenges specific to the physics classroom (*Chapter 3: What To Expect*), including grading, homework, and exams. Finally, we give time-tested solutions to these challenges (*Chapter 4: How To Succeed*), including a very important tactic called the physics team.

You should strive to absorb and use as much of the information contained within these (short) four chapters as possible. We therefore recommend that you read this part, in its entirety, *before* your class begins (or, if your class has already started, as soon as possible). If you take the extra time to prepare for your physics class early—before you get swamped with classwork—you will *save* time over long run, time that would otherwise be wasted in inefficient and ineffectual study patterns.

We know that you may be tempted to skip this part. After all, this information seems unimportant compared to all of the *physics* material that you will need to learn for exams. However, we promise that this part will make your overall physics learning experience more enjoyable, efficient, and successful. That's why, after all, these chapters are first in the book!

1
WHAT IS PHYSICS?

1.1 THE STUDY OF EVERYTHING

Physics is the study of everything.

Want to know why the earth is round? Physics can tell you. Wondering why magnets stick to your refrigerator? Physics can tell you. Wish to know why your coffee gets cool and your cola gets warm when they sit in the open air? Physics can tell you.

The origin of all science is to know causes.
WILLIAM HAZLETT
English Author, 1778 - 1830

But physics is much, much more than a collection of cold facts. What makes physics interesting is that it is also a method for *discovering* new information. Want to know why your computer crashed? Physics can help you. Wondering whether the universe will ever come to an end? Physics can help you. Wish to know if we can ever make a Star Trek-like warp drive? Physics can help you.

Thus, physics is *both* an extensive body of knowledge *and* a creative process for producing new information. In order to succeed in your physics class, then, you must master both parts of the subject: you must learn fundamental physical concepts, *and* you must be able to apply these concepts to problems. For that reason, we might say that *you have not learned physics until you can do physics.*

Surprised? That's just the beginning. There's a lot more to physics than you may realize. In this chapter, we learn more about this fascinating subject itself—so that you can more easily understand the concepts (when they come).

1.2 THE FUNDAMENTAL SCIENCE

I want to know God's thoughts... the rest are details.
ALBERT EINSTEIN
American Physicist, 1879 - 1955

Although physicists study everything, their primary concern is not detail. Instead, physicists seek to describe *fundamental* phenomena—how things work in a broad sense—while leaving particulars and specifics to others. Physics is thus the **fundamental science**: the other great modern sciences branch out from the sturdy support of physics, just as the limbs of a tree grow away from its trunk and roots (see Fig. 1.1).

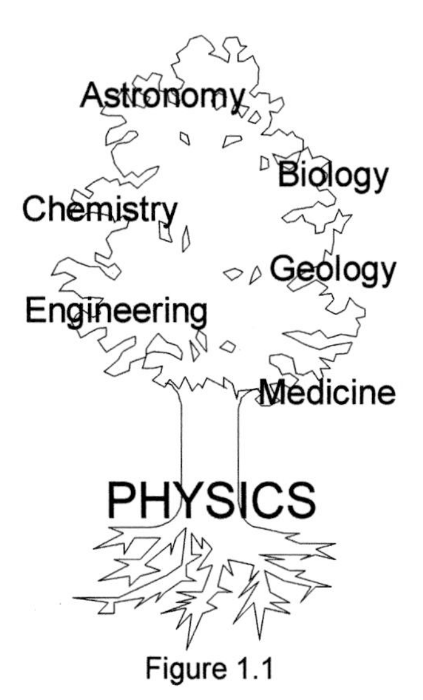

Figure 1.1

EXAMPLE 1.1: Physics: A Fundamental Science

The three greatest physicists of all time—Galileo, Newton, and Einstein—illustrate the physicists' focus on fundamentals.

Galileo, for example, concentrated on finding a clear, simple *description* of motion on earth. Newton extended Galileo's work by developing a theory that explained *why* these objects move. And Einstein furthered the effort by expanding Newton's theory to cover the motion of *all* objects, even those under extreme conditions.

While "motion" may not seem to be an exciting topic, it is in fact the foundation upon which all other knowledge about the universe rests. Thus, these three physicists can be credited with solving one of the most enduring questions about existence: *how does the universe work?* And they did it by studying the fundamentals.

The aim of science is always to reduce complexity to simplicity.
WILLIAM JAMES
American Philosopher, 1842 - 1910

Because it focuses on fundamentals, physics is a science of great simplicity and beauty. It is based on a few, *simple* laws that—given enough time and attention—*anyone* can learn to use. In this sense, physics is perhaps the most democratic of all of the sciences: by mastering a few basic principles, you too can attack the greatest problems of our time with the same authority as any Ph.D. physicist.

EXAMPLE 1.2: Albert Einstein

Albert Einstein serves as a good illustration of this aspect of physics.

Einstein began his adult life as a front desk clerk for a German patent office—far from the rarified heights of academia and professional physics. He did, however, have one thing going for him: he had a lot of time on his hands because the patent office traffic was slow!

Einstein spent his free time wisely. He studied the scientific and mathematical textbooks of his day, eventually mastering their content. In time, he became a professor of physics, a Nobel Prize winner and, arguably, the most famous scientist of all time.

Although Albert Einstein was an exceptional case, we can still draw lessons for our physics training from his example. Einstein teaches us that physics can be learned, *even if we are novices*, as long as we take the time and expend the energy needed for success.

I have no special gifts; I am only passionately curious.
ALBERT EINSTEIN
American Physicist, 1879 - 1955

Once the fundamental physical laws are learned, they are applied and reapplied to different situations. Physics, then, is not about knowing *everything*; rather, it is about knowing a *few* things that apply everywhere.

More laws are vain when less will serve.
ROBERT HOOKE
English Scientist, 1635 - 1703

EXAMPLE 1.3: The Challenger Explosion

When the space shuttle Challenger exploded in 1986, the race was on to discover the tragedy's cause. The problem was that the shuttle was and still is one of the most complicated machines on the planet. Where should scientists begin searching for clues to the disaster?

A great twentieth century physicist, Richard Feynman, went straight to the heart of the problem: What was the *main* difference between that fateful day and all other, successful launches? He quickly came to the right answer: It was cold! In fact, the day the Challenger exploded was the coldest launch day ever...so cold that icicles were photographed hanging from the shuttle casing.

Having isolated the *fundamental* problem—coldness—Feynman then thought about the *few* physical consequences of coldness, looking to see which applied to the situation at hand. For example, one thing that happens in the cold is that objects become hard and contract. Some materials, as they harden and contract, break under the stress. Feynman thought that perhaps one or more of the thousands of materials that made up the shuttle could have broken during the cold night before the launch.

Stress and strain are covered in Section 10.3.1.

In fact, this is exactly what happened. A thin rubber seal designed to keep fuel away from the ignition mechanism had frozen, contracted, and cracked during the night. As a result, fuel leaked where it shouldn't have, and an explosion resulted.

In a famous photograph, Feynman can be seen dropping a sample rubber seal into a dry ice bath. The seal cracked, as Feynman knew it would, and the case—which seemed so complex and difficult on its face—was solved. And all because a physicist (1) isolated the *fundamental* problem, and then (2) applied a *few, simple* physical laws towards its solution.

Even the few laws that govern physics can be linked by a single thread: **matter**. Just about every thing that exists in the universe—from each tiny atom to the largest galaxy—is comprised of matter.* Thus, physics is also *the study of the existence, interaction, and change of matter.*

NOTE 1.4: What is Matter?

In a nutshell, matter can be thought of as the "stuff" making up all materials, such as atoms and molecules, electrons and protons, etc.

The question "What is matter?" is of the kind that is asked by metaphysicians, and answered in books of incredible obscurity.
BERTRAND RUSSELL
English Mathematician, 1872 - 1970

* Except for light. Light does not consist of matter, and it can move through empty space. Therefore, it is just about the only physical phenomenon that doesn't (necessarily) involve matter. You will learn about light in the optics section of your course; we omit it in this book.

Figure 1.2 below shows the relationship of each major physics topic to matter. You may want to mark this page so that as you progress through the course, you can begin to see how *matter* is the common thread woven throughout physics.

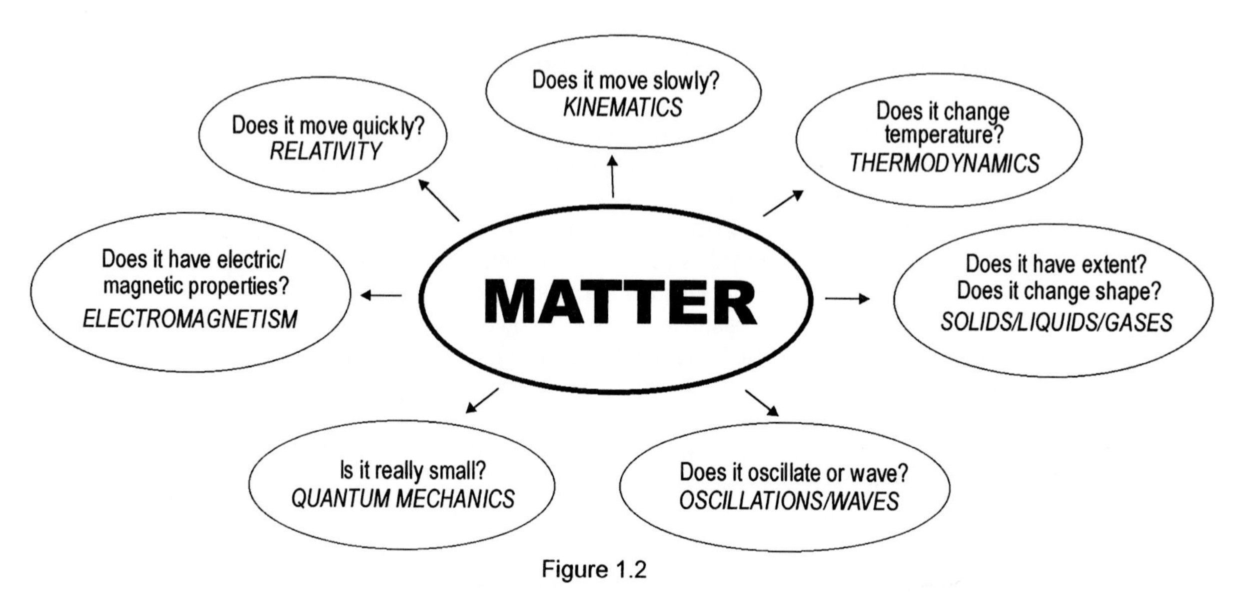

Figure 1.2

To summarize this section, then, we might say that while the scope of physics is—quite literally—universal, the *essence* of physics is minimalist: reducing all that can be known to a few fundamental principles. So take a moment to breathe a sigh of relief: you *don't* have to know everything to succeed at physics!

1.3 AN APPLIED SCIENCE

As we learned in the introduction, physics is as much about *doing* as it is about learning. Physicists use theoretical information (learned from courses like yours) to solve problems and to do experiments. Therefore, physics is an **applied science**: every physical discovery or learned fact is *applied* to problem solving or experimentation.

Chapter 4 will help you schedule your time for physics learning. 

CAUTION 1.5: Physics: An Applied Science

Because physics is an applied science, your physics course will most likely involve problem solving and lab attendance—in addition to regular reading and lecturing—so that you can learn to *use* what you have been taught.

As a result, comprehending physics takes a lot of time...maybe more time than you may want to allot to it. The good news is that if you do allow adequate time for learning physics, you should be able to master it as well as anyone else.

1.4 WHY PHYSICISTS LOVE PHYSICS, AND WHY WE THINK YOU WILL TOO

Physicists *love* physics. You may wonder why this is so...after all, there seems to be so much work involved!

Perhaps the main reason physicists love physics is because physics is *beautiful*. Physics—once really understood—displays the same aesthetic qualities as an exquisite work of art or a delicate piece of music: clarity, symmetry, and intricate simplicity. It both provokes and soothes the imagination; it excites and comforts the soul. Thus, physicists pursue physics in part because it satisfies the basic human needs for sublimity, harmony, and artistic inspiration.

The scientist [studies nature]... because he delights in it, and he delights in it because it is beautiful.
HENRI POINCARE
French Mathematician, 1854 - 1912

But physics has appeal far beyond mere aesthetics. Physics aims to answer no less than the fundamental questions of the universe, the questions humanity has been asking since the beginning of time: How did we come to be? What is our place in the universe? What does the future hold?

[People] love to wonder, and that is the seed of our science.
RALPH WALDO EMERSON
American Author, 1803 - 1882

Yet, even as physics strives to answer these and other questions, it reminds us of the elusiveness of ultimate truth. For although physics has already developed some answers to these timeless questions, the very nature of the science demands that discovery always develops and changes...even if that means overturning or revising old knowledge.

Chapter 2 discusses the nature of scientific truth in more detail.

Physics holds other attractions as well. It teaches us methods for solving problems and natural puzzles, using logic, creativity, personal intuition and insight; while rejecting rote memorization, categorization, or pigeonholing. And as long as the solutions to these problems flow logically from a few basic principles, physics allows us to entertain fantastic, outlandish, or whimsical ideas: ideas that appeal to our inner sense of playfulness and childlike wonder.

Science...uses the same kind of talent and creativity as painting pictures and making sculptures.
MAXINE SINGER
American Biochemist, 1931 -

Most of all, physics is *practical*, providing common sense information about the way the world works. When this knowledge is combined with physical problem solving techniques and the economical patterns of thinking that are the hallmark of practicing physicists (see Chapter 2), we are given a powerful tool for success in *life* as well as science. Frank Oppenheimer (1904 - 1967), a great physicist of the twentieth century, wrote of this critical aspect of physics:

The true and lawful goal of science is to endow human life with new powers and inventions.
FRANCIS BACON
English Philosopher, 1561 - 1626

> *By trying to understand the natural world around us, we gain confidence in our ability to determine whom to trust and what to believe about other matters as well. Without this confidence, our decisions about social, political, and economic matters are inevitably based entirely on the most appealing lie that someone else dishes out to us.*
>
> *Our appreciation of the noticings and discoveries of both scientists and artists therefore serves, not only to delight us, but also to help us make more satisfactory and valid decisions and to find better solutions for our individual and societal problems.*

The good of mankind may be much increased by the [scientist's] insight into the trades.
ROBERT BOYLE
English Chemist, 1629 - 1691

Concern for man himself and his fate must always be the chief interest of all technical endeavors.
ALBERT EINSTEIN
American Physicist, 1879 - 1955

Examining each of these facets of physics—beauty, knowledge, analytic prowess, and practicality—we begin to understand that learning physics is a highly personal journey: what you get out of it largely depends on what you put in. If you dare to give physics a try, you will learn new and interesting things about yourself and the world around you—things that will enlighten and inform you, and things that will be helpful in the *real world* outside of physics.

1.5 CONCLUSION

Many students wonder why physics is taught to non-physicists at all. The eternal question rings through dorm rooms and academic hallways everywhere: "If I'm studying to be a _______ (fill in the blank with your desired profession), why do I need to learn *physics*?"

Hopefully, this chapter has supplied you with many answers to this question: physics is fundamental to all other sciences; physics provides basic facts for understanding the universe, as well as techniques for discovering new information; physics is beautiful, practical, and engaging. The next chapter presents yet another answer: physics offers a new way of thinking that, once you get a hang of it, is perhaps the most efficient and effective method of analysis yet discovered.

The real voyage of discovery consists not in seeking new lands but in seeing with new eyes.
MARCEL PROUST
French Author, 1871 - 1922

For these reasons, and more that will become evident as you explore the subject in depth, it should be clear that physics is a natural prerequisite for most disciplines, and indeed, life. It is no wonder, then, that millions of students take introductory physics each year—many of whom have no intention of majoring in the subject.

If this is your situation, try to stay positive: physics asks much of you, but offers much in return. If you must take the class, you might as well do your best. We think that you will find your efforts well repaid.

2
WHAT IS SCIENCE?

2.1 INTRODUCTION

In your introductory physics course, you will spend a considerable amount of time learning material that other people discovered long ago. Over the intervening centuries, many people have worked at this body of knowledge—slowly honing, reshaping, and refining it—so that most of the little mysteries and puzzles that once piqued someone's interest have long since been smoothed away.

On the whole, this lengthy process of refinement is a net benefit for you, the introductory physics student, since otherwise you might get caught on the rough edges of reality and never be able to appreciate the grand work of art that is modern physical theory. On the other hand, when faced with such a sublime piece of work, it is easy to forget how much *time* and *effort* was expended, and how many *mistakes* were made, before the final magnificent result was achieved.

Science has always been full of mistakes....
EDWARD TELLER
American Physicist, 1908 -

The truth is that discovering the body of knowledge we call the physics canon has been a difficult and error-ridden process. That physics has come as far as it has, despite these difficulties, in just a few centuries is a testament to a revolutionary technique of intellectual investigation called the scientific method. The hallmark of this method, as we shall soon see, is "Eliminating the Unessential;" that is, *simplification*.

This method is important in part because its very existence proves that physics is *not* a seamless, perfect whole. Rather, it is a vibrant, living science full of quirks, riddles, and inadequacies; otherwise, there would be no need for such a system of simplification and idealization. Beginning physics students like you can take comfort in this fact, for it implies that this imperfect science is attainable by new learners.

But the real significance of the scientific method lies in the lessons it teaches us about *understanding* physics. Through this method, we find that the universe can *only* be understood in a simplified, idealized way. It is not just that the universe—as viewed through the lens of physics—seems to be simple and ideal; instead, we now think that the universe *is* simple and ideal. Therefore, the laws we create to describe the universe —and the techniques we use to create the laws—must also be simple and ideal. The scientific method, then, teaches us that in order to comprehend the universe (through physics), we must strive to *think* in an elemental, uncomplicated way.

Nature uses as little as possible of anything.
JOHANNES KEPLER
Italian Astronomer, 1571 - 1630

Thus, a good way to begin to learn physics is to start with the scientific method. *While initially it may seem like this chapter is irrelevant to your immediate class material, take the time to read it carefully anyway.* The scientific method offers a model for learning physics most efficiently and effectively, and so is well worth your extra time.

Learn the ABC's of science before you try to ascend to its summit.
I.P. PAVLOV
Russian Psychologist, 1849 - 1936

2.2 THE SCIENTIFIC METHOD

The scientific method allows ordinary people to do extraordinary things.
FRANCIS BACON
English Philosopher, 1561 - 1626

Galileo Galilei, a mathematician and natural philosopher in the beginning of the seventeenth century, developed a new technique for physical investigation, now called the **scientific method**.* Prior to his work, it was believed that questions about nature were best answered through reason and logic. Galileo, by contrast, insisted that theoretical arguments about the universe must be supported by *physical evidence* obtained through direct observations of nature. Galileo's scientific method, for the first time in history, *combined* observation with reason, resulting in a powerful technique for probing the secrets of the universe.

In its ideal form, Galileo's scientific method is a three-step cycle of *observation*, *theory development*, and *prediction*:

1. *Observation.* The scientist *observes* a phenomenon in nature.
2. *Theory Development.* Using reason, the scientist develops an explanation, or **theory**, to account for the observation (or, as happens more frequently, the scientist *revises* an existing theory to take the observation into account).
3. *Prediction.* The scientist uses the theory to *predict* a new, unobserved phenomenon. The prediction is called a **hypothesis**.

Reason, Observation, and Experiment; the Holy Trinity of Science.
ROBERT INGERSOLL
American Politician, 1839 - 1899

* Sir Francis Bacon (1561- 1626) slightly preceded Galileo with his own version of the scientific method. Galileo, however, was the first to also *implement* the new technique, using it to make scientific discoveries.

The cycle is completed when the scientist re-observes nature (returning to Step 1) to see if the prediction is verified. If it is, the cycle continues when the scientist thinks of a new *prediction* (Step 3) to be tested. If not, then the cycle continues when the scientist thinks of a new *theory* (Step 2). In this way, the cycle continually repeats itself (Fig. 2.1):

Thus analysis consists in making observations and experiments and in drawing general conclusions by induction....
ISAAC NEWTON
English Experimentalist, 1642 - 1727

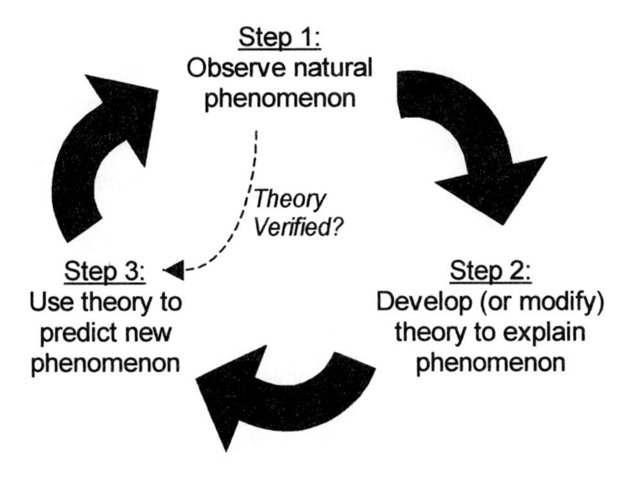

Figure 2.1

As is clear from the diagram, theories and observations are equally vital components of the scientific method: theories must be tested in nature, and observations must be guided by theory.

Sit down before fact as a little child, be prepared to give up every preconceived notion, follow humbly wherever...Nature leads, or you shall learn nothing.
THOMAS HUXLEY
English Biologist, 1825 - 1895

EXAMPLE 2.1: The Scientific Method In Your Life

Let's see how the scientific method might work in your life. For example, imagine that your grade in physics depends in part upon a series of short quizzes. Say you decide to take your first physics quiz without studying, and luckily, you pass.

Your first *observation*, then, about physics quizzes is that you did not need to study for one, and you did fine. As a result, you *formulate the theory*: "All physics quizzes are easy. I do not need to study for them." This theory explains (though not necessarily correctly!) your first experience with physics quizzes.

From this theory, you *make the prediction* that you do not need to study for any physics quizzes. You *test* this theory by not studying for the next quiz, whereupon you *observe* that you flunk. Your theory, though it worked in the first case, did not work in all cases. Thus, the theory *requires* revising. A revised theory might be: "All physics quizzes are not necessarily easy. The first quiz did not require study, but many physics quizzes will."

Now you must test this new theory, and observe *its* results.

See Section 4.3.5 to learn how best to study for physics quizzes.

2.3 ELIMINATING THE UNESSENTIAL

Galileo asserted that it is not enough to march mindlessly and mechanically through the scientific method. Instead, he maintained that scientific discovery stems from conscientious use of the method according to the guiding principle of "Eliminate the Unessential."

Science may be described as the art of systematic oversimplification.
KARL POPPER
English Philosopher, 1902 -

There are three ways to eliminate the unessential in any situation: simplification, abstraction, and idealization. To **simplify** a problem is to eliminate distracting and insignificant features from it. To **abstract** a problem is to make elements in the problem more symbolic and essential. To **idealize** a problem is to recast it in a more perfect, purer form.

See Section 7.2.1 for more on gravity. 

EXAMPLE 2.2: The Apple, Part I

Sir Isaac Newton, a great early physicist who coincidentally was born in the same year Galileo died, used Galileo's advice to "Eliminate the Unessential" when developing his revolutionary theory of gravity.

It is said that Newton's thoughts on gravity were "occasioned by the fall of an apple" (though it probably did not fall on his head, as you may have heard before!). Although the fall of an apple may seem to be a straightforward action, it is actually an extremely complex one: the apple may spin during its descent, a wind may blow it off course, etc.

When Newton saw an apple fall, he found
In that slight startle from his contemplation—
'Tis said (for I'll not answer above ground
For any sage's creed or calculation)—
A mode of proving that the earth turn'd round
In a most natural whirl, called "gravitation;"
And this is the sole mortal who could grapple
Since Adam, with a fall, or with an apple.
LORD BYRON
English Author, 1788 - 1824

So how was it–from this very complicated system—that Newton was able to develop the (rather simple) theory of gravity? Simplification, abstraction, and idealization.

Newton *simplified* the system by only considering aspects of the apple's motion that were *directly relevant* to falling. That is, he *ignored* rotation, the effects of wind resistance, and other complicating factors.

Then, Newton *abstracted* the problem by thinking about how *all* objects fall, not just apples. This released him from having to consider whether green apples fall differently than red apples, or whether apples fall differently than, say, pianos...or other similar complications.

Finally, Newton *idealized* his concept of falling apples by considering them as *point particles*: objects without shape or extent. Since the apple's shape was no longer an issue, Newton did not have to consider it when developing his theory of gravity.

What was left after Newton "Eliminated the Unessential" from the apple system was a very simple model of a point particle falling (in a straight line) because of gravity and *nothing else*. All irrelevant and troublesome aspects of the problem were eliminated, so that Newton could focus on the truly important issue: developing a theory of gravity.

We consider it a good principle to explain the phenomenon by the simplest hypotheses possible....
PTOLEMY
Ancient Roman Philosopher, c. 125 A.D.

Galileo's instruction to "Eliminate the Unessential" is the guiding principle underlying physics-style thinking, as we shall see later in this chapter (Section 2.5). But before we get to that, let's digress a bit and examine four important examples of this rule—mathematics, modeling, theories, and observation.

2.3.1 MATHEMATICS

Mathematics is a tool for *simply, abstractly,* and *ideally* describing reality. If math is difficult for you, you may find this statement hard to believe. Yet the power of mathematics is so pervasive and significant that *no* modern, legitimate science proceeds without it.

There are many reasons why mathematics is so critical to scientific discovery. The most fundamental is that mathematical language allows complicated theoretical relationships to be written more concisely than ordinary words can. This benefit goes far beyond typography, though: the clarity provided by mathematical expressions actually makes the concepts easier to understand!

When you can [express] what you are speaking about...in numbers, you know something about it; but when you cannot express it in numbers, your knowledge is of a meager and unsatisfactory kind...
LORD W. T. KELVIN
English Physicist, 1824 - 1907

EXAMPLE 2.3: Distance, Speed, and Time, Part I

You may know that the distance you travel in a certain time depends upon the speed at which you are traveling. You might express this relationship in words as "distance equals speed times time." A shorter way of writing exactly the same idea is expressed mathematically as:

$$x = v\,t,$$

where x stands for "distance," v represents "speed," and t is for "time." (Physicists use v for "velocity" rather than s for "speed" for reasons not important here.) Note how mathematical language greatly simplifies the expression of this familiar physical law. In just the same way, mathematical expressions simplify the expression of *all* important physical laws.

For more on this equation, and a definition of velocity, see Section 5.3.1.

Because mathematical expressions are so much more concise than their written-language counterparts, we can manipulate them in ways that would be difficult with words. For example, we can mathematically rearrange equations so that different, but *equivalent*, views of the same phenomenon can be obtained. This allows us to view the phenomenon from new angles, enabling us to understand *all* facets of its nature.

No human inquiry can be called true science unless it proceeds through mathematical demonstrations.
LEONARDO DA VINCI
Italian Artist, 1452 - 1519

EXAMPLE 2.4: Distance, Speed, and Time, Part II

The previous example gave an expression for finding the distance traveled, given the speed and time elapsed. But what if you had the distance and time, and wanted the speed? Just rearrange the above equation using algebra, and you have it:

$$v = \frac{x}{t}$$

Algebra is covered in Section 13.4.

All the effects of nature are only mathematical results of a small number of immutable laws.
PIERRE de LAPLACE
French Mathematician, 1749 - 1827

In some cases, mathematics allows us to manipulate theories in ways that would be *impossible* using everyday language. For instance, seemingly disparate theories can be blended together mathematically, sometimes resulting in brand new theories.

See Section 7.3.2 for more on Maxwell's Equations.

EXAMPLE 2.5: Maxwell's Equations

Consider Maxwell's Equations, four (simple!) equations describing the entire field of electromagnetism. These equations were once considered to be independent, or distinct from each other. Then Maxwell came along and mathematically blended them, yielding a *completely new* (fifth) equation that turned out to describe the nature of light.

Thus Maxwell's mathematical work led to a startling new theory that light is related to electricity and magnetism.

Mathematics has yet another advantage. Theoretical results obtained using math are often general enough to be used in other situations. Any situation that appears *mathematically* like a previously solved problem has the *same* solution, even if the *physical* aspects seem very different. This characteristic of mathematics means that not only can we copy conclusions from one problem to other without doing any new work, but we can also draw analogies between the two situations, making our understanding of both phenomena richer.

EXAMPLE 2.6: Gravity and the Electric Force

Over three hundred years ago, Isaac Newton found that the gravitational force causing planets to orbit the sun has a certain mathematical form. A century later, Charles Augustus Coulomb found that the electric force between charged particles has the same form. Thus the two physical phenomena of gravitation and electricity, which could not be more different physically, are actually mathematically *identical*.

These forces, and their mathematical relationship, are discussed further in Section 7.2.2.

The importance of this result lies in a kind of synergy between the mathematical concepts of gravitation and electricity. Discovery in both fields could increase more rapidly, for a discovery in one area also applied in the other!

Science was born of a faith in the mathematical interpretation of Nature...
JOHN RANDALL
American Historian, 1899 - 1980

Even if these four important reasons did not apply, scientists would still use mathematics in science for one simple reason: the scientific method requires that theories must be tested in nature. The results of any such test are simply *numbers*. But equations yield numbers as well! Therefore, to check theories against reality, scientists need to *compare* observational numbers with theoretical values.

To sum up, mathematics confers simplicity, abstraction, and idealization to science by its very nature. For this reason, it should be clear that one requirement for learning physics is mathematical literacy. More than that, a student of physics should strive to be *comfortable* with mathematics. Galileo, our constant guide in this chapter, said it best:

> *[Physics] is written in this grand book, the universe, which stands continually open to our gaze. But the book cannot be understood unless one first learns to comprehend the language and read the letters in which it is composed. It is written in the language of mathematics....without [which], one wanders about in a dark labyrinth.*

Part III of this book will help increase your comfort level with mathematics.

2.3.2 MODELS

A **model**, in the scientific sense, is a *simplified, abstract,* and *idealized* version of a situation, used to develop theories. Unlike everyday models—such as model airplanes—scientific models are not tiny renderings of real objects. Rather, models are conceptual representations of circumstances that, in fact, may not resemble reality at all!

EXAMPLE 2.7: The Apple, Part II

Consider the apple model used by Newton in Example 2.2 above. Its final form—a point particle falling in a straight line due to gravity—looks *nothing* like the real situation—an apple falling on an unsteady path because of gravity, wind, rotations, and so on. Yet the model meets its objective: simplifying nature so a theory can be developed.

Scientists use models because they simplify reality enough to develop workable theories. In fact, most (if not all) of the theories you will learn in your physics course are based upon models, because the corresponding real situations are too complex to be described.

However, because models, *by definition*, eliminate some aspects of a problem from consideration, theories based on models never completely describe the associated real-life phenomenon. Instead, these theories are only valid to the degree that the models are valid; that is, the degree to which the model compares favorably with reality.

If we try and understand everything at once, we often end up understanding nothing. We can either wait and hope for inspiration, or we can act to solve the problems we can...
LAWRENCE M. KRAUSS
American Physicist/Author

EXAMPLE 2.8: The Apple, Part III

If we want to find out how apples fall in real life, we must *modify* our model in Example 2.2 above. That model is *not* valid in the more realistic situation when wind, rotation, etc., are taken into account.

To a considerable degree, science consists in obtaining the maximum amount of information with the minimum expenditure of energy.
EDWARD O. WILSON
American Sociobiologist, 1929 -

Models, in general, are created by ignoring all non-essential elements of a problem, and then abstracting and idealizing what remains. This can be difficult in practice, for what may be ignorable in one situation may be vital in another. Once created, models are always checked (within a certain level of accuracy) prior to usage to ensure that they capture all of the important aspects of a problem without exhibiting any of the irrelevant ones.

EXAMPLE 2.9: The Atom Thru the Ages, Part I

Philosophers in ancient Greece thought that all matter was made of tiny particles they called "atomos." They modeled these particles as small, hard spheres. This was an effective model of matter, for it was simple, abstract, and idealized.

Of course the Greeks did not have the technology to *verify* that matter was really made of "atomos," but the theories they developed using this model matched up adequately with their view of reality. Therefore, *within the level of accuracy required by the Greeks*, the spherical "atomos" model was a good one.

Unfortunately, sometimes excellent models suddenly become unsuitable when more accurate observations reveal previously unknown facets of a phenomenon. Since the level of accuracy has changed, the model may have to change as well.

EXAMPLE 2.10: The Atom Thru the Ages, Part II

The Greek "atomos" model lasted in some form or another for almost two thousand years, largely because the predictions made by this model continued to be good enough for the level of accuracy required by the Greeks and their successors.

However, early in the twentieth century, experiments showed that atoms were *not* the smallest pieces of matter; instead, they consisted of even tinier particles eventually labeled electrons, protons, and neutrons. Thus, in light of the information gained from more accurate testing procedures, a new model of the atom was *required*.

See Section 10.2 for more on the constituents of matter.

Although a new model may be needed when more accurate observations reveal the old model to be obsolete, the old model is not necessarily useless. We do not always *need* more detail to understand the basic principles involved in a phenomenon. In fact, for most phenomena of everyday life, some of the oldest models do the best (simplest!) job of explaining reality.

EXAMPLE 2.11: The Atom Thru the Ages, Part III

While some of the most exciting physics (and chemistry) stems from the complex interactions between protons, neutrons, and electrons, there are many areas in physics in which there is no need for such a detailed model. Instead, many modern theories are based on the idea that the constituents of atoms work well enough together that physicists need only consider the *whole* atom when developing theories. And so lives on the "atomos" model!

Some models do such a good job of representing their version of reality that they are tried in other situations. This strategy can be extremely successful, but care must be taken when transplanting models to new circumstances: assumptions and approximations used in creating the original model may no longer apply. Inappropriate use of a model will result in theories that generate nonsensical predictions.

EXAMPLE 2.12: The Atom Thru the Ages, Part IV

Fairly soon after electrons, protons, and neutrons were discovered, it was shown that protons and neutrons clump together in a ball, called the *nucleus*, while electrons move around in some fashion outside of this ball. See Figure 2.2 (a).

One question that arose from this discovery was "*How* do electrons move around the nucleus?" In a first attempt to answer this question, physicists borrowed an existing, successful model of a sphere moving around another sphere: that of the earth moving around the sun! Refer to Figure 2.2 (b).

The first tests of this "planetary" model of the atom provided good, though rudimentary, results. More detailed experiments, however, revealed that the planetary model did not match overly well with experimental observations. Thus, the planetary model, while good for describing the earth's movement around the sun, did not work for describing the electron's movement around the nucleus. So, a new model for this motion was *required*, one given by modern quantum mechanics.

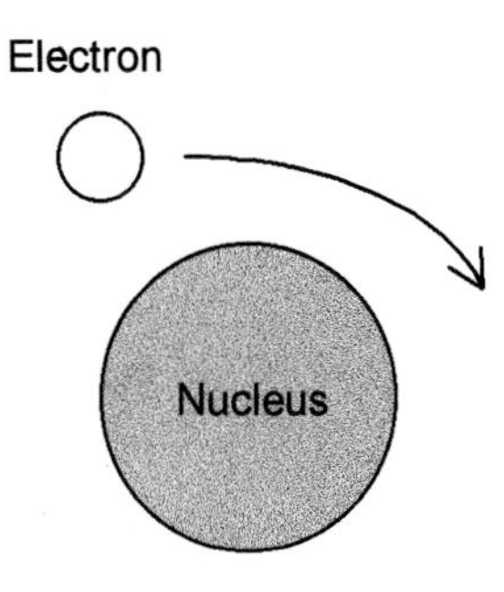

Figure 2.2 (a)

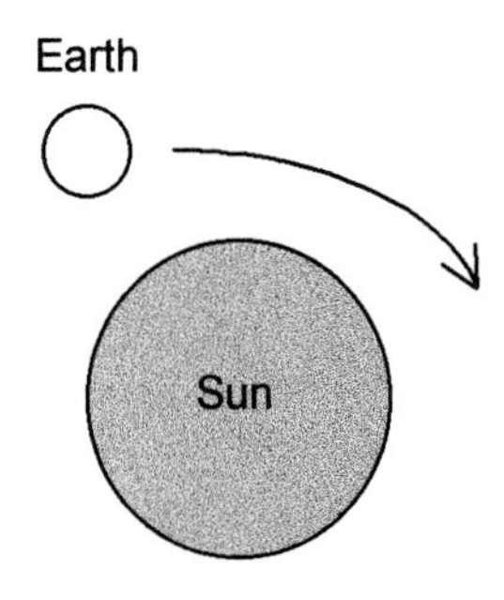

Figure 2.2 (b)

DEFINITION 2.13: Types of Models

There are four types of models.

Analysis models replace one physical problem with a similar version that has already been solved. Then, the solution to the old problem is used to answer the new one. Example 2.6 above ("Gravity and the Electric Force") is an example of this model type.

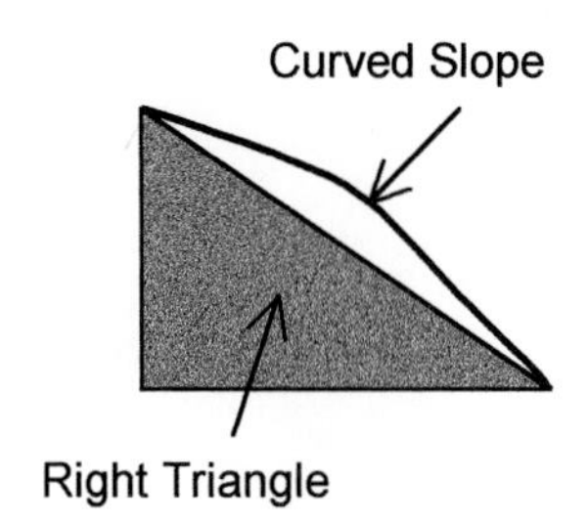

Figure 2.3

Definition 2.13 Continued...

Geometric models use geometry to simplify irregular shapes or situations, so that mathematics can be applied to the problem. For example, a hill can be modeled as a right triangle. See Figure 2.3.

Simplification models reduce a complicated real situation to a simpler, more manageable form. For example, the point particle model of the apple in Example 2.2 above is a simplification model.

Structural models are used when the scale of a problem is outside of our everyday experience, for example, in the case of very small atoms or very large galaxies. They allow us to draw analogies between these strange situations and objects with which we are familiar. The planetary model of Example 2.12 illustrates this kind of model.

2.3.3 THEORIES

[A theory's] object is to connect ...apparently diverse phenomena, and above all, to suggest, stimulate, and direct experiment.
J. J. THOMPSON
English Physicist, 1856 - 1940

A **theory** is a *simplified, abstract,* and *idealized* explanation of a phenomenon.* Theories are used in two ways: first, to understand the nature of the phenomenon itself, and second, to make predictions about the phenomenon that can be tested with observation.

EXAMPLE 2.14: Newton's Theory of Gravity, Part I

Newton's theory of gravity, developed using the point particle model of Example 2.2, is the quintessential example of a good theory.

The theory is *simple*: it is completely given by a single formula; *abstract*: the formula is a mathematical equation; and *idealized*: it deals only with point particles. So, Newton's gravity meets all of the requirements of a good theory.

This theory has been (and will be, in your class) utilized in two different ways. First, it is used to understand the motion of massive objects relative to each other, such as the falling of an apple from a tree, or the orbiting of the earth around the sun. Second, it is used to make predictions about other kinds of motions, such as movements at very high velocities, or the bending of light around very large objects.

Because of their key role in the scientific method, it is easy to believe that theories describe the "true" nature of the universe. Unfortunately, this is not the case, for the following reasons:

- Theories are often ephemeral: any theory that does not correctly explain a phenomenon, or make accurate predictions of new phenomena, *must* be revised or thrown out. Since there are always

* In fact, the simpler the better: **Occam's Razor** posits that given more than one theory to describe a situation, the simplest theory is the best one.

new phenomena being discovered, this revision process never ends.

- Theories can never be proven correct: no amount of supporting evidence can *ever* rule out the possibility that one piece of contradictory evidence still lurks unfound.
- The validity and precision of theories depends upon the validity and precision of the models on which they are based.

Consequently, theories are *not* reality; rather, they simply *describe* it. Yet, they perform this task astonishingly well: without the intellectual framework that theories provide, science as we know it could not exist.

No number of experiments can prove me right; a single experiment can prove me wrong.
ALBERT EINSTEIN
American Physicist, 1879 - 1955

The universe is not an idea of mine. It is my idea of the universe that is an idea of mine.
FERNANDO PESSOA
Portuguese Poet, 1888 - 1935

EXAMPLE 2.15: Newton's Theory of Gravity, Part II

Newton's law of gravity is a highly successful theory that describes a vast range of motions. As a result, it reigned supreme as *the* description of motion for several centuries.

However, when in the early twentieth century the theory was applied to some strange situations—such as movement at high speeds or light motion around very massive objects—Newton's theory no longer worked.

For that reason, Newton's gravity had to be replaced by a new theory that was (1) valid when Newton's gravity was valid, but (2) also valid when Newton's gravity was not. That new theory was Einstein's Theory of General Relativity.

Thus, a perfectly good theory (Newton's) had to be replaced by another (Einstein's) when new data made it obsolete. In fact, the replacement itself has been (and is) subject to revision as more information comes to light.

General Relativity is covered in Section 5.5.2.

Still, some theories, known as **laws**,# have withstood so many testing cycles without radical change, and explain such a broad range of phenomena, that scientists have come to view them as fact. New information can and does make laws obsolete, but since they have stood the test of time, any seemingly contradictory information is examined *very* closely before the law is changed.

EXAMPLE 2.16: Newton's Theory of Gravity, Part III

Even though Newton's gravity does not work in some esoteric situations, it still does a good job of describing most mundane ones. Thus, Newton's theory of gravity has been elevated to the status of a Law.

Unfortunately, the word "law" is sometimes used rather loosely in physics. Here, a **law** is considered to be a fundamental statement about the universe that is broadly applicable and extensively tested for validity. Not all "laws" meet this definition; for example, "Ohm's Law," which relates voltage to current, only applies in certain, very restricted situations.

2.3.4 OBSERVATION

Scientific observation, in its perfect form, is a *simplified*, *abstract*, and *idealized* method for validifying theories. It consists of two parts: experiments and error analysis. **Experiments** are tests of natural phenomena that result in numbers, or **measurements**, characterizing the phenomena. **Error analysis** is the procedure by which we determine our level of confidence in the measurements.

See Section 16.2.3 for more on error analysis.

The specifics of experimentation vary widely depending upon the phenomenon being investigated; however, most experiments share the following characteristics:

The test of all knowledge is experiment.
RICHARD FEYNMAN
American Physicist, 1918 - 1988

Theories are the generals; experiments the soldiers.
LEONARDO DA VINCI
Italian Artist, 1452 - 1519

Nature will reveal herself if we will only look.
THOMAS ALVA EDISON
American Inventor, 1847 - 1941

- Every effort is made to test *only* one aspect of the phenomenon at a time.
- In an experiment, this aspect of the phenomenon is varied while everything else in the surrounding **environment** is kept constant.
- An identical test—called the **control**—is also performed, but without any variation in the phenomenon.
- *All* effects of both tests—no matter how small—are carefully noted and recorded. These results should be repeatable and verifiable by any other scientist doing the same experiment.
- Measurements from the experiment and the control are compared via error analysis.
- The hypothesis is thus supported or disproven.

All well-executed experiments also share one important feature: *honesty.* Clearly, the value of any trial depends upon the truthfulness of the experimenter, both when doing the experiment itself, and when reporting its results. For this reason, *scientific integrity* is a guiding principle of the scientific method, just important as "Eliminating the Unessential."

EXAMPLE 2.17: The Zucchini Experiment

As an example of a simple experiment, consider a gardener who grows more zucchini than he can eat at one time. The gardener decides to freeze his surplus, but he is concerned about maintaining the zucchini's flavor. To determine the best route for preserving his hard-earned bounty, the gardener performs a quick experiment.

The gardener cuts one zucchini in half. He freezes one piece after only a washing. The other half he cooks first and then freezes. Then, after they have frozen, he takes both pieces out of the freezer and defrosts them. He tests each piece by tasting.

This simple example illustrates many of the main issues in experimentation. The gardener tested only *one aspect* of zucchini storage: whether or not to cook before freezing. He used a *control*, raw zucchini identical to the tested zucchini. He *measured* his results by tasting, and he *compared* the raw and cooked results. Consequently, the gardener found that cooked zucchini fares better in the freezer than raw. The happy end of this simple experiment was better frozen zucchini!

2.4 THE SCIENTIFIC METHOD IN THE REAL WORLD

Mathematics, modeling, theories, and observation exemplify how scientists strive for simplification, abstraction, and idealization in the scientific method. Yet because science is a human endeavor, it should not be too surprising that intellectual discovery does not usually progress according to a perfect plan. In fact, science in the real world often advances in ways that are highly unpredictable and, at least when viewed over the short term, do *not* conform to any particular pattern or cycle.

For one thing, it is the rare scientist that is proficient in both theory *and* observation. Rather, scientists today are usually either "theorists" *or* "experimentalists"—and these two populations may not interact with each other overly much. For another, scientific discovery usually advances in fits and starts. Sometimes discoveries develop over long periods of time and after many mistakes; and other times, a brilliant mind or raw good luck will progress science very quickly.

For these reasons, the general pace of scientific discovery can appear to be quite random and disordered when viewed over short time spans. However, when observed across the sweep of history, a "cycle" of complementary theoretical and experimental discoveries can often be seen. With the benefit of hindsight, the development of science thus can seem to have progressed according to Galileo's ideal scientific method.

The great tragedy of Science: the slaying of a beautiful hypothesis by an ugly fact.
THOMAS HUXLEY
English Biologist, 1825 - 1895

The most important of my discoveries have been suggested to me by my failures.
SIR HUMPHREY DAVY
English Chemist, 1778 - 1829

Science moves, but slowly, slowly, creeping on from point to point.
LORD ALFRED TENNYSON
English Poet, 1809 - 1892

[Science] seldom proceeds in [a] straightforward logical manner....
JAMES D. WATSON
American Biologist, 1928 -

2.5 THINKING LIKE A PHYSICIST

The scientific method provides a good introduction into the manner in which physical knowledge has been pursued since Galileo's time. But more importantly for our purposes here, the method also offers a metaphor for understanding the unique way physicists view the world—a mindset that is *critical* for all students of the subject to adopt.

The essence of thinking like a physicist is embodied in Galileo's principle of "Eliminate the Unessential." *Physicists constantly strive to view the world in the simplest way possible.* All problems—large or small—are approached in the same step-wise manner:

1. A model is created to simplify the problem;
2. An answer is developed for the modeled problem;
3. One or two of the original difficulties are added back to the model;
4. Steps 2 and 3 are repeated until a result mirrors reality.

The result of this so-called **top-down thinking** is an extraordinarily clear vision of the universe and its workings. So clear, in fact, that physics can be considered as the archetypical example of the scientific method today.

The aim of science is, on the one hand, as complete a comprehension as possible of the connection between perceptible experience... and on the other hand, the achievement of this aim by employing a minimum of primary concepts and relations.
ALBERT EINSTEIN
American Physicist, 1879 - 1955

EXAMPLE 2.18: Galileo and His Principle

Galileo himself serves as a perfect example of this unique mindset. He came to the brilliant conclusion that the state of *real* physical knowledge in his time was negligible: every so-called fact known to humanity was the product of superstition or fallacious arguments.

Understanding that the whole of physics (indeed, science) lay undiscovered before him, Galileo acted in a characteristically physicist-like fashion. Rather than attempt to answer all possible questions on the nature of the universe (and there were many!), Galileo instead distilled all of these unanswered questions into the most basic and pressing issue: "What is motion?"

Galileo understood that motion provides a framework for the study of all other questions in the universe—for *all* objects move in some fashion—and so devoted his life to its study. From this sturdy basis, Galileo's successors (like Newton) were able to build up the science we now call physics.

The whole history of science has been the gradual realization that events do not happen in an arbitrary manner, but that they reflect a certain underlying order.
STEPHEN HAWKING
English Astrophysicist, 1942 -

What is truly amazing about the "Eliminate the Unessential" principle is that the universe itself seems to abide by it! History has shown that all of the discoveries made since Galileo's time can be sorted into only a small number of categories; that is, *nature is repetitive*. It seems that the universe, with all its apparent diversity in form and function, is based on just a few, invariable laws.

These two equations are covered in Sections 6.3 and 5.5.1 respectively.

See Section 12.7 for more on symmetry.

Furthermore, these basic laws are simple! Two of the most important physical theories—Newton's Second Law ($\vec{\mathbf{F}}_{net} = m\vec{\mathbf{a}}$) and Einstein's Theory of Relativity ($E = mc^2$)—have such simple mathematical expressions that it seems hard to believe that these formulae could contain the important information about the universe that they do. In fact, physicists have come to rely on the doctrine of simplicity to such a degree that today theories are evaluated as much for their clarity and symmetry as they are for their ability to produce accurate predictions.

NOTE 2.19: A Theory of Everything

While it seems improbable that we will ever truly come to the end of physics, we *have* come to a point where we can begin to speculate on the development of a "Final Theory." Such a theory would represent the epitome of the scientific method, for it would explain all that is knowable about the universe using just one simple equation.

Achieving a "Theory Of Everything" is probably not possible—we have seen, after all, that theories are never really complete—yet physicists work towards it anyway because the process itself is vastly rewarding. The products of science enrich and uplift our society far beyond the aesthetic rewards of finding a united theory.

Science proceeds by successive answers to questions more and more subtle, coming nearer and nearer to the very essence of the phenomenon.
LOUIS PASTEUR
French Biologist, 1822 - 1895

Because physicists are trained to see the inherent simplicity in physical laws, they think physics is easy. By contrast, because non-physicists haven't had such training (yet!), they frequently think physics is not easy: they keep getting bogged down in details. However, once non-physicists understand they must make a *change in mindset*—so that they see the world and its laws as simply as possible—they begin to learn, understand, and even enjoy(!) physics.

All physics is clear, evident, or obvious.
ENRICO FERMI
American Physicist, 1901 - 1954

Not so to me.
LAURA FERMI, Enrico's wife

Therefore, *the key to becoming a successful physics student is in seeing the world like a physicist.* While it is true that other mindsets can be useful in other professions—for example, the artist may *celebrate* the unessential—in physics, you must search for that which is fundamental to every problem, and *ignore everything else.* Learn to see the simplicity inherent in the universe, and in physics. Eliminate the Unessential.

2.6 WHY PHYSICS IS TAUGHT THE WAY IT IS

Here, perhaps, we find ourselves in a paradox. If the goal of physics is to view the universe simply, then why (for heaven sakes!) does physics instruction seem so impenetrable? After all, it is easy to feel swamped by the flood of information flowing from lecture, lab, and reading.

The answer is that, traditionally, physics instruction has been based on the belief that students are able to form their own "big picture" view of main themes in physics; therefore, to "fill in the blanks," physics education must stuff as much information into an introductory course as is possible.

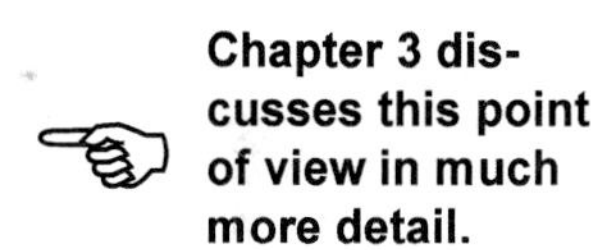

Like it or not, this is how most college-level physics is taught today. So, your job as a physics student is to tame the flood of information and continually Eliminate the Unessential: in other words, *practice* Galileo's principle in order to learn it. As the next chapter points out, this is a time-consuming task. As the last chapter showed, it is also a worthwhile one.

2.7 CONCLUSION

Galileo's development of the scientific method in the beginning of the seventeenth century was the spark that ignited the explosion of information we witness in our own time. Galileo knew that his scientific method was revolutionary and would affect scientists for generations to come. But his method also influences students of science—*us*—for once we understand how scientists go about their work, by Eliminating the Unessential, we can begin to develop a deeper and more intuitive understanding of science itself.

So that we may say the door is now opened, for the first time, to a new method...which in future years will command the attention of other minds.
GALILEO GALILEI
Italian Experimentalist, 1564 - 1642

3
WHAT TO EXPECT

3.1 INTRODUCTION

Your introductory physics course will, in many ways, be different from any other class in your past or current experience. Some of these differences—because you are not used to them—may make you uneasy or uncomfortable, and affect your ability to learn physics. However, if you knew what to expect from your physics class *before* it began, your chances for success might improve.

The purpose of this short chapter, then, is to provide you with an insider's look at the typical physics classroom. We examine many facets of physics courses, including lectures, homework, reading, quizzes and exams; your instructor and other students in your class; and why you might find physics "hard."

When I heard the learn'd astronomer where he lectured with much applause in the lecture room
How soon unaccountable I became tired and sick.
WALT WHITMAN
American Poet, 1819 - 1892

The picture we paint in this chapter—heavily based on the work of nationally recognized researcher and author Sheila Tobias*—may turn out to be nothing like your experience. After all, we are describing the *average* physics course; your particular class may be an exception. However, if you *do* recognize some of the issues discussed in this chapter (and we think you will!), remember that the next chapter provides assistance in dealing with them.

Chapter 4: How To Succeed

---

* *They're Not Dumb, They're Different; Succeed With Math;* and *Overcoming Math Anxiety.*

3.2 YOUR INSTRUCTOR

Your physics instructor is not like you, in many ways.

He* probably first became interested in physics long before high school. You, on the other hand, are likely thinking about physics for the first time in your life. He probably learned much of what he knows about physics from his early-life hobbies and reading choices. You are trying to learn physics from a college-level class. He probably decided on physics as a career very early in his life. You are likely taking physics because your (non-physics) major or program *requires* it.

There are more subtle differences between you and your instructor as well. He likes to think in terms of equations and mathematics. You like to think in terms of concepts and words. He likes to reduce things to simplicities and common denominators. You like to have all of your detailed questions answered and see the "big picture." He likes to work alone. You may prefer to work with other people. He is trying to *train more physicists*. You are trying to *learn* (but perhaps not master) a required subject.

Science is nothing but... organized common sense.
THOMAS HUXLEY
English Biologist, 1825 - 1895

And if these differences weren't enough, well, sometimes physics instructors are a touch...eccentric.

EXAMPLE 3.1: Some Physics Instructors, Part I

For example, one instructor we know is also a pig farmer. This instructor slops his pigs in the same *single* pair of jeans he wears to class each and every day of the term. Each day of the semester, then, the jeans become a little bit dirtier and stiff.

Another instructor we know insists on starting his first class of the day ten minutes *prior* to the bell, on the argument that students have no need of a "passing period" for the first hour. He honestly feels students are better served spending their time in lecture instead!

A third physics instructor we know—who comes complete with an Einstein-style shock of white hair—is mysteriously and extremely wealthy. He owns and flies his own private jet, which he uses whenever the whim strikes...even in the middle of the day!

These differences, when taken as a whole, can make up a rather large communication gulf between you and your instructor: you may literally feel you are speaking different languages. Unfortunately, to close this gap, *you* are going to have to make an effort: it is your grade on the line, after all, not his.

To build a relationship with your instructor, then, try to get to know the *person* behind the lecturer, eccentricities or other differences notwithstanding. In almost every case, the same instructors who seem odd or forbidding during lecture are actually fascinating and kind on the personal level.

---

* Most college-level physics instructors are male, so we use 'he' here.

EXAMPLE 3.2: Some Physics Instructors, Part II

You may think that one or more of the physics instructors introduced in the last example is a little peculiar. That may make you unwilling to get to know any of them better. However, each of these instructors have shown themselves to be extremely kind, intelligent, and willing to help their students in any way possible; in short, just the sort of people you want to get to know. So think twice about dismissing your instructor as "weird" and try building a friendship instead.

3.3 THE LECTURES

Science, when well digested, is nothing but good sense and reason.
STANISLAUS
Patron Saint of Poland, 1030 - 1079

Physics lectures will probably be unlike any in your previous experience. This is because the goals and traditional lecture styles of physics and non-physics courses are different.

EXAMPLE 3.3: Humanities and Physics Lectures

To see how physics and non-physics courses differ, let's compare a typical humanities lecture with one in physics. Note how the *goals* and *techniques* of the two types of lectures vary, and, subsequently, how your *level of participation* in the course may also vary.

* * *

The goal of a humanities class is to teach you to *analyze a body of work*, such as the sculpture of ancient Greece or the writings of American post-modernists. The purpose of this approach is to extract general themes and underlying meaning from the works.

Your humanities instructor will try to attain this goal by acting as a *facilitator* to your personal investigations. She will ask you to read or examine a large selection of materials; form an opinion about them; and then discuss your findings in class, in an essay, or on an exam.

Thus, the humanities instructor will often hardly lecture at all, choosing instead to let you learn for yourself from the readings, the discussions, and your own self-inquiry. Consequently, humanities classes often serve as a jumping off point: a place from which *you* can develop your own hypotheses and ideas—ideas that may sometimes be better than the instructor's own!

* * *

The goal of a physics lecture, on the other hand, is to teach you to *solve problems*. The purpose of this style is to enable you to apply general physics information to specific situations.

Your instructor will strive to attain this goal by acting as a *source of factual information* and a *modeler*. He will lecture to pass on a canon of knowledge, and he will solve problems to model necessary techniques.

Example 3.3 Continued...

Physics lectures, then, serve as a place where you *watch* a master work, while you copy down his efforts for later self-review.

Therefore, the physics instructor may spend the entire class hour lecturing—after all, he has so little time to pass on all the information *and* problem-solving tricks that you will need to do the homework—and no time answering your specific questions or helping you understand important points on a one-on-one basis. Accordingly, in this type of classroom environment, you are often *not* expected to contribute to the discussion or have ideas that differ substantially from the instructor's, since you have not (yet!) mastered the basic material.

Science is all metaphor.
TIMOTHY LEARY
American Philosopher,
1920 - 1997

Though they differ on traditional teaching styles, college-level humanities and physics instructors share the goal of teaching you *how to think.* However, the thinking skills required for each subject are *different*, so the lecture styles may also be different.

One style is not necessarily better than another, though you may feel more comfortable in one over the other. For instance, if you are used to humanities lectures, the physics style may seem uncomfortable to you.

EXAMPLE 3.4: Physics Lectures: Some Challenges

Let's take a closer look at a few aspects of physics lectures that might be disagreeable to you.

For one thing, since the instructor in a typical physics course spends most—if not all—of class time talking, you might have a hard time staying alert. This is not necessarily your fault: psychologists have shown that the average adult pays attention to lectures for only 10 to 15 minutes; physics lectures can be up to five times that long!

Furthermore, the instructor uses all that time to pass information from him to you. This may leave you feeling like he is the "keeper of knowledge," and you are nothing but a copier and a re-doer of his work. This feeling can be exacerbated when you go home and solve problems that have been solved (by someone else) a thousand times before. You may feel there is no room for your individual creativity or personal contribution in this kind of environment.

What's more, the information passed on to you can seem to have a "canned" or contrived quality to it: the equations and theories apply only to special, dry situations, or under certain, unrealistic assumptions. You may wonder how any of it applies to real life.

And to top it all off, it may seem like your instructor's lectures have a disjointed or disconnected quality; that is, they seem to lack a "storyline" that connects concepts into a "big picture." You might see this on a small scale —the instructor makes no effort to link today's lecture to yesterday's or tomorrow's work—or on a large scale—the instructor doesn't connect major concepts to preceding or following ones. Whatever the manifestation, you are left feeling that all you gained from the lecture is a confusing jumble of dry facts, absent any context or connections.

While some aspects of lecture may make you feel uncomfortable in the physics classroom, you must keep in mind that your instructor is probably trying his best. After all, he was taught in this way, and it certainly worked for him.*

If you *do* feel uncomfortable in the physics-learning environment, don't let it become an excuse to fail. Instead, strive to *change* your mindset and learning skills to fit the available classroom style. Chapter 4 provides advice on how to best meet this challenge.

3.4 YOUR CLASSMATES

Physics is experience arranged in economical order.
ERNST MACH
Austrian Physicist, 1838 - 1916

Physics students may seem different from students in other classes.

EXAMPLE 3.5: Humanities and Physics Students

Let's compare typical humanities students with those in physics. Because the approach to, and level of participation in, each type of lecture differs, students seem to behave differently in each class as well.

✷ ✷ ✷

Students in the humanities classroom must participate in graded class discussions. Therefore, the typical humanities student seems outspoken, interested in the subject matter, and respectful of classroom rules—since otherwise, discussions would never materialize or would quickly descend into chaos.

Furthermore, the typical humanities class—even at the introductory level—often only has twenty or so students. In this kind of classroom atmosphere, students can get to know each other and their instructor on a fairly intimate level, developing a warm sense of community and cooperation.

In short, humanities students and instructors—even though they might be strangers at first—*seem* approachable and friendly, so that fellow students can quickly become chummy with each other. In this type of situation, it seems easy to get help and make friends with fellow students.

✷ ✷ ✷

Physics students, on the other hand, are often expected to sit passively in class, absorbing lecture material quietly on their own. Thus, the typical physics student may appear (on the outside) to be bored, tired, uninterested, scared, or even rude.

And since most introductory physics courses enroll hundreds of students at a time, it can be hard for students to reach out to each other or the instructor on a personal level—even if they did feel welcome to do so.

So, physics students and instructors often end a class term as much as strangers as when they began. It is easy to see why, in this type of atmosphere, you may not feel connected to a community, or that anyone cares about your progress.

* Things are changing at the college level. Many physics education researchers now advocate using some humanities-style lecture techniques in physics.

It can be easy to feel isolated in a typical introductory physics course. This perception may be aggravated if you start out feeling different from the other students; for example, if you don't intend to be a physics major, or if you are not white or male.

For more on physics teams, see Section 4.2. 

Luckily, *you do not have to feel isolated for long.* All you have to do is *team up* with your fellow students. Although they may seem forbidding at first (just think, you probably look forbidding to them, too!), keep in mind that they are probably in the same boat as you are. So why not join forces? What you don't understand, someone else surely does, and vice versa.

3.5 YOUR COURSEWORK

Science starts only with problems.
KARL POPPER
English Philosopher, 1902 -

The typical introductory physics coursework usually consists of regular readings and homework sets, weekly labs and quizzes, and anywhere from one to five exams over the term.

3.5.1 GRADING

Of all of this coursework, exams usually (but not always) make up the lion's share of your grade, while homework, quizzes, and labs are given much less importance.*

NOTE 3.6: Why?

Why would your instructor put so much emphasis on test grades when he could just as well check the level of your understanding with (easier) homework, labs, or quizzes?

Of course there are as many reasons for this as there are instructors, but one very important factor is that most introductory college-level physics classes enroll *huge* numbers of students. It is much easier for the instructor to check the progress of all of these students through a few exams once or twice a term, rather than with reams of homework sets or quizzes due each week.

Because of this emphasis on test results, *success in physics classes is largely a matter of success on exams.* However, the paradox of physics grading is that you *won't* do well on the exams unless you attend lectures, do the reading, and perform well on homework, quizzes, and labs! That is, your instructor is using the exams as a *test* (literally!) of how well you have completed other aspects of the course. This part of the chapter looks at each important element of your grade.

* Your instructor will inform you of the relative importance of each of these elements in your particular course at the beginning of the term.

3.5.2 PROBLEM SOLVING

The goal of most of your coursework is to teach you to *solve problems.*

Problem solving is discussed in more detail in Section 4.4.

> **CAUTION 3.7:** Getting Wet, Part I
>
> Just as you can't learn to swim without getting wet, you can't learn physics without doing problems. So, the typical physics class is *heavy* on the problem solving.

Like it or not, you need math to solve problems. But you also need to know appropriate physical concepts, and a variety of problem solving techniques.

If your math skills need sharpening up, see Part III of this book.

> **CAUTION 3.8:** Getting Wet, Part II
>
> Each of these skills can be *learned*, so that anyone can become a good problem solver. Problem solving is *not* a function of how smart you are, or how quickly you can figure out answers. Instead, it is a step-wise (and time-consuming) process that must be done carefully and slowly each time.
>
> This does not mean, luckily, that you can't make mistakes! Mistakes are an important part of the problem solving process...as long as you take the time to figure out what went wrong and avoid it next time.

The student must remember, for his consolation...that his failures are almost as important to the cause of science...as his successes. It is as much to know what we cannot do in any direction—the first step, indeed, toward the accomplishment of what we can do.
LOUIS AGASSIZ
American Zoologist, 1807 - 1873

You will practice problem solving in many ways in your physics class, including homework sets, quizzes and exams, and lab work. So, even if problem-solving competence is not a goal for *you*, it *is* a goal of the instructor and of the course. *Problem solving is a key to understanding physics, not just a test of it.* So use your coursework to your advantage: as a tool for getting a deeper understanding of the problem-solving process, and of physics itself.

Section 4.3 discusses coursework in detail.

3.5.3 HOMEWORK

Humanities and physics homework assignments are different.

> **EXAMPLE 3.9:** Humanities and Physics Homework
>
> Typically, humanities homework consists of readings and one or more papers over the course of the term. Often there are no real due dates

Example 3.9 Continued...

on any of this work, so many students find they can get away with leaving it unfinished until the very last minute.

* * *

By contrast, physics homework is due regularly, usually at every class meeting or at least once a week. Furthermore, students that leave physics assignments to the last minute usually can't finish them! In fact, students who don't work on their physics homework every day—regardless of when it is due—will not succeed in the course.

The bad news is that despite all of the effort you put into your homework sets, you will probably not get much of a direct grade benefit from them. Instead, the largest part of your grade will likely come from exam scores. *But you will not do well on your exams unless you do well on your homework.* So you must strive for high homework scores as a down payment on your exam grade.

Science can only state what is, not what should be.
ALBERT EINSTEIN
American Physicist, 1879 - 1955

In general, physics homework consists almost exclusively of problem solving. Chapter 4 will help with this. Sometimes, though, there are conceptual, short-answer-type questions sprinkled throughout your assignment. Answer these as you would any short essay question: with concise but complete, to-the-point, sentences.

Your instructor will try to assign a good mix of problems and questions: some will be easier, some harder; some will require a lot of algebra, some will require a conceptual understanding; some you will solve *instantly*, some you will simply *never get*. Usually, the number of problems assigned per concept is a good indicator of the relative importance of a concept; however, in most cases, the total number of problems assigned is less than the number you will need to really learn the material.*

3.5.4 QUIZZES & EXAMS

Your instructor assesses problem-solving skill with quizzes and exams:

- *Quizzes* are very short tests of your understanding of simple concepts. The questions are usually brief and require little math. In most cases, then, quizzes are the instructor's way of checking that you have been paying attention...and nothing more. Therefore, you should *never* flunk a quiz: not only are you screaming to the instructor that you haven't kept up, but you are throwing away *free points*!
- *Exams*, on the other hand, are longer and harder than quizzes. They are designed to check for deep understanding of important topics. Exam problems are usually few and moderately difficult; however, they are usually *less hard* than the hardest homework problems.

* In fact, some instructors will not assign *any* homework!

EXAMPLE 3.10: Humanities and Physics Exams

Humanities and physics exams are quite different.

One obvious difference between them is that humanities tests usually consist of broad essay questions, while physics exams are composed of specific problems. Therefore, a student who doesn't study for a humanities test may actually be able to provide competent answers on their exam anyway, simply by writing well and drawing on general knowledge of the subject.

This is *absolutely* not the case in a physics exam. Many physics concepts are counter-intuitive, which means that applying "common sense" logic to a problem can result in a wrong answer. The only way to approach physics exam problems with confidence is to prepare *prior* to the test. And the only way to do this is to practice homework problems. And then practice some more.

Quizzes and exams are where all of your hard labor on homework pays off. For one thing, exam questions often look very similar (if not identical) to homework problems. But even when the exam problems do not appear familiar on the surface, experienced problem solvers (*i.e.,* YOU after doing your homework) will quickly find a solution—because of their effort on the homework.

Science is not everything, but science is beautiful.
J. ROBERT OPPENHEIMER
American Physicist, 1904 - 1967

3.5.5 THE LABORATORY

As we have seen, physics is as much about experimentation as it is about problem solving. Therefore, your physics grade will likely depend (at least to some degree) on a laboratory segment.

In Section 1.3.

EXAMPLE 3.11: Physics Laboratories

There is no analogue to the physics laboratory in the humanities-style class. This does not mean, however, that physics labs are irrelevant or skippable. On the contrary, laboratories are critical parts of the physics learning process: they are the place where you get "hands on" with many of the concepts you have learned in class. So, you must take the time to do laboratories well.

There are usually three parts to an introductory laboratory:

1. Taking measurements;
2. Analyzing data; and
3. Drawing a conclusion.

Chapter 16 will discuss each of these elements of labs in more detail. Here, we simply look at the three main types of labs, as the type of lab determines the kind of conclusion you will draw:

For more on constants of nature, refer to Section 11.5.3.

- *Verifying a constant of nature*: In this type of lab, you try to measure a value for a constant of nature, like the acceleration due to gravity. Your conclusion is the answer to the question "Is the value I have measured 'close enough' to the accepted value?" You can answer "yes" to this question if the difference between your value and the accepted value is small and lies within your error margin.

Error analysis is discussed in Section 16.2.3.

For more on the conservation of momentum, see Section 8.8.

- *Verifying a theory*: In this type of lab, you try to verify that a certain theory, like the conservation of momentum, applies in a particular situation. Your conclusion will be a "yes, the theory has been verified," or a "no, it has not been verified." To determine your answer, you must calculate a value for the quantity in question. You may then answer "yes" if the difference between your experimentally determined value and the theoretically predicted one is within your error margin.
- *Experiencing a phenomenon*: In this type of lab, you try to view or experience a certain physical phenomenon. Your conclusion will be a description of what you saw or experienced. If the experiment was done right, it should be similar to what was described in the textbook, lab manual, or by the instructor.

3.5.6 READING

With all of the emphasis on problem solving and lab work, you might feel that *reading* in your physics class is seriously undervalued. While it is certainly true that reading does not hold the same importance in a physics class that it does in a humanities course, it is still a vital part of the physics learning process.

It is such an important part, in fact, that your physics textbook will probably be *huge*: typical books weigh several pounds and run over a thousand pages. Luckily, however, physics textbook content is not all words; instead, there are plenty of activities, charts, and pictures that help illustrate and clarify concepts.

Section 4.3.3 looks at each of these in depth.

EXAMPLE 3.12: Humanities and Physics Textbooks

Your physics book will differ from humanities texts in ways other than length and content. Here we compare the two types of reading materials.

Example 3.12 Continued...

Humanities textbooks are often wordy and circular, taking numerous interesting detours this way and that, repeating or summarizing material over and over again. The purpose of this sort of writing style is to examine a topic from multiple angles, to reinforce the content and develop multiple levels of meaning. The author does not necessarily expect you to understand each new concept right away, since he or she will probably return to the idea again.

Furthermore, humanities texts almost always outline a "big picture" during topic development. This enables you to grasp an entire concept at the same time as you learn supporting details.

Finally, for obvious reasons, humanities texts almost never include mathematical content.

* * *

Physics writing, by contrast, is precise and to the point. It is organized logically and stepwise—never repeating itself or summarizing past concepts—such that each new idea builds on earlier material. The purpose of this writing style is to move logically from one assertion to the next—building up a rational argument, and supporting it with evidence—to prove a physical fact. The author will therefore expect that *you* will take the time to understand a concept before you move on, because he or she will probably not explain it again.

Because of this emphasis on stepwise, logical argumentation, physics texts hardly ever take the time to outline "big picture" concepts. Rather, they expect you to create your *own* overview during private study.

Lastly, and perhaps obviously, physics texts are highly mathematical. You will find dozens of equations in each chapter, some important, some less so, often with few words to explain them. Physicists—in their mathematical mindset—consider a topic explained once an equation has been developed; thus, the textbook author will often provide an equation and think only a few explanatory words are necessary. Therefore, it is your job to interpret these equations as "mathematical sentences," creating personal meaning with each one.

Math sentences are discussed further in Section 13.4.1.

In short, reading physics textbooks involves a different set of skills than you may have learned or used previously. Chapter 4 will provide reading tips to help you get the most out of your physics texts.

True science teaches us to doubt, and in ignorance, refrain.
CLAUDE BERNARD
French Scientist, 1813 - 1878

3.6 WHY PHYSICS SEEMS "HARD"

Many students of physics, especially non-majors, believe that physics is "hard." There are some good reasons for this belief:

- Physics relies heavily on mathematics. If you feel a bit anxious about your math skills, you will understandably feel anxious about learning physics as well.

- The typical introductory physics course crams quite a bit of information into only one or two terms of classwork. Thus, you have to learn *lots* of material, and you have to do it *fast*.
- And, unhappily, to learn all of that material, you must expend a lot of energy: solving homework problems, going to laboratories, reading a very large textbook—not to mention studying for and taking quizzes and exams—take time and effort to complete.
- In addition, physics instruction—let's admit it—can sometimes be boring. If you are having a hard time staying awake during class (or even, *getting to* class), then physics will be difficult to learn because you are missing critical instruction.

It is only in science that I find we can get outside ourselves. It's realistic, and to a great degree verifiable, and it has this tremendous stage on which it plays.
ISIDOR ISAAC RABI
American Physicist, 1898 - 1988

- Furthermore, just because you are present at, and can follow, the lecture, does not always mean you actually *understand* what is being said. That is, physical concepts often seem odd, counterintuitive, or downright unbelievable. Though your instructor may swear that this or that idea is "true," you may not believe a word of it—which can be a very uncomfortable position to be in.
- And finally, even if you do all of the work and attend all of the classes, it can still be difficult to earn a good grade in physics. There are many reasons for this: it can be hard to get help from your instructor; your classmates can be competitive about grades; and class averages on tests can be low.

NOTE 3.13: A Funny (???) and True Story

At a recent lecture to college-level introductory physics instructors, Sheila Tobias spoke on the topic of grading. She made the comment—based on her research—that at the college level, only math instructors grade harder than physics instructors. A member of the audience, presumably a physics instructor himself, quipped, "We have to try harder!"

When these points are taken together, it is easy to see why you might feel that whatever the benefits of physics, they are *not worth all of the work*! However, despite this perception, chances are that you will do well in your physics class, *as long as* you put in some honest effort. So the question then becomes, "How can I meet the challenge?" rather than, "How hard is it?" The next chapter will help you to meet and beat physics classroom challenges.

3.7 CONCLUSION

Science is organized knowledge.
HERBERT SPENCER
English Philosopher, 1820 - 1903

As we have seen in this chapter, there are many differences between the typical physics course and other, non-physics courses—including instructor-student interactions, student-student interactions, and course-work expectations.

It is true that, taken as a whole, these differences can seem overwhelming, frustrating your attempt to learn physics. However, there is a solution to this dilemma—*but* it takes some effort on your part. *To succeed in your physics class, you must (temporarily) modify the expectations, abilities, and study skills that have worked for you in other, non-physics courses.*

This **book offers an overview; see Part II for more.**

For example, no matter how much you may crave it, you are probably *not* going to get a conceptual overview from your class or textbook. Instead, you are going to have to create one for yourself. Similarly, while you may be used to doing most of your concentrated thinking during class, in a physics course, you will think hardest *outside* of the classroom, while doing homework and studying for exams.

Therefore, you need to develop a *different set* of study tools for your physics toolbox than those that have been successful for you before. These tools are absolutely *within your reach*, but you must seek them out, they will not come to you. This book can assist you: it is devoted to helping you develop and fill in your physics toolbox. The next chapter is especially useful in this regard.

In all things in nature there is something of the marvelous.
ARISTOTLE
Greek Philosopher, 384 - 322 B.C.

4
HOW TO SUCCEED

4.1 INTRODUCTION

Chapter 3: What to Expect

As we learned in the last chapter, physics classes can be very different from other kinds of courses. So, even though you may be used to getting A's and B's in your non-physics classes, you may find it tricky to earn such grades in physics.

The true delight is in the finding out....
ISAAC ASIMOV
American Author, 1920 - 1992

One reason for this regrettable situation is that the study skills, coping techniques, and classroom participation methods that have brought you success in other classes do not necessarily work in the physics learning environment. Therefore, if you want to learn physics well (or, even if you just want to get a good grade in the course!), you must be prepared to *change* how you learn, both inside and outside of the classroom.

The purpose of this chapter, then, is to provide you with time-tested advice about thriving in *physics* class. Topics include a discussion on "physics teams;" an in-depth look at succeeding with physics coursework (such as lectures, readings, homework, quizzes, exams, and laboratories); a problem-solving guide; and information on time management with respect to physics (your "physics workout").

Because of its focus on physics-specific learning methods, this chapter comprises a *critical* component of your physics toolbox. So it is worth it to take the time to incorporate this information into your regular study regimen. And what if you need more general study skills advice? Just refer to the study skills guide in the Appendix.

Appendix: Studying for Success

4.2 THE BEST ADVICE: PHYSICS TEAMS

The generalizations of science sweep in ever widening circles, and more aspiring flights, through a limitless creation.
THOMAS HUXLEY
English Biologist, 1825 - 1895

Of all the study skills advice presented in this chapter (and, indeed, the book), we offer the best and most important information here: *join a physics team*. If you get nothing else from this book, get this: physics is *much* easier to learn in a group than on your own.

A **physics team** is a group of three to five people, all taking the same physics course, who meet several times per week *outside of class*. The team may also choose to sit together during class, as well.

NOTE 4.1: Do Physics Teams Work? And How?

Physics teams *work*—according to extensive research—for two reasons. First, people learn best by *teaching* other people. By its small group nature, the physics team ensures that each student has the opportunity to teach material to other students in the group. Second, research tells us that students learn most efficiently with their peers. By allowing students to work with each other, physics teams enable them to learn more quickly than they would on their own.

In short, physics teams are effective learning tools because students *teach each other* in a fun, stress-free, small group environment. In fact, small group collaboration works so well that *most* physics in the "real world" is done in teams.

We can't stress this enough: physics teams are the *most important* element of your study plan. Their only drawback is that if you do not correctly build your team *from the start*, or if you do not put in enough effort to *maintain* it, the team may end up being more of a hindrance than a help.

EXAMPLE 4.2: Physics Team Traps

Let's look at some traps that may doom your physics team.

One way your team can break down is the "hero trap," in which one person does all of the work for the entire team. The problem with this trap is that only the "hero" learns any of the material; meanwhile, the rest of the team *thinks* they are learning physics—until a test proves them wrong!

Another way physics teams can fail is the "freeloader trap," in which one member of the group does no work of his own, but helps himself to the group's efforts. Not only does the freeloader not learn any physics, but the rest of the team may become disgruntled and dissatisfied with the group effort.

A third problem is the "party team trap," in which teammates have too much fun during meeting times. The difficulty here is that little or no actual work gets done, so that *no one* learns any physics.

The trick to making your physics team successful is to *strike a balance* between having fun and getting work done. The best way to do this is to set some ground rules for the team from the start. The following guidelines might help:

1. **Choose your teammates carefully.**

   - *Get the right people.* You want to pick a team that appropriately mixes work and fun. So with each prospective team member, ask yourself: Does this person seriously want to learn physics? Will he or she work for the *entire* semester? Is this person *too serious*, making group meetings a chore? Or is this person *too fun*, making group meetings unproductive? And think twice before asking your good friends to join the team—you want to get work done, not just have a regular excuse to party.

   - *Get different kinds of people.* The idea here is to end up with as many different viewpoints as possible, so that you can discuss physics from as many different angles as you can. Pick people from different walks of life, with different abilities and talents, and different outlooks on physics. Just be careful that you don't pick *such* different individuals that the team can't work together effectively (defeating the rule above).

2. **Decide upon a game plan.**

   - *How serious do you want to get?* Your team meetings can be as informal as a fifteen-minute review of material prior to class, or as involved as a holistic, team-based study plan for the entire term. It's all up to you. Spend a few minutes at your first meeting determining the group's goals for the term.

3. **Set regular meeting times and places...and then stick to them!**

   - *Strive for at least three times a week.* You might want to meet with your team before or after class, at lunchtime, or in the evenings. It really doesn't matter when these meetings take place; the most important thing is that they take place regularly and preferably several times a week—so that your team never gets behind.

   - *Choose a regular meeting place.* Find a quiet, comfortable, out-of-the-way location that is free of distractions and available for use at all of your meeting times. Libraries are ideal for this purpose, but dorm rooms or class rooms will also work well.

   - *Stick to your schedule.* Once a schedule of meeting times is developed, stick to it! Only miss a meeting for the most important reasons.

4. **Have teammates teach each other.**

   - *Explain concepts to the group.* Physics teams work by allowing students to become teachers. So each team member must be responsible for explaining concepts they understand to other teammates who don't. In this way, the entire group—one by one—gets a chance to verbalize concepts and correct mistakes in understanding.

   - *Everyone must participate equally.* Everybody in the team must have an equal chance to learn. So be on the lookout for teammates who hog the spotlight, or others who fade into the background. If either of these events happen, in a friendly way, remind "smarties" to let other people have their turn, and encourage "wallflowers" to ask questions and participate in discussions.

   - *Rephrase explanations.* After each explanation, and before moving on, have a different teammate rephrase the concept. Repetition doubles your chances of understanding new material, especially when the concept is said in a different way.

5. **Don't copy.**

   - *Never use the group session as an opportunity to copy work or "divide up" your workload.* Not only can copying get you in a load of trouble (you could be expelled for this!), but also it completely defeats the purpose of having a study session in the first place: when you copy, you are not *learning* anything. A better route is to have a team member explain *how* they did a problem, including the appropriate equations and concepts, and then have the rest of the team work the problem individually.

6. **Have fun.**

   - *Finally, have fun*, get to know some new people, and share your feelings about the class and the instructor. Just remember that your first priority is *learning physics*; don't waste *too* much time making plans for the weekend or insulting your instructor!

EXERCISE 4.3: Setting Up Your Physics Team

Use the following checklist to set up your physics team.

- ☐ Develop a list of possible people for your team.
- ☐ Ask the people from your list to join your team.
- ☐ Have your first meeting:
  - o Decide upon a game plan.
  - o Pick a schedule and regular meeting location.
  - o Discuss "rules of the game," including how to share talking time, what to do about copying, and so on.
- ☐ Start having regular meetings.

You may initially feel embarrassed or reluctant to start a physics team: after all, you may have to ask strangers to join your group. *Get over it!* First of all, it is very likely that your classmates are just as uncomfortable in the class as you are, and so they may actually be grateful for your offer of help. Secondly, once your team gets going, you will have friends in your physics class, making the lecture less lonely. But the most important reason to make the plunge into a physics team is that physics teams *work*: research has shown that students who study in teams *get better grades* than those who don't!

In fact, the benefits of working in teams are so great that your instructor may even help your effort by posting a notice or making a class announcement about teams. But even if *you* have to do the work to set up the team, it's worth it! The benefits of creating a physics team far outweigh any awkwardness you might feel in putting one together.

4.3 GETTING THE BEST GRADE

In Section 3.5.

As we pointed out in the last chapter, your physics class will have many elements—lectures and readings; homework, quizzes, and exams; and laboratories—some of which will be graded and some of which won't. However, as we also saw in that chapter, the scores you earn on graded assignments will be a *direct reflection* of how well you did on the non-graded portions of the course. Therefore, in order to earn the best grade in your physics class, you need to know how to excel in *all* aspects of the course. This section is designed to help you do just that.

4.3.1 LECTURES

In Section 3.3.

Physics class time, as we saw in the last chapter, is almost completely devoted to lecturing and instructor problem solving. This type of instruction requires different classroom skills than those that work in a discussion-style class. The following tips will help you use your physics lecture time most effectively:

1. **Come prepared.**

   - *Read the assigned text before class.* Even if you don't understand everything in the reading, you will get an outline of important concepts. As a result, the up-coming lecture will make more sense and you will get more out of it.

2. **Stay alert.**

   - *Do your best to follow the lecture and stay alert for the whole class period.* Admittedly, this can sometimes be a difficult task. But the more learning you do *in* class, the less you have to do outside of class. So take all necessary steps to stay alert: bring coffee or food, sit next to friends (as long as you don't have *too* much fun!), concentrate on taking good notes, and so on.

3. **Remember that lecture periods are primarily meant for you to copy down the instructor's work so that you can review it later.**

Note-taking skills are discussed in the Appendix.

- *Take good notes.* It is *critical* that you develop good note-taking skills for your physics class; otherwise, you will not be able to get down everything that is being said *and* written on the board.
- *Write down everything exactly as it is said or written, including exact representations of any figures or diagrams.* Make sure you copy down lecture material *exactly*—even if it doesn't make sense at first. Later, you will need to recreate the lecture in your mind, and your notes will be critical aids in this regard: if you take sloppy notes, you will have a poor understanding of the material. Take special care to faithfully reproduce any graphs, charts, or diagrams that appear during the class: that "stray" mark may actually be an important part of the picture.
- *But don't be just a stenographer.* Take some time during the lecture to *think* about what you are writing down.
- *Recognize that the instructor will probably make a mistake sooner or later.* Instructors are people too! However, this means that your carefully transcribed notes will also be mistaken. Don't worry, as a part of your "physics workout," you are going to fill in your notes with information from the textbook, so you can correct any errors at that time.

Physics workouts are covered in Section 4.5 below.

4. **Ask some questions during class.**

- *Don't ask too few...* Asking questions in class will help keep you alert, so feel free to raise your hand when you don't understand a concept. It is likely that you are not the only confused one!
- *...But, don't ask too many....* Your instructor (and your classmates) will probably not appreciate it if you interrupt lecture too often; after all, physics classes usually do *not* have a discussion format.
- *...And, ask smart.* The best physics students know that a well-thought-out question not only enhances their own understanding of the material, but may also result in "brownie points" from the instructor. So think a little about your question before you ask it.

The essence of Science: ask an impertinent question, and you are on your way to a pertinent answer.
JACOB BRONOWSKI
French Mathematician, 1908 - 1974

5. **Finally, try to have a little fun.**

- *The more engaged you are in the class, the better you will do.* Don't get too carried away, but do try to enjoy your physics class: there are lots of interesting things to learn!

4.3.2 MEETING WITH YOUR INSTRUCTOR OUTSIDE OF CLASS

Unfortunately, the harried pace of the typical physics lecture does not leave much time for interpersonal interaction or private question-and-answer periods with your instructor. Yet, to learn physics most effectively, you *need* just that sort of individualized assistance.

Although your physics team will help you get some of that personalized instruction, you still need to meet with your physics instructor on a regular basis. That is, *you must take time out of your busy schedule to meet with the instructor after class or during his office hours.* Fortunately, as we saw in the last chapter, the vast majority of physics instructors welcome any opportunity to help students outside of class.

So, talk with your instructor and set up a mutually convenient time to meet.* Then, use the time to:

- Ask questions about concepts,
- Go over homework problems, and/or
- Review for exams.

We promise that this extra time *will* pay off in some form, whether it is a "benefit of the doubt" grade increase at the end of the term, some "inside" information on test content, or even a friendship with a really interesting person.

EXAMPLE 4.4: A True Story

A physics student tried the above advice, and attended her instructor's weekly office hours and pre-test review sessions. During one test review session, the woman realized that she did not understand a certain topic. So she asked the instructor to provide a "mini-lecture" to shore up her knowledge.

The other students at the session scoffed, sure that the topic was so arcane that it would "never" be on the test. The instructor, however, silenced them and proceeded to give a short but careful mini-lecture on the topic.

When the test came, that very same day, the "arcane" topic made up fully 20% of the exam. Only the students that attended the review session had any idea that the instructor thought the material was important, and therefore, were the only ones to get A's on the test. And the woman got the highest grade of all.

The moral of the story? Instructors are *people.* They *want* to help students who work hard. If you show them that you are willing to put in the extra effort, they are often willing to help you out in return.

* One note of caution: Whatever you do, don't set up a private meeting with your instructor and then back out at the last minute (of course, except in the case of a true emergency). Instructors everywhere have been "stood up" by students before, and we can assure you that they do not look kindly on such transgressions.

4.3.3 READING

In Section 3.5.6. 

The best way to prepare for your lecture is to read the assigned text prior to class. However, as we discussed in the last chapter, your textbook may be intimidating: typical physics texts can have thousands of pages, and involve multiple instructional approaches, including text, examples, exercises, boxes, illustrations, graphs, charts...and so on.

With so much material to absorb, it is easy to see why many students skip their physics reading assignments. Unfortunately, this is the wrong tactic. Physics reading is an important tool in your physics toolbox.

Yet, because physics textbooks are *different* from other kinds of books, you must read them differently than other books. The following tips will help you get the most out of your physics reading:

1. **Read the text *very* slowly.**

    - *Try to understand each sentence before moving on.* Remember, the textbook will not repeat concepts or return to old material during exposition, so you must make sense of *each point* before going on to the next one. To help you read the book at the appropriate speed, think of your textbook as a reference—like a dictionary or an encyclopedia—rather than a story, novel, or essay.

For more on mathematical sentences, see Section 13.4.1. 

    - *Read equations like "mathematical sentences."* Don't skip over equations as you read. Instead, sound out each symbol with its math meaning in your head (or out loud). Then, interpret the entire equation in your own words: ask yourself, "What does the equation mean to me?"

The Appendix has more information on taking notes while you read. 

2. **Take notes while you read...***

    - *...Either in the margin of your book or on a separate piece of paper.* This step is important: research shows that the very act of writing down new information "jump starts" the learning process and increases retention.

    - *Take just enough notes to capture important points only*: if you take too many notes, you waste time better spent solving problems; if you take too few, you can't reconstruct the author's argument later.

3. **Look for a "big picture," but keep it simple.**

    - *Make connections between topics...* Ask yourself: How does this concept relate to what I have already learned?

    - *...But, don't get so carried away with details of your "big picture" that you no longer see fundamental issues.* Keep in mind that physicists

* Don't bother to outline your text with a marker, though; studies have shown that, for scientific reading, outlining the text is not as effective for learning as taking notes.

are always looking for the essence, the simplicity, of a problem. Don't get tangled up in detail. See the forest for the trees.

4. **Remember the main issue: *problem solving.***

   - *Take an interest in, and try to understand, physical concepts, but not at the expense of problem solving.* Problem solving is ultimately what physics is all about. This means that if you have to choose between time spent reading and understanding concepts, and time spent solving problems, choose problem solving!

5. **Use the text's instructional features to your advantage.**

   - *The examples, exercises, charts, graphs, and other elements of your book are there to help you learn, so use them!* Table 4.1 below lists common features of most introductory-level physics texts, and how to cope with them.#

Table 4.1: Common Features of Introductory Physics Texts

FEATURE	HOW TO COPE
TABLE OF CONTENTS & INDEX	These components of the textbook help you find topics as you "skip around" during problem solving or reference-checking. Get in the habit of using these; they will make your work much easier!
PREVIEW & SUMMARY	These usually appear at the beginning and end, respectively, of each chapter. Skim through them to get an idea of what is to come, and to review what you have just learned.
DEFINITIONS	Throughout the text, quantities and terms will be "defined," either in equation or text form. Memorize these, but don't waste time trying to figure out "where they came from." They're just names!
EXAMPLES & EXERCISES	Work out each example or exercise yourself, alongside the text. Don't try to "do it on your own;" instead, take the opportunity to use the text as your personal problem-solving guide.
BOXES	These short, informative essays on topics in the text are usually optional. However, skim over them, because they usually provide the "story" or "historical context" that makes comprehension easier.
TABLES & GRAPHS	Think of tables and graphs as text in a more concise format. Thus, always take time to familiarize yourself with titles, captions, and column or axis headings, as well as the general trend of the data.
ILLUSTRATIONS	Illustrations include photographs, drawings, diagrams, and maps. They are usually numbered so that you can refer to them from the text. Captions usually clarify their meaning. Always spend some time with illustrations so that you get a better idea how the physical concept applies to reality.
EQUATIONS	Never skip equations; read them like text. Try to interpret the meaning of the equation "sentence."
DERIVATIONS & PROOFS	Technically, a physical equation is not considered correct unless it can be derived from fundamental principles using algebra. Therefore, many texts show derivations or proofs so that you can be sure that each equation is sound. Often, these are optional; in that case, skim over them and move on. If, however, in your class derivations/proofs are required, work through them step by step.
END-OF CHAPTER PROBLEMS	Do as many as possible, even if they are not assigned! Check your work with the solutions manual.

This book attempts to get you used to these features by including them in the text. For example, you will find plenty of exercises, examples, and boxes here; be sure and read them as you would read the regular text: carefully and in sequence.

Finally, as we leave this section on physics textbooks, we remind you that—unlike the case in some humanities-type courses—physics reading is just *one* ingredient in a complicated recipe, not the entire dish. There are many other components of your physics course, and they need your time and attention too.

So, don't read your physics textbook too much or too closely: one goal of introductory physics courses is to lessen your reliance on *words* (while increasing your reliance on *numbers*). Instead, *budget* your time carefully between reading and other elements of the course, especially problem solving.

4.3.4 HOMEWORK

More on problem solving in Section 4.4.

As we have said before, and will say again, problem solving is the core of physics. The key to problem-solving mastery is practice, practice, practice; that is, *doing homework.* This section covers some general tips for earning your highest score on homework assignments. More specific information on problem solving in general comes later in the chapter.

1. **Make homework a top priority.**
   - *You cannot pass tests without problem solving, and you cannot learn problem solving without homework.* Therefore, make your physics homework your *top* priority.
2. **Give each and every problem your best effort.**
   - *Problem solving takes time*! So, make sure you allot enough time for it. A good rule of thumb is an hour of dedicated problem solving per day.
   - *Use a consistent problem-solving method.* Section 4.4 provides one; your instructor may have another: choose the one that works best for you and stick with it.
   - *Work in groups.* What you don't understand, someone else will, and vice versa. This is the perfect opportunity to get the most out of your physics team.*
3. **Get help when you need it.**
   - *Don't waste time getting frustrated.* The shame in physics classes is not in needing help, but in not getting help when you need it! Help can come from many sources: your instructor, your teaching assistant (if any), your teammates, tutors, and so on. Seek out help and get your questions answered.
4. **Work extra problems.**
   - *The more problems that you solve, the better you get at problem solving.* Therefore, take it upon yourself to work problems that

* Unfortunately, some instructors do not allow group work on homework. If this is your case, do not do anything that could jeopardize your grade or get yourself accused of cheating.

are not assigned *in addition* to the graded ones. If you need some incentive to work these extra problems, just keep in mind that most physics instructors, sooner or later, take exam questions straight out of the textbook.

We cannot stress this point enough: doing well on homework is the best way to ensure that you will do well on exams. So make the effort to excel on your homework.

4.3.5 QUIZZES & EXAMS

The best way to study for quizzes and exams is to solve problems. That is, to do your homework! Then, when you are done with the homework, solve unassigned problems. Then, when you have solved those problems, make up your own problems, and solve those. Your instructor may even be willing to provide old exams; if so, get a copy and work out those problems, too. In other words, *the key to quiz and exam success is practice, practice, practice!*

Reason rules the world.
ANAXAGORAS
Ancient Greek Philosopher, 500 - 428 B.C.

In addition to problem solving, the following tips for preparing for and taking physics quizzes and exams will also help:

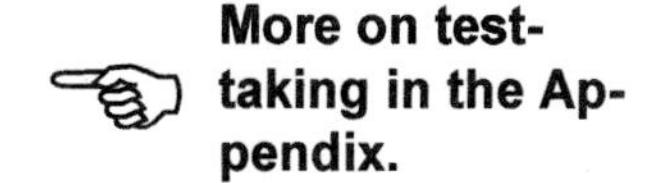
More on test-taking in the Appendix.

1. **During your pre-test practice, strive for speed *and* accuracy.**

   - *In a test situation, it is not enough to know how to do a problem; you must also do it quickly and correctly the first time.* To improve your skills, try solving a problem, putting it away for a time, then solving it again: the second attempt should be faster and more accurate.

   - *One caution with this plan*: don't get in the habit of memorizing problems and their solutions. You are trying to learn how to tackle *new* problems, not rehash old ones!

> **EXAMPLE 4.5:** Memorizing Homework Problems
>
> While it is true that homework problems sometimes show up verbatim on tests, the vast majority of exam questions will be problems you have never seen before. Therefore, it does not *help* to know how to solve problem #43 and only problem #43; rather, you need to know how to solve problems *like* problem #43 should they appear on the test.

2. **Create a review sheet.**

   - *On one 8 ½ x 11" piece of paper, write down all of the formulas needed for your quiz or exam.* But only write down the equations that will help you on the test; not everything you have ever learned!

- *If allowed by your instructor*, take your review sheet into the test and refer to it while you work. If not, take time before the exam to memorize the formulas on the sheet. Then, immediately upon receiving the test, write them down so that you won't forget them.

3. **Take it slow.**

   - *Read the instructions carefully*. Do you have to answer every question, or can you pick and choose? Do you have to do anything special with your work, like draw a box around your final answer? The only way you will know these things is if you read the instructions first.

   - *Read each problem carefully.* Does it have multiple parts? What is the final answer you have to find? The number one cause of test mistakes is students not correctly reading the problem. Don't let it happen to you.

   - *Take a few moments to think about each problem before you dive in.* There may be a shortcut if you just stop to think about it!

   - *Watch out for multiple-part questions.* If you can't answer all parts, try to solve what you can. If you need a solution from one unsolved part to solve another, make up a reasonable value and use that. Just be sure to leave a note to the grader explaining your actions.

4. **Keep at it.**

   - *If you can't start a problem, at least try writing down formulas and sketching pictures.* These might jog your memory, and, at the very least, may score you some partial credit.

   - *If you still can't solve a problem, move on to the next one.* Don't waste time! The goal is to get as much of the test done as possible.

5. **Allot enough time to check your work at the end.**

   - *Did you do the math right? Is your writing legible? Check it!*

4.3.6 THE LABORATORY

The laboratory is your chance to get "hands on" with physics. If you find the theoretical aspects of physics lectures and problem solving difficult, laboratory is your way to see (literally!) how physical concepts work in the real world. The following tips will help you get the most out of this unique opportunity:

1. **Be prepared.**

   - *Read the manual before you get to lab.* Although your instructor will provide some instruction on doing the lab, you will be better off in the long run if you have a "feel" for what to expect from the lab and the necessary equipment you will be using before you get there. So check out that lab manual *before* class.

   - *Make a plan.* Take some time to plan your experimental strategy before you jump in. The short time you spend in laying the groundwork before beginning the lab will be more than recouped during the lab (when you don't have to repeat the experiment because you did it wrong the first time!).

2. **Take careful measurements.**

   - *Familiarize yourself with the equipment before you begin.* In most cases, you may have never seen the lab paraphernalia before. So, spend a little time getting to know the devices you will be using before you start the experiment.

   - *Use the equipment correctly, to ensure you get accurate measurements.* If you have a question about how to take measurements, ask!

   - *Carefully record your measurements.* Get a notebook and reserve it just for labs. In it, neatly write down *everything* you measure or do in the experiment. Create tables for the data if necessary. But don't get carried away; the important thing is to have a neat record of your work, not a masterpiece of artistic organization.

What we call physics comprises that group of natural sciences which base their concepts in measurement....
ALBERT EINSTEIN
American Physicist, 1879 - 1955

3. **Analyze your results.**

   - *Do error analysis.* Your instructor, or Chapter 16 of this book, will provide you with more detailed instructions on this topic.

   - *Examine your results and draw a conclusion.*

Refer to Section 3.5.5 for the types of conclusions you can draw.

To summarize all of these points, we would say that the key to success in physics laboratories is in *taking care*: taking care in making your measurements, taking care in analyzing your results, and taking care in drawing conclusions. It can be tempting to rush through labs (because you can sometimes leave if you finish early), but *don't do it*! You will waste time and effort in making mistakes that have to be fixed...time and effort better spent solving homework problems.

4.3.7 A NOTE ABOUT CALCULATORS & COMPUTERS

Today, calculators and computers are just as much a part of the physics class as are pencils and paper. Unfortunately, if you are not used to using these devices, they can make your life more difficult rather than less. This section is designed to help you get the most out of these important problem-solving tools.

CALCULATORS

Calculators today are incredibly complex; even the simplest ones contain as much memory as some PC's. In general, there are two types of scientific calculators on the market:

- 3 × 5 inch-card sized calculators with a simple array of important functions, like exponents and trigonometric functions.
- Book-sized calculators/hand-held PC's with a dizzying selection of graphing functions, word processors, and other bells and whistles.

We recommend (and use ourselves!) the first kind, for two reasons. First, they are much simpler to learn to use, and the faster you can get started on the calculator, the better! Second, you simply don't *need* all of the power of the fancier calculators—your computer (see below) will do the more complicated work. Whichever style you prefer, however, take some time to pick the right calculator for you: you will be spending a *lot* of time with your calculator over the term, so select the one that *you* feel most comfortable with.

Once you buy a calculator, spend a little time reading the manual and playing around with it. Find out how to use the function keys (including the trigonometric and exponential functions), try out the graphing mode (if any), and experiment with the order of input: *it makes a difference whether you add or multiply first*! The more comfortable you get with the calculator and its functions now, the faster problem solving will go later.

Finally, get in the habit of carrying your calculator around with you. Yes, it is a touch geeky, but you never know when you might need it!

COMPUTERS

Computers are fast becoming indispensable parts of physics education. It is easy to see why: there are physics web sites that offer great instructional animations, spreadsheets that do your calculating for you, and tutorial programs to help you through rough spots in your class. Not to mention that hardly any physics discovery today happens without them.

So you need to get your hands on a computer. If you can't afford one, your school's computer lab should have free (or low cost) computer time; all you have to do is sign up for it. If you can afford a computer, buy the *best* one you can afford: it will last longer before becoming obsolete, and it will do your work faster. And don't forget to get a good printer.

Once you have access to a computer, you should spend a good amount of time learning how to use its basic elements, such as web browsing, word processing, and spreadsheet calculations. Take a class if you need to—just make sure that you can use the computer competently and with some degree of speed.

After familiarizing yourself with the basic functions of your computer, start solving problems with it. Yes, it will take a little while to get used to, but you will most likely save time over the long run. Then use your computer to analyze your lab results: spreadsheet programs are *ideal* for solving complicated lab equations and presenting your results neatly. Then visit some of the countless physics web sites that exist today: you are sure to find at least one that interests you. In short, do as physicists do: use your computer for physics!

4.4 PROBLEM SOLVING

We hope that we have finally convinced you that problem solving is an *essential* skill for your physics course. This section will help you sharpen up your problem-solving skills so that you can do your best on each assignment.

4.4.1 A PROBLEM SOLVING GUIDE

Perhaps the hardest part of problem solving is *getting started*. So, in this book, we provide the following problem-solving guide. Your textbook or instructor may also offer other problem-solving strategies; pick the method that feels the most comfortable to *you*.

When the only tool you own is a hammer, every problem begins to resemble a nail.
ABRAHAM MASLOW
American Psychologist, 1908 - 1970

The one hitch with any kind of problem-solving guide is that—by definition—it will only get you started. All physics problems are different, so no one procedure can help you in every circumstance. But, as you do more problem solving, you will develop more problem-solving techniques that will help you in more situations. The key—as always—is to practice, practice, practice!

NINE STEPS FOR EFFECTIVE PROBLEM SOLVING

1. **Read the problem carefully.**
   - Many problem-solving mistakes come from reading a problem too quickly, so that important details are missed. Eliminate this problem by reading the problem right the *first* time.

2. **Then, re-read the problem to pick out important "code words."**
   - Most physics problems are "word problems," which have a story-like format. On the second read-through, your task is to *ignore* most of the setup and hone in on the *few* important words, called **code words**, which give away the solution to the problem. Table 4.2 below lists some code words and their meanings. You should memorize this list, and then add to it as your problem solving skills improve.

Table 4.2: Physics Problem-Solving Code Words

CODE WORD	MEANING	CODE WORD	MEANING
is, are, has	equals (=)	at rest	velocity is zero
sum, more, total, altogether, greater than	add (+)	constant velocity	acceleration is zero
difference, fewer, left over, remain, less than	subtract (–)	slippery, icy, smooth surface	friction is zero
of, product, times	multiply (×)	"long" object	object has infinite length
per, percent, ratio	divide (÷)	"very small" object	consider object as a point particle
force	free body diagram, Newton's Laws	"very large" object	consider object to be infinitely big
collision	momentum, energy	objects are "close"	the field from one object affects other
free-fall	acceleration is constant: $a = -g$		

EXAMPLE 4.6: Lotus Esprit Problem, Part I

Let's consider the following problem:

> *"A Lotus Esprit super-coupe can travel* 30 m *from rest in* 3.3 s. *What is its acceleration?"*

From Step 1 of our problem-solving guide above, we know that we must read this problem carefully (which, hopefully, we just did). Now, from Step 2, we see that we must re-read the problem again, ignoring the setup and looking for code words:

> *"A Lotus Esprit super-coupe can travel* 30 m *from rest in* 3.3 s. *What is its acceleration?"*

The setup here is the super-coupe. However much we may be interested in this fast, cool car, we must ignore it (for now), and hone in on the *important* details. The code words here are:

Distance traveled $\equiv x = 30$ m
"From rest" $\Rightarrow$ Initial speed $\equiv v_i = 0$ m/s
Time traveled $\equiv t = 3.3$ s
Acceleration $\equiv a = ?$

From these code words, we immediately know that we will be looking for a quantity called acceleration, and we will use distance, initial speed, and time to find it.

3. **Draw a good picture or two.**

- A picture helps you to visualize important details of a problem, allowing you to understand the problem better. A "good" picture (in physics) is *not* a realistic illustration or drawing; rather, it is a bare bones, quick sketch (roughly to scale), of essential details and components in the problem, and *nothing* else.
- Sometimes you may need to draw more than one picture to depict different objects, situations, or times.

Particular facts are never scientific; only generalizations can establish science.
CLAUDE BERNARD
French Scientist, 1813 - 1878

EXAMPLE 4.7: Physics Sketching, Part I

With a little practice, *anyone* can learn to sketch quick and useful diagrams, no matter what their drawing skill. The trick is to take a little care in making straight lines straight, circles circular, and right angles perpendicular. For example, draw a square "squarely":

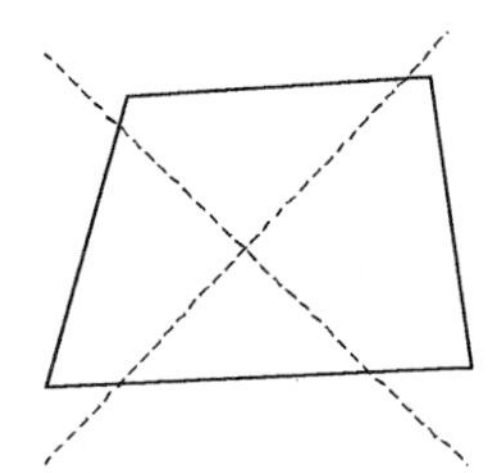

And draw a circle "circularly":

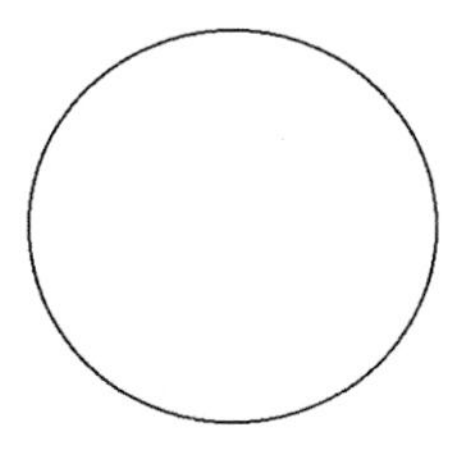

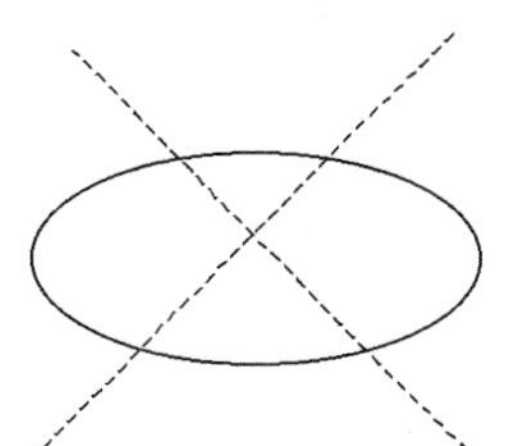

And, take care with right angles:

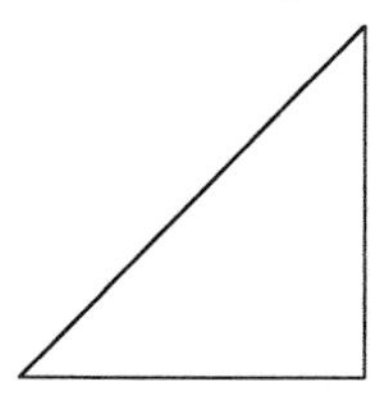

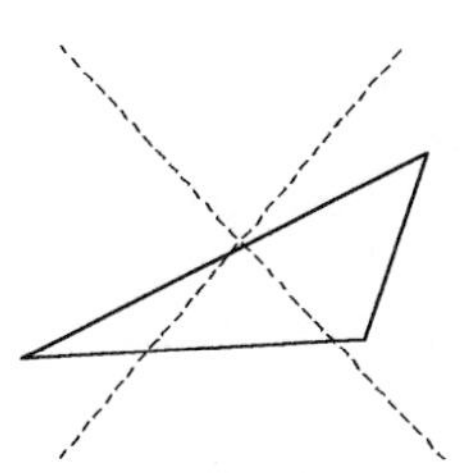

You can draw other angles too! Note that the bottom left angle in the right triangle above left is $45°$, so a $30°$ angle would be about two-thirds of that, while a $10°$ angle would be about one-quarter of it:

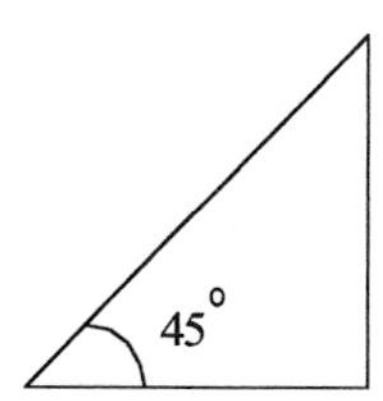

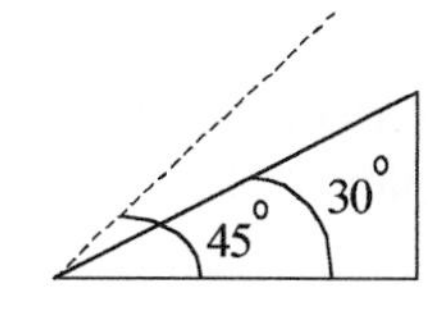

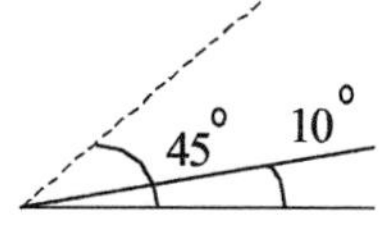

Example 4.7 Continued...

Believe it or not, you can even learn to draw in three dimensions! Just keep lines or planes vertical, horizontal, or at an angle of $45°$ relative to each other:

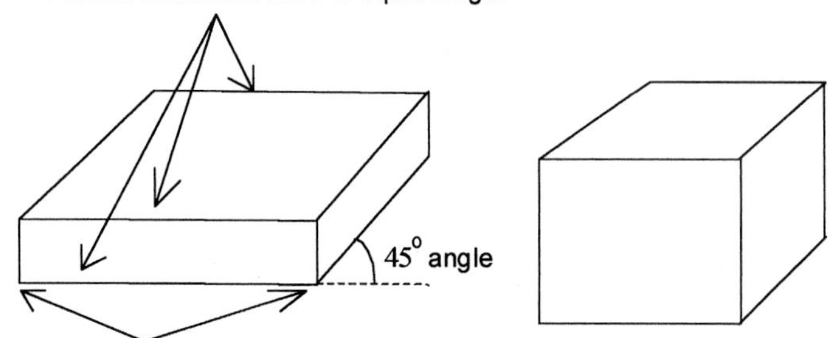

Cylinders have oval ends, rather than squares:

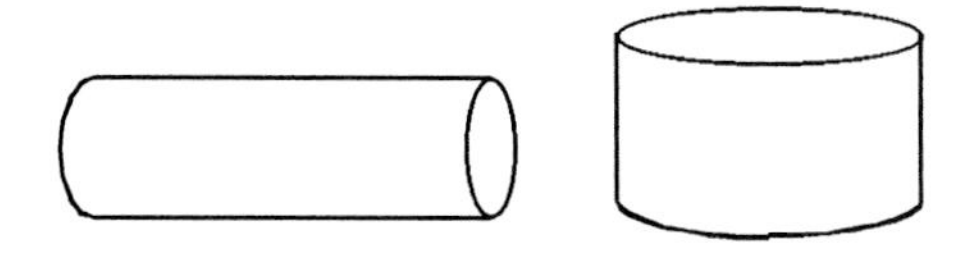

And spheres are merely circles with shading or perspective added:

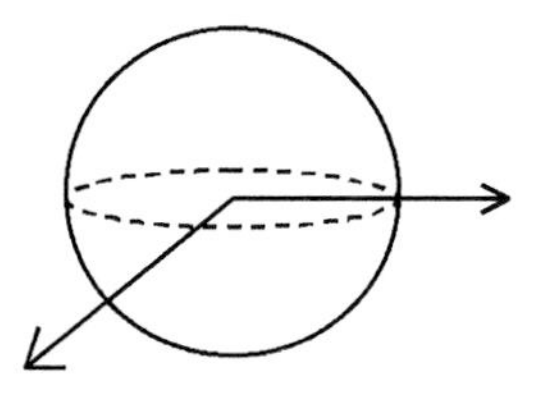

EXERCISE 4.8: Physics Sketching, Part II

Now you try! In the margins of this book, sketch the following:

- A square
- A rectangle that is twice as long as it is tall
- A circle
- An isosceles (two legs of equal length) right triangle
- A cube
- A sphere

EXAMPLE 4.9: Lotus Esprit Problem, Part II

A good physics picture consists of just the basic elements of a problem, and nothing else. For example, bodies are reduced to simple boxes or even just points. Arrows indicate motion and all quantities are labeled with symbols (not numbers!).

Example 4.9 Continued...

So, in the case of our example problem in Example 4.6 above, we might draw our Lotus Esprit as a box, and the road as a flat line. We would draw arrows to indicate the direction of motion and the distance traveled. And we would label x, v_i, and a appropriately. Thus, our picture might look like:

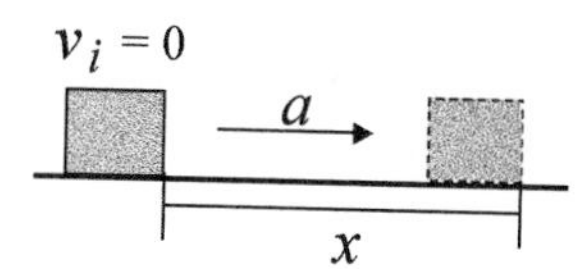

Note that the whole thing took less than thirty seconds to draw. We didn't bother to make anything look *realistic*; we only drew enough detail to clarify relationships among the critical variables.

4. **Make a list of knowns and unknowns.**

- Your ultimate problem-solving goal is to express your final answer in terms of what is known, and nothing else. So, you need a list of what *is* known and unknown.
- When making your list, use appropriate symbols and accepted SI units.

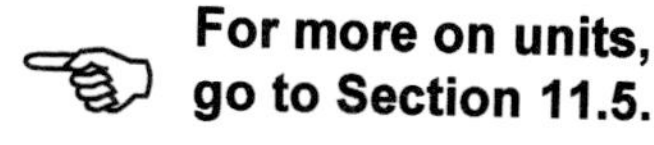
For more on units, go to Section 11.5.

EXAMPLE 4.10: Lotus Esprit Problem, Part III

In our example problem, our knowns are x = 30 m, v_i = 0 m/s, and t = 3.3 s. Our unknown is a, which has units m/s^2.

5. **Choose one or more equation(s) that fits your list.**

- You will always need the same number of pieces of information as unknowns. Most of the time, but not always, the information is given in the form of equations.
- Usually, you will be given just the right amount of information to solve the problem; sometimes, though, there will be more or less data than you need. Your task in these cases is to weed out unneeded information, or look up or assume other facts.
- Pick your equations carefully: some equations apply only to certain situations.
- Be on the lookout for "classes" of problems that have similar solutions: once you have solved one problem in a class, you can use the results of that problem to solve other, similar problems more quickly.

EXAMPLE 4.11: Lotus Esprit Problem, Part IV

We want an equation that has x, v_i, t, and a in it. We only need one equation because there is only one unknown (a). Referring to our kinematic equations from Chapter 5, we find one that fits these criteria:

$$x = v_i t + \tfrac{1}{2} a t^2$$

This equation has everything we need in it (x, v_i, t, and a), and nothing we don't. So it's the right equation.

This equation is discussed in Section 5.4.2.

6. **Solve the equation with math.**
   - Carry out the algebra slowly and carefully, choosing the easiest possible route.
   - Don't try to work out the problem in your head! Instead, write out every algebraic step so that you can check your work later.
   - Refrain from plugging in any numbers until the algebra is completely done. At that time, plug in all of your knowns at once, including the units, and solve for the unknown.

When faced with different paths toward the solution of a problem, you should seek always the beautiful one.
HENRI POINCARE
French Mathematician, 1854 - 1912

EXAMPLE 4.12: Lotus Esprit Problem, Part V

For our Lotus problem, we can do some algebra:

$$x = v_i t + \tfrac{1}{2} a t^2$$

$$\Rightarrow \tfrac{1}{2} a t^2 = x - v_i t$$

$$\therefore \quad a = \frac{2(x - v_i t)}{t^2}$$

Now that we have an equation for a that is completely in terms of knowns, we can plug in our known values, and solve for a:

$$a = \frac{2[(30\ \mathrm{m}) - (0\ \mathrm{m})(3.3\ \mathrm{s})]}{(3.3\ \mathrm{s})^2}$$

$$= 5.51\ \mathrm{m/s}^2$$

7. **Check your answer for plausibility.**
   - Ask yourself after every problem, "To the best of my understanding of the physics involved in this problem, and considering my own experiences with similar phenomenon, *does*

this answer make sense?" If your response to this question is "no," there are three possibilities:

1. Your math is wrong: go back and check it.
2. Your assumptions are wrong: go back and check them.*
3. The answer may be purposely non-intuitive. Perhaps the answer was designed to make you think a little more carefully!

The result of a mathematical development should be continuously checked against one's own intuition about what constitutes reasonable...behavior. When such a check reveals disagreement, then the following possibilities must be considered:
(a) A mistake has been made in the [math]...;
(b) The starting assumptions are incorrect and/or constitute a too drastic simplification;
(c) One's own intuition about the...field is inadequately developed;
(d) A penetrating new principle has been discovered.
HARVEY J. GOLD
American Biochemist, 1932 -

EXAMPLE 4.13: Lotus Esprit Problem, Part VI

Our answer from the previous example, $5.51\ \mathrm{m/s^2}$, is about equal to $\frac{1}{2}g$, or half the acceleration due to gravity. We would expect that a fast car, like a Lotus Esprit, would give a pretty good punch when it accelerates. Thus, we figure this answer is at least in the right ballpark.

8. Round your checked answer to the correct number of significant digits, and write it in scientific notation.

See Sections 11.2 – 11.3 for more on these techniques.

EXAMPLE 4.14: Lotus Esprit Problem, Part VII

To correctly write our answer in significant figures and scientific notation, we use the techniques of Chapter 11, obtaining $5.5\ \mathrm{m/s^2}$.

9. Check to see that you answered the question asked.

- Often a correct *result* is *not* the correct *answer*, because it does not satisfy the question posed.

EXAMPLE 4.15: Lotus Esprit Problem, Part VIII

Does our result answer the question asked? The question was:

> *"A Lotus Esprit super-coupe can travel* $30\ \mathrm{m}$ *from rest in* $3.3\ \mathrm{s}$.
> *What is its acceleration?"*

We obtained an acceleration of $a = 5.5\ \mathrm{m/s^2}$, so yes, the question is correctly answered.

Once you have completed the nine steps, you're done! A summary of these problem-solving steps is in Table 4.3 below.

* Be careful, though. Some assumptions are *required* to make the problem solvable; yet, they do not correspond to any real life situation.

Table 4.3: Nine Steps for Effective Problem Solving

STEP	PROCEDURE
1	Read the problem carefully.
2	Re-read the problem to pick out code words.
3	Draw a good picture.
4	Make a list of the knowns and unknowns.
5	Choose appropriate equation(s).
6	Solve the equation(s).
7	Check your answer for plausibility.
8	Round your answer to the correct number of significant figures, and write it in correct scientific notation.
9	Check to see if you answered the question asked.

EXERCISE 4.16 A Runner

Try out your new problem solving skills with this problem:

> *"A fast runner can run* $100\ \mathrm{m}$ *in under* $10\ \mathrm{s}$. *What is her maximum acceleration?"*

Hint: the answer is $2\ \mathrm{m/s^2}$.

4.4.2 WHAT TO DO WHEN YOU ARE STUCK

Sometimes, unfortunately, the problem-solving guide of the last section is not enough to get you through a problem. Here is a list of some ideas to keep you on track:

I think and think for months and years. Ninety-nine times, the conclusion is false. The hundredth time, I am right.
ALBERT EINSTEIN
American Physicist, 1879 - 1955

- *Try slowing down.* Instead of roaring through a problem, try to slow down a bit. Sit and *think* about the problem for a while—or even put it away for a short time—and something will come to you eventually. Problem solving is *not* a quick process. Mistakes, wrong turns, and wasted time are *part of the game*. Even though you may wish it were otherwise, you need to work at a slower pace when problem solving then you would with other types of work. But keep at it: the answer *will* eventually come to you.
- *Ask yourself questions.* When you get stuck, start asking yourself questions, like: What equation should I use? Can I use that concept here? If so, why? If not, why not? Did we cover this problem in class? And so on. One of these questions is bound to jog an idea loose in your head.

- *Find out what is holding you back.* Ask yourself, "Why is this problem hard for me?" Is there something you can do to make it easier? For example, could you break it down into parts? Choose easier numbers? Etc.
- *Look up formulas or examples in your book or your notes.* Equation hunting is a fast way to jump-start your brain.
- *Try working the problem backwards.* Believe it or not, professional physicists use this technique all of the time to solve problems! When they already have an answer (perhaps you will find yours in the back of the book or in the solutions manual), they start there and work backwards to get to the setup.
- *Guess*! Sometimes guessing can be a good way to at least get a problem started.
- *Ask someone for help.* Don't waste your time getting frustrated; instead, get help!

The best problem-solving advice we can give you, though, is the following. The main difference between "novice" and "expert" problem solvers, according to research, is that "expert" problem-solvers think about the problem physically, in terms of numbers and problem types, *before* determining the required equation or concept. "Novice" problem solvers, by contrast, do the last step first: they reach for equations and jump into algebra before they fully understand the problem. To avoid this critical mistake, then, train yourself to work problems like experts: *think* about the problem before jumping headlong into it!

4.5 STUDYING FOR PHYSICS OVER THE TERM

Although we may fervently wish it otherwise, learning physics takes a lot of time. Unlike a humanities course, in which the required work can (sometimes) all be crammed into the end of the term, physics knowledge is *built up* over time. If you try to force all of your physics material into your head all at once, you will *not* learn it, and *you will flunk your exams.* Therefore, you must study physics slowly, stepwise, and each and every day.

The good news is that if you work according to a well-thought-out plan—which we help you develop in this section—you can both *maximize* your learning and *minimize* the time spent doing it. We call our version of a study plan a **physics workout**, because it has many things in common with the kind of workout you get at a gym.

Your physics workout has two parts: what you need to do each day, and what you need to do over the term.

NOTE 4.17: Physical Vs. Physics Workouts

The purpose of a physical workout is to build muscle. You do this by exercising regularly (experts recommend a minimum of 30 minutes a day, five to six times a week), and by doing several repetitions of each exercise. Unfortunately, though, if you lower your workout frequency, or the number of repetitions of each exercise, you will not only stop making progress, but you will actually *lose* the results you have built up over time.

The physics workout operates in the same way, although you build *brain* muscle, not your body. You do this by studying regularly (we recommend a minimum of 60 minutes a day, five to six times a week), and by solving many problems. If you don't keep up with this workout, though, you will not learn new material, and you will *forget* old material.

4.5.1 YOUR DAILY PHYSICS WORKOUT

The scientific mind does not so much provide the right answers, as ask the right questions.
CLAUDE LEVI-STRAUSS
French Anthropologist, 1908 -

The best way to ensure that you will study for physics every day is to *schedule* (at least) sixty minutes of study time into each day of the week. Label this study time your daily physics workout, so that you don't "accidentally" skip it.

When you make your schedule, take your other commitments, jobs, and activities into consideration, so that your schedule is as convenient, satisfying, and easy to implement as possible. Set study times for team meetings *and* for alone study, and be sure to schedule in "buffer zones" for unexpected assignments or well-earned leisure activities.

EXAMPLE 4.18: One Student's Schedule

Anna is taking five classes, including physics, and has a part-time job. She also likes to run three times a week. The study schedule she came up with to meet all of her needs is shown below.

	Monday	Tuesday	Wednesday	Thursday	Friday	Saturday	Sunday
8		CALCULUS		CALCULUS			
9	*Eat*	↓	*Eat*	↓	*Eat*		
10	ENGLISH	CHEMISTRY	ENGLISH	CHEMISTRY	ENGLISH	*Work*	*Study*
11	HISTORY	↓	HISTORY	↓	HISTORY		
12	*Team Mtg*	*Alone Study*	*Team Mtg*	*Team Mtg*	*Team Mtg*		
1	PHYSICS	↓	PHYSICS	PHYSICS	PHYSICS		
2	*Work*	*Work*	*Work*	LAB	*Work*		
3				*Laundry/Clean*			
4							
5	↓	↓	↓	↓	↓		
6	*Run*	*Buffer!*	*Run*	*Study for Quiz*	*Run*	↓	
7	*Eat*	*Eat*	*Eat*	*Eat*	*Eat*	*Eat*	
8	*Study*	*Study*	*Study*	*Study*	*Fun*	*Fun*	↓
9							*TV Night!*
10							
11	↓	↓	↓	↓	↓	↓	

Example 4.18 Continued...

Notice that since Anna's physics class starts at one, her team has decided to meet for lunch (from 12 - 1) each day before it starts, except for Tuesdays when Anna studies alone. Also note that her physics class has a quiz every Friday, so Anna schedules alone study time each Thursday evening to study for it.

Anna also schedules in time to study for her other classes, as well as work, eating and running. She even schedules in two nights of fun, a TV night, time for her personal needs, and a buffer zone!

Divide each study session into two phases: the *review* and the *look-ahead.*

THE REVIEW

Every day, complete the following tasks:

See the Appendix for more on reviewing class notes.

1. Rewrite, or summarize, class notes from the previous lecture in a way that allows you to *review* the concepts presented.
2. Fill in any missing information in your notes from the textbook, Part II of this book, or from conversations with teammates.
3. Check to be sure that there are no holes in your knowledge of the covered concepts. If there are, mark the questionable material and get help for your question as soon as possible.
4. Work assigned homework problems. Check your progress, when you can, with the answers at the back of your book or in the solutions manual. If you make a mistake, go back and review your notes or the book for help. If you are still stuck, mark the problem and get help as soon as possible.
5. If you had success with the homework problems, work some more problems that were not assigned.

THE LOOK-AHEAD

When you are done with the review, take a short break. Then begin the look-ahead:

1. Read the assigned section in your textbook, and the associated section of Part II of this book, covering the material you will learn at the *next* lecture.
2. While you are reading, work example problems in the text alongside the given solutions. *Do not cheat yourself on this step;* you should be *working* the problems, not just following along with someone else's work.
3. Familiarize yourself with any new concepts. If your textbook offers short "self-quizzes," give these a try; if not, make up your own.

> The trick here is *not* to try to learn the material all by yourself—you will, after all, be getting a lecture on the topic—instead, you are trying to get a *feel* for what is to come and how it relates to what you have just learned.

When you are done with the look-ahead, put your physics materials away, and don't worry about them again until your next workout.

EXERCISE 4.19: Your Physics Workout

Take a few minutes now to write down *specifics* on your version of the physics workout. Include times, places, and other particulars so that you have a very clear idea of how you will implement your plan.

4.5.2 YOUR PHYSICS WORKOUT OVER THE TERM

Now that you have a plan for your daily physics workout, you need to think about how you will make this plan work over the entire class term. What happens when there is a football game you want to attend? Or a play you want to try out for? How do you plan for exams and homework due dates?

The key to sticking to your physics workout is in planning for the entire term from the *very beginning.* Invest in a scheduling calendar, the kind that is small enough to fit in your backpack, but big enough to write in all of your appointments.

At the beginning of the term, get a syllabus from the instructor. It should have the dates of each exam, as well as homework due dates. Write each of these into your calendar right away, so that when an interesting non-physics event arises, you can be sure that you will not miss an important deadline.

Then, as the term progresses, update your calendar with new information or events. Use your calendar to keep you on track when "fun stuff" threatens to distract you.

4.6 CONCLUSION

Science is built up of facts, as a house is built up of stones; but an accumulation of facts is no more science than a heap of stones a house.
HENRI POINCARE
French Mathematician, 1854 - 1912

Because the environment of a physics class is so different from that of other kinds of classes, it is important to have some tools in your physics toolbox for coping. This chapter has provided you with that kind of information. But *you* have to make the information work for you. Make a plan and then stick to it!

PART II: PHYSICS BLUEPRINT

Now that you understand what physics is, and how it is taught, it's time to tackle the material itself. This is a "bad news–good news" situation.

The bad news is that, despite your preparation from Part I, you may still find learning physics difficult. Because physics instruction is often very detailed—filled with examples, counter-examples, and more examples again—it can be tricky for beginning students to extract critical underlying themes on their own.

The good news is that this state of affairs is fixable! In fact, the purpose of this part of the book is to provide you with an *overview* of concepts typically covered in an introductory physics course. This overview can help you construct a framework for understanding physics. Then, once you have a grasp of the overall "big picture," you can fill in the details from other sources.

Nature is incomprehensible at first,
Be not discouraged, keep on,
There are divine things well envelop'd...
WALT WHITMAN
English Author, 1819 - 1892

Overviews, by definition, ignore much of the minutiae needed to completely understand physics. So, in this book, we concentrate primarily upon main ideas—drawing connections between different subjects, and presenting topics in a slightly different order than your instructor might—while leaving particulars, fine points, and more detailed subjects to your textbook and instructor.

Because it deletes so much detail, therefore, THIS PART IS *NOT* INTENDED TO BE A SUBSTITUTE FOR YOUR REGULAR TEXTBOOK

OR LECTURE. IF YOU ATTEMPT TO LEARN PHYSICS SOLELY FROM THIS BLUEPRINT, YOU WILL FAIL YOUR COURSE!

However, if you use this part of *The Physics ToolBox* appropriately—for example, like a paperback tutor or mentor, guiding you through the course, enabling you to learn details later on your own—this part of the book can help you learn physics more effectively.

To use this part successfully, you should read each chapter one by one, when the associated concept is discussed in class. For example, the material in *Chapter 5: Kinematics* will most likely be the first topic discussed in your course; therefore, you should read that chapter (and *only* that chapter) when your course begins. Read the next chapter (*Chapter 6: Newton's Laws*) when you discuss Newtonian mechanics; this will probably take place two to four weeks later. And so on.

The most incomprehensible thing about the universe is that it is comprehensible.
ALBERT EINSTEIN
American Physicist, 1879 - 1955

A caveat: this part—although not as math heavy as your textbook—assumes that you understand algebra and vector math. If you do not, or if your memory of these topics is a little hazy, be sure to refer to the appropriate parts of Chapters 13 and 15 as you need them.

Good luck! If you have any questions about the material contained within this part, see your instructor and textbook for more information.

5
KINEMATICS

5.1 INTRODUCTION

As we have seen, physics is the study of *everything*. There is nothing too big, too small, too fast, or too complex for physicists to investigate: we study galaxies, subatomic particles, supersonic jets, and nuclear weapons; black holes, superconductors, spaceships...even the human body.

In Chapter 1.

These topics are exciting and interesting! So why, then, do most introductory physics classes start so *boringly*, with "obvious" discussions of position, velocity, and acceleration? The answer, as we saw in Part I of this book, is that physicists always begin with fundamentals. And the core of physics is *motion*.

The study of motion, however, turns out to be a *huge* topic. So, to start things off, we simply try to *describe* motion, leaving for later the deeper questions of *how* things move. Our description of motion—first developed by Galileo and now called **kinematics**—is the subject of this chapter. It may seem dull at first, but stick with it! Without this description of motion, you cannot begin to understand the more interesting material that follows.

See Chapters 6 through 9 for proof of this!

[Galileo] has been the first to open to us the door to the whole realm of Physics....
THOMAS HOBBES
English Philosopher, 1588 - 1679

5.2 POSITION

The most basic information about a physical object—and the starting point of many physics problems—is its **position** in space and time.

5.2.1 LOCATION

Coordinate systems are discussed in detail in Section 13.2.

The *location* (in space) of any object is given by a set of numbers, called **coordinates**, on a pre-defined grid, called the **coordinate system**.

There are many kinds of coordinate systems, but in an introductory physics class, you will probably use only two:

- **Cartesian coordinate systems** consist of the familiar x, y, and z coordinates (usually given in meters) on three mutually perpendicular axes. One- and two-dimensional systems are also used. See Figures 5.1 (a) – (c).*
- **Polar coordinate systems** have two coordinates: the radial distance from the origin, or **radius** r (usually reported in meters), and the angle θ between the radius and a horizontal line (almost always given in radians) See Figure 5.2 below.

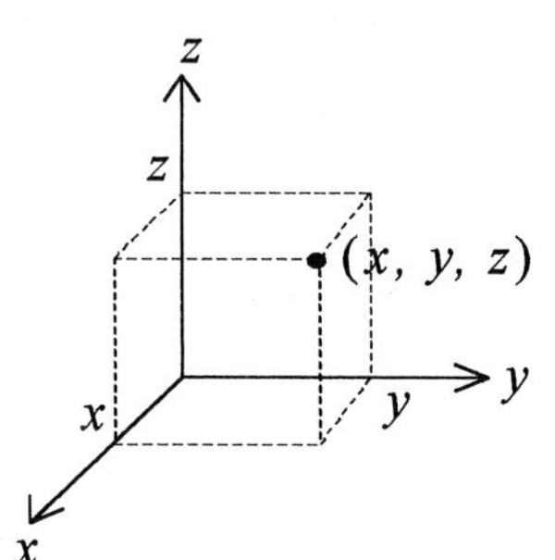

Figure 5.1 (a)

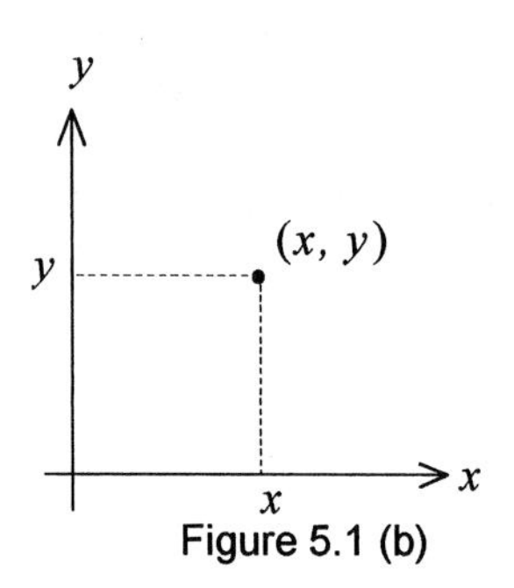

Figure 5.1 (b)

x

Figure 5.1 (c)

To analyze motion, you must *choose* the coordinate system that best describes a particular situation. Some choices are better than others—and the "right" choice in one situation may be the "wrong" choice in another—so you should try to get a feel for what works where. You will get better at this through time; and Section 13.2 can help.

> **EXAMPLE 5.1:** Choosing Coordinate Systems
>
> Cartesian coordinate systems are usually the best choice when dealing with *straight line* or *linear* motion. For example, a ball dropping vertically is best described by a Cartesian system.
>
> Polar systems, on the other hand, are best for *rotational* or *circular* motion. For instance, the tip of a clock's minute hand moves circularly, so the polar system is the best choice for describing that motion.

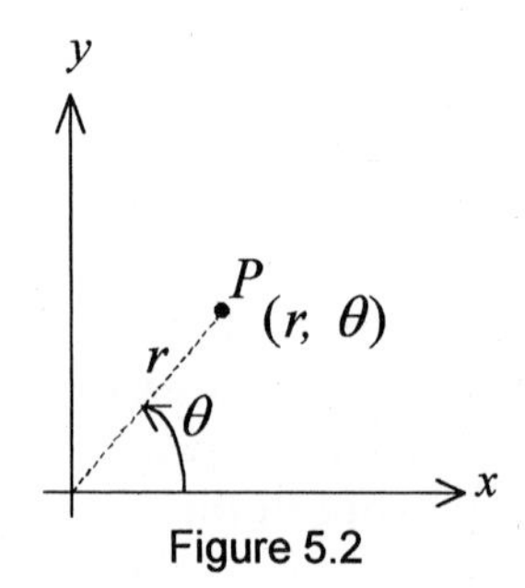

Figure 5.2

Once you have chosen a coordinate system, each point in the system is described in one of two (equivalent) ways:

- **Set notation**: Coordinates are written list-wise inside parentheses; *e.g.*: (x, y, z) for a Cartesian system or (r, θ) for polar coordinates.
- **Vector notation**: Coordinates are written as vectors; *e.g.*: $\vec{\mathbf{r}} = x\hat{\mathbf{i}} + y\hat{\mathbf{j}} + z\hat{\mathbf{k}}$ for a Cartesian vector or $\vec{\mathbf{r}} = r\hat{\mathbf{r}}$ for a polar one.

More on vectors in Chapter 15.

* Please note the dual use of the symbol x to indicate both an axis, *and* a position in space (the same is true, of course, for the y and z symbols as well). Do not be confused by this notation.

EXERCISE 5.2: Coordinate Notation

A point P is described by the Cartesian coordinates $x = 2$ m and $y = 3$ m. Draw this point on a Cartesian axis, and write the coordinates in both set and vector notation.

SOLUTION:

Figure 5.3 shows P on a two-dimensional Cartesian coordinate grid.

By plugging in our values for x and y into the equations listed above, we see that the set notation for this point is written $(2, 3)$, while the vector is notated $\vec{\mathbf{r}} = 2\hat{\mathbf{i}} + 3\hat{\mathbf{j}}$.

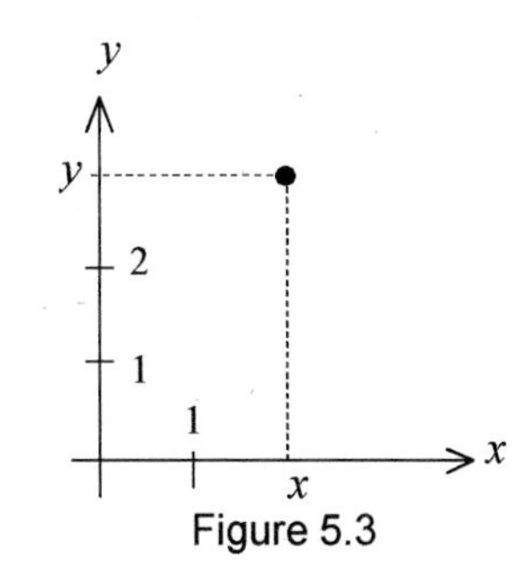

Figure 5.3

5.2.2 TIME

In order to fully describe an object's location, you must also specify the *time* (usually given in seconds) that it was there. As was the case with coordinate systems, you are free to pick the best time frame for a given situation. That is, you get to determine when to "start the clock."

See Section 11.5.2 for more on time and other fundamental quantities.

EXAMPLE 5.3: Choosing Time Frames

For instance, when would be the best time to "start the clock" for an experiment in which you drop a ball from a tall building? In principle, you could start your clock at *any* point: for example, when you first woke up that morning, or at 11:00 a.m. sharp, or even when you first arrived at the building. But in practice, the *best* choice for "starting the clock" for this experiment is the instant that you drop the ball.

"Clearly," the Time Traveler proceeded, "any real body must have extension in four dimensions...Length, Breadth, Thickness, and—Duration....[But], we incline to overlook this fact."
H.G. WELLS,
in *The Time Machine*
English Novelist, 1866 - 1946

Unfortunately, in most problems, physics textbooks often skip the time component of location. The reason for this is that it usually doesn't matter *when* an experiment takes place, so that the time measurement becomes irrelevant extra information. Nonetheless, in the "big picture," time is an intimate part of space and you should always be on the lookout for ways in which time can affect a given problem.

NOTE 5.4: Relativity and Space-Time

Albert Einstein was the first to point out that (one-dimensional) time and (three-dimensional) space are actually equivalent components of a seamless (four-dimensional) "space-time." This space-time is the fabric of our universe and the ultimate source of gravity.

5.2.3 DISPLACEMENT

If an object *moves* (through time), its position in space changes. We describe this change via the **displacement**, a simple subtraction between the first and second locations of the object. For example, the displacement of an object moving in the x-direction from x_1 to x_2 is:

$$\Delta x = x_2 - x_1$$

There are similar relations for the y, z, r, and θ directions, as well as for time t:

$$\Delta y = y_2 - y_1 \qquad \Delta r = r_2 - r_1$$
$$\Delta z = z_2 - z_1 \qquad \Delta\theta = \theta_2 - \theta_1$$
$$\Delta t = t_2 - t_1$$

A positive space displacement indicates the object is moving in the positive direction; *i.e.*, towards increasing x. Thus, a negative time displacement is not possible, because time always moves forward.

EXERCISE 5.5: A Traveling Cat, Part I

A cat moves from an initial position of $x_1 = 2$ m (at time $t = 0$ s) to a final position of $x_2 = 5$ m (at time $t = 4$ s). What is the cat's x-direction displacement? In which direction is she moving? What time elapsed between x_1 and x_2?

SOLUTION:

From our displacement equation above, we have:

$$\Delta x = x_2 - x_1 = 5\,\text{m} - 2\,\text{m} = +3\,\text{m}$$

Because the final value is positive, the cat is moving in the direction of increasing x. The total time elapsed during the cat's displacement is:

$$\Delta t = t_2 - t_1 = 4\,\text{s} - 0\,\text{s} = +4\,\text{s}$$

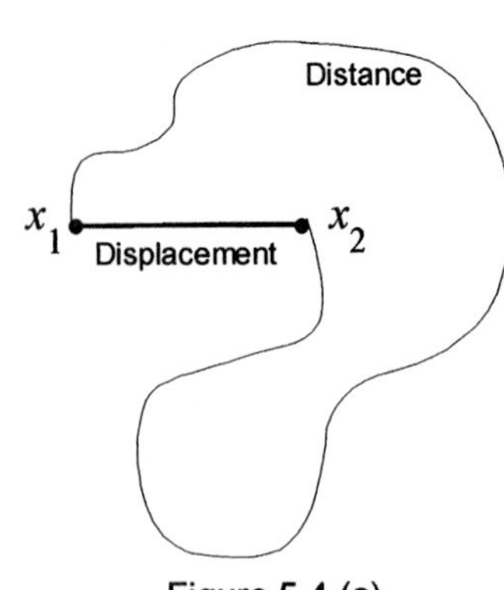

Figure 5.4 (a)

Distance
x_1 x_2
Displacement

Figure 5.4 (b)

Don't get confused. *Displacement* (net distance traveled) is *not* the same thing as *distance* (total distance traveled): an object on a winding path can travel quite a long distance between points x_1 and x_2, and still have a very small displacement (Fig. 5.4 (a)). [However, in the special but important case that all movement takes place *in one direction* on a straight line (Fig. 5.4 (b)), then the displacement equals the distance.]

The distance equation, unlike the displacement equation, is in general very complex and therefore not relevant for most introductory physics courses. However, the magnitude of the distance (in three dimensional space) is given by the equation:

Learn more about this equation in Section 13.5.1.

$$d = \sqrt{(x_2 - x_1)^2 + (y_2 - y_1)^2 + (z_2 - z_1)^2} = \sqrt{\Delta x^2 + \Delta y^2 + \Delta z^2}$$

EXAMPLE 5.6: Magnitude of the Distance

You can adapt the above equation to one- or two-dimensions by dropping one or more of the terms. For example, the distance traveled by an object moving in two dimensions from $(5, 4)$ to $(6, -2)$ (in m) is:

$$d = \sqrt{(x_2 - x_1)^2 + (y_2 - y_1)^2} = \sqrt{(6-5)^2 + (-2-4)^2} = \sqrt{37}\ \text{m} \approx 6\ \text{m}$$

5.3 VELOCITY

Once we have determined an object's position in space and time, the next quantity of interest is the object's rate of motion, usually called *speed*. Physicists, however, define a more precise, but related, quantity for rate of change of motion, known as **velocity**.

Velocity, in general, is a vector. We will discuss the directional nature of the vector soon. But first, we examine the magnitude of velocity, called the **speed** (a precisely-defined term which is not to be confused with the colloquial use of the word).

5.3.1 AVERAGE & INSTANTANEOUS SPEED

Velocity is the *rate of change of position (through time)* of an object; speed is the magnitude of its vector. Since there are two ways to calculate rates of change—using an average, and using a derivative—there are two main techniques for figuring speed.

Derivatives are covered in Section 14.2.

The **average speed** of an object is its *displacement divided by the total time elapsed* during travel. If the displacement is in the x, y, or z direction, the average speed is linear; otherwise, if the displacement is in the θ direction, the average speed is angular:

$$\bar{v}_x = \frac{\Delta x}{\Delta t} \quad (\text{unit} = \text{m/s})$$

Average Linear Speed

$$\bar{\omega} = \frac{\Delta \theta}{\Delta t} \quad (\text{unit} = \text{rad/s})$$

Average Angular Speed

EXERCISE 5.7: A Traveling Cat, Part II

What is the average linear speed of the cat in Exercise 5.5?

SOLUTION:

From that exercise, we know that $\Delta x = +3$ m and $\Delta t = +4$ s. Thus, the average linear speed of the cat is:

$$\bar{v}_x = \frac{\Delta x}{\Delta t} = \frac{+3\ \text{m}}{+4\ \text{s}} = 0.75\ \text{m/s}$$

EXERCISE 5.8: A Clock, Part I

What is the average angular speed of the tip of the minute hand of a clock?

SOLUTION:

A minute hand moves around an entire circle (2π radians) in one hour (or 3600 seconds). Thus, the average angular speed of its tip is:

$$\overline{\omega} = \frac{\Delta\theta}{\Delta t} = \frac{2\pi \text{ rads}}{3600\,\text{s}} = 1.7\times10^{-3} \text{ rad/s}$$

Note the use of ω (not w!) to differentiate angular and linear velocity.

The **instantaneous speed**, by contrast, is the *derivative of position with respect to time*. For example, the magnitude of the **instantaneous linear speed** (or, just **speed**) of an object moving in the x, y, or z directions is:

$$v_x = \frac{dx}{dt}; \quad v_y = \frac{dy}{dt}; \quad v_z = \frac{dz}{dt} \quad (\text{unit} = \text{m/s})$$

The **instantaneous angular speed** (or, **angular speed**) for an object moving in the θ direction is:

$$\omega = \frac{d\theta}{dt} \quad (\text{unit} = \text{rad/s})$$

EXERCISE 5.9: A Falling Ball, Part I

A ball falls from rest, changing position according to the equation:

$$y(t) = \tfrac{1}{2}gt^2$$

More on the gravitational field in Sections 7.3.1. and 11.5.2.

where g is the gravitational field ($= 9.8$ m/s^2). What is the ball's instantaneous linear speed after two seconds?

SOLUTION:

Instantaneous linear speed is the derivative of position. Thus, the ball's instantaneous linear speed is:

$$v_y(t) = \frac{dy}{dt} = \frac{d}{dt}(\tfrac{1}{2}gt^2) = \tfrac{1}{2}(2)gt = gt$$

After two seconds, this equation becomes:

$$v_y(2) = (9.8\text{ m/s}^2)(2\text{ s}) = 19.6\text{ m/s} \approx 20\text{ m/s}$$

Vector magnitudes are discussed in Section 15.2.1.

Note that this number is positive since speed is a vector magnitude.

DEFINITION 5.10: Average vs. Instantaneous

The average and instantaneous speeds of an object are, in general, not the same.

Think of a car trip. Your average speed for the entire trip is the total distance traveled divided by the total time elapsed. This number will almost always be different from your instantaneous speed, measured moment-to-moment by your speedometer. That is, during your trip, you may encounter both stop-and-go traffic (low instantaneous speed) and highway travel (high instantaneous speed), so your average speed for the trip will be somewhere in between (and not equal to) these two values.

The only important situation in which average and instantaneous speeds are equal occurs when an object moves at a constant speed. For example, the moving minute hand of Exercise 5.8 ticks along at a constant rate, so its average (angular) speed equals its instantaneous (angular) speed at every moment in that case.

Of the two quantities, instantaneous speed is by far the more important, and will be used almost exclusively in your introductory course.

The **total speed** of a body, or magnitude of the three-dimensional velocity vector, is given by the equation:

$$v = \sqrt{v_x^2 + v_y^2 + v_z^2}$$

The magnitude of the angular speed, by contrast, is simply ω.

EXERCISE 5.11: Magnitude of the Velocity

A jogger has a speed in the x-direction of $v_x = 4$ m/s, and in the y-direction of $v_y = 5$ m/s. What is the magnitude of his velocity vector?

SOLUTION:

The jogger's magnitude of the velocity is given by the above equation:

$$v = \sqrt{v_x^2 + v_y^2} = \sqrt{(4\text{ m/s})^2 + (5\text{ m/s})^2} = \sqrt{36\,(\text{m/s})^2} = 6\text{ m/s}$$

Table 5.1 lists some typical values for speeds.

Table 5.1: Some Speeds (linear in m/s, *angular* in rad/s)

Speed of light (the fastest speed)	2.99×10^8	Top speed of a Lamborghini	62	*Rotation of a molecule*	10^{12}
Speed of sun around galaxy	2.10×10^5	Speed of fastest 100 yd dash	10	*Rotation of a pulsar*	200
Speed of earth around sun	2.96×10^4	Speed of a flying bee	5	*Rotation of helicopter rotor*	50
Speed of moon around earth	1.00×10^3	Speed of a walking ant	1.00×10^{-2}	*Rotation of $33\frac{1}{3}$ rpm record*	5
Speed of sound (in air)	3.33×10^2	Speed of a swimming amoeba	4.50×10^{-5}	*Rotation of tip of minute hand*	10^{-3}

Linear and angular speeds are related via the equation $v = \omega r$, where r is the radius of the circle traveled.

EXERCISE 5.12: Linear and Angular Speeds

The distance from the earth to the moon is approximately 3.84×10^8 m. The moon orbits the earth at a speed of about 1.00×10^3 m/s.

(a) What is the angular speed of the moon's orbit around the earth?

(b) Through what angle does the moon travel in one month (28 days $= 2.42 \times 10^6$ s)?

You can obtain this value for seconds per month by doing a unit conversion. See Section 11.6 for more on this technique.

SOLUTION:

(a) The formula above tells us that $v = \omega r$. But in this problem we don't want the linear speed, we want the angular speed. So we must rearrange this equation to solve for ω. $\omega = v / r$.
Now, we know the value of r and v from the problem, so:

$$\omega = \frac{v}{r} = \frac{1.00 \times 10^3 \text{ m/s}}{3.84 \times 10^8 \text{ m}} = 2.60 \times 10^{-6} \text{ rad/s}$$

Notice that the radians in ω's units (rad/s) appeared when the meters in r and v cancelled out.

(b) We can use the equation $\overline{\omega} = \Delta\theta / \Delta t$ from page 71 to solve this part because the moon travels at an approximately constant rate. So, its average angular speed is equal to its instantaneous angular speed (found in part (a) above): thus, $\overline{\omega} = \omega$.
Rearranging our equation to solve for the angle, we then have:

$$\Delta\theta = \omega\Delta t = (2.60 \times 10^{-6} \text{ rad/s})(2.42 \times 10^6 \text{ s}) = 6.29 \text{ rad} \approx 2\pi \text{ rad}$$

That is, the moon travels about one full orbit per 28-day month.

5.3.2 THE VELOCITY VECTOR

Vectors are the subjects of Chapter 15.

As we said at the beginning of this section, velocity is a vector. The magnitude of the vector is the instantaneous speed (which we will now just call speed); the direction of the vector is the direction of motion.

EXAMPLE 5.13: A Velocity Vector

Consider the car from Definition 5.10. At each instant, the car has a speed specified by its speedometer. Additionally, however, at each instant the car also has a direction, indicated by the slant of the car's front tires. Thus, the velocity vector of the car is given by its speed *and* its instantaneous direction.

Since velocity is a vector, we can write it in vector notation. The linear velocity vector is written:*

Vector notation is in Section 15.3.3.

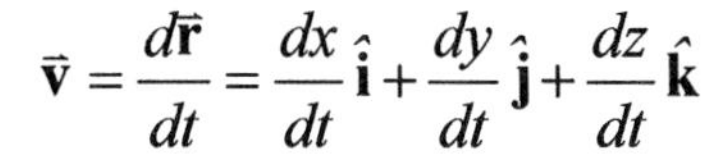

$$\vec{\mathbf{v}} = \frac{d\vec{\mathbf{r}}}{dt} = \frac{dx}{dt}\hat{\mathbf{i}} + \frac{dy}{dt}\hat{\mathbf{j}} + \frac{dz}{dt}\hat{\mathbf{k}}$$

For more on vector derivatives, see Section 15.4.1.

EXERCISE 5.14: A Falling Ball, Part II

Write the velocity of the ball from Exercise 5.9 in vector notation.

SOLUTION:

We have already found, in that case, that $v_y = dy/dt = 20$ m/s. The ball is *falling,* that is, it is moving in the negative y-direction, so the direction of the velocity vector is downwards, or $-\hat{\mathbf{j}}$. See Figure 5.5. Therefore, the vector notation describing the ball's fall is:

$$\vec{\mathbf{v}} = \frac{dy}{dt}\hat{\mathbf{j}} = (20\text{ m/s})(-\hat{\mathbf{j}}) = -20\text{ m/s}\,(\hat{\mathbf{j}})$$

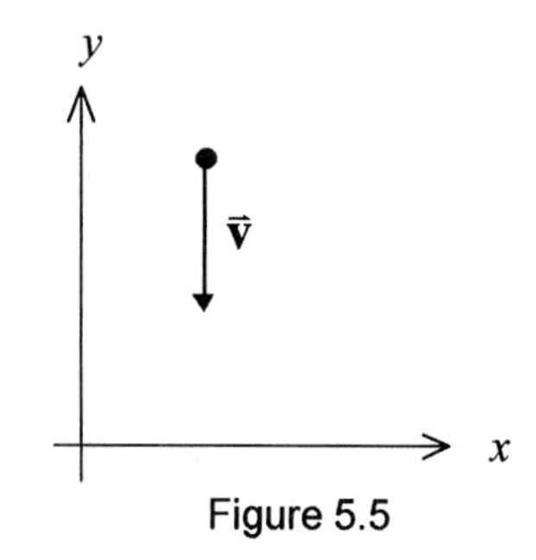

Figure 5.5

By contrast, the angular velocity vector is written: #

$$\vec{\omega} = \frac{d\theta}{dt}\hat{\mathbf{n}}$$

For more on the normal unit vector, see Section 15.3.3.

Luckily, in most introductory physics courses, you will not need to write the angular velocity in vector notation. Instead, it is usually enough to describe the direction of the vector in your work and move on.

So how do you find the direction of the angular velocity vector? With the **curl right hand rule**: Using your *right hand,* circle your fingers in the direction that θ changes; your thumb will naturally point in the direction of ω.

WARNING!

Be *very* careful with this rule! *Sloppiness results in wrong answers.* If you use your left hand (for example, if you write with your right hand and forget to put down your pencil before you do the rule), you will get the direction WRONG! Or, if you do not take extra care to line up your hand with the direction of motion *exactly*, you may get the wrong result as well.

---

* Note that because $\vec{\mathbf{r}}$ is a vector, changes in position *or* direction result in velocity!

ω is a vector, even though θ is not. To see that this is true, try this experiment. Place a book face up on a table. Rotate it 90° to the right and then rotate it 90° up to the vertical. What is the orientation of the book? Now, begin again. Place the book face up. Rotate it 90° *up* first, *then* rotate it 90° to the right. Is it in the same orientation as before? No! Why not? Because the *order of rotation,* for large angles, *matters.* But the order for vectors *doesn't* matter! Therefore, angles *cannot* be vectors. So how can ω be a vector if θ isn't? The answer is that $d\theta$ *is* a vector! The order of rotation for tiny angles *doesn't* matter. Try *this* with your book and see.

EXERCISE 5.15: A Clock, Part II

What is the angular velocity vector, including magnitude and direction, for the minute hand of the clock in Exercise 5.8? (See Figure 5.6)

Figure 5.6

SOLUTION:

As we saw in Definition 5.10, the average angular velocity of the minute hand equals its instantaneous angular velocity at every instant. Thus, the *magnitude* of its angular velocity is 1.7×10^{-3} rad/s, as found in Exercise 5.8.

The *direction* of this vector, however, must be found via the curl right hand rule described above. Let's work it out together. Using our *right* hand, we circle our fingers in the direction of the minute hand's motion; that is, *clockwise*. As a result, our thumb naturally points *into* the clock face. That means that the angular velocity is directed *inward*.

Our final answer, then, is that the angular velocity vector of the minute hand is 1.7×10^{-3} rad/s, directed inwards.

5.3.3 RELATIVE MOTION & REFERENCE FRAMES

It does not appear to me that there can be any motion other than relative....
GEORGE BERKELEY
English Philosopher, 1685 - 1753

The discussions above notwithstanding, we must be very careful when we talk of "the" velocity of an object. Different, but equally competent, observers can measure *different* speeds for the *same* object! This is known as the **principle of relative speeds**.

EXAMPLE 5.16: Relative Motion, Part I

To see that this is true, imagine the following scenario.

Two experimenters measure the speed of a rocket car moving at a high but constant speed. Stationary Stan stands on the test highway (safely behind the car!), while Moving Monica drives behind the rocket car in a second, slower (but still constant-speed) car (Fig. 5.7). Both Stan and Monica have identical, perfectly functioning laser guns with which to measure the speed of the rocket car.

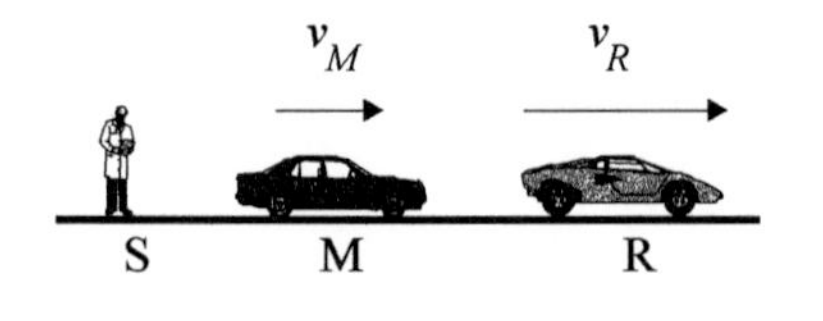

Figure 5.7
NOT TO SCALE

The experiment begins. With their laser guns, Stan measures v_R to be 400 m/s, while Monica measures v_R to be 300 m/s. But wait! They obtained different results! So who is right?

Both experimenters are correct. However, Monica's measured speed is lower than Stan's because she is *moving* relative to Stan. That is, the laser gun in her car measures the *relative* speed between her and the rocket car—which is less than the relative speed between Stan and rocket car, because Stan is not moving.

EXAMPLE 5.17: Relative Motion, Part II

To get a better feel for the principle of relative motion, try this thought experiment.

Imagine that you wake up inside of a windowless, soundproof pod. The pod has an engine and perfect shock absorbers. In short, once inside of this pod, you *cannot tell* by sight, sound, or feeling whether the pod is moving or at rest.

The only tools that you have in the pod are a perfect laser gun just like Stan and Monica's, and a radio to communicate with both of them. The gun—unlike you—can "see" the rocket car, and can measure its speed. So, having nothing else to do in the pod, you decide to take a measurement of the rocket car's speed. Then, you radio back to Stan and Monica to compare your readings.

Evaluating the three readings can tell you whether you are moving or at rest. That is, if you obtain a speed of 400 m/s, you are stationary like Stan; otherwise, you are moving like Monica. Furthermore, by comparing your reading to Stan's, you can find out how fast you are moving (relative to Stan): the difference between your measurement of the rocket car's speed and Stan's measurement is your speed.

Either way, *your measurement is correct*: the velocity measured by the perfect laser gun *is* the speed of the rocket car as it moves away from you—but it is your motion that determines the exact *value* of the measurement.

The principle of relative speeds, tricky enough to understand on its own merit, becomes trickier to grasp when we realize that *nothing* in the universe is truly stationary; that is, there is no **standard of absolute rest**. Thus, *every* measurement of velocity is *always* measured relative to another *moving* object!

EXAMPLE 5.18: Relative Motion, Part III

In Example 5.16 above, Stan was said to be stationary. This is not strictly true. Stan *is* stationary relative to another observer standing on the highway. But Stan *is actually moving*; for the earth, on which he stands, is both spinning on its axis and orbiting the sun. Meanwhile, the sun is orbiting the galaxy, and the galaxy is moving through the universe. Even the universe is not stationary, because it is expanding! In short, *nothing* (even Stan!) is truly at rest.

To keep track of all of these moving bodies, we attach **reference frames**—fancified coordinate grids—to them (Fig. 5.8). As the object moves, the reference frame moves with it. Thus, we can measure the object's motion *in its own frame*, relative to scales also in the frame.

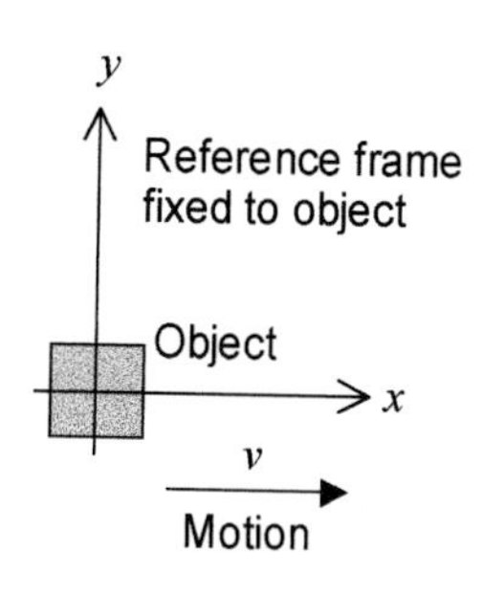

Figure 5.8

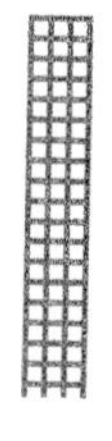

EXERCISE 5.19: Relative Motion, Part IV

Set up reference frames for Stan and Monica from Example 5.16.

SOLUTION:

To do so, we simply attach separate coordinate grids to Stan (Figure (a) below) and Monica (Figure (b)).

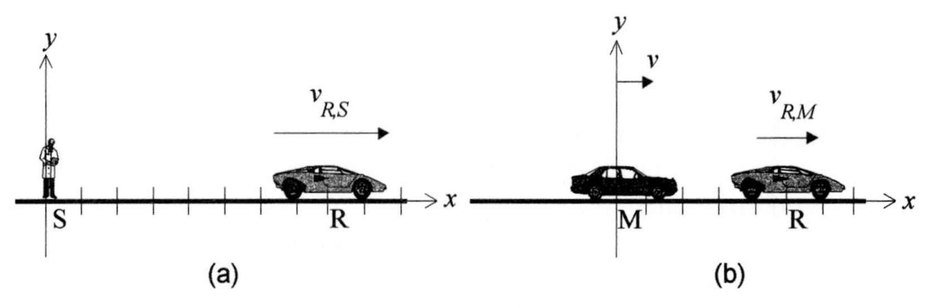

In this exercise, you should note two things. First, because Monica is moving, her *frame* is also moving, as compared to Stan's frame. We'll denote the speed of Monica's frame, as measured by Stan, as v.

Second, in keeping with the principle of relative speeds—which states that the velocities measured by Stan and Monica are different—we label the speed of the rocket car differently in each frame: in (a), it is the speed of the rocket car with respect to Stan, $v_{R,S}$, and in (b), it is the speed of the rocket car with respect to Monica, $v_{R,M}$ (such that $v_{R,S}$ is greater than $v_{R,M}$ because Monica is moving).

EXERCISE 5.20: Relative Motion, Part V

Find a relationship between $v_{R,S}$, $v_{R,M}$, and v in the previous example.

SOLUTION:

There is more on graphical addition of vectors in Section 15.2.2.

Velocity is a vector, so we can use vector addition here. Adding the vectors graphically (in one dimension), we have:

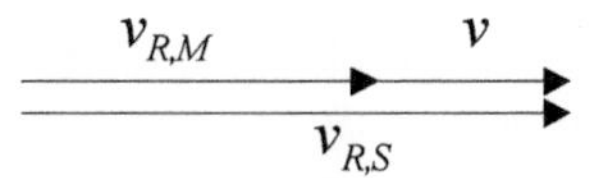

Algebraically, this is relation written $v_{R,S} = v_{R,M} + v$. This equation can be rearranged in any number of ways to get relationships more convenient for different situations; for example:

$$v_{R,M} = v_{R,S} - v \quad \text{or} \quad v = v_{R,S} - v_{R,M}$$

Reference frames that are stationary, or those that move at a constant velocity, are called **inertial frames**. These kinds of frames are very important, as we shall soon see.

EXAMPLE 5.21: Relative Motion, Part VI

Stan's reference frame, since he is "stationary" ($v_S = 0$) is inertial. However, Monica's is too, because she is traveling at a constant velocity ($v =$ constant).

5.4 ACCELERATION

[It is our purpose] to investigate and to demonstrate some of the properties of accelerated motion (whatever the cause of this acceleration may be).
GALILEO GALILEI
Italian Experimentalist, 1564 - 1642

The time rate of change of position, as we have just learned, is velocity. Likewise, the *time rate of change of velocity* is **acceleration**.*

5.4.1 AVERAGE & INSTANTANEOUS ACCELERATION

The **average** and **instantaneous linear accelerations** (unit = $\mathrm{m/s^2}$) of an object are calculated in the same way as was done above for velocity:

$$\bar{a} = \frac{\Delta v}{\Delta t}$$

Average Linear Acceleration

$$\vec{\mathbf{a}} = \frac{d\vec{\mathbf{v}}}{dt} = \frac{d^2\vec{\mathbf{r}}}{dt^2}$$

Instantaneous Linear Acceleration#

EXERCISE 5.22: A Falling Ball, Part III

Find the acceleration of the ball from Exercises 5.9 and 5.14 above.

SOLUTION:

From those exercises, we found the velocity to be $\vec{\mathbf{v}}(t) = gt(-\hat{\mathbf{j}})$. So, to find the acceleration, we must take the derivative of velocity:

$$\vec{\mathbf{a}}(t) = \frac{d\vec{\mathbf{v}}(t)}{dt} = \frac{d}{dt}[(gt)(-\hat{\mathbf{j}})] = g(-\hat{\mathbf{j}}) = 9.8\ \mathrm{m/s^2}(-\hat{\mathbf{j}}) = -9.8\ \mathrm{m/s^2}(\hat{\mathbf{j}})$$

Notice that $\hat{\mathbf{j}}$ is a constant, so that it can be taken out of the derivative. Furthermore, the negative sign in front of the $\hat{\mathbf{j}}$ indicates that the vector points *downwards*. Finally, notice that the final equation does *not* have *t* in it, so the value of acceleration is the *same* at all times; that is, a is a constant in this case.

Unit vectors, and their derivatives, are covered in Section 15.3.3.

* We could keep going with derivatives, if we wished. For example, the time rate of change of acceleration is the **jerk**. However, higher order time derivatives of position will never be seen in your introductory physics class.

Note that because $\vec{\mathbf{v}}$ is a vector, changes in speed *or* direction result in acceleration!

The **average** and **instantaneous angular accelerations** of an object are calculated similarly (unit = rad/s^2):

$$\bar{\alpha} = \frac{\Delta\omega}{\Delta t}$$

Average Angular Acceleration

$$\vec{\alpha} = \frac{d\vec{\omega}}{dt} = \frac{d^2\theta}{dt^2}\hat{\mathbf{n}}$$

Instantaneous Angular Acceleration*

The direction of $\vec{\alpha}$ is found using the same right hand rule as for $\vec{\omega}$ above (p. 75), with one caveat: If the magnitude of $\vec{\omega}$ is increasing, then $\vec{\alpha}$ points in the same direction of $\vec{\omega}$; otherwise, if the magnitude of $\vec{\omega}$ is decreasing, then $\vec{\alpha}$ points in the opposite direction to $\vec{\omega}$.

EXERCISE 5.23: A Clock, Part III

What is the angular acceleration for the minute hand on the clock of Exercise 5.8? If the clock in that exercise ran slow (that is, if its angular velocity was *not* constant and was decreasing), in which direction is the angular acceleration of the minute hand?

SOLUTION:

The angular velocity of the minute hand in Exercise 5.8 was found to be 1.7×10^{-3} rad/s, which is a *constant* value. The derivative of a constant is zero, so there is *no* angular acceleration in this case.

If the clock runs slow, however, the angular velocity of the minute hand is no longer constant; it decreases. Therefore, its angular acceleration has the opposite direction as the hand's angular velocity.

In Exercise 5.15, we found that the minute hand's angular velocity is directed *into* the clock. Thus, if the clock ran slow, the angular acceleration vector of its minute hand would point *out of* the clock.

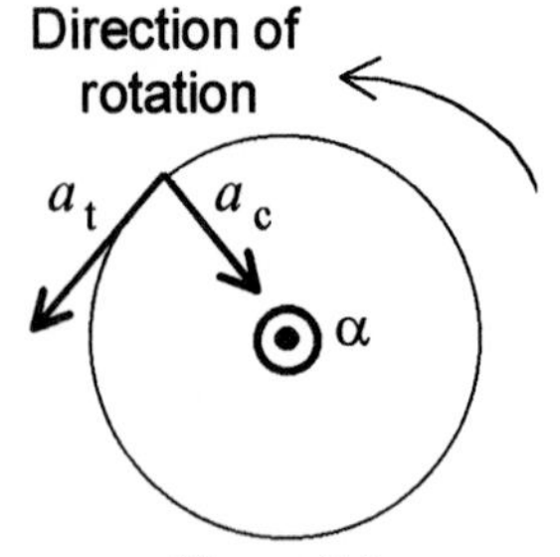

Figure 5.9
NOT TO SCALE

The angular acceleration α is related to the **tangential component** of linear acceleration a_t:

$$a_t = \alpha r$$

But, even in the case of no angular acceleration, there is still a **centripetal acceleration** a_c on rotating objects (due to the circular motion):

More on this equation in Section 7.2.9. 

$$a_c = \frac{v^2}{r} = \omega^2 r$$

See Figure 5.9.

The *total* linear acceleration of an object moving in a circle is $a = \sqrt{a_t^2 + a_c^2}$.

To see how this equation was obtained, refer to Section 15.3.2. 

* Note that because ω is a vector, changes in speed *or* direction result in acceleration!

EXAMPLE 5.24: Tangential Vs. Centripetal

An ant sitting on the tip of the minute hand of Exercise 5.8 (in the case of the perfectly functioning clock) will not feel a tangential acceleration since there is no angular acceleration in that situation (Exercise 5.23).

However, the ant will certainly feel a centripetal acceleration because of the minute hand's rotation around the circle. The value of that acceleration (assuming a hand length of ten centimeters) is:

$$a_c = \omega^2 r = (1.7\times10^{-3}\ \text{rad/s})^2(0.10\ \text{m}) = 2.9\times10^{-7}\ \text{m/s}^2$$

5.4.2 THE CASE OF CONSTANT ACCELERATION: THE KINEMATIC EQUATIONS

In the special, but frequently seen, case that acceleration is constant, the four differential equations given above [$v_x = x'(t)$, $a_x = x''(t)$, $\omega = \theta'(t)$, and $\alpha = \theta''(t)$] can be solved quite easily (your textbook or instructor will do this for you).

Differential equations, and their solutions, are covered in Section 14.4.

The results of these calculations, called the **kinematic equations**, are summarized in Table 5.2 below. These ten very important relations will solve *any* problem in which the acceleration is constant.

To use this table, choose the equation that has all of the variables in your problem, but *none* of the ones that you don't have. The i subscript on many of the quantities below indicates the value of each quantity at the beginning of the motion; *i.e.*, when $t = 0$.

Table 5.2: The Kinematic Equations

EQUATION	MISSING VARIABLE	EQUATION	MISSING VARIABLE
$x = v_{i,x}t + \frac{1}{2}a_x t^2$	v	$\theta = \omega_i t + \frac{1}{2}\alpha t^2$	ω
$x = v_x t - \frac{1}{2}a_x t^2$	v_i	$\theta = \omega t - \frac{1}{2}\alpha t^2$	ω_i
$x = \frac{1}{2}(v_{i,x} + v_x)t$	a	$\theta = \frac{1}{2}(\omega_i + \omega)t$	α
$v = v_{i,x} + a_x t$	x	$\omega = \omega_i + \alpha t$	θ
$v^2 = v_{i,x}^2 + 2a_x x$	t	$\omega^2 = \omega_i^2 + 2\alpha\theta$	t

Physicists like to think that all you have to do is say, "These are the conditions, now what happens next?"
RICHARD FEYNMAN
American Physicist, 1918 - 1988

EXERCISE 5.25: A Falling Ball, Part IV

In Exercise 5.22, we found that the acceleration of the falling ball was constant: $a = g$. Using the correct kinematic equation, find the position of the ball after 2 seconds.

SOLUTION:

In this problem, we are given two quantities, the (constant) acceleration a, and the time of fall t. We also know (from Exercise 5.9) that the ball began its fall with zero initial velocity, $v_i = 0$. Therefore, we need to choose the kinematic equation that contains a, t, and v_i, as well as the distance of fall y (the unknown quantity) but *no* other variables.

That equation is the first one listed on the left-hand side of the above table, $y = v_{i,x}t + \frac{1}{2}a_x t^2$. Note that we have changed x to y in that equation because the fall is in the vertical direction. Then we plug in:

$$y(2) = v_{i,x}t + \tfrac{1}{2}a_x t^2 = (0)(2\,\text{s}) + \tfrac{1}{2}(9.8\,\text{m/s}^2)(2\,\text{s})^2 = 19.6\,\text{m} \approx 20\,\text{m}$$

You can use the equations of Table 5.2 in more than one dimension as well. *Separately* calculate the necessary values in *each* direction, and combine the results using the sum of the squares method on page 73.

EXAMPLE 5.26: A Thrown Ball

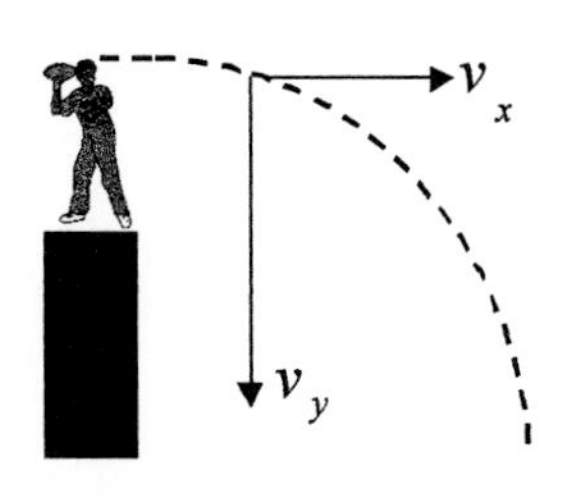

Figure 5.10
NOT TO SCALE

For example, consider the case in which a ball is thrown *horizontally* off of the tall building in Exercise 5.9 above, rather than dropped straight down. In this case, the ball will have two *distinct* motions: a horizontal (x) motion due to the throw, and a vertical (y) one due to gravity. See Figure 5.10.

So, to find the subsequent motion of the ball, we must calculate its speed in the x *and* y directions:

$$v_x = v_{i,x} + a_x t \quad \text{and} \quad v_y = v_{i,y} + a_y t$$

Then, we must combine the two answers vectorially to obtain the magnitude of the velocity vector:

$$v = \sqrt{v_x^2 + v_y^2}$$

Let's work out an example together. Assume a ball is thrown horizontally with a speed of $v_{i,x} = 5$ m/s. Therefore, $v_{i,y} = 0$. What is the speed of the ball after three seconds?

Under normal conditions, the only acceleration on the ball is gravity in the negative y-direction. Thus, $a_x = 0$ and $a_y = g$. So, our equations for the x and y components of the velocity are:

$$v_x = v_{i,x} + a_x t = (5\,\text{m/s}) + (0\,\text{m/s}^2)(3\,\text{s}) = 5\,\text{m/s}$$

$$v_y = v_{i,y} + a_y t = (0\,\text{m/s}) + (9.8\,\text{m/s}^2)(3\,\text{s}) = 29.4\,\text{m/s}$$

$$\Rightarrow v = \sqrt{v_x^2 + v_y^2} = \sqrt{(5\,\text{m/s})^2 + (29.4\,\text{m/s})^2} \approx 30\,\text{m/s}$$

5.5 RELATIVITY & QUANTUM MECHANICS

Position, velocity, and acceleration make up the sturdy foundation upon which all other physical knowledge about the universe is built. And as the most basic and fundamental aspects of physics, position, velocity and acceleration are rather *dull,* right?

If you think so, you are in lofty company: throughout most of physics history, scientists thought they had reached the end of the road with these quantities. That is, until the turn of the twentieth century, when some very bright people took a closer look.

It was a wonderful mess at that time. Wonderful! Just great! It was so confusing—physics at its best, when everything is confused and you know something important lies just around the corner.
ABRAHAM PAIS
Danish Physicist, 1918 -

5.5.1 SPECIAL RELATIVITY

In 1905, at the age of 26, Albert Einstein stated two rather plain postulates regarding physical law:

- **The Postulate of Relativity**: All observers in inertial frames see the same laws of physics.
- **The Postulate of Light Speed**: All observers in inertial frames measure the same speed of light, $c = 2.99\times10^8$ m/s.

I simply ignored an axiom.
ALBERT EINSTEIN,
on developing relativity
American Physicist, 1879 - 1955

These two postulates certainly seem reasonable on their face. In fact, we have already assumed the first principle to be true (though we never said it explicitly); otherwise, we could never hope to find one set of physical laws, because physics itself would change from place to place!

Considered separately, then, these two principles are not shocking. Taken together, however, they form the **theory of special relativity**, an extremely important new law of physics.

There was a young lady named Bright,
Who traveled much faster than light.
She departed one day
In an Einsteinian way
And returned on the previous night.
ANONYMOUS

EXAMPLE 5.27: Special Relativity

Let's see how special relativity affects our rocket car example above, redrawn in Figure 5.11.

Imagine that the rocket car can, and does, travel at the speed of light. By the Postulate of Light Speed, Stan and Monica must measure the *same* speed for the rocket car; thus, $v_{R,S} = c$ and $v_{R,M} = c$.

Now, in Exercise 5.20, we saw that the relationship between $v_{R,S}$, $v_{R,M}$, and v is given by the equation $v_{R,S} = v_{R,S} + v$. Plugging in our values for $v_{R,S}$ and $v_{R,S}$ into this equation, we get $c = c + v$.

Yikes! This equation is nonsense. What went wrong?

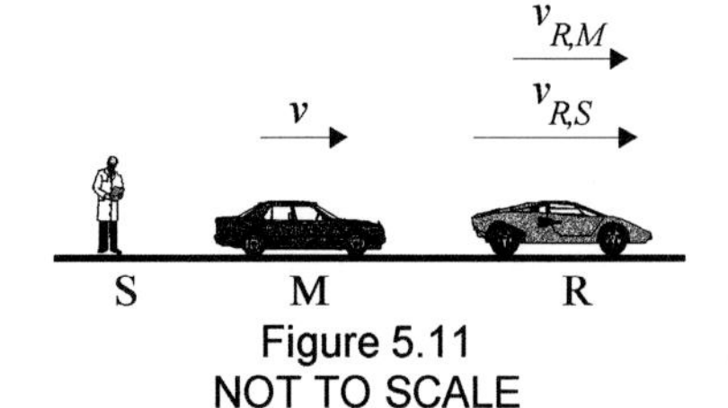

Figure 5.11
NOT TO SCALE

* * *

The problem is in our definition of speed (a law of physics which is subject to the first Postulate), and its relation to the constant speed of light (subject to the second Postulate). Speed is defined as the ratio of distance to time:

$$v = \frac{\text{distance}}{\text{time}}$$

Example 5.27 Continued...

But, in the case of a speed equal to c, this equation becomes:

$$c = \frac{\text{distance}}{\text{time}}$$

Now, different observers *must* measure the same value for c, even though they are traveling at different speeds. The only way this is possible is if the different observers measure *different* values for the *same* distance and time! That is, the speed of light remains constant in this equation because the definitions of distance and time do *not*.

In general, observers moving relative to a fixed distance in a "stationary" frame will measure different distances than do observers stationary in that frame; furthermore, observers moving relative to a fixed clock in a stationary frame will measure different elapsed times than will observers stationary in that frame. The *ratios* of the measurements, however, will *equal each other*, and they will both equal c.

* * *

In our rocket car example, this means that Stan will measure a different distance for the rocket car, in a different amount of time, than will Monica...even though Stan and Monica are equally competent experimenters. However, the ratio of each observer's measurements will still *both* be equal to c.

Where does that leave our equation $v_{R,S} = v_{R,S} + v$? This equation is *fine* for speeds low compared to c (for example, the 400 m/s in Example 5.16). But, when speeds get high (around $0.1c = 3.0\times10^7$ m/s), it is no longer adequate. We must *replace* it with a relativistic equation:

$$v_{R,S} = \frac{v_{R,M} + v}{1 + (vv_{R,M}/c^2)}$$

Notice that this equation reduces to the first one when $v << c$.

Many seemingly crazy consequences come out of special relativity, including:

Time travels in divers paces with divers persons.
WILLIAM SHAKESPEARE
English Author, 1564 - 1616

- **Time dilation**: A clock in motion with respect to an observer appears to run more slowly than an identical clock at rest with respect to the observer.
- **Length contraction**: A length in motion with respect to an observer appears to be shorter than an identical length at rest with respect to the observer.
- **Unsimultaneous simultaneity**: Two events that appear simultaneous to an observer at rest with respect to the locations of the events may not appear to be simultaneous to an observer in motion with respect to these locations.
- **Energy is mass; mass is energy**: Matter has energy simply by virtue of existing: the value of this mass-energy is given by the familiar equation $E = mc^2$.

There was a young fencer named Fisk,
Whose thrust was exceedingly brisk.
So fast was his action
Lorentz-FitzGerald contraction,
Reduced his rapier to a disk.
ANONYMOUS

5.5.2 GENERAL RELATIVITY

Einstein later extended his theory of special relativity to include the motion of *accelerating* (non-inertial) reference frames. This expanded theory, called **general relativity**, says (among other things) that the curvature of space-time is the source of gravity.

Oh, that stuff! We never bother with that in our work.
ERNEST RUTHERFORD, on relativity
English Physicist, 1871 - 1937

As it turns out, Einstein's Theory of Relativity is actually a more inclusive, better theory of motion than the more familiar version developed by Newton, discussed in Chapters 6 and 7 (Fig. 5.12). Yet it is Newton's theories of motion that we learn in introductory physics classes, not Einstein's. Why is this so?

We study Newtonian mechanics, not Einstein's relativity, because the motions of everyday life almost always occur at speeds much lower than c, and over distances in which space-time curvature is small. In this realm, there is no need for the more complex equations of relativity. So, thinking like a physicist, we begin with the *simpler* (though less accurate) theory of motion given by Newton—leaving the more precise but more difficult Einsteinian version for advanced studies.

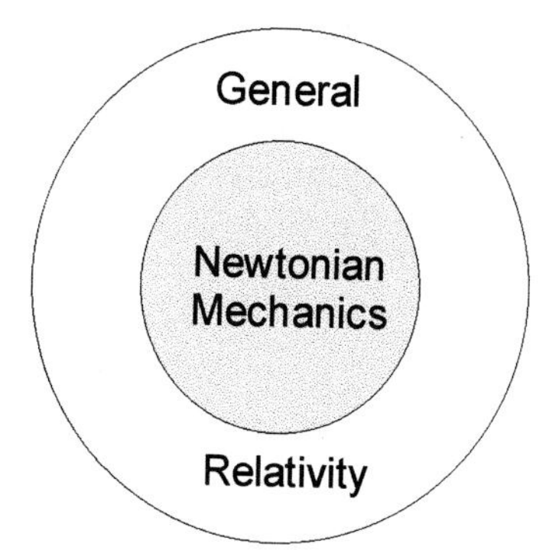

Figure 5.12

5.5.3 QUANTUM MECHANICS

Just as relativity is the physics for describing very *fast* objects, **quantum mechanics** is the physics for describing very *small* objects ($<10^{-10}$ m).

Like relativity, quantum mechanics is a theory of motion that is more inclusive (and more accurate) than Newtonian mechanics (Fig. 5.13). And like relativity, at the introductory stage, we learn Newtonian mechanics anyway, because in real life we almost never deal with objects small enough to need quantum mechanics.

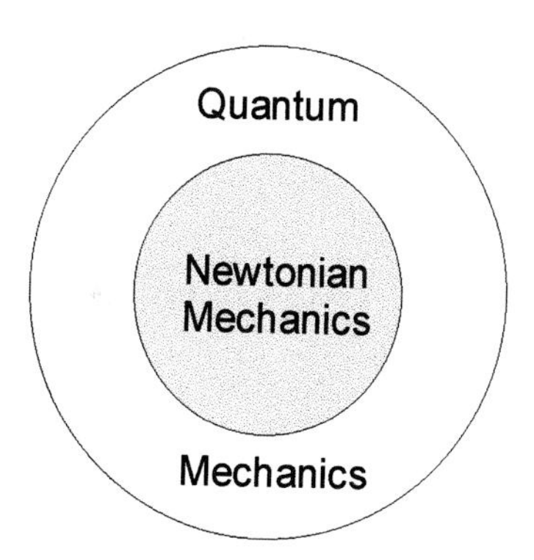

Figure 5.13

At any rate, quantum mechanics tells us that in the tiny realm of atoms and subatomic particles, many strange things happen:

- A wide variety of subatomic particles exist, such as the familiar electrons, protons, and neutrons, but also including some more esoteric ones, like quarks, neutrinos, and muons.

There's more on atomic particles in Section 10.2.

- All subatomic particles exist as *particles* or *waves*, depending upon the situation.* Each particle-wave of momentum p has wavelength $\lambda = h / p$, where h is **Planck's constant**—a fundamental constant of nature as important to physics as is c.

Waves are in Chapter 9.

The fundamental constants are in Section 11.5.3.

- Subatomic particles interact by transferring **virtual particles** back and forth between each other. The sum of all virtual particle activities is detected in the macroscopic world as forces.

Forces are covered in Chapter 7.

* In particular, they move through space as if they were waves, but they interact with matter as if they were particles.

Electromagnetism is in Section 7.3.2.

For more on energy and momentum, see Chapter 8.

Probability is covered in Section 16.4; differential equations are in Section 14.4.

- As atomic and subatomic particles jostle around in their strange little world, they sometimes *emit* energy and momentum in the form of **electromagnetic radiation**, also known as light.
- Radiation is itself both a particle and a wave, depending upon the circumstance. Its energy and momentum have specific (**quantized**) values according to wavelength: $E = hc/\lambda$ and $p = h/\lambda$.
- We interpret particle-waves of any type with the mathematics of probability and differential equations.
- One consequence of probabilistic particle-waves is that we can *never* measure their position or their velocity exactly. In fact, the uncertainties in our measurements of position and velocity are *related* by the **Heisenberg Uncertainty Principle**:

$$\Delta x \cdot \Delta p \geq \tfrac{1}{4\pi} h$$

Nobody understands quantum mechanics.
RICHARD FEYNMAN
American Physicist, 1918 - 1988

So, just as different observers, in the relativistic domain, can measure different values for distance and time—as a result of the constancy of c—it is not possible, in the quantum realm, to measure both position and velocity exactly—as a result of the constancy (and small size) of h. In both cases, the troubles are *not* caused by instrument or user error; instead, they are physical limits imposed on reality by natural law.

5.6 CONCLUSION

Every great advance in Science has issued from a new audacity of imagination.
JOHN DEWEY
American Philosopher, 1859 - 1952

The kinematical description of motion both effectively describes (low speed, large size) motion, and hints at discoveries in relativity and quantum mechanics (high speed, small size motion). These physical laws—which supersede Newtonian mechanics, even though we do not study them that way—have changed forever our perception of space, time, and the universe around us.

But kinematics serves us in other ways, as well. We need this basic, fundamental description of motion in order to even ask questions about *why* objects move; questions that are best answered, in real life, by Newtonian mechanics. And so we leave kinematics, and turn now to the heart of the introductory physics class: analysis of motion through Newton's Laws and their consequences.

6
NEWTON'S LAWS

6.1 INTRODUCTION

In Chapter 5, we learned that Galileo was the first to quantify and describe motion, using the kinematic equations. However, Galileo did not determine *why* objects move. This very important subject, called **dynamics**, was left to his successor, Sir Isaac Newton.*

See Section 5.4.2 for more on the kinematic equations.

Newton developed three laws—now bearing his name—that show that accelerated motion is due to the presence or absence of *forces*. A **force** is a vector quantity that describes the action of one body on another. In remembrance of Newton's contribution to the subject of forces, the SI unit of force was named after him (newton = $\mathrm{N} = \mathrm{kg{\cdot}m/s^2}$).

See Chapter 7 for more on forces.

In this chapter, we introduce Newton's Laws. *These Laws are so fundamental to the study of physics, that with the exception of a small number of topics usually covered late in the course (or ignored altogether), <u>everything</u> <u>else</u> in your physics class will be an expansion on, or a restatement of, these laws.* You may find this comforting: if you understand these three laws (and you will!), you understand the vast majority of classical physics.

Mechanics is the paradise of the mathematical sciences because in it we come to the fruits of mathematics.
LEONARDO DA VINCI
Italian Experimentalist, 1452 - 1519

---

* The subjects of kinematics (Chapter 5) and dynamics (this chapter) are collectively known as **mechanics**; or, more specifically, **Newtonian mechanics**.

6.2 THE FIRST LAW: FORCES & INERTIA

Once moving, [bodies] are never at rest unless impeded. Once in motion, only an outside force can restrain them.
G.B. BENEDETTI
Italian Experimentalist, c. 1585

Newton's First Law (which we will call N1 for brevity) defines forces. This law comes first in the sequence because, in typical physicist fashion, Newton recognized that further statements about forces—that is, the Second and Third Laws—could only be developed *after* the meaning of a force is made precise. N1 is most simply stated as follows:

A moving body will continue its motion unless acted upon by a force.

As is true for all fundamental laws about the universe, there are many subtleties about N1 that must be clarified in order to fully understand it.

EXAMPLE 6.1: Newton's First Law: Subtle Point #1

The definition of a "moving" body in N1 is less obvious than it may seem at first glance.

First of all, the word "moving" in the statement of this law means that the body is traveling in a straight line at a *constant velocity*; *i.e.*, it is neither speeding up, nor slowing down, nor changing direction.

Secondly, the body's constant velocity can have *any* value, including *zero*. That is, the special case in which the body is at rest—when the body is not "moving" in the colloquial sense of the word—is also incorporated into this law.

In Section 5.3.3. 

(As an aside, note that these two points are directly related to the issue of relative motion, as discussed in detail in Section 5.3.3. In the language of that section, we would say that the body referred to in N1 is located in an *inertial frame*. Indeed, N1 *defines* the notion of inertial frames.)

If we wanted to make these two ideas about motion more explicit in the statement of N1, we could rewrite the law as:

A body in constant velocity motion, or at rest, will continue its constant velocity motion, or remain at rest, unless acted upon by a force.

N1, also known as the **law of inertia**, tell us that objects *keep doing* whatever it is they were doing (whether moving at constant velocity or sitting at rest), unless a force gets involved. That is to say that without a force, there is no change in motion...and *no stopping.*

EXAMPLE 6.2: Newton's First Law: Subtle Point #2

N1 is an extremely counterintuitive statement about the universe: in real life, bodies in motion *always* stop moving sooner or later. In fact, prior to Newton, people believed that the "natural state" of any object was at rest; that is, moving bodies *must* eventually stop their motion and return to their natural resting state.

Example 6.2 Continued...

The "natural state" hypothesis does fit with everyday experience. To see that this is so, try a simple experiment. Push this book across your desk. The book moves while you push it, but, once you stop pushing, the book stops. It seems as though the book's "natural state" is to be at rest: without your pushing force, the book "naturally" comes to a stop.

* * *

The "natural state" hypothesis is not correct, however, as Newton makes explicit in his first law. When in motion, the book is actually experiencing *two* forces: the force of your hand in the forward direction, *and* the force of friction (between the book and the table) in the backward direction. The book comes to a stop once the pushing force is gone not because rest is its "natural state," but because the frictional force is *still there*.

Friction is discussed further in Section 7.2.6.

What would happen, then, if you pushed the book along a surface with less friction, for example, along the surface of an ice rink? We know from experience that the book would travel much farther on the ice than on the tabletop (even with the same push) before it came to a stop.

It is easy to imagine the ideal case (Eliminating the Unessential) in which the book slides on a surface with *no* friction: the book would literally travel forever (at a constant speed!) as a result of the one push from your hand.

Eliminating the Unessential is in Section 2.3.

* * *

Newton's true genius is therefore evident in his statement of N1. He wrote this law in order to directly address the question of the "natural state" of the object—a notion that had been around since the time of Aristotle!

DEFINITION 6.3: Inertia

The word *inertia* describes the tendency of *all* objects to continue with their previous motion, in the absence of a force.

Today, we correlate inertia with the presence of mass. Thus, an object with a lot of inertia (*i.e.*, a lot of mass) is harder to deviate from its trajectory (*i.e.*, it takes a lot of force to change its motion) than an object with less inertia.

If a force is present, a body will *alter* its motion—either by speeding up, slowing down, changing direction, or some combination of these—according to N1.

EXAMPLE 6.4: Newton's First Law: Subtle Point #3

To understand this point, think of a drifting spaceship, moving through space in a straight line at a constant speed. This spaceship will *only* change its motion if a force acts upon it: for example, if the engines are turned on (creating a forwardly-directed engine force), or if it is hit by

Example 6.4 Continued...

an asteroid (pushing the ship away from the asteroid), or if it encounters some sort of matter cloud that slows its progress (by creating friction or drag on the ship).

So, N1 really says two things:

- In the *absence* of a force, a body keeps moving as it was moving.
- In the *presence* of a force, a body *changes* its motion.

NOTE 6.5: Newton's First Law: A Paradox?

N1 is paradoxical in many ways.

On the one hand, it is very easily stated—so easily, in fact, that you have surely heard this law said before. On the other hand, it is vastly counterintuitive—in real life, bodies do *not* seem to move forever; rather, they eventually come to a stop.

Furthermore, while the importance of this law (in a theoretical sense) cannot be overstated—without it, forces cannot be defined—in a practical sense, you will almost never *use* this law for any constructive purpose. Instead, you will use the Second and Third laws for the bulk of your problem solving.

6.3 THE SECOND LAW: $\vec{\mathbf{F}}_{net} = m\vec{\mathbf{a}}$

According to N1, if one or more forces are applied to a body, the body's motion changes. But *how* does the motion change, exactly? Newton's Second Law (N2) tells us.

See Chapter 15 for more on vectors.

N2 says that the *net* force on a body *accelerates* it. That is, according to N2, the vector sum of all forces on a single object is equal to the product of the object's mass and its acceleration:

$$\vec{\mathbf{F}}_{net} = m\vec{\mathbf{a}}$$

EXAMPLE 6.6: Force & Acceleration

According to N2, forces provide acceleration to a body. According to kinematics (Chapter 5), accelerations change velocity through time. Thus, forces change a body's motion through a time derivative of velocity. Since velocity is a vector, changes in velocity can affect the velocity's magnitude (the object changes its speed), its direction (the object alters its bearing), or both.

N2 implies that bodies with different masses will have different accelerations *even if* they experience the same force. Conversely, it also tells us that objects of different masses can achieve the same acceleration only if proportionally different forces are applied to them.

EXAMPLE 6.7: Car-Truck Accidents

Accident statistics prove that in car-truck accidents, the car is likely to sustain more damage than the truck. Why would this be so?

The answer lies in N2. Because the car is less massive than the truck, it will experience a proportionally higher acceleration than the truck, *even when subject to the same force.* This higher acceleration translates into more chances for the car to be damaged in an accident.

EXAMPLE 6.8: Moving Day

Imagine that you are helping a friend move. You push a heavy box along the ground while your friend pushes a lighter box immediately next to you. The two of you start to race.

Being stronger than your friend, you are able to accelerate your box as fast as your friend. However, because your box is more massive, according to N2, you must apply a bigger force on the box in order to keep the accelerations the same. You may tie the race, but you'll get the better workout!

To calculate the acceleration of an object using N2, we vectorially sum *all* of the forces acting *on* the object. It is the *total* force that equals the mass times the acceleration in $\vec{\mathbf{F}}_{net} = m\vec{\mathbf{a}}$, not any single force.

We use a **free body diagram** to help us with this task. A free body diagram is a picture in which an object is represented by a single point.* All forces on the object are drawn as vector arrows radiating away from the point. The vectors have lengths proportional to the forces' strength, and directions in the direction of the forces.

See Section 15.2.1 for more on visualizing

EXERCISE 6.9: Free-Body Diagram, Part I

An elephant, a horse, and a bear—working for a circus—each pull a rope that is attached to a barrel of food. See Figure 6.1. The elephant pulls twice as hard as the horse and three times as hard as the bear. Find the direction of the barrel's acceleration using a free body diagram. The directions in Figure 6.1 are correctly drawn.

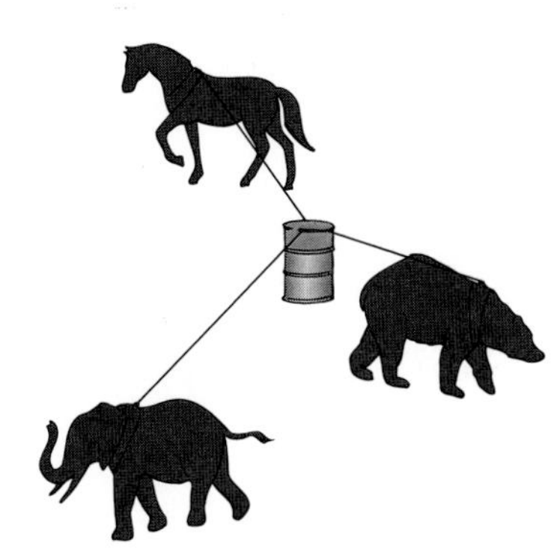

Figure 6.1

* We can represent a body by a point because there is a place in each body, called the **center of mass**, where the body moves as if all of its mass was concentrated there. Your textbook will have more information on how to calculate the center of mass.

Exercise 6.9 Continued...

SOLUTION:

To work this problem, we must draw a free body diagram for the barrel. This means we need to make a *new* sketch in which the barrel is represented by a dot, and the forces on the barrel are represented by vector arrows coming out from the dot. We do *not* need to redraw the animals.

There are three forces acting on the barrel: the force due to the elephant, F_e, the force due to the horse, F_h, and the force due to the bear, F_b. From the statement of the problem, we know that F_e is twice as long as F_h and three times as long as F_b. From the diagram in Figure 6.1, we also know the directions of the vectors.

So, from this information, we can draw a free body diagram:

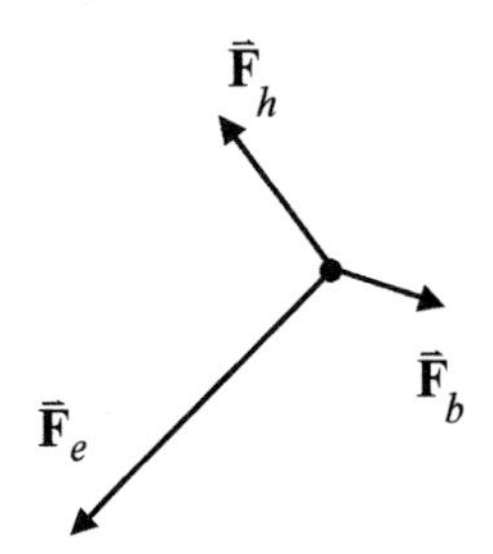

See Section 15.2.2 for more on graphically adding vectors.

To solve the problem, we must add these forces vectorially:

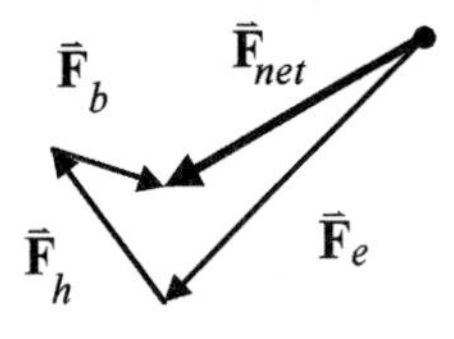

Refer to Section 15.3.5 for more on scalar multiplication.

Since mass is a positive scalar, the acceleration of the barrel points in the *same* direction as the total force:

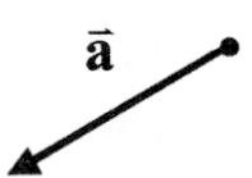

Thus, the barrel will move in the general direction between the elephant and the horse.

No body begins to move or comes to rest of itself.
AVICENNA
Ancient Persian Philosopher, 980 - 1037

When the total force on an object is not zero, the acceleration is also non-zero. So, the object will *move*. This situation is analyzed by means of a division of mechanics known as **dynamics**. On the other hand, when the total force on the object *is* zero (for example, when there are no forces on the body, or when all of the forces cancel each other out), the acceleration *must* be zero. In this case, the object will move at a *constant velocity* or *remain at rest*. This situation is analyzed via the branch of mechanics called **statics**.

EXERCISE 6.10: Free-Body Diagram, Part II

Three children fight over a toy, as sketched in Figure 6.2. Each child pulls on the toy, but it does not move. If Andy pulls twice as hard as Larry, with what force does Marty pull? Assume that the directions of Amanda and Larry are correctly drawn in Figure 6.2.

Figure 6.2

SOLUTION:

To solve this problem, we must draw a free body diagram.

We redraw the object experiencing force (the toy) as a dot. Then, we sketch the forces due to the children as arrows coming out from the dot. The length of the vector arrows must be proportional to the forces exerted, as given in the statement of the problem. Notice that we do *not* know the length or direction of Marty's force arrow at this time, so we draw it in arbitrarily:

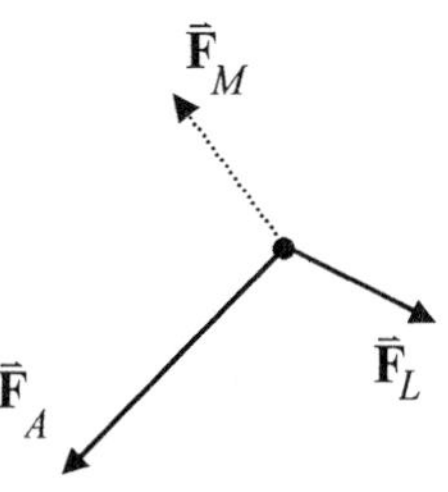

We now add up the forces exerted by Larry and Andy:

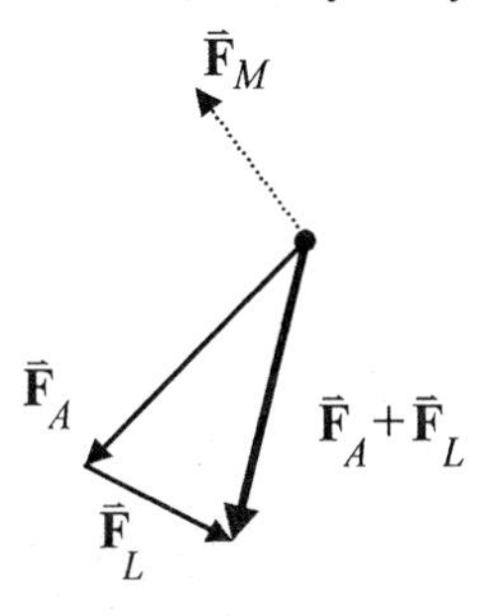

Since the toy does not move, its acceleration is zero. Therefore, the vector sum of the forces must also be zero. This means, in this static situation, that Marty's force arrow must have the same length as $\vec{F}_A + \vec{F}_L$, but point in the *opposite* direction. Therefore, the complete free body diagram must look like:

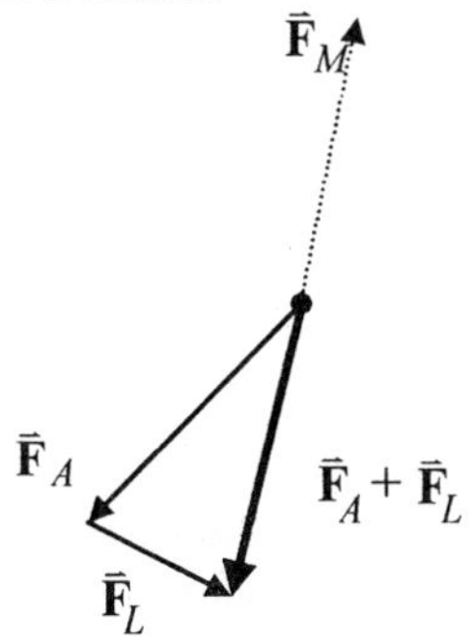

6.4 THE THIRD LAW: ACTION & REACTION

An object offers as much resistance to the air as the air does to the object. You may see that the beating of an eagle's wings against the air supports its heavy body in the highest and rarest atmosphere...
LEONDARDO DA VINCI
Italian Experimentalist, 1452 - 1519

Newton's Third Law (or N3) is perhaps the most difficult of the three laws to understand. Whereas N2 describes the effect of forces *on* a *single* object, N3 examines the forces *between multiple* objects.

This law is most simply stated as:

If object *A* experiences a force from object *B*, then *B* will experience a reciprocal force from *A*, such that:

The force on *A* due to *B*,
is *equal* in magnitude, but *opposite* in direction, to
the force on *B* due to *A*.

The force from *B* is the **action**, while the force from *A* is the **reaction**.*
Hence, N3 is sometimes known as the **law of action and reaction**.

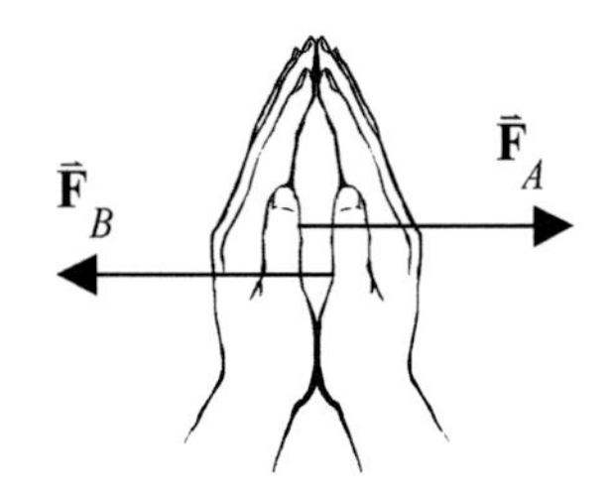

Figure 6.3

EXAMPLE 6.11: The Force of Prayer?

To *feel* the effect of Newton's third law, tightly press your hands together in the praying position.

Notice that your left hand (object *A*) presses on your right hand (object *B*) with a force equal to the force your right hand (*B*) exerts on your left (*A*). However, the directions of these two forces are different: the force from your left hand is directed rightwards, while the force from your right hand is directed leftwards. Figure 6.3 shows these forces.

Thus, as N3 dictates, the force on each of your hands is *equal* in magnitude, but *opposite* in direction.

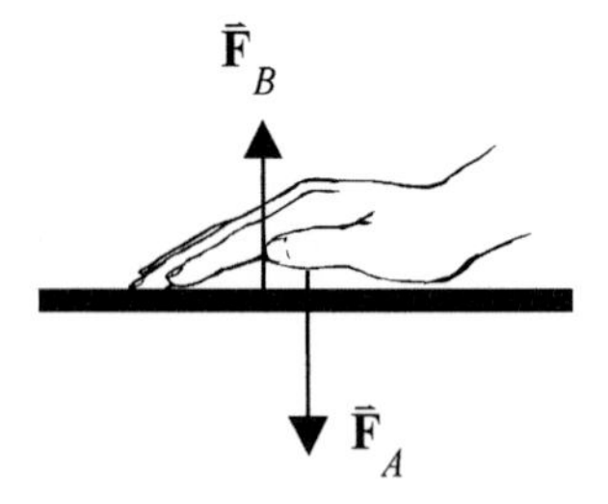

Figure 6.4

EXAMPLE 6.12: Laying On of Hands

Let's investigate N3 further. Try pressing one of your hands (object *A*) onto a hard surface, like a table (object *B*).

Notice that the force you apply to the table is equal in magnitude, but opposite in direction, to the force you feel from the tabletop (Fig. 6.4). This is similar to the case in the previous example.

Now, try *increasing* the force you apply against the table. The harder you push, the harder it pushes back on you! In other words, the force from the table seems to *adjust* as force from your hand changes, in order that N3 remains valid.

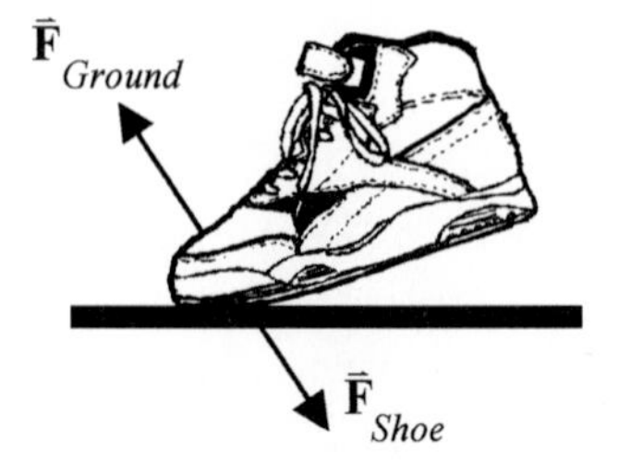

Figure 6.5

EXAMPLE 6.13: Walking

Walking is a good example of an application of N3. Every step you take, your foot pushes into the ground. At the same time, the ground pushes back on you, propelling you forward (Fig. 6.5).

* Though, perhaps confusingly, these labels can be reversed.

EXAMPLE 6.14: The Earth and the Moon

The earth and the moon exert a force (**gravitation**) on each other. These forces are equal in magnitude but opposite in direction. The earth constantly pulls on the moon, keeping it in orbit. Meanwhile, the moon constantly pulls back on the earth, causing the tides.

See Section 7.2.1 for more on gravitation.

6.5 NEWTON'S LAWS AND ROTATION

In most physics classes, you will learn about Newton's Laws quite early on in the course, using them to solve many straight-line motion problems. Some time later, you will apply these laws to *rotational* motion, and solve more problems. However, there is no fundamental difference between the physics of Newton's Laws for straight line or rotational motion, so, there is no physical reason to separate the two lessons. In this book, we discuss them together.

Section 6.5.3 below summarizes Newton's Laws for rotation. They are very similar to the laws we have just learned—with two exceptions. Whereas, according to N2, we calculate the linear acceleration with the *mass* of an object and the total *force* upon it ($\vec{\mathbf{a}} = \vec{\mathbf{F}}_{net} / m$), we calculate *rotational* acceleration using the *rotational inertia* (a quantity that is related to but not the same as mass) and the total *torque* on the object (a quantity related to but not the same as force).

6.5.1 ROTATIONAL INERTIA

As we have seen, the *inertia* of an object is its tendency to resist changes in velocity. The **rotational inertia** (or **moment of inertia**, or even just **inertia**, though this name can obviously be confused with the previous usage of the term) of a rotating object is a measure of its tendency to resist changes in *rotational* velocity. The value of rotational inertia depends on *both* the total mass of an object *and* on the distribution or shape of the mass relative to the axis of rotation.

EXAMPLE 6.15: Rotational Inertia, Part I

The best real-life example of rotational inertia is the dumbbell: a long metal tube with weights on each end (Fig. 6.6).

Imagine holding a dumbbell at the center of the tube and rotating your wrist so that the dumbbell remains horizontal. The resistance you feel to the rotation is the dumbbell's rotational inertia. You can *change* the dumbbell's inertia by increasing the weight (making the dumbbell harder to rotate) or by moving the weights along the tube (moving the weights closer to your wrist makes the dumbbell easier to rotate).

Figure 6.6

The value of the rotational inertia is *different* for different shapes, and can also be different for the *same* shape on different axes of rotation.

Figure 6.7

EXAMPLE 6.16: Rotational Inertia, Part II

Imagine that a blizzard has dumped three feet of snow on your driveway. You decide to shovel it off. On each scoop of the shovel, you gather snow onto the shovel. Then, you rotate the shovel to unload the snow on the lawn. See Figure 6.7.

The value of the shovel's rotational inertia depends on two things: how much snow is in the scoop (affecting the *mass* of the shovel), and where you place your forward hand on the shaft (affecting the *axis of rotation* of the shovel).

You, of course, want to make shoveling as fast and easy as possible. You can speed up the process by getting a lot of snow into each scoop. But, this increases the shovel's rotational inertia, making it harder for you to rotate the shovel.

However, you can also decrease the shovel's rotational inertia by moving your hand closer to the scoop. This moves the axis of rotation closer to the bulk of the mass, lowering the inertia back down to a reasonable value.

To sum up, the rotational inertia of the shovel depends both on the mass of the snow, and where the mass is located relative to the axis of rotation. (By the way, the safest way to scoop snow is to push it. Picking up and rotating a heavy shovel can lead to back pain.)

You will learn to calculate rotational inertias for various shapes and axes in your physics class; your textbook or instructor will provide the procedure for doing so. It will help to know that rotational inertias of all symmetric objects have the same format:

$$I = kmr^2$$

Your textbook will have a much longer list of these. 

where k is a constant (less than or equal to one) that varies depending on the object's shape, m is the object's mass, and r is some defining length of the object. Figure 6.8 lists the rotational inertias for some simple objects.

Figure 6.8: Rotational Inertias

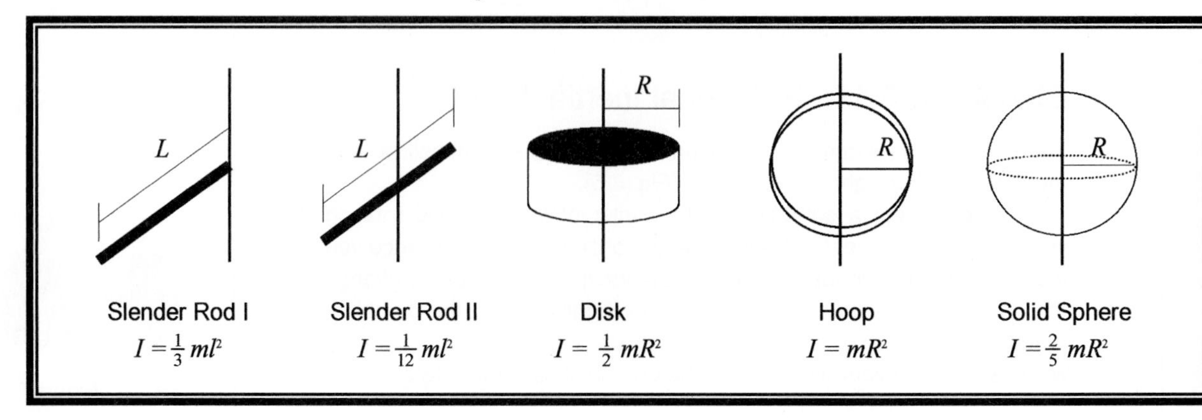

6.5.2 TORQUE

When a force is applied to a rotating object, we must consider the location of the force on the body *relative to* the axis of rotation. A force applied *at* the axis will not change the motion of the rotating body; only a force located a distance away from the axis (but still on the body) will cause angular acceleration. We call the combination of a force, acting at a distance from the axis of rotation, the **torque**.

EXAMPLE 6.17: Torque and Doors, Part I

To get a feel for torque, stand in front of the nearest hinged door. This kind of door rotates along an axis that runs through the hinges attached to the wall.

Try pushing the door. Note that your hand naturally applies the pushing force at a location far from the hinge axis of rotation, *i.e.*, at the unhinged side of the door. This is because, through trial and error, you have learned that pushing a door near the hinges will not result in the door properly closing.

To prove to yourself that this is true, try pushing the door close to the hinges. In this case, notice that you need to apply a greater pushing force on the door in order to rotate it shut. Also note that the closer to the hinges you push, the greater the force that is required. In fact, if you choose to push *on* the hinges, the door will not move at all.

Your pushing force, applied at a distance from the axis of rotation, torques the door into closing.

We call the (shortest, straight line) distance between the axis and the location of the applied force the **moment arm** $\vec{\mathbf{r}}$. The moment arm is a vector that points *from* the axis of rotation *to* the location of the applied force. Torques are maximized when the moment arm is *perpendicular* to the direction of the applied force.

EXAMPLE 6.18: Torque and Doors, Part II

Try pushing the door again. Notice that you push in a direction roughly perpendicular to the plane of the door itself. As a result, the door rotates closed without much fuss. Now try pushing the door at an angle significantly less than ninety degrees. It takes much more effort to close the door this way, doesn't it?

The results of the three door trials of this example and the last are summarized in the figure below:

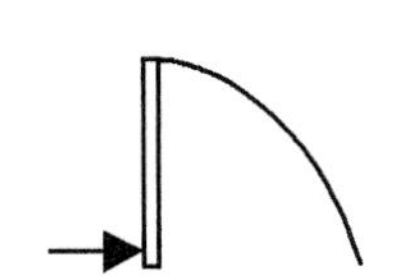
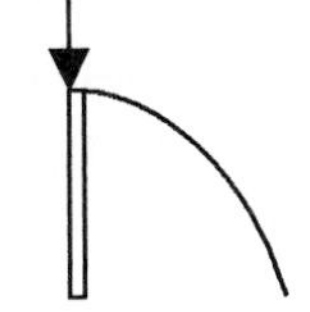
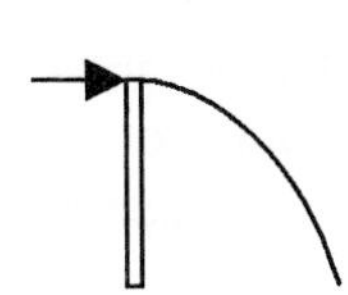

A force at the hinge does nothing...Neither does a force parallel to door...This force maximizes the torque.

See Section 15.3.5 for more on cross-multiplication. 

The mathematical expression for torque is given by the equation:

$$\vec{\tau} = \vec{r} \times \vec{F}$$

where $\vec{r}$ is the moment arm of the rotating object, $\vec{F}$ is the applied force, and $\vec{\tau}$ is the symbol for the vector quantity torque.

DEFINITION 6.19: Torque & The Cross Product

Notice that, in the torque equation above, $\vec{r}$ and $\vec{F}$ are multiplied as a cross product. The magnitude of the torque is thus given by the cross product equation $\tau = rF \sin \phi$, where r and F are the positive magnitudes of their respective vectors, and ϕ is the smallest angle between the vectors.

Because of this relation, we see that the cross product is maximized when the vectors are perpendicular to each other; that is, when $\sin \phi$ is one. Therefore, when the moment arm is perpendicular to the applied force, we obtain the highest possible value for the torque.

Furthermore, this relation tells us that cross products are zero when vectors are parallel; *i.e.*, when $\sin \phi$ is zero. Thus, if the force is parallel to the moment arm, there won't be *any* torque.

Moreover, according to this equation, if the moment arm has zero length (the force is applied *at* the axis of rotation), the torque will be zero as well.

The direction of the torque is given by a cross product right hand rule. Thus, the torque always points perpendicular to the plane formed by $\vec{r}$ and $\vec{F}$, and is *not* in the direction of motion. See Chapter 15 for more on vectors, cross products, and cross product right hand rules.

In short, the mathematical expression for torque exactly corresponds to our experience of it!

6.5.3 NEWTON'S LAWS FOR ROTATIONAL MOTION

Newton's First Law for rotation is almost identical to that for straight-line motion:

STRAIGHT LINE MOTION	ROTATIONAL MOTION
1. An object in constant straight-line motion (or at rest) remains in motion (or at rest) until acted upon by a force.	1. An object in constant rotational motion (or at rest) remains in motion (or at rest) until acted upon by a torque.

Angular acceleration is in Section 5.4.1. 

Newton's Second Law for rotational motion, however, is slightly different than the Second Law for straight-line motion. For the latter, we use $\vec{F}_{net} = m\vec{a}$; for the former, we use $\vec{\tau}_{net} = I\vec{\alpha}$, where $\vec{\alpha}$ is the vector quantity rotational acceleration.

6.6 STATIC EQUILIBRIUM

A special case of Newton's Laws that is dear to the hearts of engineers everywhere is **static equilibrium**. In this case, the net force *and* the net torque on a body is zero:

$$\vec{\mathbf{F}}_{net} = \mathbf{0} \text{ and } \vec{\boldsymbol{\tau}}_{net} = \mathbf{0}$$

EXAMPLE 6.20: Static Equilibrium

If it were somehow possible for the three children of Example 6.10 to pull on the bear, such that the bear did not move laterally ($\vec{\mathbf{F}}_{net} = \mathbf{0}$) nor did it rotate ($\vec{\mathbf{F}}_{net} = \mathbf{0}$), then the bear would be in static equilibrium.

6.7 CONCLUSION

The real beauty of Newton's Laws is that they are so *general*. We can use these laws in an infinite number of situations—and we do—because Newton phrased them in terms of *forces*. It doesn't matter whether the force is due to an elephant pulling on a rope or the earth pulling on the moon, we need only the magnitude and direction of the force to find the object's acceleration. And once we have the acceleration, we can predict the future motion of the object using kinematics.

See Chapter 5 for more on kinematics.

In the next chapter, we provide a listing and description of many natural forces. All of them obey Newton's Laws, and therefore, they can be analyzed in exactly the same way as we have just done. The only real difference between the analysis of this chapter and the next, however, is that in the next, we use vector mathematics a little more heavily than we have done here, as we shall see.

7
FORCES & FIELDS

7.1 INTRODUCTION

In Chapter 6, we introduced Newton's three Laws, which describe forces and their influence on motion. These laws are very general—that is, they work for *any* and *all* forces—making Newton's Laws useful in a wide variety of situations.

In this chapter, we examine many different kinds of forces, including their causes and mathematical expressions. Some of these forces are easy to understand and widely used in an introductory physics course—such as the familiar force of weight—and others are more esoteric and utilized less often at the beginning level—for example, the buoyancy force. As you learn about each of these forces, however, keep in mind that all of them are intended for use with Newton's Laws.

We also discuss the new, but related, concept of the **force field**. Force fields represent a link between Newton's description of motion, discussed in Chapter 6 in this book, and the energy description of motion, discussed in Chapter 8.

[My goal is] to derive two or three general principles of motion for phenomenon, and afterwards, to tell how the properties and action of all corporeal things follow from these manifest principles.
ISAAC NEWTON
English Experimentalist. 1642 - 1727

For more on forces, see Chapter 7; for more on energy, see Chapter 8.

7.2 FORCES

In this rather long section, we look at several natural forces important in the typical introductory physics course. In your class, these forces will not appear all at the same time; rather, you will learn about them as the class progresses through the term. In this book, however, we gather them all together so that you can begin to recognize similarities and differences between forces.

7.2.1 GRAVITY

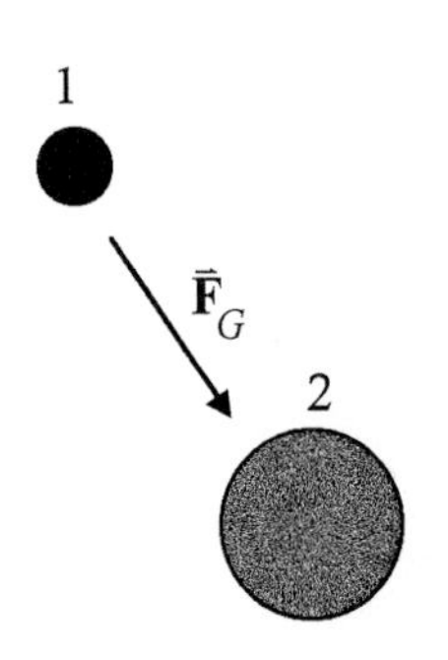

Figure 7.1

See Section 11.5.3 for more on constants of nature.

Perhaps the most fundamental and important force in our lives is the force of **gravity**: the attraction between objects because they have mass. It is not surprising that this most basic of forces was first described by Newton, who found that gravity depends both upon the masses of *and* the distance between two objects.

That is, the magnitude of the gravitational force can be expressed as:

$$F_G = \frac{Gm_1m_2}{r^2}$$

where m_1 and m_2 are the masses of the two objects, r is the distance between them (measured from the center of one object to the center of the other), and G—the **Gravitation constant**—is a constant of proportionality having a value of 6.672×10^{-11} N·m² / kg².

The direction of the gravity vector, as depicted by Figure 7.1, is *from* one mass *toward* the other (that is, from body 1 to body 2 as sketched, but also, from body 2 to body 1).* The gravitational force is thus considered *attractive*: gravity always causes bodies to accelerate toward each other.

The masses in this equation can be *anything*: for example, they can be the earth and the moon; two people; or two protons. No matter what they are, according to N3, the force on each object has the *same* magnitude.

See Section 6.4 for more on N3.

For my part, I think that gravity is nothing but a certain natural striving with which parts have been endowed...[so that] they may join together in their unity and wholeness.
NICOLAUS COPERNICUS
Italian Astronomer, 1473 - 1543

EXERCISE 7.1: Gravity

What is the gravitational force between:

(a) The earth and the moon?

(b) Aaron (having a mass of 70 kg) and Rochelle (with a mass of 62 kg), if they are separated by 0.75 m?

(c) Two protons, separated by 10^{-10} m?

---

* This *radial direction* is given by the unit vector $-\hat{\mathbf{r}}$. More on unit vectors in Section 15.4.3.

Exercise 7.1 Continued...

SOLUTION:

(a) To solve for the force of gravity for *any* two objects, we apply our formula $F_G = Gm_1m_2/r^2$, and plug in appropriate numbers. In the case of the earth and the moon, we have:

You can find the mass of the earth and the moon, and the distance between them, in the appendix of your textbook.

$$F_G = \frac{(6.672\times10^{-11}\ \text{N}\cdot\text{m}^2/\text{kg}^2)(5.98\times10^{24}\ \text{kg})(7.36\times10^{22}\ \text{kg})}{(3.84\times10^{8}\ \text{m})^2} = 1.99\times10^{20}\ \text{N}$$

This is a *huge* force, which is why the earth and the moon stay in orbit around each other.

(b) The procedure in this case is exactly the same as for part (a):

$$F_G = \frac{(6.672\times10^{-11}\ \text{N}\cdot\text{m}^2/\text{kg}^2)(70\ \text{kg})(62\ \text{kg})}{(0.75\ \text{m})^2} = 5.1\times10^{-7}\ \text{N}$$

This is a very *small* force, which is why you never feel a gravitational attraction to the people around you.

(c) Repeating the same procedure as above:

You can find the mass of a proton in the appendix of your textbook.

$$F_G = \frac{(6.672\times10^{-11}\ \text{N}\cdot\text{m}^2/\text{kg}^2)(1.67\times10^{-27}\ \text{kg})^2}{(10^{-10}\ \text{m})^2} = 1.86\times10^{-44}\ \text{N}$$

Why do you think physicists consider the gravitational force between protons to be negligible (at least, for most applications)?

EXERCISE 7.2: Gravity and Acceleration

We know from N3 (and Example 6.14) that the earth and the moon experience the same strength gravitational force, which, in part (a) above, we found to be 2×10^{20} N. Do they, as a result, also experience the same acceleration?

SOLUTION:

For more on N2, see Section 6.3.

No. The earth and the moon have very different masses. By N2, we know that the acceleration of a body depends on *both* the force it experiences *and* its mass. Because the moon is about 100 times *less* massive than the earth, it will be accelerated about 100 times *more* than the earth.

When calculating the gravitational force between a planet (such as the earth) and a small object near the surface of that planet, we often rewrite the gravity equation as:

$$F_G = mg$$

where m is the mass of the smaller object, and $g = GM/r^2$, such that M is the mass of the planet and r is the distance between the objects (measured center to center). We call g the **gravitational field** or the **acceleration due to gravity**; for earth, g is about $9.8\ \text{m/s}^2$.

See Section 11.5.2 for more on the gravitational field.

It is common to call F_G the **weight** of the object. Notice, however, that weight depends on which planet you are on!

All Celestial Bodies whatsoever have an attraction or gravitating power...whereby they attract not only their own parts...but that they do also attract all other Celestial Bodies that are within the sphere of their activity.
ROBERT HOOKE
English Scientist, 1635 - 1703

EXAMPLE 7.3: The Weight of Your Cat

What if you wanted to calculate the gravitational attraction between the earth and your cat—or, in simpler terms, the cat's weight? Use the formula given above to make a quick calculation.

If your cat has a mass of 3 kg:

$$F_G = mg = (3\ \text{kg})(9.8\ \text{m/s}^2) = 29.4\ \text{N} \approx 30\ \text{N}$$

Thus, your cat weighs almost 30 newtons.

EXERCISE 7.4: Your Weight on Mars

If you weigh about 120 lbs on earth, what do you weigh on Mars?

SOLUTION:

First, we find the value of your mass:

$$\frac{F_{G,earth}}{F_{G,Mars}} = \frac{mg_{earth}}{mg_{Mars}}$$

Now, m in the right hand side of the above ratio cancels out, because we are trying to find the weight of the *same* object (you!) on the two planets. Thus, we have:

$$\frac{F_{G,earth}}{F_{G,Mars}} = \frac{g_{earth}}{g_{Mars}}$$

$$\Rightarrow F_{G,Mars} = \frac{g_{Mars}}{g_{earth}} F_{G,earth}$$

Next, we must calculate g_{Mars}, using M and r from your textbook:

See the appendix of your textbook to find the mass and radius of Mars.

$$g_{Mars} = \frac{(6.672\times10^{-11}\ \text{N}\cdot\text{m}^2/\text{s}^2)(5.98\times10^{23}\ \text{kg})}{(3.38\times10^{6}\ \text{m})^2} = 3.49\ \text{m/s}^2$$

Plugging in the rest of our numbers into our rewritten ratio, we get:

$$F_{G,Mars} = \frac{g_{Mars}}{g_{earth}} F_{G,earth} = \left(\frac{3.49\ \text{m/s}^2}{9.8\ \text{m/s}^2}\right)(120\ \text{lbs}) = 43\ \text{lbs}$$

On Mars, you would weigh just one-third of your weight here. That doesn't sound too bad!

7.2.2 THE ELECTRIC FORCE

The **electrical force** between two objects with charge is given by an equation—called **Coulomb's Law**—that is mathematically very similar to the equation for gravity:

See Example 2.6 for more on their similarity.

$$F_E = \frac{kq_1q_2}{r^2}$$

where q_1 and q_2 are the net electrical charges of the two objects (measured in coulombs (C)), r is the distance between them (measured from the center of one charge to the center of the other), and k is an electrical constant which has value 8.988×10^9 N·m²/C² $\approx 9 \times 10^9$ N·m²/C². This constant is also written (for complex reasons that are not important here) as $1 / 4\pi\varepsilon_0$.

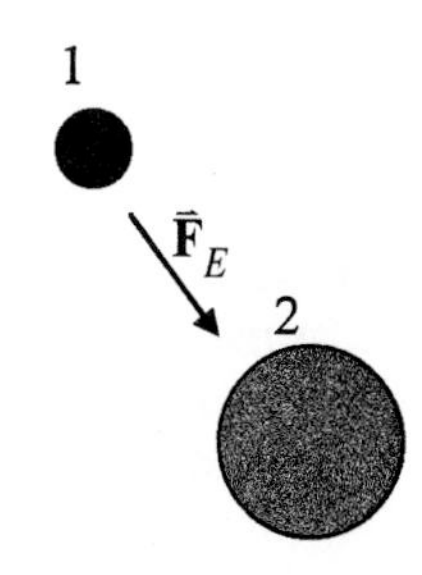

Figure 7.2 (a)

Notice that if you exchange G for k, m_1 for q_1, and m_2 for q_2 in the above equation, Coulomb's Law for electricity is *identical* to Newton's Law for gravity. The only real difference between these two formulas is that whereas the masses in the gravity equation always have positive values, the charges in Coulomb's Law can be positive *or* negative.

This means that the Coulombic force may be negative (attractive) *or* positive (repulsive), depending on the combination of q_1 and q_2. That is, the sign of the electric force indicates whether the force on q_1 points *toward* q_2 (attractive) or *away* from it (repulsive). See Figures 7.2 (a) and (b). So, the old saying "opposites attract" describes the rule that when q_1 and q_2 have *opposite* signs, the electrical force will be *attractive*.

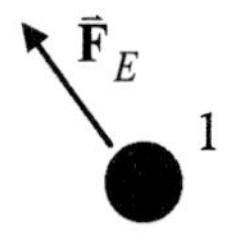

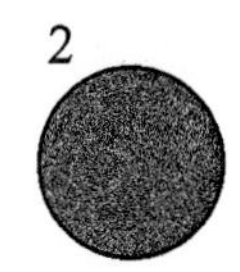

Figure 7.2 (b)

EXERCISE 7.5: Getting the Sign Right

In the chart below, fill in the sign of the resulting electrical force, given charges of the designated signs. Then determine if the force will be attractive or repulsive.

q_1	q_2	F_E	Description
+	+	?	?
+	–	?	?
–	+	?	?
–	–	?	?

SOLUTION:

q_1	q_2	F_E	Description
+	+	+	repulsive
+	–	–	attractive
–	+	–	attractive
–	–	+	repulsive

Electricity is of two kinds, positive and negative. The difference is, I presume, that one comes a little more expensive, but is more durable; the other is a cheaper thing, but the moths get into it.
STEPHEN LEACOCK
Canadian Humorist, 1869 - 1944

EXERCISE 7.6: Coulomb's Law and the Proton

What is the Coulombic force between two protons separated by 10^{-10} m? Compare this to the gravitational force found in Exercise 7.1 (c).

SOLUTION:

We solve this problem in the same manner as in Exercise 7.3; namely, by applying the formula $F_E = kq_1q_2/r^2$ and plugging in values:

See the appendix of your textbook to find the charge of a proton.

$$F_E = \frac{(9\times10^9\ \text{N}\cdot\text{m}^2/C^2)(1.6\times10^{-19}\ \text{C})^2}{(10^{-10}\ \text{m})^2} = 2.4\times10^{-8}\ \text{N}$$

This is a repulsive force because it is positive. Note that the electrical force between the protons, though still small, is *much* larger (10^{36} times larger) than the gravitational force between them. Therefore, we can conclude that the electrical force is a predominant factor in atomic interactions.

EXAMPLE 7.7: Changing Things A Bit

What happens if we replace one of the protons in the previous example with an electron? Electrons have the *same* amount of charge as protons, but they are *negatively* rather than positively charged.

We should reason, then, that the magnitude of the electrical force between an electron and a proton will be the *same* as the magnitude of the electrical force between two protons, but the force will be *attractive* rather than repulsive.

7.2.3 THE MAGNETIC FORCE

Charges produce an electrical force on other charges. But when they *move*, they also produce a **magnetic force** on other charges. The magnetic force may be negative (attractive) or positive (repulsive), depending on the combination of moving charges and materials involved.

The mathematical form of the magnetic force *varies* depending upon many circumstances. Therefore, physicists usually use an easier version of the magnetic force equation, called the **magnetic field**, to solve problems. The magnetic field will be discussed later in this chapter.

See Sections 7.3.1 & 7.3.2.

7.2.4 TENSION

Tension F_T is the force exerted by ropes or strings when they are *taut*. If the rope or string is very light (or "massless"), then the tension force has the same value all along its length. This force is always directed *along* the rope or string, *away* from an object connected to it (Fig. 7.3).

See Section 4.4.1 for more on problem solving "code words."

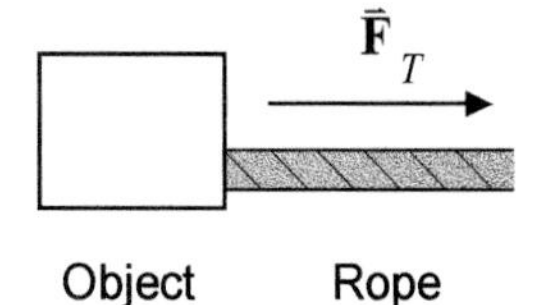

Figure 7.3

EXAMPLE 7.8: Gravity, Tension, and the Shoe

Dangle your shoe from your shoelace, as in Figure 7.4. What is going on in this situation?

Gravity is pulling down on the shoe. But the shoe is not falling (*i.e.*, acceleration in the y-direction is equal to zero), so there must be a force pulling upwards on the shoe, by N2. What is this force? The tension in the shoelace! We know that the tension is equal in magnitude, but opposite in direction, to the force of gravity. Therefore, the shoe stays put.

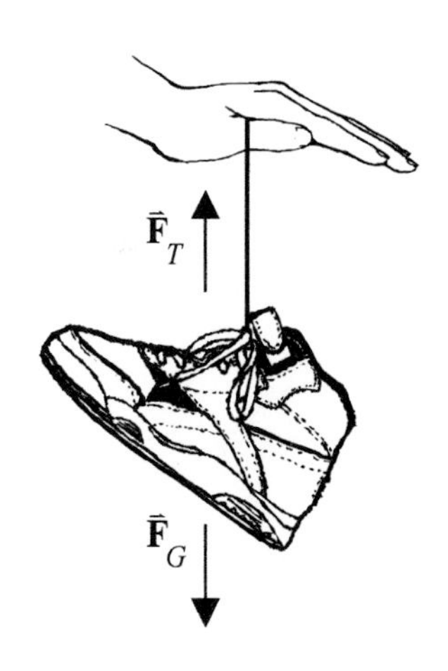

Figure 7.4

EXAMPLE 7.9: Pulley With Masses

Consider the following setup often seen in introductory physics classes. Two boxes are connected by a massless rope over a pulley, as sketched in Figure 7.5. Let's analyze this situation.

Because of gravity, the hanging block will fall. So, the tension in the rope will pull the block on the table to the right. But, the inertia of the block on the table will resist this gravity-induced rightwards motion, causing the rope to pull upward on the hanging block, slowing the fall.

Eliminating the Unessential, we assume that the rope remains taut throughout the motion. Since the rope is massless, then, the *numerical value* of the tension will be the same everywhere along the rope.

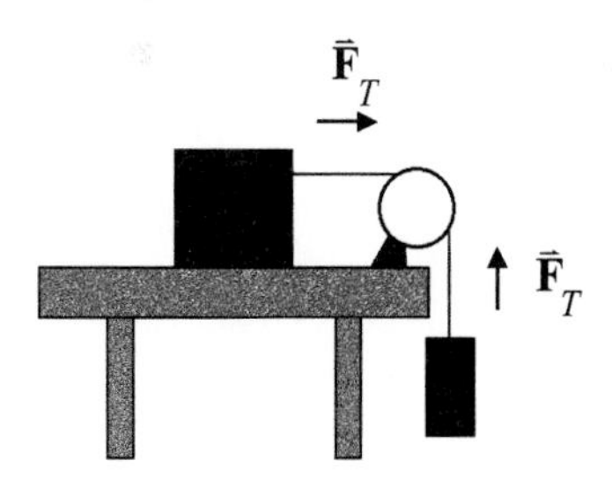

Figure 7.5

7.2.5 THE NORMAL FORCE

The **normal force** F_N is the force exerted on objects by surfaces. It is always directed perpendicularly, and outwards, from a surface (Fig. 7.6).

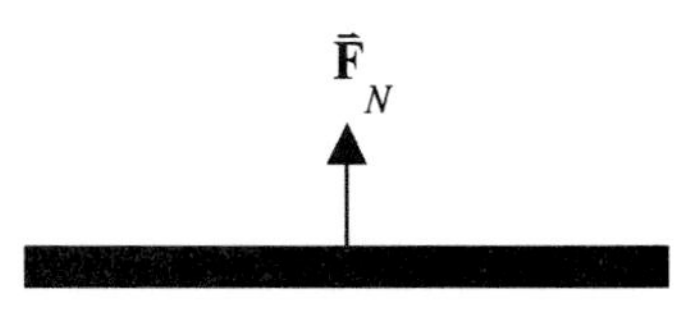

Figure 7.6

EXAMPLE 7.10: The Normal Force, Part I

You're studying physics right now, so presumably there is a big cup of coffee sitting nearby. We know from Section 7.2.1 that gravity is pulling down on your coffee cup. So why isn't the cup falling to the ground? Just as the shoe lace in Example 7.8 prevents the shoe from falling, the normal force from your table prevents the cup from falling.

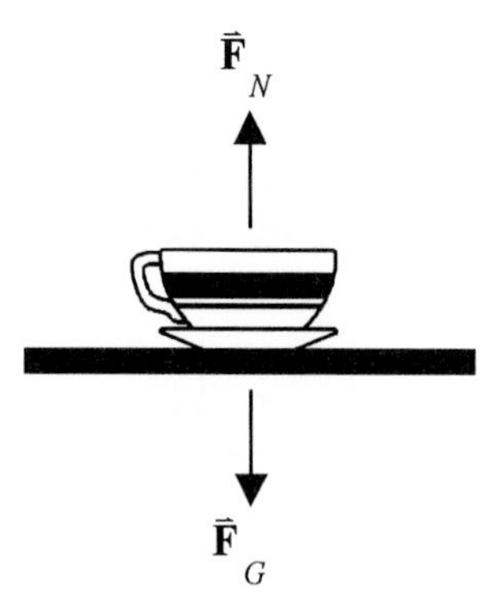

Figure 7.7

Example 7.10 Continued...

Because the cup isn't moving ($a_y = 0$), we reason from N2 that there *must* be another force on the cup, which has the same magnitude as the gravitational force but which acts in the opposite direction. This force is provided by the *surface* of the table. That is, there is a *normal force* acting perpendicular to, and upwards from, the table that exactly counteracts gravity. See Figure 7.7. So, the cup stays put.

EXAMPLE 7.11: The Normal Force, Part II

Okay, so we understand why the cup stays put. What would happen, though, if you put a heavier object on the table, like a textbook?

In this case, because the book is heavier than the cup, gravity pulls harder on it. Yet we know that the book, like the cup, doesn't sink through the table. So, the table *must* be exerting a stronger normal force on the book than on the cup!

How can the table possibly know to push harder on the book than on the cup? The answer comes from electrical interactions at the atomic level. By Coulomb's Law (Section 7.2.2), the electrons at the surface of the cup repel the electrons at the surface of the table. As gravity tries to pull the cup through the table, the electrons in both bodies arrange themselves in such a way as to prevent this from happening.

The same thing occurs in the case of the book. But, since the book is heavier than the cup, the electrons in the book and the table "make more effort" to counteract the effects of gravity.

What would happen, however, if you put a very heavy object—say, an elephant—on your table? The electrons in the elephant and in the desk could not possibly configure themselves in an efficient enough manner to counteract the huge weight of the elephant, and the electrical bonds in the table would collapse. The table would crack, and the elephant would end up on the floor.

CAUTION 7.12: The Normal Force, Part III

In the previous examples, the normal force was always equal in magnitude, but opposite in direction, to the weight. You might reasonably conclude that the normal force *always* does this. *This is NOT necessarily true!* This situation only occurs in special cases. We will return to this point in Example 7.25 below.

7.2.6 FRICTION

Friction arises because no surface is perfectly smooth. Therefore, when two surfaces rub against each other, they catch, causing the objects to slow down. The force of friction is thus directed *parallel* to the surface and *opposite* the *relative* direction of motion.

WARNING!

Be *very* careful with the direction of the friction vector: it is not always in the obvious direction. The general rule is that friction is pointed opposite to the direction of *relative* motion between the two surfaces.

EXAMPLE 7.13: Direction of Friction, Part I

When one of the surfaces is stationary, the direction of motion of the moving object *is* the direction of relative motion for the system. For example, a box is pushed to the right in the bed of a stationary truck. Therefore, the friction vector points opposite to the (only) direction of motion. So, the direction of friction here is to the left. See Figure 7.8.

Figure 7.8

EXAMPLE 7.14: Direction of Friction, Part II

When both surfaces move, the relative motion is the motion *between* the objects. In this case, the friction vector points opposite to the direction of the relative motion.

For example, imagine that the truck suddenly accelerates forward (to the right in Figure 7.9). As a result of the truck's abrupt motion, then, the box slips backwards. Thus, the relative motion between the box and the pickup truck in this case is in the backwards direction. So, the frictional force—which points opposite to the direction of relative motion—is actually directed forwards (in the same direction as the truck's motion).

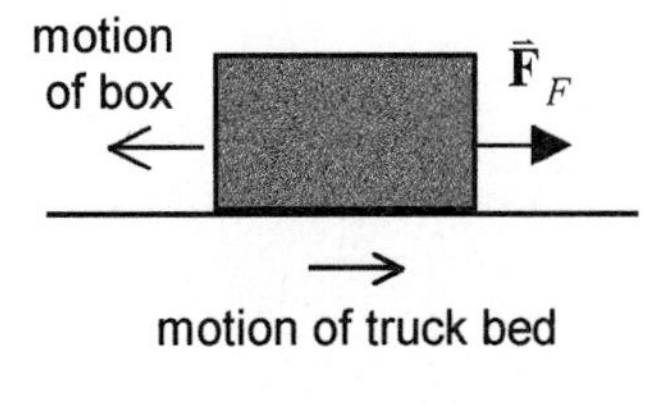

Figure 7.9

NOTE 7.15: Friction at the Microscopic Level

Like the normal force, friction is a *macroscopic* description of interactions that really take place on the *microscopic* level. Recall our experiment in the last chapter in which we pushed a book along a tabletop. The book slowed to a rest because of friction.

In Example 6.2.

We can describe this action in terms of electrons in the book and in the table, using the same model as in Example 7.11 above. The electrons at the surface of each object repel each other with the Coulombic force. The result of all of these electrical repulsions *is* friction.

There are two types of friction:

- **Static friction** is the frictional force between two objects that are *not* moving relative to each other.
- **Kinetic friction** is the frictional force between two objects that *are* moving relative to each other.

Generally it is easier to further the motion of a moving body than to move a body at rest.
THEMISTIUS
Ancient Greek Philosopher, 390 - 320 B.C.

Static friction *varies* from zero to a maximum value, according to the external force on the object. The value of kinetic friction stays roughly *constant*, but is always *less than* the maximum value of static friction.

As long as a force on a stationary body is less than the maximum value of static friction, the body will not move. Once a force overcomes this value, however, the object will move and kinetic friction begins.

EXAMPLE 7.16: Static & Kinetic Friction, Part I

To see the difference between static and kinetic friction, try this experiment.

Place your textbook on your desk. Now lightly push it with your finger...but don't push so hard that you move the book. In this case, static friction is perfectly balancing the force from your finger, so that the textbook doesn't move.

Now try pushing harder, but don't move the book yet. The static frictional force between the book and the table has *increased* to compensate for your increased force. This happens in just the same way that the normal force increases to compensate for larger weights; *i.e.*, through changing electromagnetic interactions between electrons on the surface of the table and the book.

Finally, push the book hard enough to move it. In this case, your finger force overcomes the maximum value of static friction between the book and the table. So, the book moves, and, at this point, kinetic friction takes over. Notice that now that the book is moving, you don't have to push as hard to *keep* the book moving as you did to *get* the book moving. This is because the value of kinetic friction is always less than the maximum value of static friction.

EXAMPLE 7.17: Static & Kinetic Friction, Part II

Let's take a closer look at static and kinetic friction with another experiment.

Pile several books into a tidy stack on your desk. Now push the stack using the bottom book only. Gradually increase your force until this book barely moves. Notice that the stack of books moves as one object! How can this be?

In this experiment, both static and kinetic frictions are present. As the stack moves, there is kinetic friction between the bottom book and the table. But, there is *also* static friction between the bottom book and the next bottom book, between the next bottom book and the one

Example 7.17 Continued...

above it, and so on. Thus, the books in stack itself do *not* move relative to each during the motion; they are "stuck together" by static friction.

Now push on the bottom book hard enough so that the stack topples. What happened in this case? The force you applied to the bottom book overcame the static friction between the books in the stack. This caused slipping in the stack, and ultimately, its collapse.

The magnitude of a frictional force between two objects depends upon four things:

1. Whether or not the objects move relative to each other (*i.e.,* whether the situation involves static or kinetic friction);
2. Whether the objects are sliding or rolling;
3. The materials making up each of the two rubbing surfaces; and
4. The magnitude of the normal force between the two surfaces.

EXAMPLE 7.18: Causes of Friction

We have already looked at the effects of motion on frictional forces. Now, let's take a closer look at the other three factors listed above. Reconsider the case of the book sliding along a table (Example 7.16).

Imagine that you apply a layer of grease to the table. The book now slides more easily. This is because the book is no longer rubbing against the table, it's rubbing against the slick grease! In general, frictional forces between materials *change* when materials change.

Now, let's imagine that our book has been cleaned of grease and firmly attached to a rolling platform. How much force would we need to get the book moving now? Even less than with the grease! This is because, as a general rule, the force of friction during *rolling* is less than the force of friction during *slipping*.

Finally, take the book off of the rollers and place it next to the stack of books of the Example 7.17. Push on both the book and the stack. Which requires more force to get it moving? The stack, of course, because the normal force on the stack is greater than the normal force on the single book. In general, friction depends upon the normal force.

We characterize the level of friction between two materials via the **coefficient of friction** μ. This is a positive number found in charts that gives the importance of the frictional force for a variety of materials.

A good chart will have several columns, depending on whether there is static or kinetic friction, whether there is sliding or rolling, etc. That is, for any two materials, there will be at least three coefficients of friction: the coefficient of static friction (maximum value) $\mu_{s,max}$, the coefficient of kinetic friction μ_k, and the coefficient of rolling μ_r. For any two surfaces, $\mu_{s,max} > \mu_k > \mu_r$ always.

EXAMPLE 7.19: Coefficients of Friction

If you tilt the end of your physics textbook up slightly, creating a ramp, you can do a little experiment with the coefficient of friction.

Place a coin on your book. As long as the ramp angle is small, the coin will not move. This is because, in this situation, the coin is held fast to the book by static friction.

See Section 15.3.1 for more on vector components.

However, once the angle of your ramp exceeds a certain value, the weight component of the coin along the ramp overcomes the frictional force between the book and the coin, so that the coin will start sliding. At this point, kinetic friction takes over.

With careful measurements of the coin's weight and the ramp angle, you can calculate μ for the book-coin system.

Mathematically, the frictional force is written as:

$$F_F = \mu F_N$$

where μ is the appropriate coefficient of friction, and F_N is the normal force between the two objects. Note that this equation, when used for the static frictional force, only returns a maximum value for $F_{F,s}$ because μ is $\mu_{s,max}$ in that case.

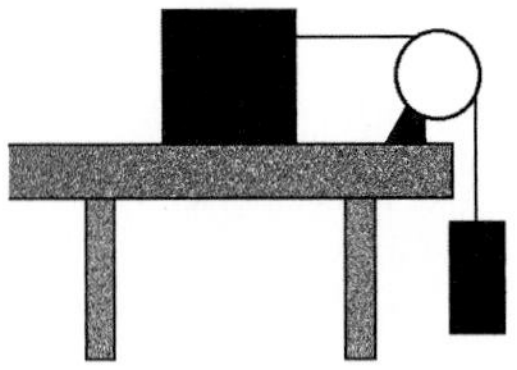

EXERCISE 7.20: Putting It All Together

Figure 7.10

Recall the diagram of Example 7.9, redrawn in Figure 7.10. Draw complete free body diagrams for each block, assuming they are stationary and that the hanging block is just about to break free. Find a relationship between the weights of each block. What will change in these diagrams if the blocks move at a constant velocity? Accelerate?

SOLUTION:

Let's begin with the simpler case of the hanging block, which we will call block 1.

There are only two forces acting on this block: the tension from the rope, directed straight up, and the weight of the block, directed straight down. Thus, the free body diagram for this block looks like:

Exercise 7.20 Continued...

Notice that each arrow in the diagram has the same length—that is, $F_T = F_{G1}$—because the block is not accelerating, so neither force can dominate. Also note that we did not label the force due to tension by a subscript indicating the block number. This is because the tension on both blocks is the *same* since the cord is massless.

* * *

Now let's look at the block on the table, which we will call block 2. This block has four forces on it: tension from the rope, directed to the right; the weight of the block, directed straight down; the static friction from the table, directed to the left; and the normal force from the surface, directed straight up. Thus, the free body diagram for this block looks like:

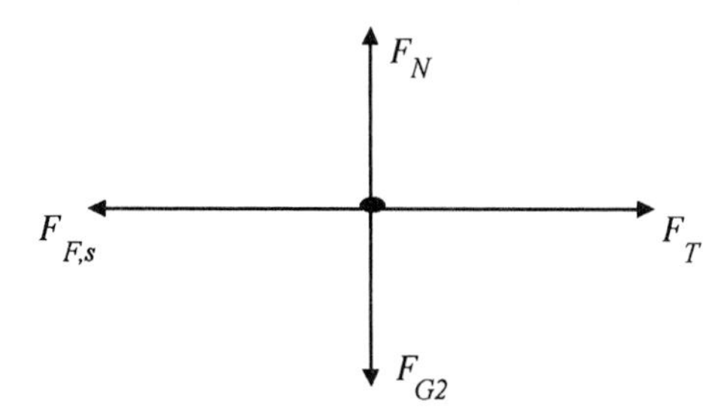

Notice that we drew each vertical arrow the same length, and each horizontal arrow the same length. This is because the block is not accelerating in either direction, so by N2 we know that $F_N = F_{G2}$ and $F_{F,s} = F_T$.

* * *

Now, we want to find a relationship between the weights of the two blocks. If we combine the results of both free body diagrams, we can conclude that $F_T = F_{G1} = F_{F,s}$.

But, according to the equation on the previous page, $F_{F,s} = \mu_s F_N$. Furthermore, we know from the analysis on block 2 that $F_N = F_{G2}$. Therefore, $F_{F,s} = \mu_s F_{G2}$. Combining this result with the relation $F_{G1} = F_{F,s}$, we have our final answer:

$$F_{G1} = \mu_s F_{G2}$$

* * *

If the blocks were moving at a constant velocity, *everything* we have just said would be the same, except that we would replace static friction $F_{F,s}$ with kinetic friction $F_{F,k}$.

If the blocks were accelerating, though, we would no longer have $F_{F,s} = F_T = F_{G1}$. Instead, we would know $F_{F,k} < F_T < F_{G1}$; that is, the vectors no longer have the same lengths in the free body diagrams above.

7.2.7 DRAG

Drag is the force on objects moving through a viscous medium, such as air. The value of drag depends on many factors, including the *shape* and *speed* of the object, and the *density* of the medium. The drag vector points *parallel* and *opposite* to the relative direction of motion.

> **WARNING!**
>
> Be *very* careful with the direction of the drag vector: it is not always in the obvious direction. The general rule is that drag is pointed opposite to the direction of *relative* motion between the object and the medium.

$\bar{\mathbf{F}}_D$

$\bar{\mathbf{F}}_G$

(a)

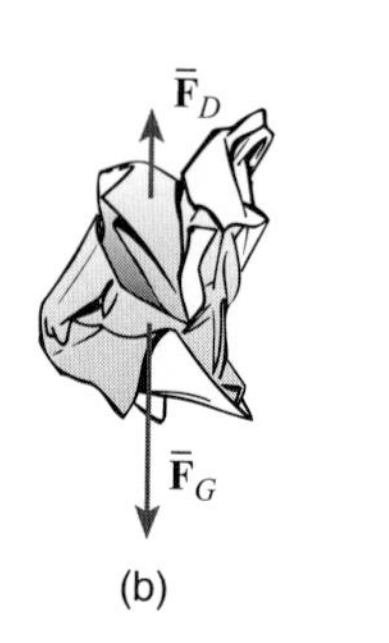

(b)

Figure 7.11

EXAMPLE 7.21: Drag on Paper, Part I

To see how *shape* affects the drag on a body, try this experiment.

Take a flat piece of notebook paper and drop it. Notice that it gently floats to the ground at an almost constant velocity. Now, take that same piece of paper and wad it up into a tight ball and drop it again. See any difference? This time, the paper drops straight down, accelerating almost at g.

What happened? In both cases, a drag force created by the air acted on the paper to slow it down. However, the *shape* of the paper changed in the two cases, causing the magnitude of the drag force to change as well. In fact, in the first case, the drag force was quite strong, so that the paper has $a \approx 0$; while in the second case, the force was very weak, so that the paper had $a \approx g$. See Figures 7.11.

EXAMPLE 7.22: Drag on Paper, Part II

Drag also depends upon the speed of the object. Take out another piece of notebook paper, and hold it steady. How much drag force is on this paper? None! The paper isn't moving, so there isn't any drag.

EXAMPLE 7.23: Drag and Paper Clips

To see how the drag force is affected by the medium in which the body is moving, think about this experiment. Drop a paper clip in air. Now drop the same clip in a jar of vegetable oil. In which case will the paper clip drop at a faster rate? The first, because the density of air is much less than the density of vegetable oil. Thus, the drag force in the first case will be less than in the second.

Mathematically, we write the drag force as:

$$F_D = \tfrac{1}{2} D\rho A v^2$$

Your textbook should have a short table of drag coefficients.

where D is the **drag coefficient**—a number that depends upon the shape of the object (you look this up in a table each time you need it)—ρ is the density of the medium (look this up as well), A is the cross-sectional area of the object, and v is the object's speed.

When a drag force acts to cancel out all other forces on a body, such that the body's total acceleration is zero, the body will move at a constant velocity (by N2). We call the constant velocity due to the drag force the **terminal velocity**.

See Section 6.3 for more on N2.

EXAMPLE 7.24: Terminal Velocity

During the Second World War, a Soviet fighter pilot was attacked at almost 22,000 feet. His plane was badly damaged, and the pilot ejected. However, there were still German warplanes flying around, so the pilot (wisely?) decided to take the fastest route away from the scene: a free-fall, with unopened parachute.

The pilot planned to open his chute at 1000 feet, and thereby make a safe landing. However, he lost consciousness during the fall and never did open the chute. Yet, this pilot not only survived the 22,000-foot fall, but only three months later, was back at work! How could this happen?

Terminal velocity (and, the lucky break of landing in three feet of snow). The pilot, during his fall, was subject to two forces: gravity, pointing down, and drag, pointing up. Because his fall was so long, the pilot was able to get to a point where these forces became equal. So, his acceleration was zero. He therefore fell the remaining distance at a constant velocity—a velocity much lower than in the case of free fall—enabling him to survive the fall when he landed in the soft snow.

(By the way, parachutes work by providing a drag force on falling objects, so that the body reaches a safe terminal velocity well prior to landing. For short falls, like those out of a plane flying at typical parachuting altitude, in fact, there is not enough time to reach terminal velocity *without* a parachute.)

EXERCISE 7.25: Skiing: What a Drag!

A woman skis down a hill (Figure 7.12). The woman weighs 600 N and the hill has an incline of $20°$. The coefficient of kinetic friction between skis and ice is 0.4.

(a) Draw a free body diagram for the woman.

(b) What is the normal force on the skier?

(c) Assuming the skier is traveling at terminal velocity, find the magnitude of the drag force on her.

Figure 7.12

Exercise 7.25 Continued...

SOLUTION:

(a) The woman experiences four forces: weight, directed straight down; the normal force, directed perpendicular to and away from the hill; kinetic friction, directed opposite to the direction of motion; and drag, also directed opposite to the direction of motion. We sketch these as:

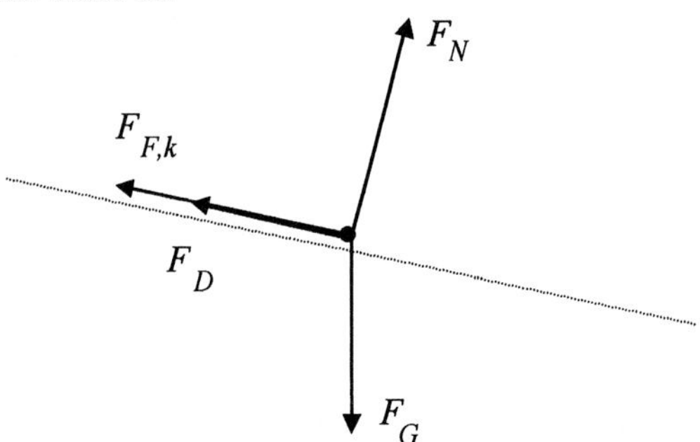

Notice that the forces of friction and drag, although sketched in the same direction, are not sketched with the same length arrow. We do this because *at this time* we do not know if they are equal, and therefore, we assume they are not.

(b) We can use the free body diagram from part (a) to find the normal force on the skier. But before we can begin, we must choose an appropriate coordinate system. There are two basic options. We can place our x-y axes in the usual up/down-left/right position, as sketched in Figure (a) below, or we can rotate them so that the x-axis is parallel to the hill, as sketched in Figure (b):

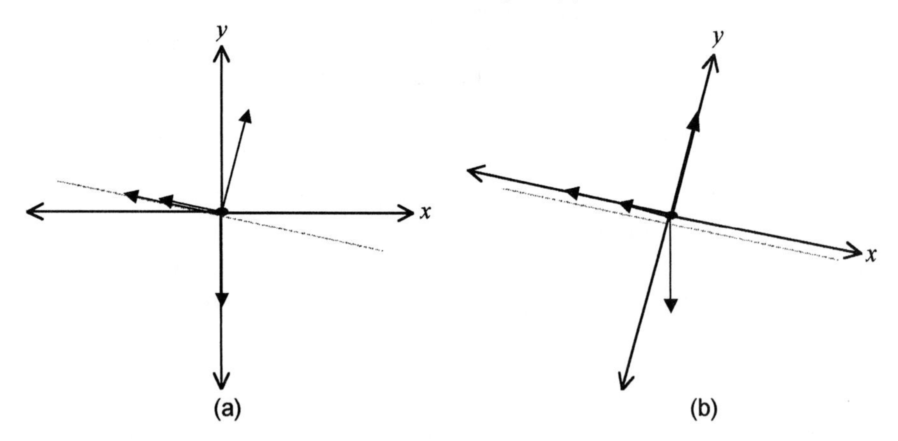

Which one would you choose? The first option certainly seems more familiar, and therefore the simpler choice. But take a closer look. Only one of the four forces in (a) lies along an axis. Therefore, we will have to break down the remaining three forces into components.

On the other hand, choice (b) looks very strange. Yet, in this case, *three* of the four forces lie along axes. Therefore, we only have to break down *one* force into components. Despite its odd appearance, then, we should choose coordinate system (b) to solve this problem because it is easier mathematically.

So, for this choice of coordinate system, there is one force—the weight—that needs to be resolved into components. Let's do that now. In order to do it, we need to know the angle between the weight vector and the negative y-axis. Using a little trigonometry, we can prove that this angle is the *same* as the slant of the hill. Thus, our sketch is:

Exercise 7.25 Continued...

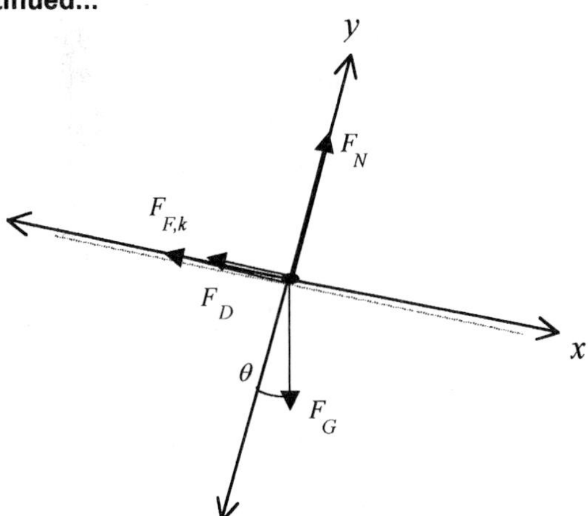

where $\theta = 20°$. Using our vector math, then, we have:

$$F_{G,x} = mg\sin\theta \text{ and } F_{G,y} = -mg\cos\theta$$

Notice that the right hand term in the y-equation is negative. This indicates the downward direction of the weight vector. However, the same term in the x-equation is positive! Check the free body diagram to see why this is.

Finally, we are now in a position to answer the original question: what is the normal force on the skier? By applying N2 in the y-direction, and vector math, we can obtain the following equation:

$$F_y = F_N - mg\cos\theta = ma_y$$

Now, the skier is *not* moving in a direction *perpendicular* to the hill (that is, she is not jumping *up* or sinking *down* into the ground). Thus, the acceleration in the y-direction is zero: $a_y = 0$. From this piece of information, and the equation above, we know that:

$$F_N - mg\cos\theta = 0$$
$$\Rightarrow F_N = mg\cos\theta = (600\text{ N})\cos(20^\circ) = 564\text{ N}$$

Notice that the normal force—in this case—does *not* equal the skier's weight!

(c) To find the magnitude of the drag force on the skier, we must apply N2 and vector math to the x-direction of the figure above:

$$F_x = -F_D - F_{F,k} + mg\sin\theta = ma_x$$

If the skier is moving at terminal velocity, she is no longer accelerating downhill. Thus, the acceleration in the x-direction is also zero: $a_x = 0$. So, the above equation becomes:

$$-F_D - F_{F,k} + mg\sin\theta = 0$$
$$\Rightarrow F_D = -F_{F,k} + mg\sin\theta = -\mu F_N + mg\sin\theta$$

Note that we replaced $F_{F,k}$ by its appropriate equation in terms of the normal force. Now we can plug in numbers into this equation, including the result of part (b) above, and get:

$$F_D = -(0.4)(564\text{ N}) + (600\text{ N})\sin(20^\circ) = 183\text{ N}$$

This magnitude is positive, which means that the vector is in the negative x-direction (as it is drawn). So, we guessed the right direction when we drew the diagram!

7.2.8 THE SPRING FORCE

The **spring force** is a *variable, restoring* force: the more you try to compress (or stretch!) a spring, the more the spring pushes out (or in) on you. That is, *springs always try to return to their normal, relaxed state.* The equation describing the spring force is called **Hooke's Law**.

[The force] of any Spring is in the same proportion with the [extension] thereof; That is, if one [force] stretch or bend it one space, two will bend it two, three will bend it three, and so forward.
ROBERT HOOKE
English Scientist, 1635 - 1703

EXAMPLE 7.26: The Spring Force & Rubber Bands

Rubber bands (partially) obey Hooke's Law. See this for yourself: get a rubber band and pull on it. The longer the rubber band stretches, the harder you must pull to overcome the *increasing* force the band exerts on you.

Hooke's Law, in equation form, is:

$$F_S = -kx$$

where k is the **spring constant**[*]—a number that characterizes how "strong" the spring is—and x is the distance that the spring is stretched (or compressed), *as measured from the natural unstretched position.* The negative sign in this formula indicates that the spring force always *opposes* stretching or compression.

EXERCISE 7.27: A Bathroom Scale

A bathroom scale measures your weight via a spring.

At its simplest, this kind of scale is simply a platform on a spring (attached to a meter to read your weight). An object of mass M is placed on a bathroom scale with spring constant k. In terms of these variables, how far will the spring compress? See Figure 7.13.

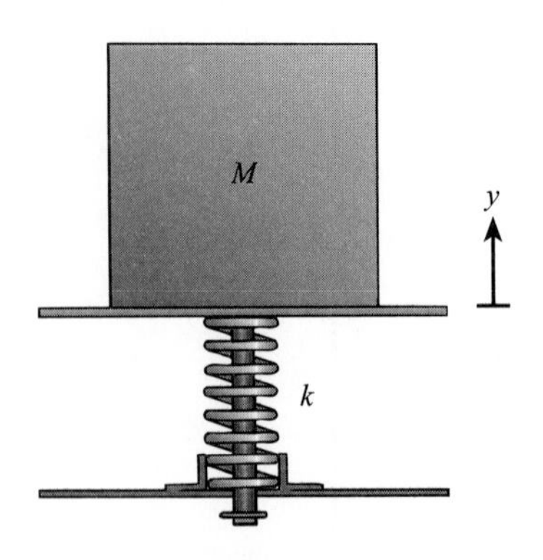

Figure 7.13

SOLUTION:

An object on a bathroom scale compresses the spring until its weight $F_G = mg$ equals the scale's spring force $F_S = -ky$. Thus, $F_G = F_S$, or:

$$mg = -ky$$

$$\Rightarrow \; y = -\frac{mg}{k}$$

Note that y is negative in this answer, meaning that the scale will come to rest *below* its initial position.

[*] Do not confuse the spring constant k with the k in the equation of the electric force.

7.2.9 CENTRIPETAL ACCELERATION

Objects that move at a constant speed v in a circle of radius r are said to move in **uniform circular motion**. They will experience a **centripetal acceleration** that is directed towards the *center* of the circle:

$$a_c = \frac{v^2}{r}$$

WARNING!

The direction of the centripetal acceleration can be confusing, because when you are traveling in a circle, it *feels* as though you are being pulled *outward* rather than inward.

The outward feeling is called the **centrifugal acceleration**, and it is an artifact of the fact that spinning bodies are not in inertial frames. This means that even though the centrifugal acceleration *feels* real, it isn't. So, don't let your intuition about the centrifugal acceleration trick you into getting the direction of the centripetal acceleration wrong.

There's more on inertial frames in Section 5.3.3.

EXERCISE 7.28: The Centripetal Acceleration

You round a curve with radius 150 m on a highway off-ramp at 75 kph (= 21 m/s). What centripetal acceleration do you experience, direction and magnitude?

SOLUTION:

$$a_c = \frac{(21\,\text{m/s})^2}{150\,\text{m}} = 2.9\,\text{m/s}^2$$, directed into the center of the circle

Technically, the centripetal acceleration multiplied by the mass of the accelerating body is a force, called the **centripetal force**. However, *be very careful with the centripetal force*: you do *not* sum up the centripetal force with all of the other forces. Rather, you identify which force or forces *provide(s)* the centripetal force, and then set the other force or forces equal to the product of the mass and the centripetal acceleration.

EXERCISE 7.29: The Loop-De-Loop

Occasionally in circuses you will see the "loop-de-loop" trick, in which a motorcycle rider drives through a vertical loop, such that at the top of the loop, the rider is completely upside down. The trick seems very

Exercise 7.29 Continued...

difficult on its face (after all, the rider is upside down!), but in actuality getting around the loop safely is only a matter of going fast enough.

Assuming the loop track is a circle with a radius of three meters (about one story high), what is the minimum speed the motorcycle must have at the top of the loop, in order to stay on the track?

SOLUTION:

In this case, we apply N2 to the rider when he is at the top of the loop. At that position, there are two forces on the rider: the normal force from the track (pointing down!), and gravity. Together, these forces equal the product of the total mass and the centripetal acceleration. This "centripetal force" is *not* included in the free body diagram shown below:

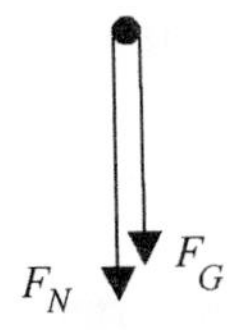

Applying N2 to this diagram (and defining the positive y-direction to be *downwards*) gives us:

$$F_N + F_G = ma_c = m\frac{v^2}{r}$$

Now, we want to know the minimum speed such that the rider just barely stays on the track; that is, when F_N is zero:

$$F_c = 0 + F_G = mg = m\frac{v^2}{r}$$

$$\Rightarrow v = \sqrt{gr} = \sqrt{(9.8\,\text{m/s}^2)(3\,\text{m})} = \pm 5.4\,\text{m/s}$$

We take the positive root as our final answer; thus $v = 5.4\,\text{m/s}$. This works out to a speed of about twelve miles per hour—about the pace of a brisk bike ride!

7.2.10 BUOYANCY

Refer to Section 10.4 for more on fluids.

Buoyancy is an upwardly directed force exerted by *fluids* on completely or partially submerged bodies. The magnitude of this force is always equal to the weight of the fluid *displaced* by the object.

EXERCISE 7.30: Fish

Imagine a $6\ \text{kg}$ tropical fish swimming in the Atlantic Ocean ($\rho_{water} = 10^3\ \text{kg/m}^3$). The fish has an approximate volume of $2 \times 10^{-3}\ \text{m}^3$. Will the fish sink or swim?

SOLUTION:

When the fish is totally submerged (as fish usually are), the volume of water that is displaced by the fish is *equal* to the volume of the fish itself: $V_{water} = V_{fish} = 2 \times 10^{-3}\ \text{m}^3$. Furthermore, volume is related to mass via the equation $m = \rho V$. So the water mass displaced is:

$$m_{water} = (10^3\ \text{kg/m}^3)(2 \times 10^{-3}\ \text{m}^3) = 2\ \text{kg}$$

The corresponding weight of the water is:

$$F_{G,\ water} = m_{water}g = (2\ \text{kg})(9.8\ \text{m/s}^2) = 19.6\ \text{N}$$

This must equal the buoyant force. So F_B is $20\ \text{N}$ also. The fish weighs:

$$F_{G,fish} = m_{fish}g = (6\ \text{kg})(9.8\ \text{m/s}^2) = 58.8\ \text{N}$$

So, we can sketch a free body diagram for the fish:

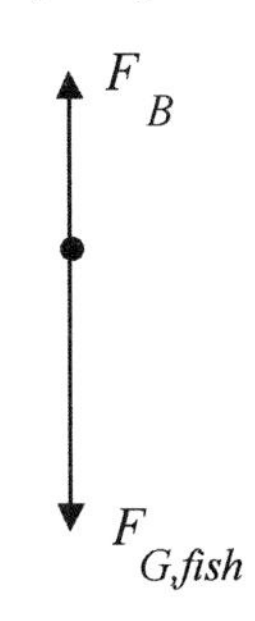

Using N2, and this free body diagram, we know that:

$$F_y = F_B - F_{G,fish} = m_{fish}a_{y,fish}$$

Solving this equation for acceleration, we get:

$$a_{y,fish} = \frac{F_B - F_{G,fish}}{m_{fish}} = \frac{19.8\ \text{N} - 58.8\ \text{N}}{6\ \text{kg}} = -6.3\ \text{m/s}^2$$

This is a negative acceleration! So the fish *should* sink! But fish don't sink. How can this be?

* * *

Fish have a *swim bladder*. Fish can change the size of this bladder by filling it with air. This changes the size and mass of the fish, which in turn changes the volume of water they displace and their weight. The more water that is displaced, the greater the buoyant force (which, of course, must also counterbalance the greater weight). With the right adjustments, the fish can make its acceleration in the y-direction zero, causing flotation at a wide range of depths.

7.2.11 LIFT

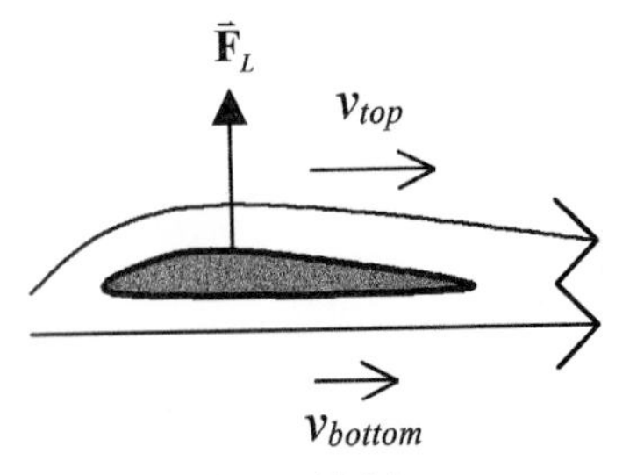

Figure 7.14

Lift is an upwardly directed net force exerted by *fluids* on bodies in motion relative to the fluid.

Lift results when an appropriately shaped body cuts through the flow of the fluid, such that some of the fluid is forced over the top of the body. See Figure 7.14. The fluid flowing over the top of the body moves *faster* than the fluid under the body, so that the fluid pressure on top of the body is less than the pressure below. The greater pressure under the body pushes the body upwards, resulting in the lift force.

7.2.12 THRUST

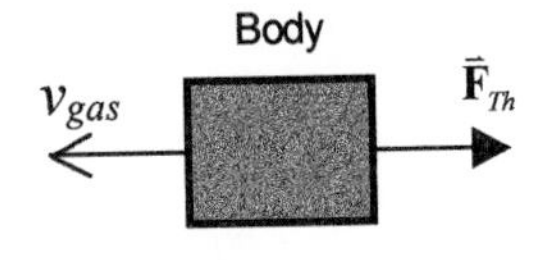

Figure 7.15

Thrust is a net force arising when *gases* are expelled from a body at high speed (*i.e.*, in a rocket engine). Thrust is directed opposite to the gaseous outflow. See Figure 7.15.

7.2.13 OTHER FORCES

There are two other fundamental forces in nature: the **strong force** and the **weak force**. These forces hold the atom together. When combined with gravity, electricity, and magnetism, the strong and weak forces make up the five most important forces in the universe.

See Section 2.5 for more on grand unification.

One goal of modern physics is to link these forces into one "super-force." Some of this work has already been done: the electric and magnetic forces have been linked together as the electromagnetic force (the **Lorentz Force**), and the weak force has been connected to the electromagnetic force as the electroweak force. Figure 7.16 below summarizes this process:

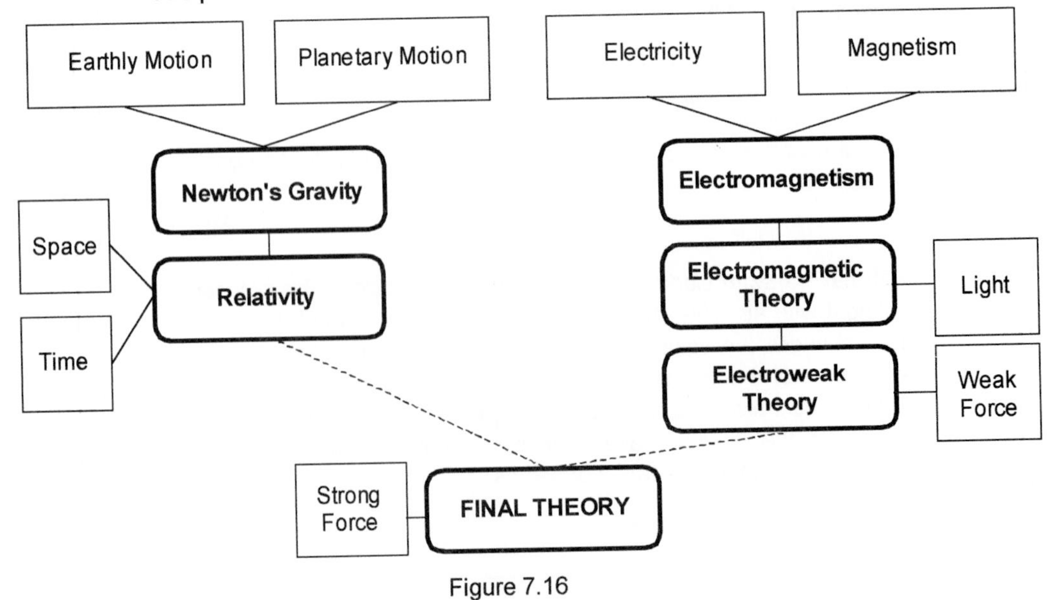

Figure 7.16

All natural forces are due to one of the five fundamental forces. These can be classed into two broad categories:

- **Contact forces**, like the normal force and friction, occur when two bodies are in *contact*. These kinds of forces, as discussed in Examples 7.11 and 7.13, are due to electromagnetic interactions between atoms at the surface of each body.
- **Action-At-A-Distance forces**, like gravity and the electric force, arise between two *separated* bodies. These kinds of forces seem more mysterious than contact forces, because it is not readily obvious how the force is *transmitted* across space. One way of explaining these forces is with *fields*, the subject of the next section.

If you insist upon a precise definition of force, you will never get it!
RICHARD FEYNMAN
American Physicist, 1918 - 1988

7.3 FORCE FIELDS & MAXWELL'S EQUATIONS

Although forces provide an immensely useful way of thinking about the causes of motion, they do not provide ready explanations for action-at-a-distance forces. Further, as forces get more complex, their equations lengthen, and they often become cumbersome to use. Therefore, physicists often desire an alternative formulation to forces—one found in **force fields**.

7.3.1 FORCE FIELDS

Force fields are *not* new quantities, nor do they represent any new physics. Instead, they are simply mathematical reconfigurations of forces.

Section 2.3.1 discusses the importance of mathematical reconfigurations in more detail.

EXAMPLE 7.31: The Gravitational Field, Part I

Let's look at one example of a force field: the *gravitational field*.

We know that the gravitational force between any two objects can be found via the equation $F_G = Gm_1m_2/r^2$. That is, no matter where m_1 is located in space, we can always figure out its gravitational effect on m_2 using this formula.

In some cases, however, we wish to figure the effect of m_1 *without regard* for m_2. We do this by *dividing out* m_2 in the gravity formula; that is, by *defining* a new quantity:

$$g(r) \equiv \frac{F_G}{m_2} = G\frac{m_1}{r^2}$$

We call this new quantity the **gravitational field**. The units of this field are N/kg or m/s^2.

There's more on the gravitational field in Examples 7.32 and 7.34.

In their defining equations, force fields contain only fundamental parameters (for *one* body) and some constants. Therefore, fields describe the influence of a single body on the space around it.

EXAMPLE 7.32: The Gravitational Field, Part II

For example, the gravitational field given above is only a function of m_1 and the distance r from it. Therefore, the field describes how m_1 affects its surrounding space, regardless of the presence or absence of other masses. If another mass is placed in the field, the field can be measured by noting the force on the second mass.

For instance, the field near m_1 is strong, so other masses experience a strong gravitational force in that location. Conversely, the field far from m_1 is weak, so other masses will experience a weak force there.

One advantage of force fields is the ability to *visualize* the effect of bodies on their surrounding space. There are two ways to do this: a **vector field map**, consisting of vectors whose lengths are proportional to the vector field at each point, and directions parallel to the direction of the field there; and a **line of force map**, consisting of lines tangent to those vectors.

EXAMPLE 7.33: Field Maps

Below are examples of a vector field map (Figure (a)) and a line of force map (Figure (b)) for the gravitational field of point mass m_1:

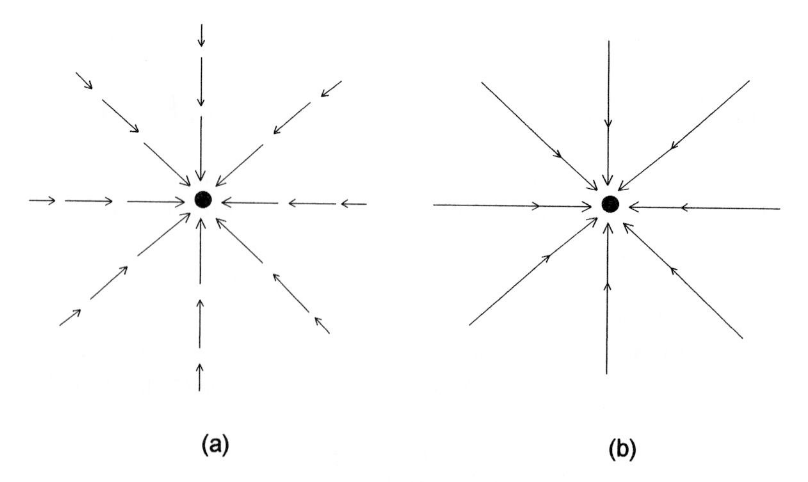

Notice that in (a), the *length* of the vectors decrease away from the mass to indicate how the field of m_1 lessens with distance. Likewise, notice that in (b), the *space between* field lines increases away from the mass for the same reason.

It is easier to draw line of force maps, so these are more frequently seen in an introductory course.

Another advantage of force fields is that they often simplify the solution of force problems.

EXAMPLE 7.34: The Gravitational Field, Part III

For example, once we have calculated the gravitational field for an object, say the earth, we can use just the *one* equation to calculate the gravitational field for a *variety* of objects.

To see that this is true, let's calculate the gravitational field of the earth. To do so, we simply plug in the mass of the earth into m_1:

$$g(r)_{earth} = G\frac{m_1}{r^2} = (6.672\times10^{-11}\,\mathrm{N\cdot m^2/kg^2})\frac{(5.98\times10^{24}\,\mathrm{kg})}{r^2} = \frac{3.99\times10^{14}\,\mathrm{N\cdot m^2/kg}}{r^2}$$

This field, once calculated, works for *any* distance r (measured from the center of the earth). For example, on the surface of the earth, r is just the earth's radius R:

$$g(R) = \frac{3.99\times10^{14}\,\mathrm{N\cdot m^2/kg}}{(6.37\times10^6\,\mathrm{m})^2} = 9.8\,\mathrm{m/s^2}$$

(By the way, this number should be familiar to you; what is it?)

Now, to calculate the force of gravity between the earth and *any* object on its surface, we simply need to multiply this field by the object's mass. For instance, the force of gravity on an apple would be...

$$F_G = m_{apple}\,g(R) = (1\,\mathrm{kg})(9.8\,\mathrm{m/s^2}) = 9.8\,\mathrm{N}$$

...and the force of gravity on a car would be...

$$F_G = m_{car}\,g(R) = (900\,\mathrm{kg})(9.8\,\mathrm{m/s^2}) = 8800\,\mathrm{N}$$

...and so on.

We could follow the same procedure with *any* mass at *any* distance from the field-producing object. For example, consider a government weather satellite has a mass of about 2100 kg, and a maximum orbit of approximately 3.6×10^7 m above the earth's surface. Thus, the earth's gravitational field on the satellite at maximum orbit is about:

$$g(r_{satellite}) = \frac{3.99\times10^{14}\,\mathrm{N\cdot m^2/kg}}{(6.37\times10^6\,\mathrm{m} + 3.6\times10^7\,\mathrm{m})^2} = 0.22\,\mathrm{m/s^2}$$

[Notice that we had to add the radius of the earth to the distance of orbit, because r is always measured from the *center* of the field-producing object.]

Finally, the gravitational force on the satellite is:

$$F_G = m_{satellite}\,g(r_{satellite}) = (2100\,\mathrm{kg})(0.22\,\mathrm{m/s^2}) = 470\,\mathrm{N}$$

Although force fields can theoretically be developed for most forces, traditionally, the most important ones in an introductory physics class are the electric and magnetic fields.

EXAMPLE 7.35: The Electric Field

We define the electric field E in a similar way as we did the gravitational field: that is, by beginning with the formula for the electric force, and then dividing away the effect of one of the charges:

$$F_E = k\frac{q_1 q_2}{r^2} \Rightarrow E(r) \equiv \frac{F_E}{q_2} = k\frac{q_1}{r^2} \text{ (electric field for a point charge)}$$

The units of the electric field are $\mathrm{N/C}$.

The line of force map for the electric field (of a point charge) looks very much like that of the gravitational field (of a point mass), with one exception: the lines of force point towards *or* away from the charge, depending on whether the charge is negative (a) or positive (b):

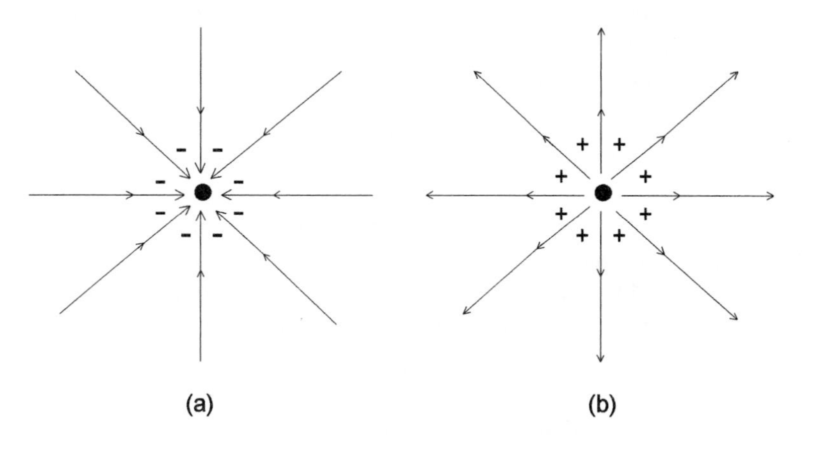

EXAMPLE 7.36: The Magnetic Field

The magnetic field $B(r)$ is defined in a similar way as the electric field. It has units of the Tesla ($\mathrm{T = kg/C{\cdot}s}$).

The magnetic field is different from the electric field in an important way: whereas positive and negative electric charges can exist separately from each other, positive and negative magnetic "charges" *cannot*. Therefore, the lines of force for a magnet must pass through both positive (south) and negative (north) poles on each body (*i.e.*, the lines must loop back on themselves):

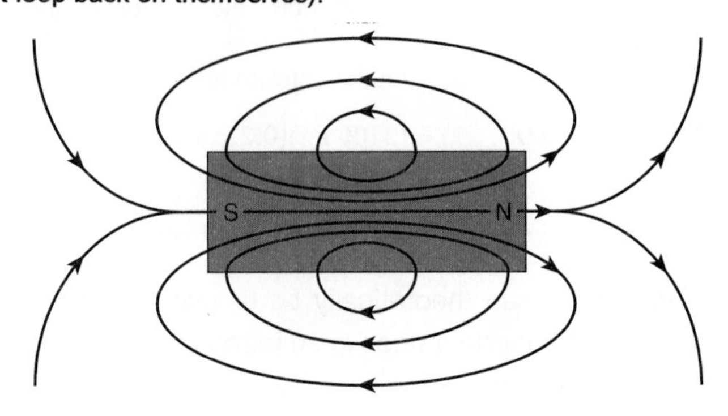

The actual formulas for electric and magnetic fields vary depending upon the distribution of charges. Table 7.1 gives the electric and magnetic field formulas for the charge distributions sketched below the grid. Your textbook or instructor will show you how to derive these equations.

Table 7.1: Electric and Magnetic* Field Formulas

SOURCE	FORMULA	SOURCE	FORMULA	SOURCE	FORMULA
POINT CHARGE	$E_{pc} = \frac{kq}{r^2}$	RING OF CHARGE	$E_{ring} = \frac{kqz}{(z^2+R^2)^{\frac{3}{2}}}$	INFINITE WIRE	$B_{wire} = \frac{2k'i}{r}$
DIPOLE	$E_{dipole} = \frac{2kqd}{z^3}$	DISK OF CHARGE	$E_{disk} = \frac{2\pi kq}{A}\left(1-\frac{z}{\sqrt{z^2+R^2}}\right)$	CURRENT LOOP	$B_{loop} = \frac{2k'NAi}{z^3}$
INFINITE LINE OF CHARGE	$E_{line} = \frac{2kq}{Lr}$	SHELL OF CHARGE	$E_{shell} = \begin{pmatrix} \frac{kq}{r^2} \text{ for } r > R \\ 0 \quad \text{for } r < R \end{pmatrix}$	SOLENOID	$B_{solenoid} = \frac{4\pi k'Ni}{L}$
INFINITE, FLAT CHARGED SURFACE	$E_{surface} = \frac{2\pi kq}{A}$			TOROID	$B_{toroid} = \frac{2k'Ni}{r}$

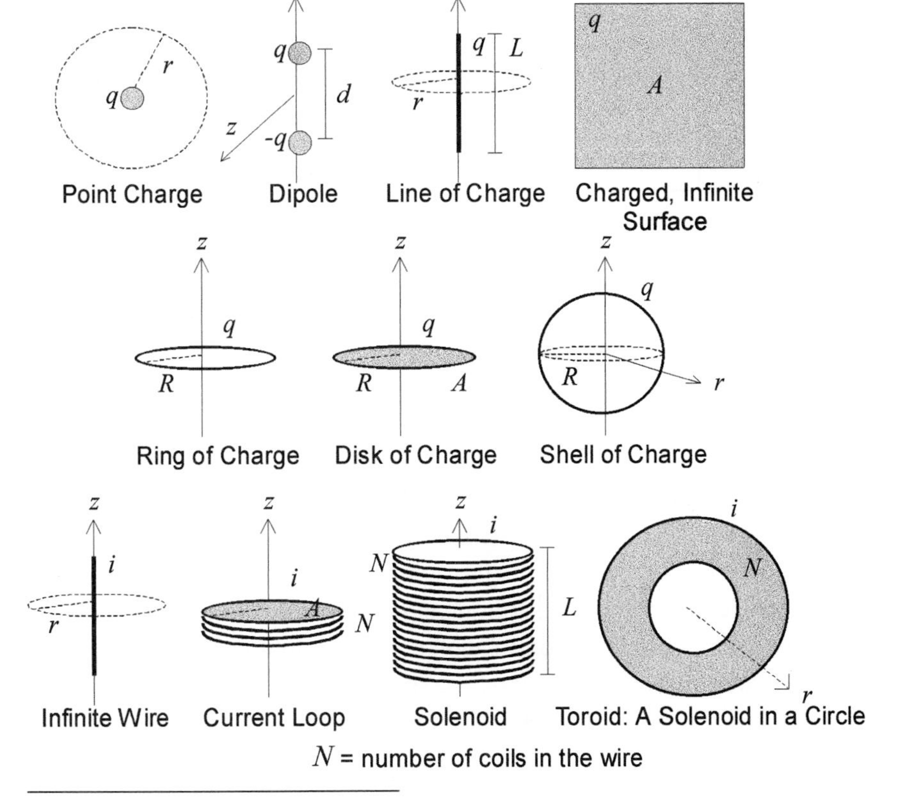

N = number of coils in the wire

* The k' in the magnetic field equations is a magnetic constant, similar to k in Coulomb's Law. It has value $k' = 10^{-7}$ T·m/A. It is sometimes written $\mu_0 / 4\pi$.

The field formulation in electromagnetism is so useful that the fields E and B are used far more often than the forces F_E and F_B in calculations. However, it is always understood (even if unsaid), that the corresponding forces can be found, when needed, with the equations:

See Section 15.3.5 for more on vector multiplication.

$$\vec{\mathbf{F}}_E = q\vec{\mathbf{E}} \text{ and } \vec{\mathbf{F}}_B = q\vec{\mathbf{v}} \times \vec{\mathbf{B}}$$

The Lorentz Force is also discussed in Section 7.2.13.

The combined effect of the electric and magnetic forces (on a point charge) is called the **Lorentz Force**:

$$\vec{\mathbf{F}}_{Lorentz} = \vec{\mathbf{F}}_E + \vec{\mathbf{F}}_B$$

7.3.2 MAXWELL'S EQUATIONS

Just as all of classical mechanics can be summarized into three laws—Newton's Laws—all of electromagnetism is summarized in four equations—Maxwell's Equations. These very important equations are stated in terms of the electric and magnetic fields E and B. We state these laws, and their "translations" in Table 7.2 below.

Table 7.2: Maxwell's Equations

NAME	EQUATION	WHAT IT MEANS	HOW YOU CAN USE IT
GAUSS' LAW FOR ELECTRICITY	$\oint \vec{\mathbf{E}} \cdot d\vec{\mathbf{A}} = \Phi_E = 4\pi kq$	The electric field is produced by positive and negative charges.	To calculate the electric field, given the charge distribution.
GAUSS' LAW FOR MAGNETISM	$\oint \vec{\mathbf{B}} \cdot d\vec{\mathbf{A}} = \Phi_B = 0$	The magnetic field is produced by *dipoles*: inseparable combinations of north and south poles.	Not used in introductory classes.
FARADAY'S LAW	$\oint \vec{\mathbf{E}} \cdot d\vec{\mathbf{s}} = -\frac{d\Phi_B}{dt}$	A changing (over time) magnetic flux produces an electric field.	To calculate the electric field, given the value of the (changing) magnetic flux.
AMPERE-MAXWELL LAW	$\oint \vec{\mathbf{B}} \cdot d\vec{\mathbf{s}} = \frac{\mu_o}{4\pi k}\frac{d\Phi_E}{dt} + \mu_o i$	A changing (over time) electric flux and/or current i produces a magnetic field.	To calculate the magnetic field, given the value of the (changing) electric flux, or a current distribution.

The wave equation is in Section 14.4.2.

Maxwell found that the mathematical combination of these four equations yielded a fifth equation (the **wave equation**) that describes the nature of light. So the field formulation, in this case, actually provided information that the force formulation could not: that is, light is a wave!

For more on this, refer to Example 2.5.

There are, in fact, *many* kinds of electromagnetic waves important to modern life. Figure 7.17 below summarizes each of these.

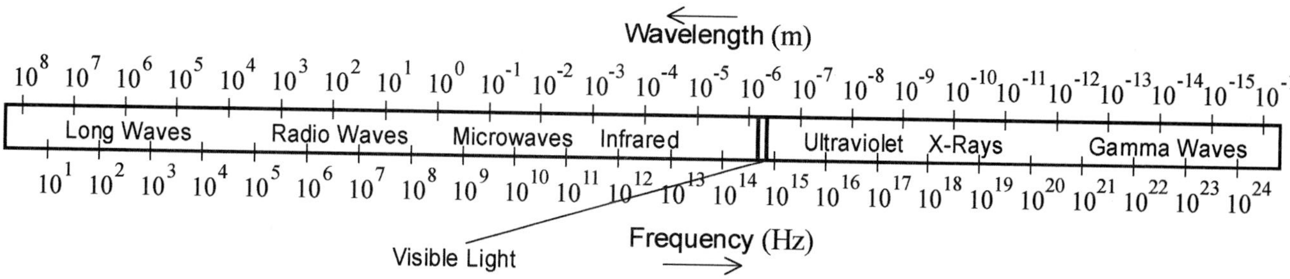

Wavelength (m)

10^{8} 10^{7} 10^{6} 10^{5} 10^{4} 10^{3} 10^{2} 10^{1} 10^{0} 10^{-1} 10^{-2} 10^{-3} 10^{-4} 10^{-5} 10^{-6} 10^{-7} 10^{-8} 10^{-9} 10^{-10} 10^{-11} 10^{-12} 10^{-13} 10^{-14} 10^{-15} 10^{-16}

Long Waves | Radio Waves | Microwaves | Infrared | Ultraviolet | X-Rays | Gamma Waves

10^{1} 10^{2} 10^{3} 10^{4} 10^{5} 10^{6} 10^{7} 10^{8} 10^{9} 10^{10} 10^{11} 10^{12} 10^{13} 10^{14} 10^{15} 10^{16} 10^{17} 10^{18} 10^{19} 10^{20} 10^{21} 10^{22} 10^{23} 10^{24}

Frequency (Hz)

Visible Light

Figure 7.17

7.3.3 FORCE FIELDS & A FINAL THEORY

As we saw in Chapter 2 and Section 7.2.12 above, a major goal of physics today is to unite the five fundamental forces—gravity, electricity, magnetism, and the strong and weak forces—into one "super-force" that describes all phenomena with one "final" theory.

Despite the importance of forces and Newton's Laws for introductory-level physics, this cutting-edge research is done using *fields*. Thus, field formulations are important not just because they make some force problems easier to solve, but because they provide insight into one of the deepest physics questions of our time.

7.4 CONCLUSION

This long chapter gave you an introduction into all of the forces you will encounter in a typical introductory physics course. You use all of these forces—different as they are from each other—in precisely the same way: with Newton's Laws. And when this process becomes too complex, you switch to the field formulation. Either way, you can discover *why* objects move as they do, and *predict* their future motion using the kinematic equations.* *Viva la forces*!

Man masters nature not by force, but by understanding. This is why science has succeeded where magic failed: because it looked for no spell to cast on nature.
JACOB BRONOWSKI
English Mathematician, 1908 - 1974

* Assuming the acceleration is constant.

8
ENERGY AND MOMENTUM

8.1 INTRODUCTION

In the last two chapters we have outlined the principles and uses of Newton's Laws. We found that just about *any problem* (at least theoretically) can be solved by means of these laws.*

However, Newton's Laws do not always provide the easiest way to solve problems. In fact, many otherwise simple problems become quite intricate and convoluted when analyzed with forces. We therefore desire one or more alternate methods for solving mechanics problems—methods that are *equivalent* to Newton's Laws but *easier* to use in certain situations.

Physical concepts are the free creations of the human mind and are not, however it may seem, uniquely determined by the external world.
ALBERT EINSTEIN
American Physicist, 1879 - 1955

In this chapter, we will explore two such methods: *energy* and *momentum*. These concepts do *not* represent any new physics—they are simply reconfigurations of Newton's Laws—yet they provide new ways to solve problems without the involvement of forces. But these two techniques do more than just make problem solving easier: when fully understood, they offer us a broader understanding of the underlying principles that govern the workings of the universe.

---

* Subject to certain conditions. For example, sufficient information about the mathematical form of applicable forces is necessary to use Newton's Laws, information that is not always easy to come by. Also, objects need to be of reasonably large size and travel at reasonably slow speeds. See Section 5.5 for more on quantum mechanics, relativity, and Newton's Laws.

8.2 ISOLATED SYSTEMS & CONSERVATION

To begin a Newton's Law problem, we must draw a free body diagram. To begin an energy or momentum analysis, on the other hand, we must first sort the problem elements into two categories: those that are in the *system*, and those that are not.

A **system**, in physics, is a carefully defined object, collection of objects, or region of space. There are two kinds of systems:

- An **isolated system** does *not* interact in any way with its surroundings, or **environment**.
- A **non-isolated system** interacts with its environment.

Isolated systems are the preferred type for energy and momentum analyses. We discuss them, and their uses, in the next few sections. We will return to non-isolated systems later in the chapter.

EXAMPLE 8.1: Isolated Systems, Part I

In your laboratory and/or classroom, you will—sooner or later—study collisions between two objects. Because this situation is so common, it provides us with a good way to introduce isolated systems.

Consider a typical case in which two carts (cart 1 and cart 2) are on a frictionless, horizontal track. Cart 1 is pushed towards cart 2. Cart 2 is at rest (Fig. 8.1). What is the isolated system for this case?

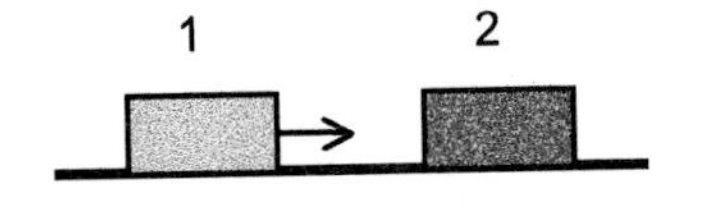

Figure 8.1

We may be initially tempted to pick cart 2 as our system, since it is at rest. What could be more "isolated" than a lone cart sitting quietly on the end of the track? Unfortunately, though, this choice would be incorrect, because in this definition of the system, cart 1 is in the environment. Yet, since cart 1 eventually *collides* with cart 2, the environment will interact with the system. *By definition*, isolated systems do not interact with their environments.

In a similar way, we might have guessed that cart 1 is our isolated system, since it is the only moving object in the problem. But this would also be an incorrect choice, for the same reason as above: the carts will collide and the system will interact with the environment.

So, we have to think of another choice. What if we define the system to be cart 1 *and* cart 2? This time, we get it right. This *is* an isolated system because the only interaction—the eventual collision between the carts—occurs between elements of the system and *nothing else*.

But what about the surface of the earth? Or the track? Don't we need to include these in our system, too? The answer is no. Because the earth and the track are oriented *horizontally* (in this problem), and the track is frictionless, we do not need to consider them in the system. In fact, they do not matter at all in this problem. In the next example, things will be different.

EXERCISE 8.2: Isolated Systems, Part II

A person drops a pencil (Fig. 8.2). Choose an isolated system.

SOLUTION:

We know that the pencil must be a part of the system, since it is doing the falling. But what else, if anything, is also in the system?

For one thing, is the person (or person's hand) part of the system? Once the falling action has begun, the person becomes irrelevant to the problem. Therefore, the person does *not* have to be considered in the system. Eliminating the Unessential, then, we ignore the person.

So, the system is just the pencil, right? Unfortunately, no. In this example, unlike the preceding one, the action (the dropping of the pencil) occurs *vertically* rather than horizontally. Vertical actions—or, more strictly, actions that are not exactly horizontal—involve interaction with the earth (through gravity). Thus, the isolated system in this case *must* include the earth.

To summarize, then, our isolated system for this example is the pencil and the earth.

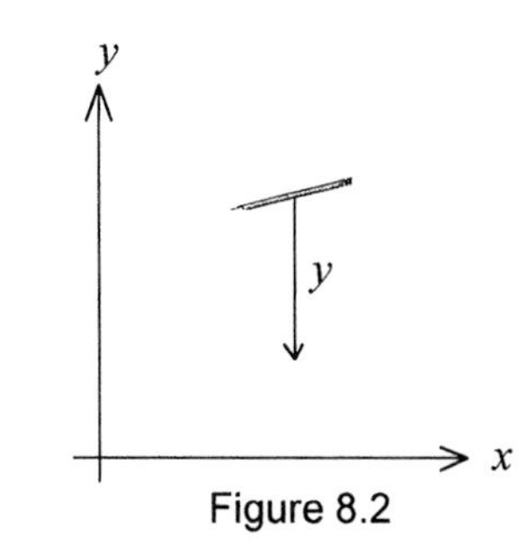

Figure 8.2

Eliminating the Unessential is in Section 2.3.

In a carefully defined, isolated system, some quantities, called **conserved quantities**, do not increase or decrease, *no matter what happens* inside of the system. Some physical processes are **conservative**—they allow these quantities to be conserved—while others are not.

EXAMPLE 8.3: Conserving Potatoes

To illustrate conservation, consider a 1 kg potato in your kitchen. The isolated system for this problem is the kitchen, and the conserved quantity in question is mass.

Say you decide to make hash browns. You shred the potato into small pieces. Shredding is a *conservative physical process* because even though the potato is now in small shreds, there is still 1 kg worth of them in the kitchen (weigh them to see). So, the potato's mass is *conserved* during shredding.

Now, imagine that you get a phone call and can't finish the hash browns. So, you put the potato shreds in the refrigerator. Nonetheless, there is still 1 kg of potato in the kitchen—as you can prove by weighing—so that the mass of the potato has been conserved. Thus, refrigeration is also a conservative physical process.

Finally, after finishing your call, you cook the hash browns. Unfortunately, you turn the heat up too high, and burn the potatoes into soot. Now, you no longer have 1 kg worth of potato; instead you have a considerably smaller amount of black dust. Consequently, burning hash browns is *not* a conservative physical process, and mass is not conserved in this case.

Another way of thinking about conservation is that a conserved quantity does not change through time—even though other aspects of the system may change. Hence, *time* becomes an important consideration in conservation problems. You will often find yourself comparing "before and after" situations when analyzing these types of problems.

Mass-energy is in Section 5.5.1.

EXAMPLE 8.4: Conserved Quantities

One example of a conserved quantity, as we just saw, is mass. In any isolated system, for any conservative physical process, mass (or more precisely, mass-energy) does not change through time. We will see other examples of conserved quantities in the following sections.

Mathematically, conservation means that you can:

1. Set the conserved quantity equal to a constant; *or*
2. Set the value of the quantity at an earlier time equal to the value of the quantity at a later time.

EXAMPLE 8.5: Writing Conserved Quantities

For instance, in the above two examples, our conserved quantity was mass. So, you could write $m = \text{constant}$, $m_i = m_f$, and/or $m_1 = m_2$, where i and f and 1 and 2 denote values of the mass at initial and final times, respectively.

8.3 ENERGY

The beginning of wisdom is the definition of terms.
SOCRATES
Ancient Greek Philosopher, c. 470 - 399 B.C.

Many terms in physics are words that we use every day. Take, for example, the word "energy": in ordinary language, to have energy is to be perky, quick, and active; to lack energy is to be sluggish, slow, and idle. The physics definition of *energy*, while hinting at the everyday usage of the term, is more strictly defined.

Scalars and vectors are discussed in Chapter 15.

In physics, **energy** (a scalar) is a measure of the *state* of a system. The unit of energy is the joule $(\text{J} = \text{kg}\cdot\text{m}^2/\text{s}^2 = \text{N}\cdot\text{m})$. There are many types of energy, corresponding to several parameters for measuring the state of the system. Table 8.1 below lists the four main categories of these.

Table 8.1: Some Types of Energy

TYPE OF ENERGY	MEASURABLE ASPECT OF SYSTEM
KINETIC ENERGY	How fast are things moving?
POTENTIAL ENERGY	Where are things positioned? How are they arranged?
INTERNAL ENERGY	How much energy is in the bodies of the system?
MASS-ENERGY	How much mass is in the system? How much energy?

See Special Relativity in Section 5.5.1 for more on mass-energy.

All energies in the real world can be sorted into one of these four types. Of these four, however, kinetic and potential energies are the most important for your introductory physics class.

8.3.1 KINETIC ENERGY

Kinetic energy K is perhaps the easiest energy to understand. There are three basic principles for kinetic energy:

1. If a mass is moving, it has kinetic energy; if it's not, it doesn't.
2. The faster the mass moves, the more kinetic energy it has.
3. The more massive it is, the more kinetic energy it has.

EXAMPLE 8.6: Superman's Kinetic Energy

Superman, when flying, has kinetic energy. Since he is "faster than a speeding bullet," (and, presumably more massive than one, too), Superman has more kinetic energy than a bullet—by virtue of his greater speed *and* his greater mass.

The equation for kinetic energy takes these three ideas into account:

$$K = \tfrac{1}{2}mv^2$$

Note that in this equation, velocity is squared, while mass is not. Thus, changes in velocity affect the value of kinetic energy more than changes in mass do. Also note that kinetic energy is always *positive*, because the mass m, and the magnitude of the velocity v, are each always positive.

EXERCISE 8.7: Jogging and Kinetic Energy

How much kinetic energy is involved in jogging? Assume that you have a mass of 70 kg (the mass of an average person), and that you jog about 2 m/s (about 4 mph; a nice, slow jog).

SOLUTION:

Applying the kinetic energy formula, $K = \frac{1}{2}mv^2$, we have:

$$K = \tfrac{1}{2}mv^2 = \tfrac{1}{2}(70\ \text{kg})(2\ \text{m/s})^2 = 140\ \text{J}$$

Notice that this energy has nothing to do with how long you have jogged, but rather it is an inherent quality of the act of jogging itself.

Objects that spin have kinetic energy, too. The kinetic energy during rotational motion is given by the equation:

Refer to Section 5.3.1 for more on angular velocity, and Section 6.5.1 for information on rotational inertia.

$$K_{rot} = \tfrac{1}{2}I\omega^2$$

where I is the object's rotational inertia, and ω is its angular velocity.

EXERCISE 8.8: 45's and Kinetic Energy

Music singles used to be released on small vinyl records called 45's. These records were named for the fact that they rotated at 45 revolutions per minute—about 5 rad/s. Assume that a 45 has a moment of inertia of about $6 \times 10^{-3}\ \text{kg} \cdot \text{m}^2$, and calculate the record's rotational kinetic energy.

SOLUTION:

Applying the rotational kinetic energy formula, $K_{rot} = \frac{1}{2}I\omega^2$, we have:

$$K_{rot} = \tfrac{1}{2}I\omega^2 = \tfrac{1}{2}(6 \times 10^{-3}\ \text{kg} \cdot \text{m}^2)(5\ \text{rad/s})^2 = 0.075\ \text{J}$$

8.3.2 POTENTIAL ENERGY

Potential energy U is energy that is *stored* in a system by virtue of the arrangement of the system's elements; thus, this kind of energy represents energy with a "potential" for other use.

EXAMPLE 8.9: A Dart Gun, Part I

A child's spring-loaded dart gun provides a good illustration of potential energy (Fig. 8.3). In these sorts of guns, a dart compresses a spring. The "cocked" spring holds energy; thus, in the spring + dart system, potential energy is stored. When the spring is released (via a trigger), the potential energy in the spring is converted into kinetic energy in the dart, causing the dart to fly away.

Figure 8.3

There are many types of potential energy, and therefore, many different formulas for it. This is because there are several types of forces which can act between elements of a system.

Furthermore, *there is no absolute value for the potential energy.* Potential energies are always defined in relation to an arbitrary **zero point**—a place where *you* decide where the potential energy is zero. Once you make this decision, the values of all other potential energies in the system are figured relative to this point.

The three most important types of potential energy*—and suggested zero points for each—in an introductory physics course are listed in Table 8.2 below. Your instructor or textbook will derive these for you.

Table 8.2: Potential Energies

TYPE	EQUATION	SUGGESTED ZERO POINT
SPRING POTENTIAL	$U_S = \frac{1}{2}kx^2$	Where the spring is relaxed, neither stretched nor compressed
HEIGHT POTENTIAL	$U_H = mgh$	At the lowest point in the problem
GRAVITATIONAL POTENTIAL	$U_G = -\frac{GMm}{r}$	M at a distance infinitely far away from m.

EXERCISE 8.10: A Dart Gun, Part II

Where would you set the zero point of the potential energy for the dart gun example above?

SOLUTION:

The best place for the zero point in this example, according to Table 8.2, is where the tip of the spring rests when it is uncocked; that is, where the spring is neither compressed nor stretched. In this way, when the spring is cocked, there *will* be a potential energy.

* Other kinds of potential energy include the **nuclear potential energy** due to the nuclear strong force, and the **electric** and **magnetic potential energies** due to the electric and magnetic forces, respectively. Equations for the electric and magnetic energies differ for different situations, and so will probably be discussed during the second half of your course.

EXERCISE 8.11: A Dart Gun, Part III

A dart gun is loaded. The spring has a spring constant of $k = 1$ N/m, and is compressed 3.5 cm. Find the potential energy stored by the dart + spring system.

SOLUTION:

First, we must identify our system. We did this in Example 8.9: it is the spring + dart. The gun itself does nothing to move the dart, so we can ignore it (and we do). We can also ignore the earth (at least at the beginning of the motion); why?

Next, we need to decide upon a zero point for the potential energy. We have done this in Exercise 8.10: it is where the spring is relaxed. We *define* this position to be $x = 0$; thus, $U_S = 0$ when $x = 0$.

Now, we need to think about the arrangement of the system. The potential energy arises when the spring is *cocked* to 3.5 cm. Without the cocking, there is no potential energy in this system. Thus, the difference between the cocked and uncocked location is $x = 3.5$ cm. Now we can find U_S at this x. Choosing the spring potential energy formula, $U_S = \frac{1}{2}kx^2$, we have:

$$U_S = \tfrac{1}{2}(1\,\text{N/m})(0.035\,\text{m})^2 = 6 \times 10^{-4}\ \text{J}$$

EXERCISE 8.12: Potential Energy on the Earth

What is the potential energy (of the system) when you are (a) at sea level, and (b) on a hill 200 m high? Assume a mass of 70 kg.

SOLUTION:

First, we choose a system. The system here is you + earth. Because you are moving *upwards* from sea level to the top of a hill (*i.e.*, not horizontally), you must include the earth in your system.

Next, we choose a zero point. According to the table above, the best zero point for this type of problem is at the lowest position of the problem. In this case, sea level is the lowest we can go. Therefore, we set potential energy equal to zero at sea level.

Good! This means that not only have we set a zero point, but we have also answered part (a) above: the potential energy at sea level is zero: $U_H = 0$.

Now, to answer part (b), we need the formula $U_H = mgh$:

$$U_H = mgh = (70\,\text{kg})(9.8\,\text{m/s}^2)(200\,\text{m}) = 1.4 \times 10^5\ \text{J}$$

EXERCISE 8.13: Potential Energy Between Earth and Moon

What is the potential energy stored in the earth + moon system?

SOLUTION:

First, we choose a system: it is the earth + moon. Then we pick a zero point. If the earth + moon were infinitely far apart, according to Table 8.2, $U_G = 0$. But since they are not, we will figure U_G below relative to this zero point.

Finally, we consider the arrangement of the system: the earth and moon are separated by a certain distance, the value of which we look up in our text.

Now we apply the right formula from the above list, $U_G = -GMm/r$:

Exercise 8.13 Continued...

$$U_G = -\frac{(6.67 \times 10^{-11}\ \text{N} \cdot \text{kg}^2/\text{m}^2)(5.975 \times 10^{24}\ \text{kg})(7.35 \times 10^{22}\ \text{kg})}{3.85 \times 10^8\ \text{m}} = -7.61 \times 10^{28}\ \text{J}$$

The negative sign in the answer means that *less* energy is stored in the earth + moon system than would be present if these two bodies were infinitely far apart. (It does *not* mean that energy has a direction.) That is, it would take a *lot* of energy (7.61×10^{28} J worth, in fact) to make the moon leave this system.

Your text should have the values for the mass of the earth and moon, as well as their distance.

We said at the beginning of this chapter that forces and energies are related. We can see that this is true with the following equations between the net force in the system and the potential energy:

$$F_x = -\frac{dU}{dx}, \quad F_y = -\frac{dU}{dy}, \text{ and } F_z = -\frac{dU}{dz}$$

Force is in Chapter 7; derivatives are covered in Section 14.2.

That is, for example, the force component in the x-direction is equal to the negative derivative of the potential energy (with respect to x).*

EXERCISE 8.14: Force and Potential Energy

In Table 8.2 above, the spring potential was given as $U_S = \frac{1}{2}kx^2$. Find the corresponding force. What is the general name for this force?

SOLUTION:

The force is the negative derivative of the potential energy:

$$F_x = -\frac{dU_S}{dx} = -\frac{d}{dx}\left(\tfrac{1}{2}kx^2\right) = -\tfrac{1}{2}k(2x) = -kx$$

This force is known as Hooke's Law.

Find out more about Hooke's Law in Section 7.2.8.

8.3.3 INTERNAL ENERGY

Internal energy E_{int} is the energy associated with atomic and molecular components of a system, and is especially important in thermodynamics. Like potential energy, there are many types of internal energy, such as molecular bond energy, molecular vibrational energies, energy due to fuel within a system, and energies associated with temperature.

In Section 10.5.

* Because force is a vector, there are three derivatives of potential energy, corresponding to each of the three dimensions. Together, the three derivatives make up the three components of the force vector. We can write these three equations more succinctly using a vector derivative. See Section 15.4.1 for more on vector derivatives.

Ideal gases are in Section 10.4.3. ☞

For example, the internal energy due to temperature of an ideal gas is:

$$E_{\text{int},T} = nCT$$

n and T are discussed in Section 11.5.2; heat capacity is in Section 8.5.2 below. ☞

where n is the number of moles, C is the heat capacity, and T is the temperature. Similarly, the internal energy due to temperature of a monatomic ideal gas is written:

$$E_{\text{int},T} = \tfrac{3}{2}nRT$$

R is in Section 11.5.3. ☞

where R is the gas constant, 8.31 J/mole-K.

EXERCISE 8.15: Internal Energy

What is the internal energy of one mole of an ideal monatomic gas at 0° C?

SOLUTION:

This equation is in Section 11.5.2. ☞

First, we must convert the given Celsius temperature to kelvins, as is our normal procedure in thermodynamics. The relation between the two temperatures is $T_K = T_C + 273$. Thus, 0° C is 273 K. Now, we apply the formula for an monatomic ideal gas:

$$\begin{aligned} E_{\text{int},T} &= \tfrac{3}{2}nRT \\ &= \tfrac{3}{2}(1\text{ mole})(8.31\text{ J/mole}\cdot\text{K})(273\text{ K}) \\ &= 378\text{ J} \end{aligned}$$

8.3.4 MASS ENERGY

Mass energy E_{mass} is the energy all massive objects have by virtue of their mass. Einstein found that this energy is given by the relation:

$$E_{mass} = mc^2$$

Mass energy is essentially ignored in introductory physics classes, because it is not relevant except in quantum mechanical discussions.

8.4 CONSERVATION OF ENERGY: A FIRST LOOK

So far in this chapter, we have defined two important concepts: isolated systems and energy. Now we will combine these two ideas into one of the most important statements about the universe you will learn in your physics course: *Energy, in an isolated system, is conserved.* That is, the sum of the energies in an isolated system can be set equal to a constant, which we (arbitrarily) call E.

If there are only kinetic and potential energies in a system—which is a very common situation in the introductory physics course—then, by **conservation of mechanical energy**, we can write:

$$K + U = E$$

This equation says that while K and U may change in a system, they do so in such a way that the *sum* $K + U$ remains constant (and equal to E). In this case, E is known as the **mechanical energy**.

EXAMPLE 8.16: Conserving Mechanical Energy, Pt. I

In the figure below, we see the oscillation of a pendulum as it swings back and forth. Notice how the kinetic and potential energies change as the motion changes, while the *total* energy of the pendulum system remains the same.

Although the conservation of mechanical energy equation is not the most precise version of the physical *law* of Conservation of Energy, it is valid in the vast majority of problems seen in an introductory physics class. On the rare occasions that it doesn't apply, it is usually enough to determine what missing energy caused the discrepancy.

For an exact expression of Conservation of Energy, see Section 8.6 below.

EXERCISE 8.17: Conserving Mechanical Energy, Part II

Recall the falling pencil from Exercise 8.2 above. Assume that the pencil has a mass of 0.30 kg, and that it is dropped from a height of 1.2 m. What is the speed of the pencil as it hits the ground? Use conservation of mechanical energy.

SOLUTION:

We choose the earth + pencil system, which was isolated in Exercise 8.2, so we can use conservation of mechanical energy to solve this problem:

$$K + U = E$$

In order to use this equation, though, we must find the value of E. At the very instant the pencil is dropped, the pencil has no speed. Thus, its kinetic energy K is zero there. However, the pencil *does* have a height; therefore, it will have a height potential energy (if we set the zero point to be at ground level, the lowest point in the problem, as we should):

$$U_H = mgh = (0.30 \text{ kg})(9.8 \text{ m/s}^2)(1.2 \text{ m}) = 3.5 \text{ J}$$

Thus, the total energy of the pencil system—at the beginning of the fall—is:

$$E = K + U = 0 \text{ J} + 3.5 \text{ J} = 3.5 \text{ J}$$

This energy—although calculated at the *beginning* of the pencil's fall—*does not change* throughout the fall, because the system is isolated. Therefore, we can use this value for E, plus the original equation $K + U = E$, to solve for the speed of the pencil when it hits the ground (or, more precisely, at the exact instant before it hits the ground—so that we don't have to worry about complications involved with the pencil's collision with the ground).

At that time, the pencil will have kinetic energy (because it is moving), but no potential energy (because it is at ground level). Thus, we can write the energy equation at that time as:

$$E = 3.5 \text{ J} = K + U = \tfrac{1}{2}mv^2 + 0$$

Using a little algebra, and plugging in for mass, we can solve for v:

$$v = \sqrt{\frac{2E}{m}} = \sqrt{\frac{2(3.5 \text{ J})}{0.30 \text{ kg}}} = 4.8 \text{ m/s}$$

This value represents the (scalar) speed of the pencil. To determine the direction of the (vector) velocity, we must think about the motion of the pencil a little more closely. The pencil begins its fall with $v_{i,y} = 0$ at $y_i = 1.2$ m, and falls *downwards* to $y = 0$ m. Thus, it's velocity component *must* be negative so that it will correctly fall downwards.

* * *

We could, if we had wished, have used the kinematic equations for free-fall from Chapter 5 to solve this problem, where $v_{i,y} = 0$, $y_i = 0$ m, $y = 1.2$ m, and $a = g$:

$$\begin{aligned} v_y^2 &= v_{i,y}^2 + 2a(y - y_i) \\ &= 0^2 + 2(9.8 \text{ m/s}^2)(1.2 \text{ m} - 0) \\ &= 23 \text{ m}^2/\text{s}^2 \\ \Rightarrow v_y &= \pm 4.8 \text{ m/s} \end{aligned}$$

We choose the negative answer because the pencil is falling. Notice, therefore, that we obtain the same result as above. This is because the physics of the problem hasn't changed, only the math!

NOTE 8.18: Potential Energy Curves

One way to visualize the effect of changing energies in an isolated system is with **potential energy curves**. In these graphs, the potential energy of the system is plotted as a function of position: $U = U(x)$. An example is shown below:

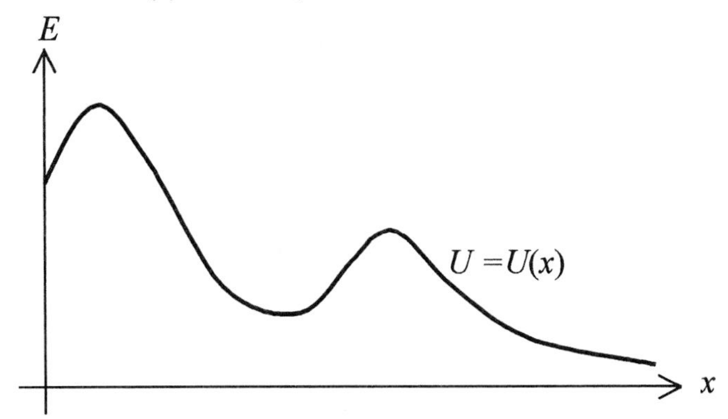

We can label some interesting points on this graph. For example, the highest point occurs at x_1, as sketched below. Here, the maximum energy $E_{max} = U$ (i.e., $K = 0$ there).

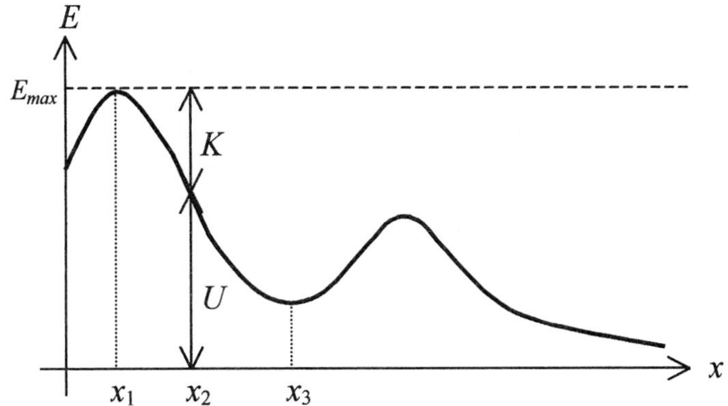

At each point along the graph, the vertical distance from the x-axis to the curve is the potential energy U. The remaining vertical distance from the curve to the horizontal line E_{max} is the kinetic energy K. We have illustrated this relationship at the point x_2 on the graph above. As you move along the x-axis, the relative values of U and K change, but the total energy E_{max} remains the same.

So, one advantage of potential energy curves is that you can "see" how an object behaves at each point in time, as a result of its energy. For example, at the top of the first hill (x_1), the object will only have potential energy. At the very next instant, however, it will start to "slide" down the hill, losing potential energy, but gaining kinetic. The object will speed up as its kinetic energy increases (and its potential energy decreases), until it reaches the bottom of the first trough (x_3). There, it will have its highest speed (because it has the greatest K there). Immediately after that point, however, the object will start to slow down—losing kinetic energy but gaining potential—as it "climbs" the second hill. And so on.

Potential energy curves also have another, completely different, practical use. As we saw in Section 8.3.2 above, the (negative) derivative of U is F. In graphical terms, this means that the (negative) slope of the $U(x)$ graph gives the net force on the object. So, using this kind of graph, we can determine the value of the net force at any point by finding the slope of the curve there. For example, what is the force at point x_3? At this point, the graph has a minimum. We know from calculus that a minimum corresponds to a zero derivative. So, the net force on the object at x_3 is zero.

8.5 ENERGY TRANSFORMS & TRANSFERS

One implication of conservation of energy is that energy can and does *change form*. The two mechanisms for change are **energy transformation**—energy conversions from one type to another, for example, from kinetic to potential energy–and **energy transference**—energy flow between a system and its environment.

8.5.1 WORK

The most important form of energy change is *work*. Work is a special physics term, which, like energy, also has a colloquial meaning. However, unlike energy, the physics definition of work is *very different* from its everyday usage.

Force is in Chapters 6 and 7; displacement is covered in Section 5.2.3.

Dot products are in Section 15.3.5.

WORK & CONSTANT FORCES

Work—in the physical sense—is a mathematical concept related to force $\vec{\mathbf{F}}$ and displacement $\vec{\mathbf{r}}$. If a force is constant, the work is:

$$W = \vec{\mathbf{F}} \cdot \vec{\mathbf{r}}$$

Work is a scalar quantity. The unit of work, like for energy, is the $\text{kg·m}^2/\text{s}^2 = \text{N·m}$, or joule. To ensure we don't confuse the two, however, we reserve joule for energy, and use the equivalent unit N·m for work.

EXERCISE 8.19: Work and Constant Forces

Recall the 1 kg potato from Example 8.3.

(a) How much work do you do to lift the potato one meter, if you use a constant lifting force?

(b) How much work does gravity do in the same situation?

(c) How much work is done by gravity if you carry the potato horizontally one meter?

(d) How much work is done by gravity as the potato slides down a $30°$ ramp of length one meter?

SOLUTION:

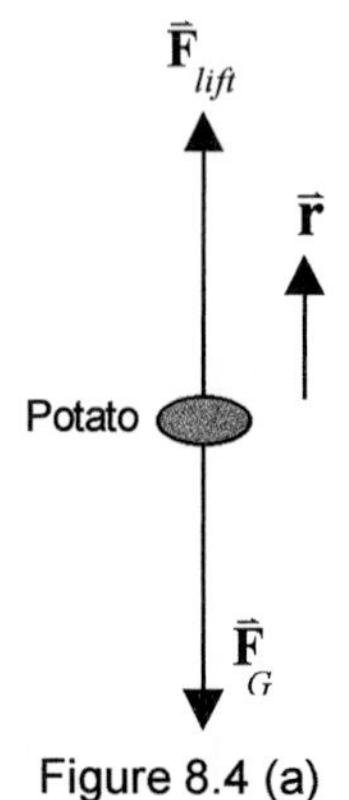

Figure 8.4 (a)

(a) Because the lifting force is constant, we can use the constant-force work equation:

$$W = \vec{\mathbf{F}} \cdot \vec{\mathbf{r}}$$

"Undoing" the dot product, we get:

$$W = Fr\cos\phi$$

where F and r are the magnitudes of the vectors $\vec{\mathbf{F}}$ and $\vec{\mathbf{r}}$, respectively, and ϕ is the smallest angle between then.

Now, the lifting force has the same magnitude, but opposite direction, as the weight of the potato (in Fig. 8.4 (a)). That is, $F_{lift} =$

Exercise 8.19 Continued...

$F_G = mg$, except that F_{lift} points upward. Now, from the problem statement we know that the displacement has a magnitude of 1 m, also in an upward direction, since the potato is being *lifted*.

We conclude that the force and the displacement have the *same* direction (upward). Thus, the angle between them is zero.

Plugging all of this information into the work equation, we have:

$$W = F_{lift} r \cos\phi = m_{potato} g r \cos\phi = (1\,\text{kg})(9.8\,\text{m/s}^2)(1\,\text{m})\cos(0°) = 9.8\,\text{N}\cdot\text{m}$$

(b) To find the work done by gravity, we do *everything* we did in part (a) above the same, *except* that now we note that the applicable force (weight) points downward. Thus, the angle between it and the displacement is 180°:

$$W = F_G r \cos\phi = m_{potato} g r \cos\phi = (1\,\text{kg})(9.8\,\text{m/s}^2)(1\,\text{m})\cos(180°) = -9.8\,\text{N}\cdot\text{m}$$

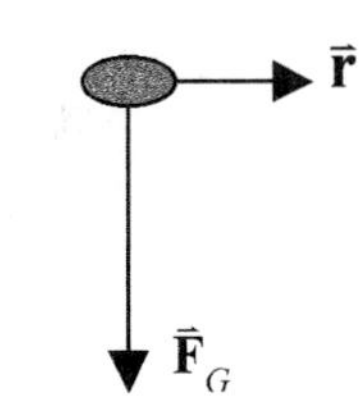

Figure 8.4 (b)

(c) Since you are carrying the potato horizontally, the potato's displacement is in the horizontal direction. However, gravity still points straight down (as in part (b) above). See Figure 8.4 (b).

Therefore, the angle between the potato's displacement and the gravitational force upon it is 90°. The cosine of 90° is zero, so the work of carrying the potato horizontally is zero.

(d) This situation is sketched in Figure 8.4 (c). Note that though gravity still points downward, the component of the weight in the direction of the displacement (i.e., along the ramp) is $F_x = mg\sin\theta$, where θ is the ramp angle. *Do not confuse* θ with the angle between F_x and the displacement $\vec{\mathbf{r}}$, *which is zero!* (Refer to the diagram to be sure.) So, the force component in the x-direction is:

$$F_x = mg\sin\theta = (1\,\text{kg})(9.8\,\text{m/s}^2)\sin(30°) = 4.9\,\text{N}$$

Thus, the work on the potato is:

$$W = F_x r\cos\phi = (4.9\,\text{N})(1\,\text{m})\cos(0°) = 4.9\,\text{N}\cdot\text{m}$$

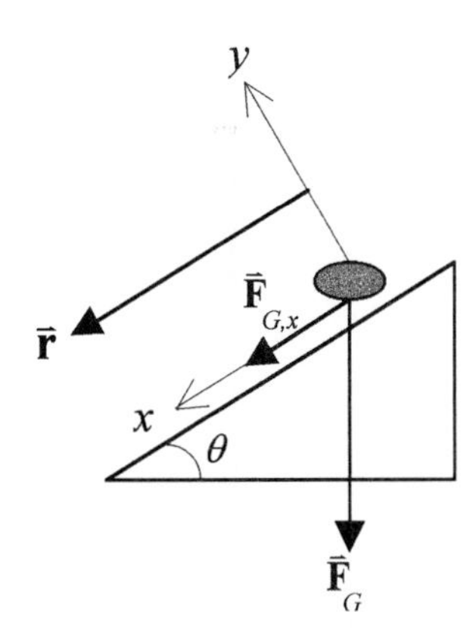

Figure 8.4 (c)

Take careful note of the dot product in the expression for work. This dot product does two things. First, it requires that a component of the force be applied parallel (or anti-parallel) to the displacement in order for work to be done.

Second, it provides a sign to work: work is positive when $\cos\theta$ is positive, and negative when $\cos\theta$ is negative. This sign is *not* directional in nature (after all, work is a scalar); rather, it indicates whether work is transferring energy into or out of the system. In the case of positive work, the *environment* does work on the system so that the energy of the system *increases*; in the case of negative work, the *system* does work on the environment, so that the system *loses* energy.

Section 15.3.1 covers vector components.

See Section 13.6.2 for more on when the cosine function is positive and negative.

EXERCISE 8.20: Taking A Closer Look At Work

Compare the four values for the work obtained in the previous exercise. Discuss their relative amounts as well as their signs.

SOLUTION:

(a) The work to lift a potato in part (a) above was found to be $+9.8\ \text{N·m}$. This is the maximum possible value for work in this example (irrespective of sign). A maximum value was obtained because the lifting force in this case was in exactly the same direction as the displacement.

The positive sign in the answer indicates that the environment (you) did work *on* the system (potato + earth); that is, without your work, the potato would not have moved.

(b) The work done by gravity in part (b) above was found to be $-9.8\ \text{N·m}$. We obtained the same answer as in part (a) above—irrespective of sign—because in this example, the lifting force was directed exactly anti-parallel to the displacement: that is, parallel and anti-parallel directions both result in maximum values for work.

The negative sign in the answer, which arose because the two vectors *were* anti-parallel, indicates that gravity did work *within* the system.

(c) In part (c) above, we found that the work to carry a potato horizontally was zero. Because the force, in this instance, was precisely perpendicular to the direction of the displacement—that is, there is no component of the force in the direction of the displacement—we found that no work was done in this case (even though your arms certainly exerted effort in the process).

(d) Finally, in part (d) above, we found that the work done by gravity was $+4.9\ \text{N·m}$. This value, midway between the maximum result of parts (a) and (b), and the minimum result of part (c), arises because the force had a non-zero component that lies somewhat, but not completely, in the direction of the displacement.

The positive sign in this case indicates that gravity transformed potential energy to kinetic, via the transfer mechanism of work.

DEFINITION 8.21: Work and Conservative Forces

If we could sum up the results of the last Exercise, it is that when it comes to work, the direction of force *matters*. Now we shall see that the *type* of force also matters.

We can sort forces into two classes: those that are **conservative** and those that are not (**non-conservative**). A conservative force is one in which the total work is zero for an object traveling on a path that begins and ends at the same point.

For example, in Exercise 8.19 (b), we raised a potato one meter into the air. As a result, gravity did $-9.8\ \text{J}$ of work on the potato. Likewise, if we were to lower the potato back down to its original position, gravity would do $+9.8\ \text{J}$ of work on it (check it!). So, if we were to take the potato through the entire path (up into the air and then back down to its starting position), the total work done on the potato is zero: $-9.8\ \text{J} + 9.8\ \text{J} = 0$. This is because the force involved in this situation—namely, gravity—is conservative.

On the other hand, one example of a non-conservative force is friction. If we were to push the potato in a closed circle on top of a table, the total work done by friction on the potato would *not* be zero because energy would be lost to friction throughout the motion.

Box 8.21 Continued...

Work done via a conservative force conserves mechanical energy; work done via a non-conservative force does not. *However,* as long as the energy lost to the non-conservative force is included in the energy conservation equation, and the table top is included in the system, the *total* energy of an isolated system is always conserved.

More in Section 8.6 below.

WORK & VARIABLE FORCES

If the force varies (in space), work has a more complex representation:

$$W = \int \vec{\mathbf{F}} \cdot d\vec{\mathbf{r}}$$

Integration is covered in Section 14.3.

EXERCISE 8.22: Work and Non-Variable Forces

In Exercise 8.14, we saw that the spring force (Hooke's Law) was $F_S = -kx$, a force that varies with x, the distance of stretch. What is the work done by a spring on an object that stretches (or compresses, for that matter) the spring a distance x?

There's more on Hooke's Law in Section 7.2.8.

SOLUTION:

The spring force is not a constant force. To calculate the work of the spring, then, we must use the non-constant work equation above:

$$W = \int \vec{\mathbf{F}} \cdot d\vec{\mathbf{r}} = \int -kx \cdot dx$$

where the force is only in the x-direction, so the dot product is reduced to the above equation. Adding limits of integration to the integrand, and doing the integral, we have:

$$W = \int_0^x -kx \cdot dx = -k\left[\tfrac{1}{2}x^2\right]_0^x = -\tfrac{1}{2}k[x^2 - 0] = -\tfrac{1}{2}kx^2$$

Notice that the result is negative. This is because the spring force is *always* oriented opposite to the direction of x. Think about it! If the spring is stretched (x is directed forwards), then the spring force pulls backwards; however, if the spring is compressed (x is directed backwards), then the spring force pushes forwards. See Figures 8.5.

(a) Spring in unstretched position

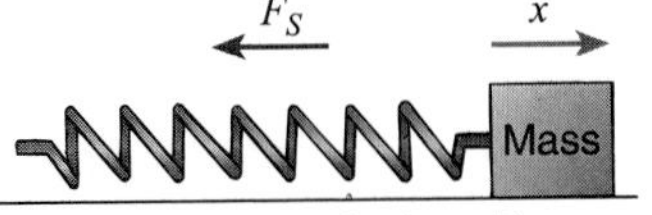

(b) Spring in stretched position

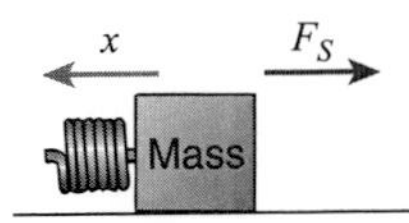

(c) Spring in compressed position

Figure 8.5

WORK – ENERGY THEOREM & POWER

The **Work-(Kinetic) Energy Theorem** says that if work is done on a system (and no other energy transfer is occurring), then the kinetic energy of the system changes:*

$$W = \Delta K = K_f - K_i$$

* This formula is only true in the case that kinetic energy is the only type of energy that is changing. That is, the Work-Energy Theorem is *not* a law; it is not applicable to every situation.

EXERCISE 8.23: The Work-Energy Theorem

Return to the falling pencil from Exercise 8.2 above. Pick your system to *just* be the pencil, so that the earth is in the *environment*. As a result, energy is no longer conserved in this situation. Find the work done by the earth on the pencil. Then, use the work-energy theorem to find the speed of the pencil right before it hits the ground.

SOLUTION:

In this case, the force in the work equation is the weight of the pencil. This force is directed downwards. Furthermore, the pencil is falling, so the direction of displacement is also downwards. The work equation then becomes (with information from Exercise 8.17):

$$W = Fr\cos\phi = mgr\cos\phi = (0.30\text{ kg})(9.8\text{ m/s}^2)(1.2\text{ m})\cos(0°) = 3.5\text{ N}\cdot\text{m}$$

The work is positive, as it should be for this situation. (Why?)

Now, at the top of its fall, the pencil has no speed, and thus no kinetic energy. Thus, $K_i = 0$. But when the pencil hits the floor, it *does* have kinetic energy. So, $K_f = \frac{1}{2}mv^2$. The increase in energy is due to the work done by the earth. Thus:

$$W = 3.5\text{ J} = K_f - K_i = \tfrac{1}{2}mv^2 - 0$$

$$\Rightarrow v = \sqrt{\frac{2W}{m}} = \sqrt{\frac{2(3.5\text{ J})}{0.30\text{ kg}}} = 4.8\text{ m/s}$$

This is exactly the same answer as we got in Exercise 8.17! We expect this to happen because work is just a type of energy transfer.

* * *

By solving this problem in three different ways (kinematics, conservation of energy, and the work-energy theorem), you should begin to see how these different concepts work together and complement each other. There's always another way to solve a problem!

The rate of work done through time is the **power**:

$$P = \frac{W}{\Delta t} = \vec{\mathbf{F}}\cdot\vec{\mathbf{v}}$$

The unit of power is the watt ($1\text{ W} = \text{J/s}$).

EXERCISE 8.24: Power

Say it takes you three seconds to raise the potato in Exercise 8.19. At what rate do you do work on the potato?

SOLUTION:

The rate of work is the power:

$$P = \frac{W}{\Delta t} = \frac{9.8\text{ N}\cdot\text{m}}{3\text{ s}} = 3.3\text{ W}$$

8.5.2 OTHER TRANSFORM & TRANSFER MECHANISMS

As we have seen, the most important mechanism for energy transformation is work. Other types of energy transformation include chemical and nuclear reactions. In an *isolated* system, however, the *total* amount of energy—regardless of the types in the system or how each type is being transformed—is always conserved.

In a *non-isolated* system, by contrast, energy can be *transferred* across a system boundary to and from its environment; thus, total energy is *not* conserved in this case. Work is the most important form of energy transfer (as well as for energy transformation); however, three other important energy transfer mechanisms exist: *waves* (*i.e.,* sound or light waves), *electrical transmissions* (*i.e.,* in circuits), and *heat*. We discuss waves in detail in Chapter 9, and since this book does not deal with circuit theory, here we concentrate on heat.

Waves are covered in Chapter 9.

Heat Q is the energy that is transferred between a system and its environment because of a temperature difference between them. Heat is positive when energy is brought into the system from the environment and negative otherwise.

There are two types of heat:

- **Sensible heat** is the heat associated with temperature changes:

$$Q_S = C\Delta T$$

  where C is the **heat capacity**, a number that you look up in charts for different materials, and ΔT is the temperature change.

Your textbook should have a list of heat capacities.

- **Latent heat** is the heat required to change the *phase* of a material, for example, from a solid to a liquid, or from a liquid to a gas:

$$Q_L = Lm$$

  where L is the **heat of transformation**—a number also found in charts that depends upon the material and on the type of phase change—and m is the mass of the sample.

Your textbook should have a list of transformation heats.

There are three ways that heat energy can be transferred in a system, depending on the type of material involved. Table 8.3 summarizes them.

Table 8.3: Methods of Heat Transfer

NAME	MATERIAL	METHOD
CONDUCTION	Solids	Atoms in solids vibrate because of their temperature: the hotter they are, the more they move. As they vibrate, atoms jostle their neighbors, forcing them to move in a similar fashion. Thus, heat is passed, atom by atom, through a solid material.
CONVECTION	Fluids	Hot fluids rise; cold fluids sink. Heated fluids, then, set up a **convection cell** that circulates heat away from the heat source.
RADIATION	All	All bodies with a temperature above absolute zero emit heat energy, via electromagnetic waves, into the atmosphere.

8.6 CONSERVATION OF ENERGY: A CLOSER LOOK

In the energy equation of Section 8.4 above, we only took into account the conservation of mechanical energy. This method is fine for simple problems that only have kinetic and potential energies, but in more realistic scenarios, we must consider other kinds of energy—along with energy transfers and transformations—in our energy equations.

However, because such situations involve non-isolated systems (by virtue of allowing energy transfers), *the total amount of energy in these systems may no longer be constant*. Thus, we find that it is better to deal with energy *changes* in our equation, rather than the energies themselves.

To sum up, a better conservation equation would:

1. Include *all* energies in a system, not just kinetic and potential;
2. Include energy transfers and transformations as well; and
3. Express the equation in terms of energy *changes*, not energies.

To create such an equation, we realize that the sum of all energy changes in a system *is equal to* the sum of the energy transfers. Thus, we can write our more precise expression for **conservation of energy** as:

$$\text{Energy Changes} = \text{Energy Transfers}$$

For more on the change symbol, Δ, refer to Section 5.2.3.

If we recall that Δ is a mathematical indication of a change (*i.e.,* for a change in kinetic energy $\Delta K = K_2 - K_1$), then we could write this expression in equation form as:

$$\Delta K + \Delta U + \Delta E_{int} + \Delta E_{mass} + \ldots = W + Q + \ldots$$

Now, this equation looks forbidding because it is so long. However, in most cases (especially in introductory physics courses), many of these terms are zero so that the equation becomes quite neat and short.

EXERCISE 8.25: Energy Conservation, Part I

Imagine that a book sits on the table next to you, and that you give it a good push. Write the equations of conservation of energy (a) while you are pushing the book, (b) while it slides on the table after leaving your hand, and (c) while it drops through the air after falling off of the table.

SOLUTION:

(a) The system in this case is the book. The only terms in the energy equation, then, are the kinetic energy of the book as it moves because of your push, and the work you do on the book to push it:

$$\Delta K + \Delta U + \Delta E_{int} + \Delta E_{mass} + \ldots = W + Q + \ldots$$
$$\Rightarrow \Delta K = W$$

which is just the work-energy theorem of Section 8.5.1 above.

Exercise 8.25 Continued...

(b) The system in this case is the book + surface, since the book will lose internal energy to friction. The only terms in this case are kinetic and internal (frictional) energies:

$$\Delta K + \Delta U + \Delta E_{int} + \Delta E_{mass} + \ldots = W + Q + \ldots$$
$$\Rightarrow \Delta K = -\Delta E_{int}$$

(c) Here, the system is the book + earth. As it falls, the book has potential energy (due to its height above the ground) and kinetic energy (because it is moving), so:

$$\Delta K + \Delta U + \Delta E_{int} + \Delta E_{mass} + \ldots = W + Q + \ldots$$
$$\Rightarrow \Delta K + \Delta U = 0$$

which is just the equation of conservation of mechanical energy.

(But wait! This equation doesn't look anything like the mechanical energy equation $K + U = E$. How do we get from $\Delta K + \Delta U = 0$ to $K + U = E$?

If we "undo" the delta signs in $\Delta K + \Delta U = 0$, our equation becomes:

$$(K_2 - K_1) + (U_2 - U_1) = 0$$
$$\Rightarrow K_1 + U_1 = K_2 + U_2$$

where we rearranged the equation algebraically. Now, since $K + U$ is a constant, which we can call E, this equation becomes $K + U = E$. We have recreated the equation of conservation of mechanical energy!)

EXERCISE 8.26: Energy Conservation, Part II

In the curve of Note 8.18 above, a teddy bear of mass = 2.0 kg sits on the top of the hill at x_1, at a height of 3.0 m. After a tiny nudge, the teddy bear slides down the hill, obtaining a maximum speed of 6.0 m/s at the bottom of the first trough (at x_3), which has a height of 1.0 m. What is the change in internal energy of the teddy? Ignore any energies that may reside in this isolated system other than potential, kinetic, and internal.

SOLUTION:

Our conservation of energy equation, applied to this situation, says that:

$$\Delta K + \Delta U + \Delta E_{int} + \Delta E_{mass} + \ldots = W + Q + \ldots$$
$$\Rightarrow \Delta K + \Delta U + \Delta E_{\text{int}} = 0$$

where we have ignored other energies and energy transfers as the problem statement indicated. We can rewrite this equation as:

$$\begin{aligned}\Delta E_{\text{int}} &= -(\Delta K + \Delta U)\\ &= -[(K_2 - K_1) + (U_2 - U_1)]\\ &= (K_1 - K_2) + (U_1 - U_2)\end{aligned}$$

Now we can plug in what we know:

$$\begin{aligned}\Delta E_{\text{int}} &= (0 - \tfrac{1}{2}mv^2) + (mgy_1 - mgy_2)\\ &= [0 - \tfrac{1}{2}(2.0\,\text{kg})(6.0\,\text{m/s})^2] + [(2.0\,\text{kg})(9.8\,\text{m/s}^2)(3.0\,\text{m} - 1.0\,\text{m})]\\ &= 3.2\,\text{J}\end{aligned}$$

Thus, the increase in internal energy is +3.2 J.

It is important to realize that in physics today, we have no knowledge of what energy is.
RICHARD FEYNMAN
American Physicist, 1918 - 1988

As we cannot give a general definition of energy, the principle of conservation of energy simply signifies that there is something *which remains constant.*
HENRI POINCARE
French Mathematician, 1854 - 1912

If you feel a little confused about the concept of energy, don't despair: energy is a nebulous concept even to experienced physicists! However, just because the idea of energy may not be completely evident or obvious to you, does not mean that it will not be a useful quantity when you try to solve problems.

Think of energy this way: All systems have a quantity of *something* we call energy. The energy resides in the system in various forms: some of it is kinetic; some of it is potential; and so on. Things "happen" in the system when energy is transformed from one type of energy to another—for example, when potential energy is converted to kinetic energy—or when energy is transferred in or out of the system.

Moreover, in an isolated system (and with a little jiggering, *any* system can be made isolated), the total amount of energy in the system *stays the same*—so that energy is conserved—even as things are transpiring in the system.

8.7 MOMENTUM

"'When I use a word," Humpty Dumpty said, in a rather scornful tone, "it means just what I choose it to mean—neither more nor less."
LEWIS CARROLL,
Through The Looking Glass
English Author, 1832 - 1898

You have heard the term *momentum* before. When we think of a body with 'momentum,' we imagine the body is moving and would be hard to stop. The physical meaning of momentum brings to mind this connotation as well, via a specific mathematical definition: **momentum** (in the physical sense) is a vector created by the product of mass and velocity:

$$\vec{\mathbf{p}} = m\vec{\mathbf{v}}$$

The unit of momentum is the $\mathrm{kg \cdot m/s}$.

EXERCISE 8.27: Momentum

What is the momentum of the pencil in Exercise 8.23 just before it hits the ground?

SOLUTION:

We know the pencil's mass and velocity (at the bottom of its fall) from that exercise. Thus, we can solve for the magnitude of momentum:

$$p = mv = (0.30\ \mathrm{kg})(4.8\ \mathrm{m/s}) = 1.4\ \mathrm{kg \cdot m/s}$$

Since momentum is always parallel to velocity, and because the pencil is *falling*, the direction of the momentum vector is *downwards*. So, we can write the momentum vector of the pencil as:

$$\vec{\mathbf{p}} = (1.4\ \mathrm{kg \cdot m/s})(-\hat{\mathbf{j}}) = -1.4\ \mathrm{kg \cdot m/s}\ \hat{\mathbf{j}}$$

As we did with energy, we want to connect the definition of momentum with force. We do this via a derivative: the time rate of change of momentum is the force:

$$\vec{\mathbf{F}} = \frac{d\vec{\mathbf{p}}}{dt}$$

Force is covered in Chapters 6 and 7.

EXERCISE 8.28: Force & Momentum

A 4.0 kg mass moves via the position vector $\vec{\mathbf{r}} = (5t^2 + 3t - 2)\hat{\mathbf{i}} + 4\hat{\mathbf{j}}$, where t is time in seconds and r is in units of meters. (a) Find the velocity of the body. (b) Calculate its momentum. (c) Find the force on it.

SOLUTION:

(a) Velocity is the time derivative of position: $\vec{\mathbf{v}} = d\vec{\mathbf{r}}\,/\,dt$:

$$\vec{\mathbf{v}} = \frac{d\vec{\mathbf{r}}}{dt} = \frac{d}{dt}\left[\left(5t^2 + 3t - 2\right)\hat{\mathbf{i}} + 4\hat{\mathbf{j}}\right] = (10t + 3)\hat{\mathbf{i}} \text{ m/s}$$

where the y-component of the derivative is zero because the y-component of the position is a constant.

This definition of velocity is in Section 5.3.2.

(b) The momentum is just the body's mass times its velocity:

$$\vec{\mathbf{p}} = m\vec{\mathbf{v}} = (4.0\text{ kg})(10t + 3)\hat{\mathbf{i}} = (40t + 12)\hat{\mathbf{i}}\ \text{kg·m/s}$$

(c) Now, the force is the time derivative of the momentum. But since $\vec{\mathbf{p}}$ is in the x-direction only, $F_x = dp_x\,/\,dt$:

$$F_x = \frac{dp_x}{dt} = \frac{d}{dt}\left(40t + 12\right) = 40\text{ N}$$

$$\Rightarrow \vec{\mathbf{F}} = 40\hat{\mathbf{i}}\ \text{N}$$

Objects that spin have a different kind of momentum, called **angular momentum**. Angular momentum (units $\text{kg·m}^2\text{/s}$) is defined as the cross product of the position and the linear momentum of a body:

$$\vec{\mathbf{L}} = \vec{\mathbf{r}} \times \vec{\mathbf{p}}$$

Cross products are covered in Section 15.3.5.

EXERCISE 8.29: Angular Momentum, Part I

Find the angular momentum around the origin of the body above.

SOLUTION:

Angular momentum is a cross product. Refer to Section 15.3.5 for more information on the following calculation:

$$\vec{\mathbf{L}} = \vec{\mathbf{r}} \times \vec{\mathbf{p}} = \begin{vmatrix} \hat{\mathbf{i}} & \hat{\mathbf{j}} & \hat{\mathbf{k}} \\ 5t^2 + 3t - 2 & 4 & 0 \\ 10t + 3 & 0 & 0 \end{vmatrix} = \left[(5t^2 + 3t - 2)(0)\right] - \left[(4)(10\text{t} + 3)\right]\hat{\mathbf{k}}$$

$$= (-40t - 12)\hat{\mathbf{k}}\ \text{kg}\cdot\text{m}^2\text{/s}$$

Rotational inertia is discussed in Section 6.5.1; angular velocity in Section 5.3.1.

We can also define linear momentum in terms of rotational inertia I and angular velocity ω:

$$\vec{\mathbf{L}} = I\vec{\boldsymbol{\omega}}$$

The direction of the angular momentum is parallel to the direction of the angular velocity. We find the direction of the angular velocity using the **curl right hand rule**: Using your *right hand*, circle your fingers in the direction that θ changes; your thumb will naturally point in the direction of ω.

WARNING!

Be *very* careful with this rule! *Sloppiness results in wrong answers.* If you use your left hand to do the right hand rule (for example, if you write with your right hand and forget to put down your pencil before you do the rule), you will get the direction WRONG! Or, if you do not take extra care to line up your hand with the direction of motion *exactly*, you may get the wrong result as well.

EXERCISE 8.30: Angular Momentum, Part II

In Exercise 8.8, we saw that a 45 record rotates at about $5\ \text{rad/s}$, and has a rotational inertia of $6\times10^{-3}\ \text{kg}\cdot\text{m}^2$. What is the record's angular momentum, including magnitude and direction?

SOLUTION:

The magnitude of the angular momentum is the product of the angular velocity and the rotational inertia:

$$L = I\omega = (6\times10^{-3}\ \text{kg}\cdot\text{m}^2)(5\ \text{rad/s}) = 3\times10^{2}\ \text{kg}\cdot\text{m}^2/\text{s}$$

The direction of the angular velocity is found via the curl right hand rule. We circle the fingers of our right hand in the direction that the record plays; that is, counter-clockwise. Our thumb naturally points upwards; thus, the angular velocity points upwards. See Figure 8.6.

The angular momentum direction is parallel to the angular velocity, so it also points upwards. Therefore, our angular momentum vector is:

$$\vec{\mathbf{L}} = (3\times10^{2}\ \text{kg}\cdot\text{m}^2/\text{s})\hat{\mathbf{k}}$$

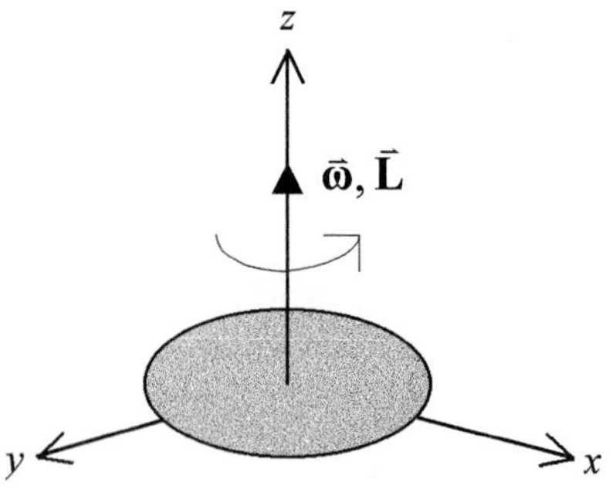

Figure 8.6

Linear momentum is connected to force via a time derivative. Likewise, angular momentum is connected to torque via a time derivative:

Torques are covered in Section 6.5.2.

$$\vec{\boldsymbol{\tau}} = \frac{d\vec{\mathbf{L}}}{dt}$$

EXERCISE 8.31: Torque & Angular Momentum

Find the torque on the 4.0 kg body of Exercise 8.29.

SOLUTION:

Torque is the time derivative of the angular momentum:

$$\vec{\tau} = \frac{d\vec{L}}{dt} = \frac{d}{dt}[-40t - 12]\hat{\mathbf{k}} = (-40 \text{ kg}\cdot\text{m}^2/\text{s})\hat{\mathbf{k}}$$

8.8 CONSERVATION OF MOMENTUM

We saw in Section 8.4 above that energy conservation is a very important law about the universe. There is a similarly significant statement about the conservation of momentum in isolated systems.

In an isolated system upon which no net external forces act ($\vec{\mathbf{F}}_{ext} = \mathbf{0}$), *linear momentum is conserved.* That is, momentum may be changed/ exchanged within an isolated system, but the *total* amount of momentum does not increase or decrease with time.

EXAMPLE 8.32: Momentum Conservation, Part I

For a classic example of momentum conservation, think of an astronaut in space. The astronaut holds a large mass—say, a communications satellite. The astronaut and the satellite are initially at rest. See Figure 8.7 (a).

Figure 8.7 (a)

The system for this problem is the astronaut + satellite. Because neither the astronaut nor the satellite is moving, the system initially has zero momentum. Suppose that, suddenly, the astronaut tosses the satellite away. This motion consists of an *internal* force (the astronaut is doing the throwing), so that no *external* forces act on the system. Therefore, *momentum in the system is conserved.*

After the throw, the satellite's momentum is no longer zero; it is—after all—moving. Therefore, the astronaut's momentum *must also* be non-zero, in order to cancel out the momentum of the satellite. That is, since the total momentum of the astronaut + satellite system started out as zero, it must still be zero, because momentum is conserved in this system. See Figure 8.7 (b).

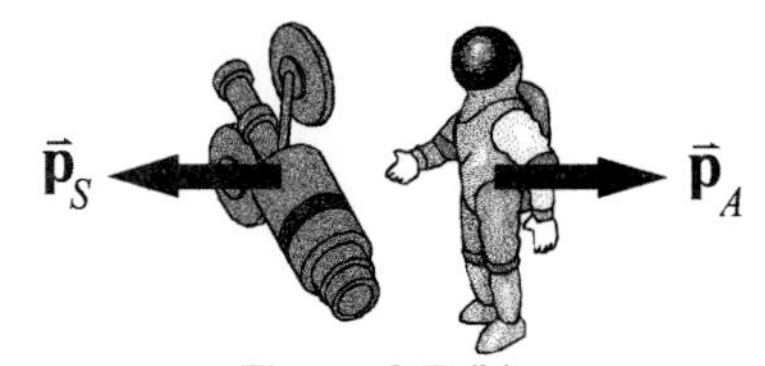

Figure 8.7 (b)

In fact, to cancel the satellite's momentum, the astronaut's momentum must be equal in magnitude, and opposite in direction, to the satellite's. Does this mean the astronaut will have the same *speed* as the satellite? Not necessarily: it will only happen if she has the same mass as the satellite; otherwise, their relative speeds will be proportional to their relative masses.

Unlike the case of mechanical energy conservation, in which we set the total amount of energy in a system equal to a constant, in momentum conservation problems, the usual procedure is to set the momentum at one time equal to the momentum at another.

EXERCISE 8.33: Momentum Conservation, Part II

In a James Bond-like spoof movie, our hero (with a mass of 75 kg) is lowered by a helicopter into a driverless golf cart (with a mass of 150 kg), which is careening "dangerously" out of control at 2.5 m/s (or about 6 mph). What is the cart's speed immediately after "James Bond" enters it, and before he has a chance to save the day by bringing the golf cart under control?

SOLUTION:

This is a conservation of momentum problem. The system is the cart.

The initial situation, in which the cart is driverless, has a momentum of $p_i = m_{cart}v_i$. The final situation, in which the cart is manned by "James Bond," has a momentum of $p_f = (m_{cart} + m_{JB})v_f$.

By conservation of momentum, we know that $p_i = p_f$, or:

$$m_{cart}v_i = (m_{cart} + m_{JB})v_f$$

$$\Rightarrow v_f = \frac{m_{cart}}{m_{cart} + m_{JB}} v_i = \left(\frac{75\,\text{kg}}{75\,\text{kg} + 150\,\text{kg}}\right)(2.5\text{ m/s}) = 1.7\text{ m/s}$$

An analogous conservation law holds for angular momentum. *In an isolated system upon which no net external torques act ($\vec{\tau}_{ext} = \mathbf{0}$), angular momentum is conserved.* That is, angular momentum may be changed/exchanged within an isolated system, but the *total* amount of angular momentum does not increase or decrease with time.

EXAMPLE 8.34: Angular Momentum Conservation, Part I

Ice skaters know conservation of angular momentum well.

When in a good spin, ice skaters can slow down their rate of spin (lowering their ω) by sticking their arms out (raising their I). Or, they can speed up their spin rate (raising their ω) by bringing their arms in tight (lowering their I).

In both cases, though, the skater's total angular momentum $L = I\omega$ is the *same* because there are no *external* torques on her. That is, the only torque on her body is from her own *internal* muscle force. Thus, the system ice skater + ice is isolated and angular momentum is conserved.

EXERCISE 8.35: Angular Momentum Conservation Part II

In the sequel to the movie in Exercise 8.33, "James Bond" is trapped on a whirling torture device known as a "merry-go-round". The MGR, as Headquarters calls it, has an rotational inertia of I_{MGR}, and is spinning with angular velocity ω_i.

Luckily, "James Bond," who has a moment of inertia while on the MGR of I_{JB}, is lifted out of his predicament by helicopter. How fast is the MGR spinning after his daring escape?

SOLUTION:

This is a conservation of angular momentum problem. The system is the MGR.

The initial situation, in which "James Bond" is on the MGR, has an angular momentum of $L_i = (I_{MGR} + I_{JB})\omega_i$. The final situation, in which the MGR is empty, has an angular momentum of $L_f = I_{MGR}\omega_f$.

By conservation of angular momentum, we know that $L_i = L_f$, or:

$$(I_{MGR} + I_{JB})\omega_i = I_{MGR}\omega_f$$

$$\Rightarrow \omega_f = \frac{I_{MGR} + I_{JB}}{I_{MGR}}\omega_i$$

8.9 COLLISIONS

Like the situation with energy, a system can transfer momentum between its parts. It does this through *collisions*.

A **collision**, in physics terms, is any situation in which two (or more) objects *interact*. Notice that this definition does not rule out interactions at a distance; *i.e.*: 'collisions' in which the objects never touch!

EXAMPLE 8.36: Types of Collisions, Part I

A familiar example of a collision is the interaction between a car and a truck when they *hit*. This is obviously a collision because the car and the truck touch each other.

But a not-so-familiar example of a collision is the interaction between two charged particles when they get close enough to *repel* each other. This is not so obviously a collision, because the charged particles in this situation will never get close enough to touch.

However, we consider both scenarios to be "collisions" because the interactions involved cause the bodies in each system to *change* their original course.

There are three types of collisions:

- In a **perfectly elastic collision**, the original, colliding objects *remain completely separate* after the collision. Kinetic energy is conserved in such a collision.
- In a **perfectly inelastic collision**, colliding objects *stick* after the collision. Kinetic energy is *not* conserved in this case, since energy is lost as the objects deform and stick together.
- In an **inelastic collision**, the objects do not stick together (though there may be some mass exchanged between the bodies), and some kinetic energy is lost.

EXAMPLE 8.37: Types of Collisions, Part II

A good example of a perfectly elastic collision is when two billiard balls strike each other in a pool game. In this case, the balls hit each other and then move away without sticking together at all.

An example of a perfectly inelastic collision is the case of a soft ice cream scoop falling to the ground. The scoop hits the ground and stays there, completely stuck to the cement.

One example of an (imperfect) inelastic collision is a crash between a car and a truck. The vehicles strike each other, cause damage, and bounce apart. Some of the car has been transferred to the truck (like some paint or a bumper), and likewise, some of the truck has been transferred to the car.

Science is the knowledge of consequences....
THOMAS HOBBES
English Philosopher, 1588 - 1679

Inelastic collisions are the most common type of collision in real life, but they are also the hardest to model: there is no way to calculate the loss in kinetic energy during the collision without a direct measurement. Therefore, most physics problems *assume* that the collision was either perfectly elastic or perfectly inelastic.

Regardless of the type of collision, however, *momentum is conserved,* so long as no external forces act on the system.

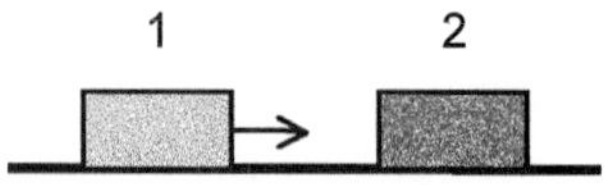

1 2

Figure 8.8

EXERCISE 8.38: Collisions, Part I

A collision takes place between cart 1 and cart 2 from Example 8.1, redrawn in Figure 8.8. Initially, cart 1 is moving horizontally towards cart 2, which is at rest. The carts collide and stick together. If cart 1 was initially moving with speed v_i, and has twice the mass of cart 2, find the speed v_f of the combined cart after the collision.

SOLUTION:

As we saw in Example 8.1, the system consists of both carts and nothing else. No external forces are involved (the only force being the collision force, which is within the system). So, the system is isolated. Thus, momentum is conserved in this situation.

Exercise 8.38 Continued...

The problem statement says that the carts collide and stick together. So we know that this is a perfectly inelastic collision. As a result, kinetic energy is *not* conserved.

Now let's get some equations. Prior to the collision, only cart 1 is moving. Thus, the initial momentum is completely due to cart 1: $p_i = m_1 v_i$. After the collision, however, both carts move *together*, so that the total momentum is the sum of the masses of the (now combined) cart times v_f:

$$p_f = (m_1 + m_2)v_f$$

We have already stated that momentum is conserved in this case, so:

$$p_i = p_f$$
$$\Rightarrow m_1 v_i = (m_1 + m_2)\, v_f$$
$$\Rightarrow v_f = \frac{m_1 v_i}{(m_1 + m_2)}$$

Now, since cart 1 has twice the mass of cart 2, we know that $m_1 = 2m_2$:

$$\therefore v_f = \frac{m_1 v_i}{(m_1 + m_2)} = \frac{2m_2 v_i}{(2m_2 + m_2)} = \frac{2m_2}{3m_2} v_1 = \tfrac{2}{3} v_i$$

Thus, the final velocity of the stuck-together carts is two-thirds of the initial velocity of cart 1. Furthermore, because the final velocity is positive, the stuck-together carts move in the same direction as cart 1 traveled initially.

EXERCISE 8.39: Collisions, Part II

Cart 1 and cart 2 from Exercise 8.38 above collide elastically. Find the speeds v_{1f} and v_{2f} of the carts after the collision.

SOLUTION:

As in the previous problem, the system consists of both carts. No external forces are involved so momentum is conserved, again, like the previous problem. But, since this exercise consists of an elastic collision, kinetic energy is also conserved!

Prior to the collision, only cart 1 is moving, so: $p_i = m_1 v_i$ and $K_i = \frac{1}{2} m_1 v_i^2$. After the collision, cart 1 and cart 2 *both* move, each at their own velocity: so $p_f = m_1 v_{1f} + m_2 v_{2f}$ and $K_f = \frac{1}{2} m_1 v_{1f}^2 + \frac{1}{2} m_2 v_{2f}^2$.

Now momentum is conserved...

$$p_i = p_f$$
$$\Rightarrow m_1 v_i = m_1 v_{1f} + m_2 v_{2f} \qquad (1)$$

...and so is kinetic energy:

$$K_i = K_f$$
$$\Rightarrow \tfrac{1}{2} m_1 v_i^2 = \tfrac{1}{2} m_1 v_{1f}^2 + \tfrac{1}{2} m_2 v_{2f}^2 \qquad (2)$$

There are two equations, (1) and (2), so we can solve for the two unknowns, v_{1f} and v_{2f}. Unfortunately, these are tricky equations to solve. We list here some of the major steps; you fill in the rest.

- (1) becomes: $m_1(v_i - v_{1f}) = m_2 v_{2f}$ (3)
- (2) becomes: $m_1(v_i - v_{1f})\,(v_i + v_{1f}) = m_2 v_{2f}^2$ (4)

Exercise 8.39 Continued...

- Dividing (4) by (3) yields: $v_i + v_{1f} = v_{2f}$ (5)

- Solve (5) for v_{1f} gets: $v_{1f} = v_{2f} - v_i$ (6)

- Plug (6) for v_{1f} into (3). After some algebra, in which we combine like terms and solve for v_{2f}, we get:

$$v_{2f} = \frac{2m_1}{(m_1 + m_2)} v_i$$

- Plugging this equation into (5), we can solve for v_{1f}:

$$v_{1f} = \frac{(m_1 - m_2)}{(m_1 + m_2)} v_i$$

Now, the equations for v_{1f} and v_{2f} above are general, meaning that we have not yet put in our problem statement information for the carts. Therefore, we can use these equations *anywhere* that the assumptions of this problem apply! In *this* problem, though, $m_1 = 2m_2$. Substituting this fact into our equations:

$$v_{1f} = \frac{(m_1 - m_2)}{(m_1 + m_2)} v_i = \frac{(2m_2 - m_2)}{(2m_2 + m_2)} v_i = \frac{m_2}{3m_2} v_i = \tfrac{1}{3} v_i$$

$$v_{2f} = \frac{2m_1 v_i}{(m_1 + m_2)} = \frac{2(2m_2)v_i}{(2m_2 + m_2)} = \frac{4m_2}{3m_2} v_i = \tfrac{4}{3} v_i$$

So, cart 1 has slowed to one-third of its initial value, while cart 2 is now traveling at four-thirds the speed at which cart 1 initially moved.

v_i

Alpha Oxygen

Initial Situation
Figure 8.9 (a)

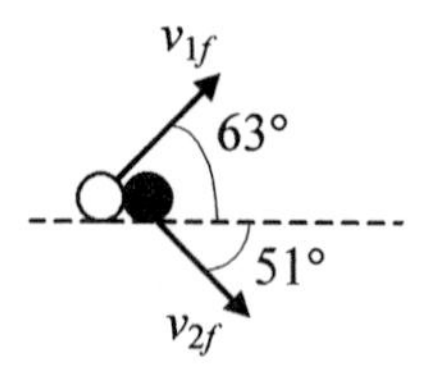

Final Situation
Figure 8.9 (b)

EXERCISE 8.40: Collisions, Part III

Collisions can also take place in more than one dimension. Consider this situation.

An α particle moves in the positive x-direction. It elastically collides with an oxygen nucleus at rest on the x-axis (Fig. 8.9 (a)). Following the collision, the α particle moves away at an angle of $63°$ above the x-axis, while the oxygen nucleus moves away at a velocity of 1.2×10^5 m/s, at an angle of $51°$ below the x-axis. See Figure 8.9 (b).

Assume the oxygen nucleus is four times as massive as the α particle, and find the initial and final velocities the α particle.

SOLUTION:

In this case, we must use momentum conservation in the x- *and* y-directions. In the x-direction, the momentum conservation equation is:

$$p_{i,x} = p_{f,x}$$
$$m_1 v_i = m_1 v_{1f} \cos 63^o + m_2 v_{2f} \cos 51^o$$

where we have used the facts that (1) only the α particle moves in the initial situation (and therefore, it is the only particle to contribute to the initial momentum), and (2) after the collision, the particles move away at angles (and therefore, the momenta must have components in the x-direction).

Exercise 8.40 Continued...

Similarly in the y-direction, we have:

$$p_{i,y} = p_{f,y}$$
$$0 = m_1 v_{1f} \sin 63^o - m_2 v_{2f} \sin 51^o$$

where we have used the facts that (1) neither particle moves in the y-direction initially, (2) after the collision, there are components of momentum in the y-direction for both particles, and (3) the oxygen atom moves downwards, so we must subtract its component.

Now, since the oxygen has four times the mass of the α particle, we know that $m_2 = 4m_1$. Thus, our two equations become:

$$m_1 v_i = m_1 v_{1f} \cos 63^o + 4m_1 v_{2f} \cos 51^o$$
$$\Rightarrow v_i = v_{1f} \cos 63^o + 4v_{2f} \cos 51^o$$

$$0 = m_1 v_{1f} \sin 63^o - 4m_1 v_{2f} \sin 51^o$$
$$\Rightarrow v_{1f} = 4v_{2f} \frac{\sin 51^o}{\sin 63^o} = 3.5(1.2 \times 10^5 \text{ m/s}) = 4.2 \times 10^5 \text{ m/s}$$

Plugging this result into the equation for v_i, we get:

$$v_i = (4.2\times10^5 \text{ m/s})\cos 63^o + 4(1.2\times10^5 \text{ m/s})\cos 51^o = 4.9\times10^5 \text{ m/s}$$

Summarizing our results, we see that the α particle approached the oxygen nucleus at almost $500{,}000$ m/s. As a result of their collision, the α particle slowed by almost 20%, to around $400{,}000$ m/s. However, the momentum the α particle lost went to the oxygen nucleus: it sped up from rest to over $100{,}000$ m/s.

The change in momentum of a body due to a collision is the **impulse**:

$$\vec{\mathbf{J}} = \Delta\vec{\mathbf{p}}$$

EXERCISE 8.41: Impulse

What is the impulse cart 1 experiences in Exercise 8.39?

SOLUTION:

Cart 1 has an initial velocity of v_i, and a final velocity of $\frac{1}{3}v_i$. Thus, its initial momentum is $p_i = m_1 v_i$, and its final momentum is $p_f = \frac{1}{3}m_1 v_i$. The magnitude of the impulse is, then:

$$J = \Delta p = p_f - p_i = \tfrac{1}{3}m_1 v_i - m_1 v_i = -\tfrac{2}{3}m_1 v_i$$

This was the impulse transferred to cart 2.

8.10 CONCLUSION

Science is a redirection of the bewildering diversity of unique events to manageable uniformity within one of a number of symbol systems.
ALDOUS LEONARD HUXLEY
English Novelist, 1894 - 1963

A conserved quantity (in the appropriately defined isolated system) *does not increase or decrease,* no matter what happens in the system, and no matter what the quantity is. Mathematically, we say that conserved quantities are *constant*. In this chapter, we discussed energy and momentum conservation, and their application to problem solving of many types, including collisions. Table 8.4 summarizes these.

Table 8.4: Conservation Laws

NAME	EQUATION
CONSERVATION OF MECHANICAL ENERGY	$K + U = E$
CONSERVATION OF ENERGY	$\Delta K + \Delta U + \Delta E_{int} + \Delta E_{mass} + \ldots = W + Q + \ldots$
CONSERVATION OF LINEAR MOMENTUM	$\vec{\mathbf{F}}_{ext} = \frac{d\vec{\mathbf{p}}}{dt} = 0$
CONSERVATION OF ANGULAR MOMENTUM	$\vec{\boldsymbol{\tau}}_{ext} = \frac{d\vec{\mathbf{L}}}{dt} = 0$

Because the energy and momentum conservation techniques of this chapter allow us to look at problems from a new vantage point—one that is free of forces—they are vastly useful for solving certain difficult problems. But they also allow us to gain a deeper, more subtle understanding of the way the universe works; an understanding that grows out of, and expands upon, the Newtonian force analysis. Therefore, energy and momentum are critical tools for your physics toolbox.

9
OSCILLATIONS & WAVES

9.1 INTRODUCTION

So far, in this part of the book, we have learned how to *describe* motion (Chapter 5) and *analyze* it (Chapters 6 through 8). In this chapter, we conclude our look at motion by examining two specific types of it: **oscillations** and **waves**.

EXAMPLE 9.1: Some Oscillations and Waves

In real life, there are many things that oscillate or wave. For instance, your heart is a complex but regular oscillator: it pumps blood through your veins with each beat. The pendulum of a grandfather clock is another oscillator: in fact, its gentle back-and-forth motion keeps time! As an example of waves, think of the ripples in water that appear when you drop a stone. But, even though you can't see them, light and sound exhibit wave motion, too!

Nor ever yet
The milky rainbow's vernal-
tinctured hues
To me have shone so pleasing,
as when first
The hand of science pointed out
the path...
MARK AKENSIDE
English Poet, 1721 - 1770

In general, oscillations and waves have highly complex forms. Your physics class (and this chapter), however, will focus only on mathematically *simple* forms of these motions: those exhibited by **simple harmonic oscillators** and **sinusoidal waves**. We shall find—in addition to learning many things about these interesting phenomena—that simple harmonic motion and waves can be physically *linked* together!

Eliminating the Unessential, Section 2.3.

9.2 PERIODIC MOTION

Oscillations and waves are both **periodic** motions; that is, they *repeat* themselves after a certain time interval. The time it takes for an oscillation or wave to repeat itself is called the **period** T. A related measure is the number of repetitions of the motion in a certain time interval, known as the **frequency** f, or alternatively, **angular frequency** ω. These three parameters are related by the equation:

$$T = \frac{1}{f} = \frac{2\pi}{\omega}$$

The units of period, frequency, and angular frequency are, respectively, seconds (s), hertz ($1\ \text{Hz} = 1\ \text{s}^{-1}$), and radians per second (rad/s).

EXAMPLE 9.2: Period and Frequency

For example, the *frequency* of beating for a heart at rest is about one time per second, or:

$$f = \frac{1}{1\ \text{s}} = 1\ \text{Hz}$$

Therefore, the *period* of this motion is...

$$T = \frac{1}{f} = \frac{1}{1\ \text{Hz}} = 1\ \text{s}$$

...and its *angular frequency* is:

$$\omega = 2\pi f = 2\pi(1\ \text{Hz}) \approx 6.3\ \text{rad/s}$$

NOTE 9.3: Why *Angular* Frequency?

You may be wondering why angular frequency ω is applied to periodic motion; after all, isn't angular frequency characteristic of *circular* motion? The answer is that periodic motion is related to circular motion, and therefore, the same notation is used for each.

To see that this is true, examine the top half of Figure 9.1. In this figure, the "shadow" of a body moving in (uniform) circular motion is projected onto a horizontal line. On this line, you can see the "back-and-forth" motion characteristic of periodic motion. Since periodic motion is clearly related to circular motion, then, we include the angular velocity ω in our equations for periodic motion.

If we now project the circular motion onto a "moving" axis (notice that the grid on the bottom half of the figure moves downward), the result is a sinusoidal oscillation. We shall soon see that periodic motions are best described by sinusoids. And, as you might guess, the corresponding sinusoidal equations will contain the angular frequency ω.

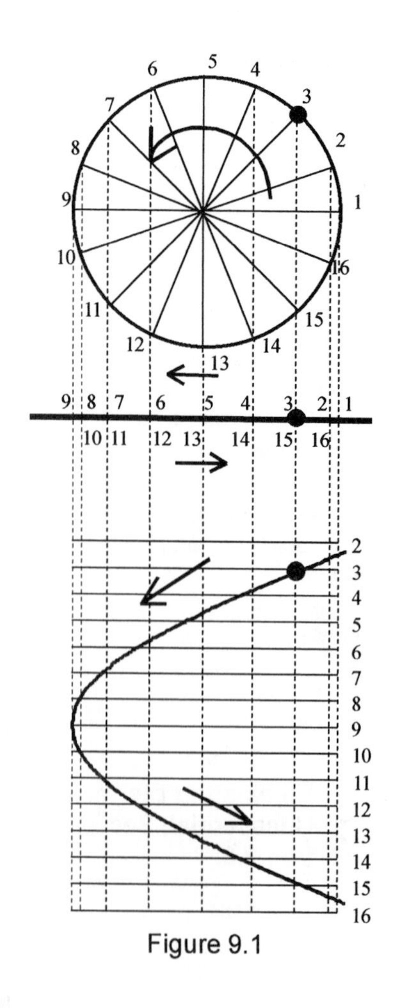

Figure 9.1

9.3 SIMPLE HARMONIC OSCILLATORS

A **simple harmonic oscillator** is any object that oscillates in such a way that its acceleration is proportional to its displacement, but with the opposite direction:

$$a \propto -x \equiv -\omega^2 x$$

For more on proportionalities, see Section 13.4.9.

where ω^2 is the constant of proportionality in this relation. This equation corresponds to a simple back-and-forth, repetitive motion or swing.

9.3.1 TYPES OF SIMPLE HARMONIC OSCILLATORS

There are two classes of simple harmonic oscillators important in introductory physics: pendulums and spring oscillators.

EXAMPLE 9.4: The Pendulum

Let's look at three different types of pendulums.

*   *   *

The easiest to understand, the **simple pendulum**, is merely a long, taut string (or light bar) hanging from a pivot. At the bottom of the string is a mass (called the **bob**). See Figure 9.2 (a). A real life example of the simple pendulum is the pendulum of a grandfather's clock.

When the bob is pulled away from its natural hanging position and released, the simple pendulum begins to swing back and forth. The resulting motion is regular and repetitive; in fact, in the case of a clock's pendulum, you can literally hear a steady "tick-tock" that marks out each constant cycle. In the ideal case (neglecting friction), the simple pendulum would continue in this even motion forever.

The angular frequency of the simple pendulum depends *only* on the length of the pendulum's string (and the gravitational field g)—and nothing else. Thus, simple pendulums are just that: simple to describe mathematically!

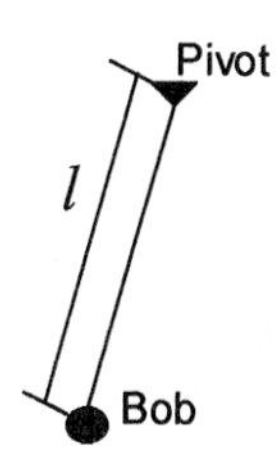

Figure 9.2 (a)

*   *   *

A more complex type of pendulum is the **physical pendulum** (Fig. 9.2 (b)). *Any* body suspended from a pivot, which moves in a back-and-forth motion, is a physical pendulum; for instance, your keys hanging from your car's ignition, or a calendar swinging by a nail in the wall, are both good examples of physical pendulums. However, the motion of these pendulums is still simple harmonic.

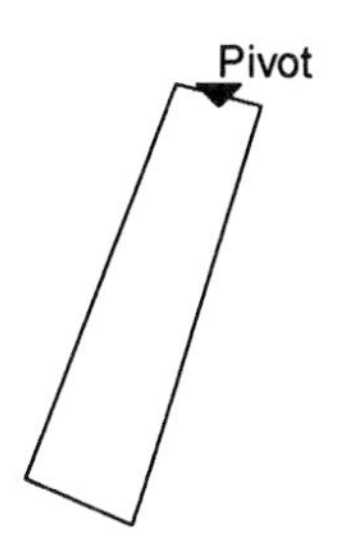

Figure 9.2 (b)

*   *   *

A third kind of pendulum is the **torsion pendulum** (Fig. 9.2 (c)). Here, a mass hangs from the end of a taut string (or light bar)—as in the case of the simple pendulum—except that the pendulum *rotates* in simple harmonic motion rather than swings. Examples of torsion pendula include a tire swing rotating around its thick center rope, and the tiny spring inside of an antique pocket watch.

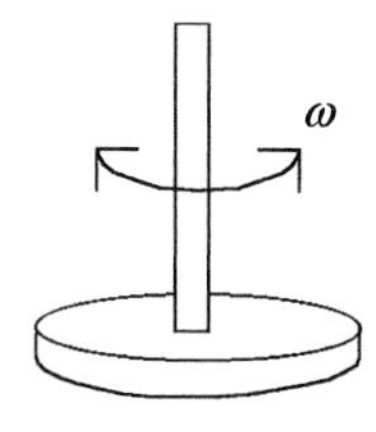

Figure 9.2 (c)

The three different types of pendulums, and their equations for angular frequency, are summarized in Table 9.1 below.

Table 9.1: Pendulums*

NAME	DESCRIPTION	EXAMPLE	EQUATION	TERMS
SIMPLE PENDULUM	A long, taut string (or light bar) with a mass at the end.	The pendulum of a grandfather's clock.	$\omega_S^2 = \frac{g}{l}$	g = Gravitational field l = Length of string or bar
PHYSICAL PENDULUM	Any body swinging from a pivot.	Car keys hanging from the ignition or a wall calendar hanging from a nail.	$\omega_P^2 = \frac{I}{mgl}$	I = Body's rotational inertia m = Mass of body g = Gravitational field l = Distance from pivot to center-of-mass
TORSION PENDULUM	Any body making small back-and-forth rotations about an axis.	A tire swing or the workings of an antique pocket watch.	$\omega_T^2 = \frac{\kappa}{I}$	κ = Body's **torsion constant** I = Body's rotational inertia

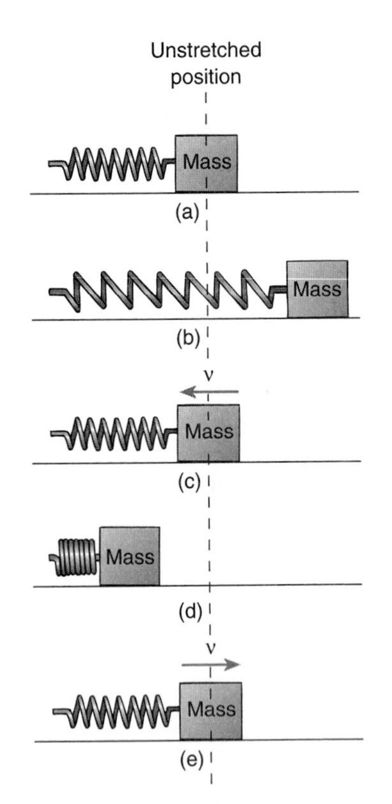

Figure 9.3

EXAMPLE 9.5: Spring Oscillator, Part I

A second important type of simple harmonic oscillator is the **spring oscillator**. This type of oscillator consists of a mass attached to a spring, which slides on a frictionless surface (Fig. 9.3 (a)).

The oscillation of this system begins when the mass is pulled away from the spring and released, so that the spring stretches (Fig. 9.3 (b)). The spring, according to Hooke's Law, opposes the stretch. So, it pulls back on the mass. The mass slides towards the spring, passing through its original location (Fig. 9.3 (c)).

But, because the mass was accelerated by the spring, it overshoots its original, unstretched location and begins to compress the spring (Fig. 9.3 (d)). This causes the spring to *push back* on the mass, moving it back the other way (Fig. 9.3 (e)). Again, the mass overshoots its original location, sliding away from the spring (and returning to the situation of Fig. 9.3 (b)).

And so the cycle continues. Since there is no friction in the system, the mass + spring will oscillate in this repetitive, regular, side-to-side (*simple harmonic*) motion forever.

(By the way, what would happen if the spring were originally *compressed* rather than stretched? The cycle would start at Figure 9.3 (d), and then proceed through the series of pictures (e)-(a)-(b)-(c), etc.)

EXERCISE 9.6: Spring Oscillator, Part II

If the mass in Example 9.5 above travels from the fully stretched position (Fig. 9.3 (b)) to the fully compressed position (Fig. 9.3 (d)) in 1.5 s, find its (a) period, (b) frequency, and (c) angular frequency.

* In each of these cases, it is assumed that the oscillatory motion is small. The equation $a = -\omega^2 x$ does not apply otherwise, and the situation becomes more complex than we care to discuss here.

Exercise 9.6 Continued...

SOLUTION:

(a) The period of an oscillation is the time it takes for the motion to complete an *entire* cycle; that is, in this case, to go from fully stretched, to fully compressed, back to fully stretched again. Thus, the period of this motion will be *twice* the time it takes to go from fully stretched to fully compressed (assuming the system does not slow due to friction or drag): that is, $T = 2\ (1.5\ \text{s}) = 3.0\ \text{s}$.

(b) Since $f = 1/T$, $f = 1/(3\,\text{s}) \approx 0.33\ \text{Hz}$.

(c) Now, since $\omega = 2\pi/T$, $\omega = 2\pi\ /(3.0\ \text{s}) \approx 2.1\ \text{rad/s}$.

EXERCISE 9.7: Spring Oscillator, Part III

Find a theoretical formula describing ω^2 for *any* mass + spring system, using the defining equation for simple harmonic oscillators, $a = -\omega^2 x$.

SOLUTION:

To find this relation, we need to place the restoring force due to the spring (Hooke's Law), $F_S = -kx$, into Newton's Second Law for the oscillator, $F = ma$:

$$F_S = -kx = ma = -m\omega^2 x$$
$$\Rightarrow k = m\omega^2$$
$$\therefore \omega^2 = \frac{k}{m}$$

Note that this equation for ω^2 *only* depends on the "strength" of the spring (given by k) and the mass m of the object—and *not* on the distance of stretch x for the spring!

Hooke's Law is covered in Section 7.2.8; Newton's Second Law is in Section 6.3.

9.3.2 DYNAMICS OF SIMPLE HARMONIC OSCILLATORS

The defining equation for the simple harmonic oscillator, $a = x'' = -\omega^2 x$, is a differential equation that can be solved for position, velocity, and acceleration as sinusoidal[#] functions of time. Your textbook or instructor will derive these for you. The final results are:

Differential equations, and their solutions, are discussed in Section 14.4.

$$x(t) = x_{\text{max}} \cos(\omega t + \phi)$$
$$v(t) = x'(t) = (-\omega x_{\text{max}})\sin(\omega t + \phi)$$
$$a(t) = x''(t) = (-\omega^2 x_{\text{max}})\cos(\omega t + \phi) = -\omega^2 x(t)$$

where ϕ is the **phase angle** of the motion, measured in radians.

[#] The term **sinusoid** is a "catch-all" name for *both* cosine and sine functions.

To see that this is true, refer to Section 13.6.2.

EXAMPLE 9.8: Examining Equations of Motion, Part I

The equations of motion for simple harmonic oscillators contain cosines and sines. These trig functions have an interesting property: when the cosine equals one (its maximum value), the sine equals zero, and when the cosine is zero, the sine is one (its maximum value).

So, in the above equations, we see that the position function (which has a cosine in it) is at a maximum when the velocity function (which contains a sine) is zero. At that point, the acceleration function (which has a cosine) also has a maximum.

Conversely, when the position is zero, the acceleration will also be zero, but the velocity will have its maximum value.

In the case of the spring oscillator, this means that at the very ends of the mass' motion (Figs. 9.3 (b) and (d))—where the position is maximum—the acceleration will have its largest magnitude, but the velocity will be zero. However, in the middle of the motion (Figs. 9.3 (c) and (e)), position and acceleration are zero, but the mass has its highest velocity.

How does this compare to the motion of a simple pendulum?

EXAMPLE 9.9: Examining Equations of Motion, Part II

The phase angle ϕ in each of the above equations is an additive factor that shifts the curve along the x-axis. The following figures illustrate this. In each case, the same sine curve has been drawn with different phases: Figure (a) has a zero phase angle; Figure (b) has a phase angle of $\frac{\pi}{2}$, Figure (c) has a phase angle of π, and Figure (d) has a phase angle of $-\frac{\pi}{2}$.

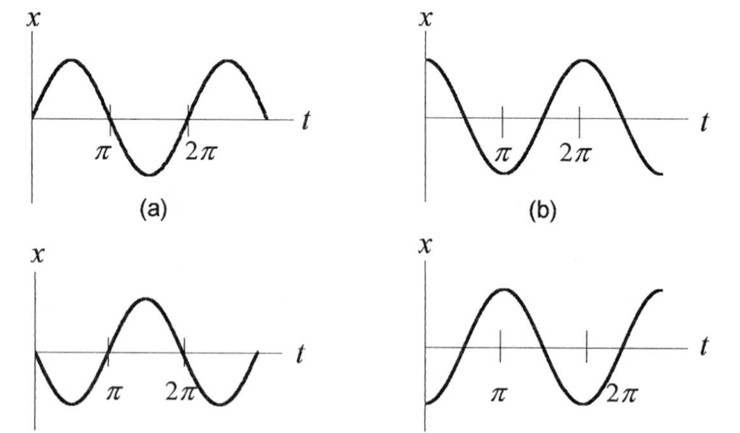

The position, velocity, and acceleration relations above can be used to calculate kinetic, potential, and total energies—as well as forces—of particular oscillators.

EXERCISE 9.10: Spring Oscillator, Part IV

Find equations for the kinetic, potential, and total energies, as well as the force, for the mass + spring system of Example 9.5. What is the force exerted by the spring 7.0 s after the mass is released? Assume the phase angle is zero radians, the mass is 1.0 kg, the maximum displacement of the mass is 5.0 cm, and the spring constant is 2.0 N/m.

SOLUTION:

The spring force is covered in Section 7.2.8; spring energy is in Section 8.3.2.

From Chapters 7 and 8, we find the appropriate formulas for kinetic, potential, and total energies—as well as the equation for force—for a spring. We simply plug into these formulas the functional relations for position, velocity, and acceleration for the spring oscillator:

$$
\begin{aligned}
K &= \tfrac{1}{2}mv^2 = \tfrac{1}{2}m[(\omega^2 x_{\max}^2)\sin^2(\omega t + \phi)] \\
U_S &= \tfrac{1}{2}kx^2 = \tfrac{1}{2}k[(x_{\max}^2)\cos^2(\omega t + \phi)] \\
E_{tot} &= K + U \\
&= \tfrac{1}{2}m[(\omega^2 x_{\max}^2)\sin^2(\omega t + \phi)] + \tfrac{1}{2}k[(x_{\max}^2)\cos^2(\omega t + \phi)] \\
&= \tfrac{1}{2}kx_{\max}^2[\sin^2(\omega t + \phi) + \cos^2(\omega t + \phi)] \\
&= \tfrac{1}{2}kx_{\max}^2 \\
F &= -kx = -k(x_{\max})\cos(\omega t + \phi)
\end{aligned}
$$

Now, to find the specific value for the force of the spring after seven seconds, we must plug in our known values into the above force equation. We know from Example 9.3 that the angular frequency of this system is 2.1 rad/s. We further know from the statement of the problem the values of t, ϕ, m, x_{max} and k. Thus, the force exerted by the spring is:

$$
\begin{aligned}
F &= -kx_{\max}\cos(\omega t + \phi) \\
&= -(2.0\ \mathrm{N \cdot m})(0.050\ \mathrm{m})\cos\left[(2.1\ \mathrm{rad/s})(7.0\ \mathrm{s}) + 0\right] \\
&= 5.3 \times 10^{-2}\ \mathrm{N}
\end{aligned}
$$

9.4 SINUSOIDAL WAVES

Waves come in many shapes and sizes, from the simple sinusoidal pattern of a sine wave, to the complex waveform of a heart beat (as seen on a heart monitor).

Despite these many types of waveforms, however, your physics class will almost certainly limit the discussion to *sinusoidal* forms only. There are two reasons for this: first, sinusoidal waves are the easiest to describe; and second, (and most importantly), we shall soon see that *all* wave shapes can be created using just sinusoids!

Eliminating the Unessential in Section 2.3.

9.4.1 WAVE PROPERTIES

In this section, we discuss basic wave properties. The most fundamental wave property is that waves move, or **propagate**, in space *and* time.

EXAMPLE 9.11: Waves on a Wet Towel

Imagine a wet towel held horizontally at one end. A wave is created on the towel by a flick of a wrist. Because of the wave, each segment of the towel then moves up and down (in *space*), even as the wave travels down the towel's length (as *time* passes).

Mathematically, we write the y position of a segment of a sinusoidal wave as a function of position x and time t:*

$$y = y_{\text{max}} \sin(kx - \omega t)$$

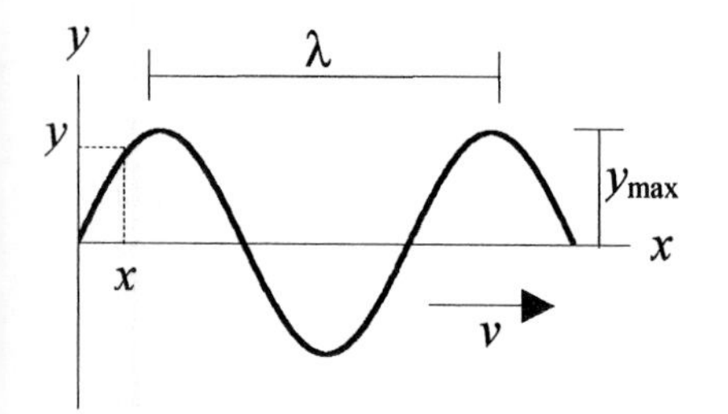

Figure 9.4

where y_{max} is the wave's maximum height, or **amplitude**, and k is the **angular wave number**, with units of rad/m.# The wave number is related to the length of one wave cycle (measured from crest to crest), or the **wavelength** λ, and to the angular frequency ω:

$$k = \frac{2\pi}{\lambda} = \frac{\omega}{v}$$

where v is the speed of wave propagation. See Figure 9.4.

EXERCISE 9.12: A Sinusoidal Wave

One sinusoidal wave is described by the function $y = 5\sin(2x + 3t)$, in appropriate SI units. Find the (a) amplitude, (b) frequency, and (c) wavelength of this wave.

SOLUTION:

(a) To solve this problem, we match the function $y = 5\sin(2x + 3t)$ to our defining equation $y = y_{\text{max}} \sin(kx - \omega t)$. From this, we see that the amplitude y_{max} of this function is 5 m.

(b) Now, the defining equation $y = y_{\text{max}} \sin(kx - \omega t)$ does not have the frequency f in it, but it does have angular frequency ω, as we can see by matching the equations. Thus, $\omega = 3$ rad/s. Since we know that $f = \omega / 2\pi = (3 \text{ rad/s}) / 2\pi \approx 0.5$ Hz.

* Be careful! Waves move in *all* directions. Therefore, the x and y notation in this formula does not necessarily mean the "left-right" and "up-down" directions sketched in Figure 9.2. As you first learn about waves, though, feel free to assume that waves *do* move as sketched; but later, begin to think more generally: let x and y come to mean *any* two perpendicular directions.

Some texts use a cosine in this formula rather than a sine. The only difference is a phase change of $\pi / 2$, so the different notation should not shake you.

Exercise 9.12 Continued...

(c) By matching the function with the wave equation, we know that $k = 2\ \text{rad/m}$. We also know $k = 2\pi/\lambda$, so:

$$\lambda = \frac{2\pi}{k} = \frac{2\pi}{2} = \pi\ \text{m} \approx 3\ \text{m}$$

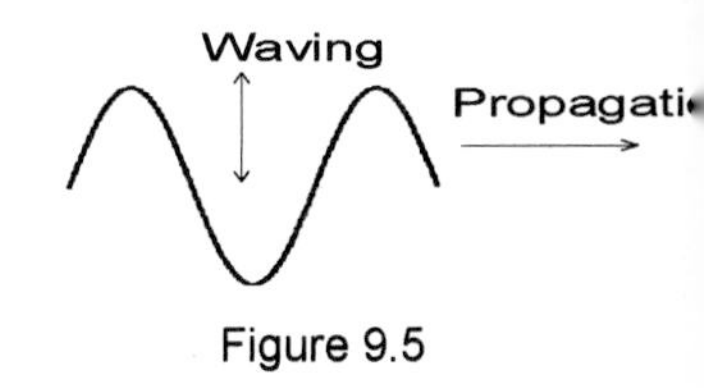

Figure 9.5

Waves carry energy from the wave source to the wave's end. The rate of energy transfer per area for a wave is the **intensity** I (unit = W/m^2).

EXAMPLE 9.13: Why a Wet Towel Burns

Why do wet towel burns—the kind you get when some evildoer flicks a long wet towel at you—hurt? Because the wave created by the flicked wrist passes energy down the towel to you.

How does this happen? At the beginning of the motion, the evildoer's wrist imparts kinetic energy to the handheld tip of the towel. This kinetic energy moves along the length of the towel, carried by the wave, until it reaches the end of the towel (and hits you).

Thus, energy is *transferred* from the handheld end of the towel to your rear *by the wave*. And you probably have the bruise to prove it!

There are two types of waves: **transverse** and **longitudinal**.

9.4.2 TRANSVERSE WAVES

Transverse waves propagate through space in a direction *perpendicular* to the waving of the wave (Fig. 9.5). *Sinusoidal* transverse waves can be created by simple harmonic oscillations.

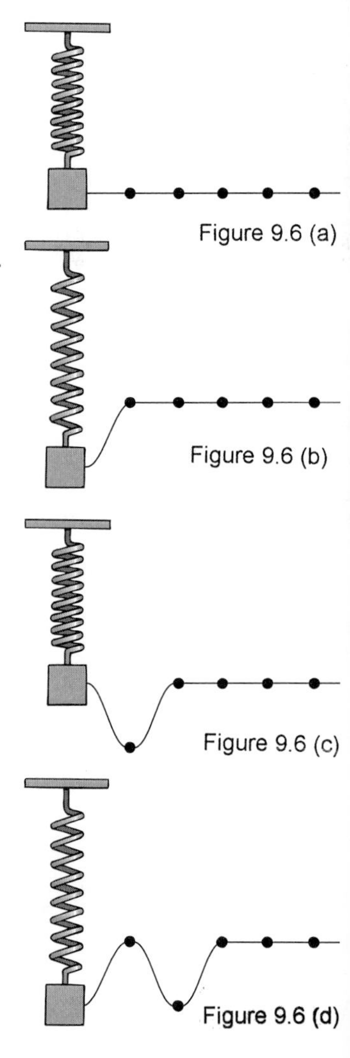
Figure 9.6 (a)
Figure 9.6 (b)
Figure 9.6 (c)
Figure 9.6 (d)

EXAMPLE 9.14: Making the Connection, Part I

To see that this is true, imagine that a mass + spring oscillator hangs vertically from a ceiling. Further imagine that a rope has been attached to the mass and held horizontally. The rope itself has been specially prepared: small beads have been glued to it at regular locations along its length. See Figure 9.6 (a).

If the mass is pulled downwards and released, it begins a vertical simple harmonic oscillation. The attached rope moves downwards with it (Fig. 9.6 (b)). When the mass moves back up, so will the rope (Fig. 9.6 (c)). The motion keeps going as long as the oscillator oscillates (Fig. 9.6 (d)). When we add up the motion of all the beads, a sinusoidal pattern appears. Thus, the simple harmonic motion of the mass + spring oscillator *created* the transverse sinusoidal wave in the rope. Simple harmonic motion and sinusoidal waves are connected!

There are many important examples of transverse waves, including waves on strings and the "Wave" seen in the stadium at football games!

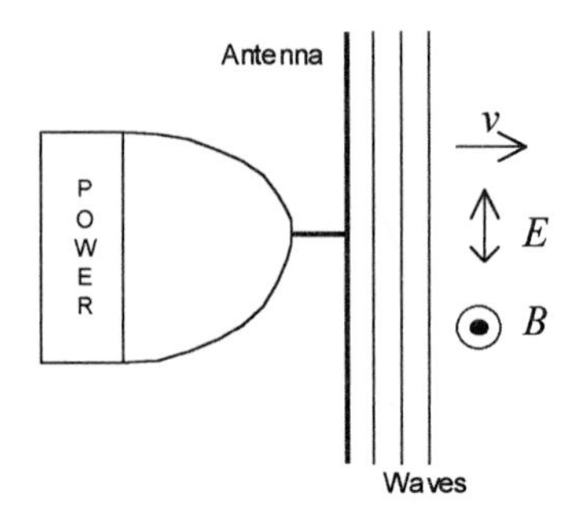

Figure 9.7

Figure 9.8

NOTE 9.15: Electromagnetic Waves

Another very important example of transverse waves are electromagnetic (EM) waves. These waves are made of oscillating electric and magnetic fields oriented perpendicularly to each other, propagating in the third perpendicular direction.

EM waves are generated by *antennas*, which, in their simplest form, are nothing more than long straight pieces of wire connected by a circuit to an A.C. power source (Fig. 9.7). The power source causes electrons in the wire to move up and down in simple harmonic motion.

As the electrons move up and down in the wire, they create electric and magnetic fields. Each position of the electrons in the wire produces a different set of fields, but their sum total are sinusoidal EM waves.

Depending upon the frequency of electron oscillation, different EM waves are created: high frequencies create x-rays and gamma rays; low frequencies make radio and TV waves. And at just the right range of frequencies (about 9×10^{15} Hz), *light* waves are created.

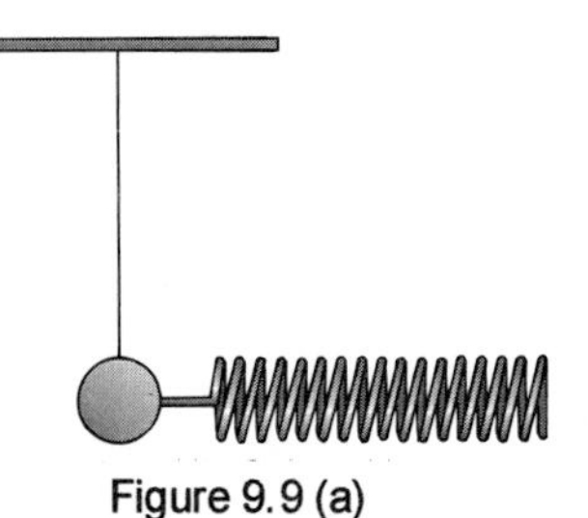

Figure 9.9 (a)

9.4.3 LONGITUDINAL WAVES

Longitudinal waves propagate in a direction *parallel* to the direction of waving (Fig. 9.8). *Sinusoidal* longitudinal waves can be created by simple harmonic oscillations.

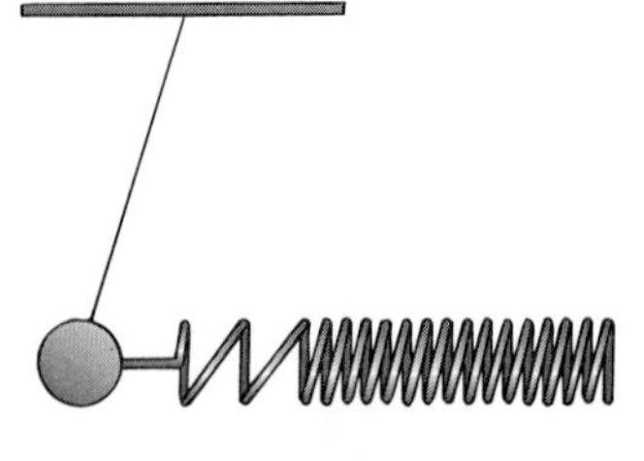

Figure 9.9 (b)

EXAMPLE 9.16: Making the Connection, Part II

To see that this is true, imagine that a pendulum hangs vertically from a ceiling, as in Figure 9.9 (a). Further imagine that a slinky has been attached to the bob and held horizontally.

If the pendulum is pulled to the left and released (Fig. 9.9 (b)), it will begin a horizontal simple harmonic motion. The end of the attached slinky will also move left, leaving a stretched region known as a *rarefaction.* As the pendulum then swings right, it contracts the slinky, creating a *compression* (Fig. 9.9 (c)). The alternation between compressions and rarefactions will continue for as long as the pendulum swings.

The changing coil density of the slinky can be described by a sinusoid, just like the y position of the rope could in Example 9.14 above. Thus, simple harmonic motion and longitudinal sinusoidal waves are connected, too!

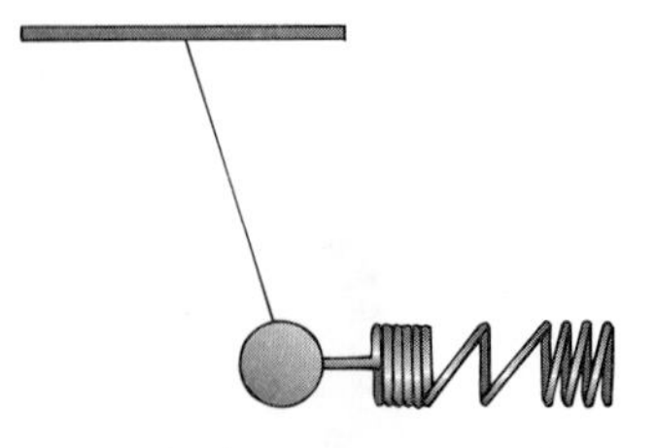

Figure 9.9 (c)

There are many important examples of longitudinal waves, including shock waves, sonar waves, and some kinds of earthquake waves.

NOTE 9.17: Sound Waves

Another very important example of longitudinal waves are sound waves. These waves are made of alternating compressions and rarefactions of air.

Sound waves are generated when a sounding device, such as a stereo speaker, causes the air molecules near it to oscillate. When the speaker moves forward, a compression is formed in front of the cone; when it moves backward, a rarefaction is formed. The moving molecules then bump into molecules that are a little farther away from the speaker, causing *them,* in turn, to oscillate.

The process continues, moving the wave further and further away from the original source. See Figure 9.10. Taken together, the compressions and rarefactions caused by the speaker make longitudinal pressure waves in air.

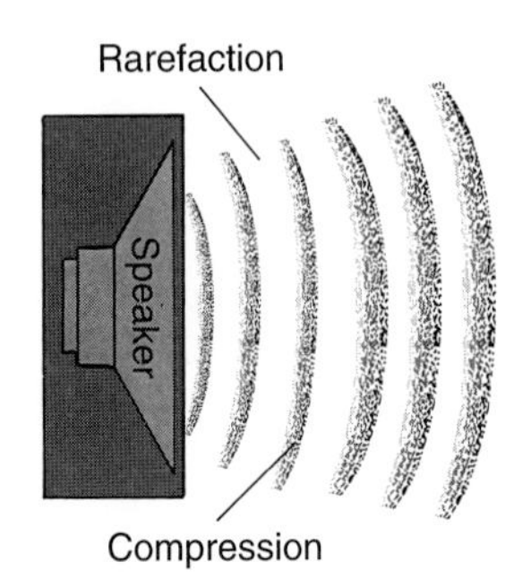

Figure 9.10

9.5 WAVE PHENOMENA

This section introduces you to interesting phenomena associated with waves. These phenomena occur for *all* types of waves—light or sound; sinusoidal or otherwise—but, for clarity, we will limit our discussion to the simpler kinds of sinusoidal waves that we have been discussing.

9.5.1 THE DOPPLER EFFECT

Whenever an observer moves relative to a source of waves, *or* the source moves relative to an observer, the frequency of the wave emitted by the source appears *different* to the observer than it does to the source. This is known as the **Doppler effect**.

EXAMPLE 9.18: The Doppler Effect and the Police

A very good example of the Doppler effect can be heard in the siren of a police car. As the car approaches you, it seems that the siren has a high pitch. Then, as it moves away, the siren seems to have a low pitch. Only when the police car is right beside you (and moving at the same speed as you) do you hear the actual (medium) tone of its siren.

DEFINITION 9.19: Other Doppler Effects

In the Doppler effect, waves bunch up or get spread out when the wave source and the observer move relative to each other. As a result, the observer detects a different wave pattern than was emitted by the source. As we saw in the police siren example above, sound waves are a good example of this: you can actually *hear* the Doppler effect.

However, it is not the only example. Another example is the **sonic boom**: when a source of sound moves faster than the speed of sound, sound waves "bunch up" against the object, causing outside observers to hear the "bunching" as a *sonic boom*.

A similar thing happens when electrons, traveling in a medium other than vacuum, move faster than the speed of light for that medium. The waves in this case are light waves, and the "bunching" is *seen* as **Cerenkov radiation**.

For more on relative motion and frames of reference, see Section 5.3.3.

The Doppler effect occurs because waves do not travel at infinite speeds; rather, it takes time for waves to move through space. Therefore, the *net* speed of a wave as *detected* by the observer must take into account the original speed of the wave and the speed of the observer. Furthermore, we must take into account that the speed of the source changes the wavelength of the wave. Mathematically, we write this as:

$$f_{observer} = f_{source}\left(\frac{v_{wave} \pm v_{observer}}{v_{wave} \mp v_{source}}\right)$$

where you choose the top signs in the formula (*i.e.,* the + in the numerator and the - in the denominator) if the source is moving *toward* the observer; otherwise, you choose the bottom ones.

EXERCISE 9.20: A Police Chase

A speeder, traveling at 145 kph (90 mph), approaches (from behind) a cop traveling at 95 kph (60 mph) with his siren on. The speeder recognizes the cop car and slows to the same speed, just as he moves alongside the cop car. At that moment, though, an even *faster* car whizzes by. The cop forgets the first speeder, and chases the new one.

Assuming that a standard police siren has a frequency of 500 Hz, and that the speed of sound in air is 330 m/s, what frequency does the first speeder hear:

(a) As he approaches the cop from behind?
(b) When both the speeder and the cop are traveling at 95 kph?
(c) When the cop accelerates to 145 kph?

SOLUTION:

(a) As he approaches the cop from behind, the speeder has velocity $v_{observer} = 145$ kph, and the cop has velocity $v_{source} = 95$ kph. The speed of sound in air is $v_{wave} = 330$ m/s $= 1190$ kph, and the siren's frequency is $f_{source} = 500$ Hz. So:

Exercise 9.20 Continued...

$$f_{\text{observer}} = f_{\text{source}}\left(\frac{v_{\text{wave}} \pm v_{\text{observer}}}{v_{\text{wave}} \mp v_{\text{source}}}\right) = (500\text{ Hz})\left(\frac{1190\text{ kph} + 145\text{ kph}}{1190\text{ kph} - 95\text{ kph}}\right) = 670\text{ Hz}$$

Notice that we used the top sign in the formula, since the cop is moving *toward* the speeder. Also note that the frequency the speeder hears is *higher* than the frequency that the cop hears. This is because the cop and speeder are approaching each other.

(b) In this case, both the source and the observer are moving at the *same speed*, so:

$$f_{\text{observer}} = (500\text{ Hz})\left(\frac{1190\text{ kph} - 145\text{ kph}}{1190\text{ kph} - 145\text{ kph}}\right) = 500\text{ Hz}$$

They hear the same sound because there is no *relative* motion between the cars.

(c) Now the cop races ahead at 145 kph:

$$f_{\text{observer}} = (500\text{ Hz})\left(\frac{1190\text{ kph} - 95\text{ kph}}{1190\text{ kph} + 145\text{ kph}}\right) = 410\text{ Hz}$$

In this case, the observer hears a *lower* frequency siren.

Just as a stone thrown into water becomes the cause and center of various circles, sound spreads in circles in the air.
LEONARDO DA VINCI
Italian Experimentalist, 1452 - 1591

Relative motion is in Section 5.3.3.

9.5.2 SUPERPOSITION

What happens when more than one wave travels through the same region of space at the same time? A *new* wave results: a wave that is the combination or **superposition** of the originals.

There are two outcomes of superposition. **Constructive interference** occurs when the original waves *add* together so the new wave is *bigger* than the originals; **destructive interference** occurs when the original waves *cancel out* so the new wave is *smaller* than the originals.

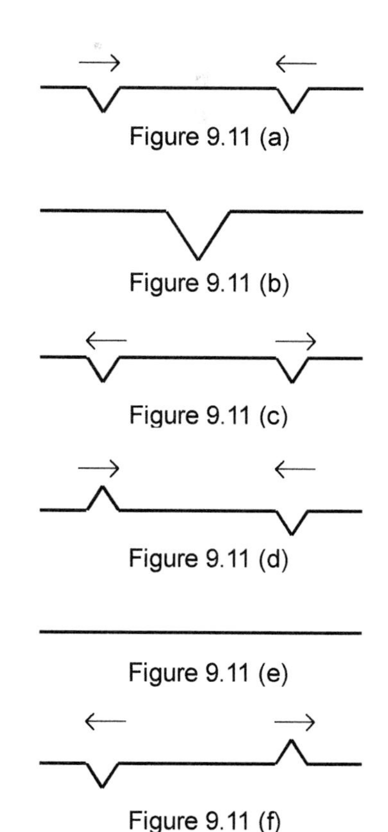

Figure 9.11 (a)

Figure 9.11 (b)

Figure 9.11 (c)

Figure 9.11 (d)

Figure 9.11 (e)

Figure 9.11 (f)

EXAMPLE 9.21: Superposition

To see constructive interference, examine Figures 9.11 (a) through (c). In Figure (a), two triangular wave pulses travel toward each other on a rope. When the pulses meet (Fig. (b)), superposition creates a wave with greater amplitude than the originals had. But, the waves *continue to move*, so, the original waves return, albeit in reverse positions (Fig. (c)).

Destructive interference is seen in Figures 9.11 (d) through (f). Here, the pulses travel towards each other (Fig. (d)), but since one pulse is directed upwards and the other downwards, when they meet (Fig. (e)), they cancel each other out! However, just like in the case above, the original waves will keep moving and eventually reappear in opposite locations (Fig. (f)).

Mathematically, we write superposition as a sum of the original waves:

$$y_{\text{new}}(x,t) = y_1(x,t) + y_2(x,t) + \cdots$$

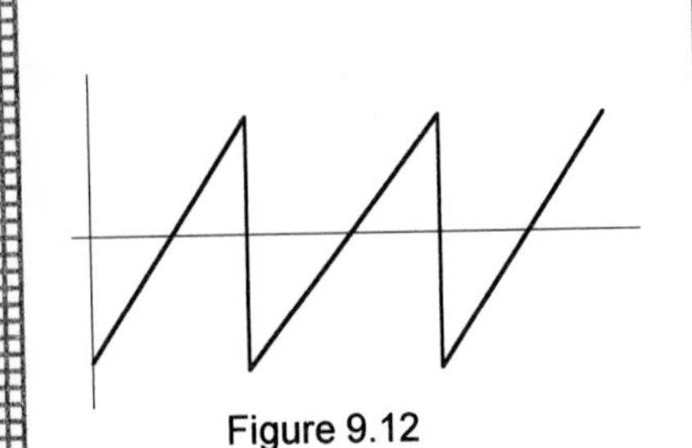
Figure 9.12

NOTE 9.22: Fourier Synthesis

One interesting effect of superposition is the ability to recreate *any* wave, no matter how complex or strange-looking it may seem, simply by superimposing a series of *sinusoidal* waves. This is known as **Fourier synthesis**.

For example, the saw-tooth wave in Figure 9.12 can be recreated by adding together the following (infinite) series of sine waves:

$$y(t) = -\frac{1}{\pi}\sin\omega t - \frac{1}{2\pi}\sin 2\omega t - \frac{1}{3\pi}\sin 3\omega t - \ldots$$

In practice, the superposition sum can be quite difficult to do. So, in introductory physics classes, we often limit ourselves to relatively simple cases, like the three important examples below.

BEATS

Beats occur when two waves, identical except for slightly different frequencies, superimpose each other (Fig. 9.13). The resultant wave has a frequency equal to the average of the two component frequencies and an amplitude which varies in time.

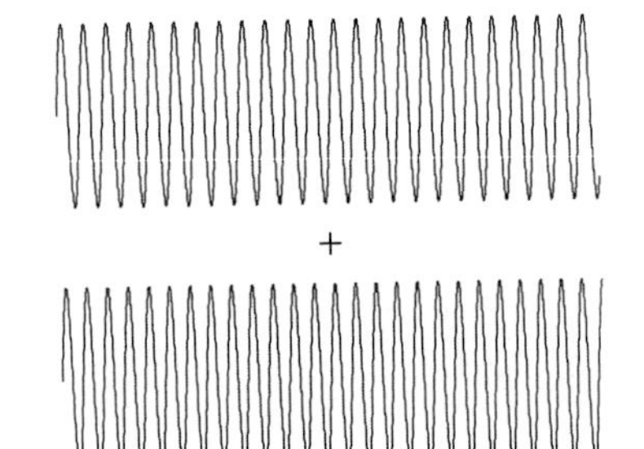

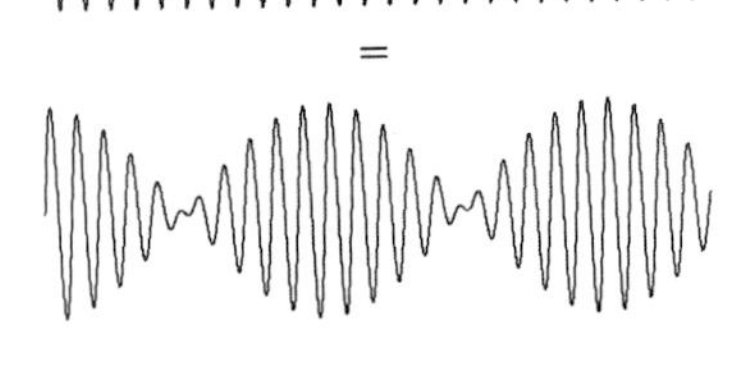
Figure 9.13

EXAMPLE 9.23: Beats

In the case of sound waves, you can actually *hear* beats. This is, in fact, how piano tuners tune pianos: when a note is struck on a piano at the same time as a tuning fork of similar frequency is rung, beats begin to sound. As the piano gets closer to the correct tone of the fork, the frequency of the beats gets progressively lower. The piano is tuned when the beats disappear completely.

Mathematically, we write the superposition (of the time dependent portions only) of the waves as:

$$\begin{aligned} y_{\text{new}} &= y_1 + y_2 \\ &= [y_{\max}\sin(\omega_1 t)] + [y_{\max}\sin(\omega_2 t)] \\ &= \{2y_{\max}\cos[\tfrac{1}{2}(\omega_1 - \omega_2)t]\}\sin[\tfrac{1}{2}(\omega_1 + \omega_2)t] \end{aligned}$$

where we have skipped some algebra that will be covered in more detail in your textbook.

A close look at this rather ugly equation reveals two distinct parts: a sinusoidal term, $\sin[\frac{1}{2}(\omega_1 + \omega_2)t]$, which has a high frequency equal to the average of the two original frequencies ($\frac{1}{2}(\omega_1 + \omega_2)$), and a beat "envelope" $2y_{max}\cos[\frac{1}{2}(\omega_1 - \omega_2)t]$, which oscillates at a much lower frequency equal to ($\frac{1}{2}(\omega_1 - \omega_2)$).

STANDING WAVES

Standing waves occur whenever two identical waves travel in opposite directions, but meet in the same location. The standing wave does not propagate; hence, it seems to *stand* in place.

Standing waves—which can be transverse or longitudinal—are made in two ways. We can visualize these methods by using waves on a rope:

- If a rope is attached at *both* ends to identical oscillators, identical waves will be created on the ends of the rope (Fig. 9.14 (a)). When the waves meet in the middle, a standing wave is created.
- If a rope is attached to an oscillator at one end, and a barrier at the other, waves generated at the oscillator will move down the rope, hit the barrier, and reflect back (Fig. 9.14 (b)). The reflected waves will then meet the original waves (still being generated), creating a standing wave.

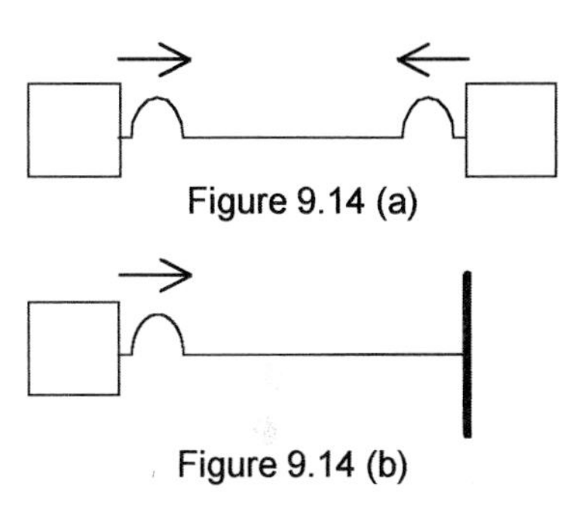
Figure 9.14 (a)

Figure 9.14 (b)

EXAMPLE 9.24: A Jumping Rope, Part I

Two rope holders (the oscillators) can generate a standing wave on a jumping rope (Fig. 9.15). To do so, each holder moves her arm in a large circle, creating a wave that is identical to the wave created by her partner. These two waves, moving in opposite directions, meet in the middle of the rope, creating one standing wave.

Figure 9.15

The expression for a standing wave is found via a superposition sum:

$$\begin{aligned} y_{new} &= y_1 + y_2 \\ &= [y_{max}\sin(kx - \omega t)] + [y_{max}\sin(kx - \omega t)] \\ &= [2y_{max}\sin kx]\cos\omega t \end{aligned}$$

where, again, we have skipped some algebra that will be in your textbook.

Notice that this wave does *not* propagate, because it does not have the form $y = y_{max}\sin(kx - \omega t)$. Instead, it is a stationary wave that has a varying amplitude equal to $2y_{max}\sin kx$.

At certain values of x, this amplitude equals zero. These points, called **nodes**, are places where the rope doesn't move at all (Fig. 9.16). There is always at least one node in a standing wave—the one created at the oscillator—other nodes may be at other oscillators or fixed barriers. Depending upon the rope's frequency of oscillation, there may also be other nodes. Figure 9.16, for instance, has four nodes.

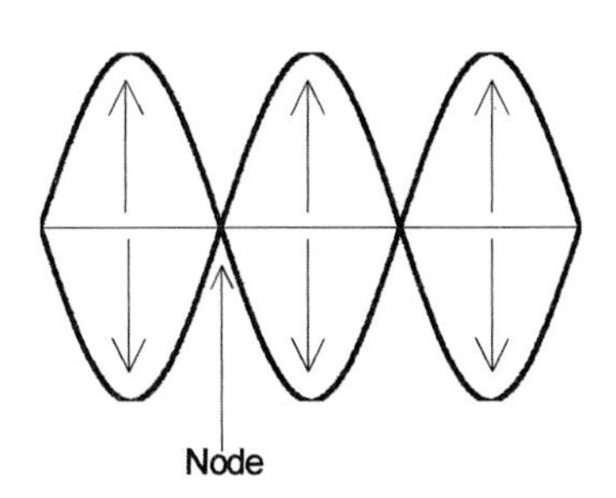

Figure 9.16

> ### EXAMPLE 9.25: A Jumping Rope, Part II
>
> A standard jumping rope has two nodes, one at each rope holder.

> ### EXAMPLE 9.26: Musical Sound
>
> Music results from standing waves created in instruments. For example, guitar strings, when strummed, set up transverse standing waves. Each note on the guitar is formed by a different pattern of standing waves. You change notes by moving your finger to different frets, *or* by strumming at different speeds, thereby setting up various standing wave patterns in the strings.
>
> Wind instruments, like oboes or trumpets, work in just the same way, only with longitudinal air pressure waves. Tone holes or valves effectively alter the pipe length, thereby moving node locations, which changes notes.

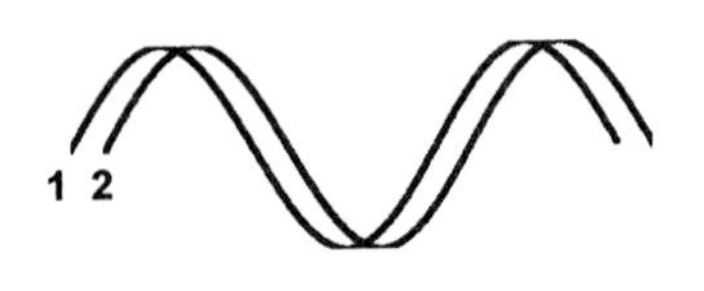

Figure 9.17

INTERFERENCE AND DIFFRACTION*

Two interesting phenomena arise when identical waves—except for their phase—are superimposed (Fig. 9.17).

Interference occurs when the two identical waves are generated separately, and then allowed to mix. The result is a pattern of alternating constructive and destructive interference "rays" spreading out from the sources' location (Fig. 9.18).

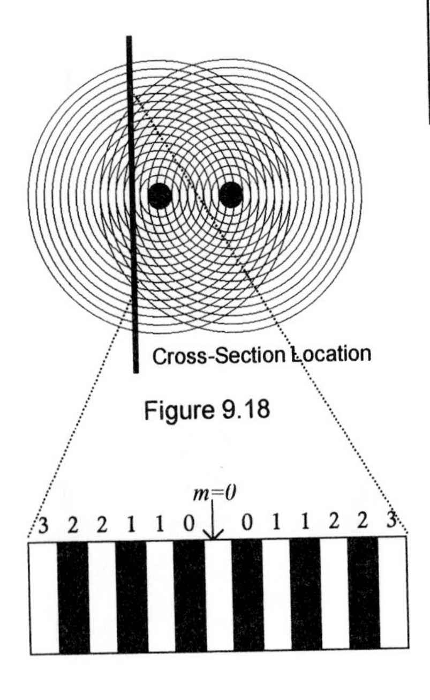

Figure 9.18

Figure 9.19

A cross-sectional cutaway of the "rays" reveals a pattern of equally spaced, equally sized light and dark bands, corresponding to constructive and destructive interference (Fig. 9.19), respectively. The angular position of each dark interference band is given by the formula:

$$\sin\theta = (m + \tfrac{1}{2})\frac{\lambda}{d}$$

where λ is the wavelength of (identical) source waves, m is an integer that numbers the bands from the center outwards (Fig. 9.19), and θ and d are geometrical parameters as shown in Figure 9.20.

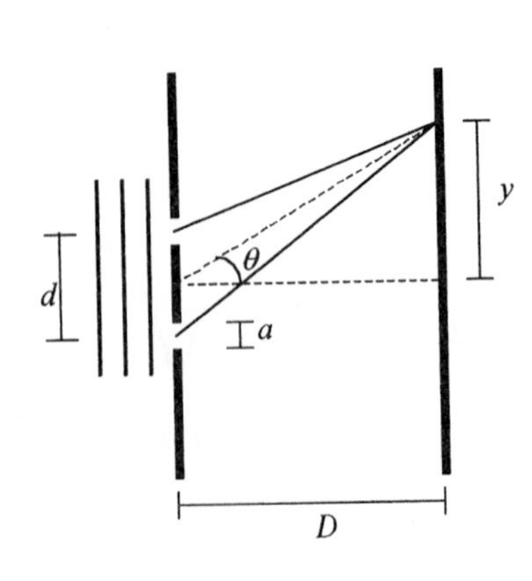

Figure 9.20

Diffraction appears when different portions of the *same* wave are allowed to mix. The result is a pattern similar to that created by interference, *except* that the bands are not equally spaced or sized; rather, they are larger and closer together nearer the center (Fig. 9.21). The angular position of each dark diffraction band is given by the formula:

$$\sin\theta = \frac{m\lambda}{a}$$

* The phenomena discussed in this section only occur when the waves propagate in more than one dimension, and are **coherent**: that is, when they *begin* in phase. Only when the waves get out of phase do interference and diffraction effects arise. Waves become out of phase when they travel different distances to get to the same final location.

where all parameters are the same as above, *except* for a, which is the width of the aperture through which the wave passes (refer to Fig 9.20).

In each case (interference and diffraction), the angular position θ can always be converted to a linear distance by $y = D\tan\theta$ (Fig. 9.20).

When diffraction occurs in *each* aperture of the two needed for interference, the result is known as **Young's Experiment**. We see a series of equally spaced and -sized bands (created by interference) inside another series of unequally spaced and -sized bands (created by diffraction) (Fig. 9.22).

The more apertures that are added to Young's Experiment, the sharper and more spaced apart each band becomes. This is desirable, since physicists choose to study each **diffraction line** separately.

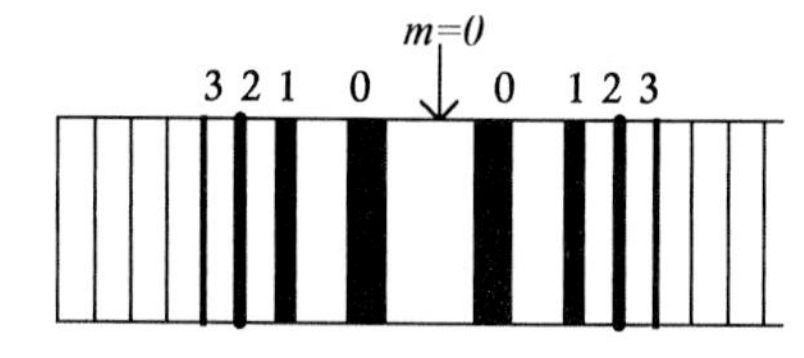

Figure 9.21

Figure 9.22

EXERCISE 9.27: Young's Experiment

In a Young's Experiment, light of wavelength 480 nm is shown through two slights on a screen 52 cm away. The slit separation is 0.12 mm, and each slit has a width of 0.025 mm.

(a) What is the angular position of the first dark interference band?

(b) What is the angular position of the third dark diffraction band?

(c) Where on the screen, above the location of the center band, is the dark band from part (b)?

SOLUTION:

(a) To find the angle of a dark interference band, we use the equation:

$$\sin\theta = (m+\tfrac{1}{2})\frac{\lambda}{d}$$

$$\Rightarrow \theta = \sin^{-1}\left[(m+\tfrac{1}{2})\frac{\lambda}{d}\right]$$

Now, the first dark interference band, according to Figure 9.18 know that $\lambda = 4.80\times10^{-7}$ m , and $d = 1.2\times10^{-4}$ m. Thus:

$$\theta = \sin^{-1}\left[(m+\tfrac{1}{2})\frac{\lambda}{d}\right] = \sin^{-1}\left[(0+\tfrac{1}{2})\frac{4.8\times10^{-7}\text{ m}}{1.2\times10^{-4}\text{ m}}\right] = 0.0020\,\text{rads}$$

(b) To find the angle of a dark diffraction band, we use the equation:

$$\sin\theta = \frac{m\lambda}{a}$$

$$\Rightarrow \theta = \sin^{-1}\left[\frac{m\lambda}{a}\right]$$

Now, the third dark diffraction band, according to Figure 9.20, has $m = 2$. Furthermore, from the statement of the problem above, we know that $a = 2.5\times10^{-5}$ m. Thus:

Exercise 9.27 Continued...

$$\theta = \sin^{-1}\left[\frac{m\lambda}{a}\right] = \sin^{-1}\left[\frac{(2)(4.8\times10^{-7}\text{ m})}{(2.5\times10^{-5}\text{m})}\right] = 0.038\,\text{rads}$$

(c) We know that the linear distance, on the screen, is given by the equation $y = D\tan\theta$ from above, where $D = 0.52$ m. Thus:

$$y = D\tan\theta = (0.52\text{ m})\tan(0.038\text{ rads}) = 0.020\text{ m}$$

9.5.3 REFLECTION & TRANSMISSION

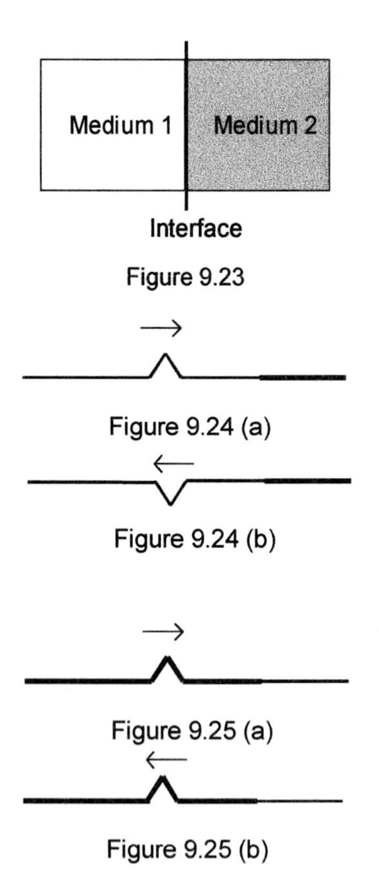

Figure 9.23

Figure 9.24 (a)

Figure 9.24 (b)

Figure 9.25 (a)

Figure 9.25 (b)

Whenever a wave passes from one medium to another—through an **interface** (Fig. 9.23)—part of the wave is **reflected** *from* the interface, and part of the wave is **transmitted** *through* it. The percentage of the wave that is reflected (compared to the percentage that is transmitted) depends greatly on the type of wave and the nature of the two media.

REFLECTION

The effects of reflection depend on the relative densities of the initial and final media through which the wave travels. We can visualize these effects with waves on a rope:*

- If the second medium is *denser* than the first, the reflected wave will have a phase change of π rads compared to the initial wave. For example, in Figures 9.24 (a) and (b), a wave on a rope approaches a denser region and is reflected upside down.
- If the second medium is *less dense* than the first, the reflected wave will make no phase change. For example, in Figures 9.25 (a) and (b), a wave on a rope approaches a less dense region and is reflected right side up.

TRANSMISSION

If some of the wave is transmitted into the second medium, the wave passes into the second medium, and is changed. We can visualize the effects of transmission with waves on a rope:#

* For clarity, the transmitted wave is not shown in these diagrams.

For clarity, the reflected wave is not shown in these diagrams.

- If the second medium is *denser* than the first, the transmitted wave will have a shorter wavelength and a slower speed than it did in the first. For example, in Figures 9.26 (a) and (b), a wave on a rope is absorbed by a denser region and becomes a smaller, slower wave.
- If the second medium is *less dense* than the first, the transmitted wave will have a longer wavelength and a faster speed than it did in the first medium. For example, in Figures 9.27 (a) and (b), a wave on a rope is absorbed by a less dense region and becomes a faster, longer wave.

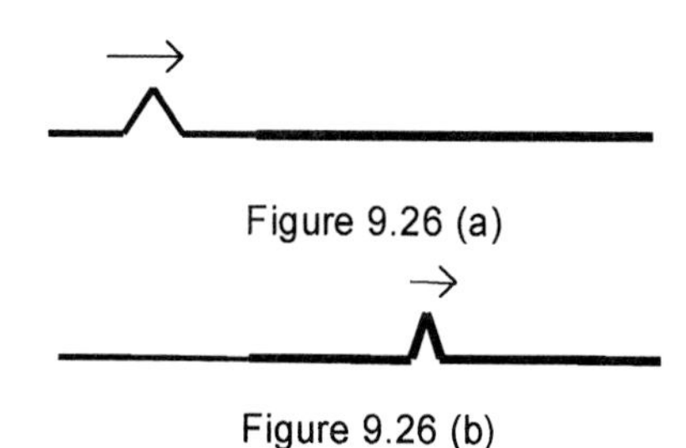

Figure 9.26 (a)

Figure 9.26 (b)

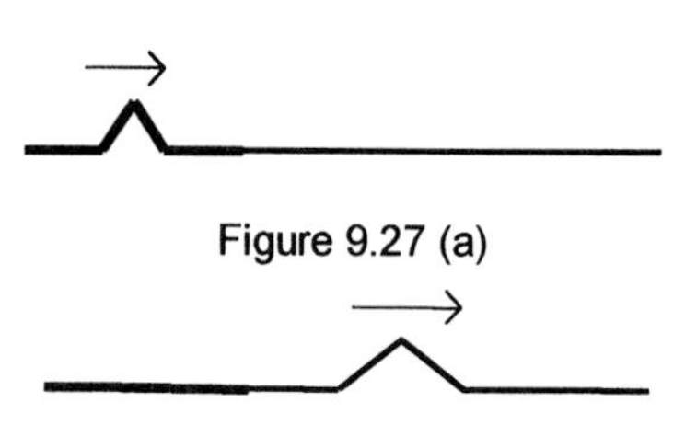

Figure 9.27 (a)

Figure 9.27 (b)

Two- or three-dimensional waves *bend* when they enter a new medium, as a result of their changing speed. This is known as **refraction**.

EXAMPLE 9.28: Light Waves

Light waves (in the regime of geometric optics, when the wavelength of the light is much less than the size of the reflecting object) reflect and refract in particular ways. The **law of reflection** says that a light wave, incident at an angle upon an interface, will reflect from the interface at the same angle at which it came in (Fig. 9.28 (a)):

$$\theta_{\text{inc}} = \theta_{\text{refl}}$$

The **law of refraction**, or **Snell's Law**, tells us that an absorbed light wave will *bend* in the second medium (Fig. 9.28 (b)) according to:

$$n_1 \sin \theta_{\text{inc}} = n_2 \sin \theta_{\text{refr}}$$

where n is the **index of refraction**, a number that you look up for each material in tables. Your textbook should have a list of these.

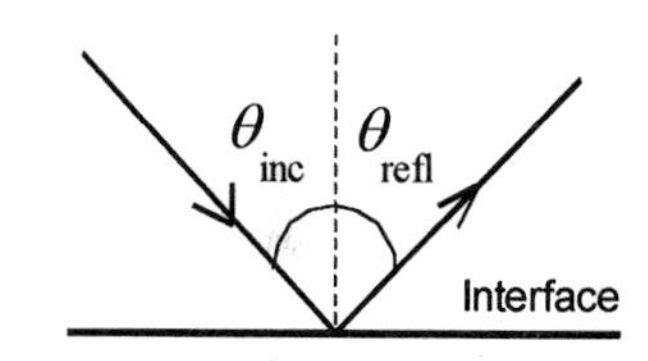

Figure 9.28 (a)

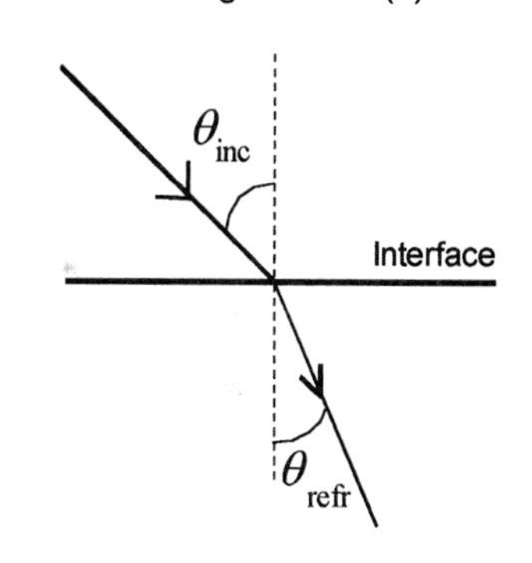

Figure 9.28 (b)

9.5.4 DAMPING & RESONANCE

Although we have been assuming otherwise up until now, in real life, waves do not propagate forever. In time, because of friction and drag in the environment, wave amplitudes and frequencies begin to decrease, and eventually, the waves fade away. This is known as **damping** (Fig. 9.29 (a)).

If we wish to forestall damping, we must *supply* the wave with more energy, **forcing** the wave to continue its original motion. The forcing energy must oscillate *with* the wave in order for this to work. If, however, the forcing oscillator has *exactly* the same frequency as the frequency of the original, undamped wave, the wave's amplitude and frequency will *increase* uncontrollably. This is known as **resonance** (Fig. 9.29 (b)).

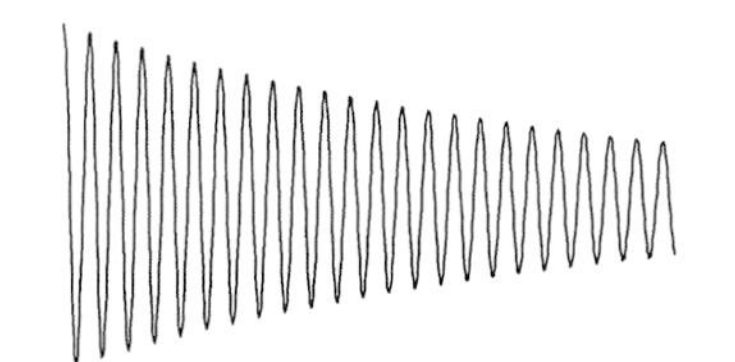

Figure 9.29 (a)

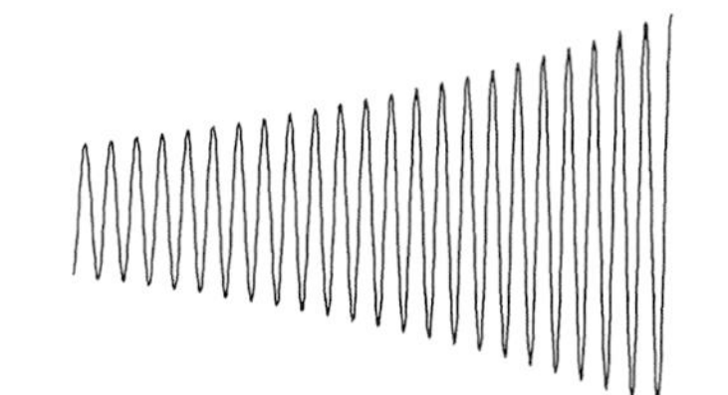

Figure 9.29 (b)

9.6 CONCLUSION

The goal is to understand the plan of creation, period.
JOHN WHEELER
American Physicist, 1911 -

This chapter introduced the concept of periodic motion, as illustrated in simple harmonic oscillations and waves. We learned how these motions are related to each other, and we took a closer look at some interesting wave phenomena.

And so, we now leave our discussion of motion, turning now to the last topic of this part: materials and their deformation.

10
MATERIALS & THERMODYNAMICS

10.1 INTRODUCTION

In Chapters 6 through 9, we learned many important theories of motion, from Newton's Laws, to energy and momentum, to waves and oscillations. Most of these theories were developed without considering the shape or makeup of the moving bodies.

We initially ignored these issues (Eliminating the Unessential) because the *fundamentals* of motion do not depend on them. However, for a more detailed and accurate picture of reality, we must examine the form and composition of bodies more closely. In this chapter we do.

Eliminating the Unessential is in Section 2.3.

EXAMPLE 10.1: Tires

Consider the tires on your car. First, their shape. For many problems, it is both legitimate and desirable to consider your tires as perfectly round and rigid disks. However, in reality, tires are somewhat flattened on the side that rests on the ground. So, a perfect tire model that works for us in a simplified analysis does not apply in reality. We must take the *exact* form of the tire into account if we desire a comprehensive physical analysis of the tire's motion.

Now think about the material that makes up the tire. Why is it rubber? Why not steel? Or water? Different materials affect the tire's motion in different ways, and we need to consider these if we want a better picture of the physics behind the tire's motion.

The gods did not reveal all things to men at the start; but as time goes on, by searching, they discover more and more.
XENOPHANES
Ancient Greek Poet, c. 560 - c. 478 B.C.

In this chapter, we examine real materials—such as solids, liquids, and gases—because their shape and makeup affect our more detailed physical descriptions.

10.2 ATOMS & MATERIALS

To begin understanding materials better, we start by looking at the building blocks of all things: atoms.

If I could remember the names of all these particles, I'd be a botanist.
ENRICO FERMI
American Physicist, 1901 - 1954

NOTE 10.2: A Quick Look at the Atomic Model

An **atom** is a very small unit of matter: so small in fact, that even just a drop of water contains trillions of them. *Everything* that exists in the universe—from cars, to galaxies, to your body—is made up of atoms. In short, the atom is the basic building block of every thing you can imagine or see around you.

* * *

Each atom consists of three parts: **protons**, or positively charged particles; **electrons**, or negatively charged particles; and **neutrons**, or particles with no charge. The protons and neutrons in each atom are bound tightly together in a ball called the **nucleus**, while electrons "circulate" around the ball at a great distance from it. See Fig. 10.1 (a).

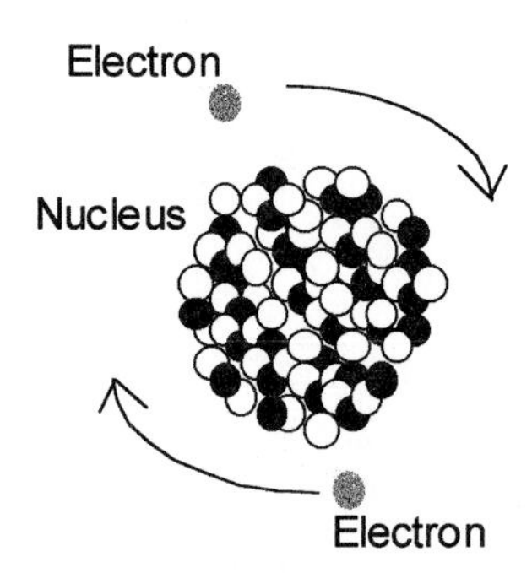

Figure 10.1 (a)
NOT TO SCALE

Now, the electrons "move" around the nucleus in a very odd way, the exact details of which you can ignore for now. Suffice it to say that they act a little like "clouds" of charge, rather than hard spheres or balls of charge (unlike the protons). So, a good way to visualize the atom is as a small, tight ball of protons and neutrons, surrounded by a rather large, loose "haze" of electrons (Fig. 10.1 (b)). The ball (protons and neutrons) has a net positive charge, and the haze (electrons) has a negative charge.

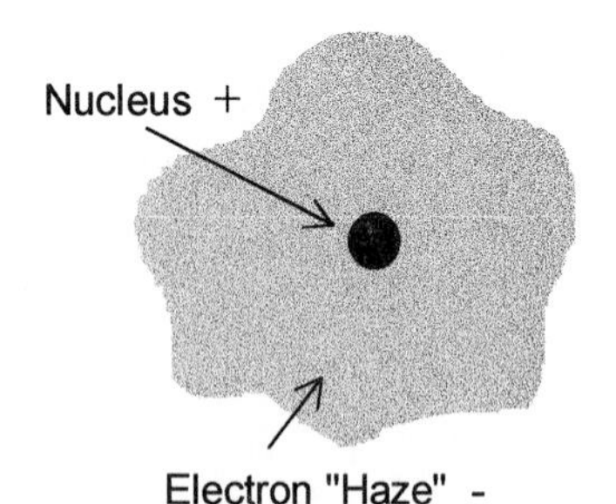

Figure 10.1 (b)
NOT TO SCALE

* * *

If there are equal numbers of protons and electrons in an atom, the atom as a whole is **neutral**: that is, it has no net charge. On the other hand, if the atom (now called an **ion**) has more protons than electrons (or vise versa), it will have a *net* positive (or negative) charge

However, even neutral atoms can *seem* to have a little bit of charge, if "viewed" from just the right angle. This is because the proton ball and the electron haze are *distinct* from each other—in an atom, the nucleus is on the inside, and the haze is on the outside. Because the positive protons and negative electrons do not mix, therefore, even the neutral atom, from just the right perspective, can seem to have a little bit of charge.

Refer to Section 7.2.2 for more on Coulomb's Law.

Because of their real *or* apparent charge, atoms interact with each other electromagnetically via Coulomb's Law. One electromagnetic "connection" between two (or more) atoms is called a **bond**; two (or more) atoms in a bond make up a **molecule**. The electromagnetic bonds that hold atoms together are *not* rigid (like glue), they are flexible (like springs).

NOTE 10.3: Atomic Bonds

Each atom, in a given material, interacts electromagnetically with its neighbors. Each atom feels *both* an attraction *and* a repulsion from these other atoms, as their positively charged nuclei and negatively-charged electron hazes come in and out of range.

Thus, bonded atoms "giggle" around each other, constantly being pushed and pulled by changing attractions and repulsions. The amount of "giggle" in a given bond depends on many things, including the number of positive and negative charges in the atoms, and the density of atoms in the material.

A material with a lot of "giggle" will deform easily (like a rubber band); a material without much "giggle" will not (like a wooden desk).

The smallest units of matter are not physical objects in the ordinary sense; they are forms, ideas which can be expressed unambiguously only in mathematical language.
WERNER HEISENBERG
German Physicist, 1901 - 1976

We sort materials into two basic types:

- **Solids** have relatively little "giggle" in their structure because their constituent atoms are tightly packed into extremely regular (three-dimensional) arrays.
- **Fluids** have relatively high "giggle" in their structure because their constituent atoms are less densely packed. As a result, the atoms are not very well (if at all) ordered into regular arrays. Fluids with relatively less "giggle" are called **liquids**; fluids with relatively more "giggle" are called **gases**.

The differences between solids, liquids, and gases can be very slight indeed.*

EXAMPLE 10.4: Glass: A Solid? Or a Fluid?

For example, window glass appears to be a solid: you can tap on it, shatter it, etc. Yet glass is actually *fluid*, albeit a very slow flowing one.

To see that this is true, visit an old building (say, a century or more) and examine the windows. (Make sure the windows are originals; new windows will not show this effect!) You should find that the glass in each window is thicker at the bottom of the pane than at the top. This is because, through time, the (fluid) window glass has flowed downward because of gravity.

* A gas made of ions, rather than of neutral atoms, is called a **plasma**. Plasma is a fourth state of matter: it is not a solid, liquid, or gas. The sun is made of plasma. We will not discuss plasmas here, and they will most probably not appear in your introductory physics course either.

That there are just over a hundred different elements can be proved by a look at the periodic table in the back of your textbook.

Every material in the universe is made of atoms and atomic bonds. Yet, as we look around, we see an infinite variety of different substances. How do we get so many different materials from just a hundred or so different atoms? The answer is that atoms and molecules can be arranged in many different ways within a basic plan.

EXAMPLE 10.5: Crude Oil: Not So Crude

Crude oil, the kind pumped straight from the ground in countless Texas oil fields, is an amazing example of this ability for identical atoms to be rearranged into a myriad of different materials.

The basic chemical structure of crude oil is rather complex, but it essentially consists of a certain combination of carbon atoms. Take away or rearrange some of the carbon atoms in the crude oil, and you can get coal, latex, aspirin...even ice cream!

In short, by shuffling around some atoms, you can get all sorts of interesting—and completely different—products from crude oil. This is despite the fact that there are only a few kinds of atoms making up the oil itself: it is the *arrangement* of the atoms that creates the infinite diversity around us, not just their individual compositions.

See Chapters 6 - 8 for more on how to analyze motion.

One job of the physicist is to characterize the properties of different materials. In principle, this could be done by examining the motion of individual atoms making up a particular substance. In practice, because even a small sample of a material contains very many atoms, this method is impossible. Therefore, physicists instead learn about the properties of solids and fluids via macroscopic, measurable parameters.

EXAMPLE 10.6: Two Parameters

Two such macroscopic parameters are important to the introductory physics class.

The first is **temperature**. Temperature is a measure of the average energy of atoms in a material. Because atomic energies are related to the speeds at which the atoms move, in principle we could calculate atomic energies in a similar way as we figured the speeds of falling balls in Chapter 5; that is, by using the kinematic equations. But in practice, there are simply too many atoms to make this calculation. So, we substitute a measurable, macroscopic quantity—temperature—that gives us *one* number representing the average effect of all of the moving atoms.

Refer to Section 5.4.2 for this kind of calculation.

Another example of a macroscopic parameter describing microscopic interactions is **pressure**. Like temperature, pressure is due to the motions of billions of atoms; unlike temperature, however, pressure is related to the total force the atoms provide to the walls of a container when they collide with it. Since we simply cannot measure all forces from each and every atom in a sample, we substitute one number—pressure—that represents the net effect of all of the atomic motions.

Force is in Chapters 6 and 7.

In summary, atoms are the tiny building blocks of all materials, their arrangement and electromagnetic interactions determining the precise nature of each particular substance. However, to understand the main classes of materials—solids, liquids, and gases—more fully, we do not examine the atoms themselves, but instead look to macroscopic parameters—such as temperature and pressure—that speak for the *net* actions of all atoms in each material.

By convention, there is color.
By convention, there is sweetness.
By convention, there is bitterness.
But in reality, there are atoms and space.
DEMOCRITUS, Ancient Greek Philosopher
c. 400 B.C.E.

10.3 SOLIDS

The main characterizing parameters associated with solids are those due to change of shape. Solids change shape, or deform, when they (1) experience a force, or (2) are heated.*

10.3.1 STRESS & STRAIN

When under force, solids deform. The force F per unit area A causing the deformation is called the **stress** σ. Stress is proportional to the fractional change in the shape of the object, known as **strain** ε:

For more on proportionalities, see Section 13.4.9.

$$\sigma = \frac{F}{A} = \mu \cdot \varepsilon$$

where μ (called a **modulus**), is the constant of proportionality in the strain equation. The units of stress are $\mathrm{N/m^2}$; strain is unitless.

As Table 10.1 below shows, a solid can be stretched or compressed in a variety of ways; each type of deformation results in a different strain and a different modulus μ.

Your textbook should have a chart of different moduli.

Table 10.1: Stresses

STRESS	DESCRIPTION	MODULUS	EQUATION	EXAMPLE
TENSILE STRESS	A force *perpendicular* to an area, which deforms object along one axis	Young's Modulus Y	$\sigma = Y\varepsilon = Y\frac{\Delta L}{L}$	$\vec{F}$; L; $L + \Delta L$
SHEAR	A force *parallel* to an area, which deforms an object in more than one dimension	Shear Modulus G	$\sigma = G\varepsilon = G\frac{\Delta d}{L}$	$\vec{F}$; L; Δd
HYDRAULIC COMPRESSION	A pressure# applied over an entire surface, compressing the material	Bulk Modulus B	$p = B\varepsilon = B\frac{\Delta V}{V}$	V; $\vec{F}$; $V - \Delta V$

* A high enough force or temperature will destroy a solid.

EXERCISE 10.7: Stress and Strain

A 15 cm steel rod is stretched 0.030 mm. Calculate the stress and strain on the rod.

SOLUTION:

To solve this problem, we must first calculate the strain. In this case, strain is the ratio of the change in length to the original length of the rod:

$$\varepsilon = \frac{\Delta L}{L} = \frac{0.030\,\text{mm}}{150\,\text{mm}} = 2.0 \times 10^{-4}$$

Notice that we converted the rod's length to millimeters to make the units cancel. So, strain is unitless.

Now, a stretching rod is under tensile stress, which corresponds to the first entry in Table 10.1. According to that table, the stress on the rod is the product of Young's Modulus for steel and ε:

Look in your textbook for the value of Young's Modulus for a steel rod. 

$$\sigma = Y\varepsilon = (2.0 \times 10^{11}\ \text{N/m}^2)(2.0 \times 10^{-4}) = 4.0 \times 10^{7}\ \text{N/m}^2$$

10.3.2 THERMAL EXPANSION

Solids expand when their temperature increases. The fractional change in each length of a solid due to a temperature increase is given by:

$$\frac{\Delta L}{L} = \alpha \cdot \Delta T$$

Your textbook should have a chart of these coefficients. 

where α is the **coefficient of linear expansion**, a number you look up in charts for different materials, and ΔT is the temperature change.*

EXAMPLE 10.8: Thermal Expansion

How much would a rod like the one in the exercise above (pre-stretch) expand if its temperature rose ten degrees Celsius?

Note that the first two entries in the table on the previous page have a stress equation, while the last has a relation for pressure. Both stress and pressure are calculated via force per unit area, so technically they both have SI units N/m^2. To keep these quantities separate, however, we call the unit for pressure the pascal: $1\ \text{Pa} = 1\ \text{N/m}^2$.

* Liquids expand as well. The relation for the change in liquid *volume* when heated is $\Delta V / V = 3\alpha\Delta T$.

Example 10.8 Continued...

SOLUTION:

To find out, we simply plug our numbers into the formula:

$$\frac{\Delta L}{L} = \alpha \Delta T$$
$$\Rightarrow \Delta L = L\alpha\Delta T$$
$$= (15\,\text{cm})(11 \times 10^{-6} /^\circ\text{C})(10^\circ\text{C})$$
$$= 1.6 \times 10^{-3}\,\text{cm}$$

Find the coefficient for steel in your textbook.

10.4 FLUIDS

Fluids—which can be liquids *or* gases—can flow. When they do, fluids are called **dynamic**; when they don't, they are called **static**. The principles that describe fluids depend on whether the fluid is static or dynamic.

10.4.1 STATIC FLUIDS

There are three important principles regarding static fluids.

First, pressure—once applied to a static fluid—is equally distributed to all points in its container. This is known as **Pascal's Principle**.

EXAMPLE 10.9: Pascal and Toothpaste Tubes

Pascal's Principle is why you can squeeze a tube of toothpaste at the bottom and still get your toothpaste out of the top!

Second, for a static fluid, pressure increases with depth.

EXAMPLE 10.10: Fluid Pressure and Depth

For example, the pressure of air at the top of a mountain is *lower* than air pressure at sea level. Likewise, the pressure of water at the bottom of the ocean is *higher* than water pressure at sea level.

Mathematically, we write the second principle as:*

$$p = p_o + \rho g h$$

where p_o is the pressure at a reference point, ρ is the density of the fluid, and h is the depth in the fluid relative to the reference point. Note that h is *negative* if you travel upwards; that is, if your final height is higher than your initial height.

EXERCISE 10.11: Under Pressure

What is the air pressure on a snorkeler's lungs? The density of water is $1000\ \text{kg/m}^3$, and the typical length of a snorkel is twenty centimeters.

SOLUTION:

To solve this problem, we apply our above formula, plugging in appropriate numbers:

$$\begin{aligned} p &= p_o + \rho g h \\ &= (1.01\times10^5\ \text{Pa}) + (1000\,\text{kg/m}^3)(9.8\ \text{m/s}^2)(0.20\,\text{m}) \\ &= 1.03\times10^5\ \text{Pa} \end{aligned}$$

This is about two percent more pressure than when the snorkeler is above water.

This fact was used in Section 7.2.10. 

The third rule for static fluids is **Archimedes' Principle**: a body wholly or partially submerged in a fluid will be supported by a buoyant force equal to the weight of the fluid that the body displaces.

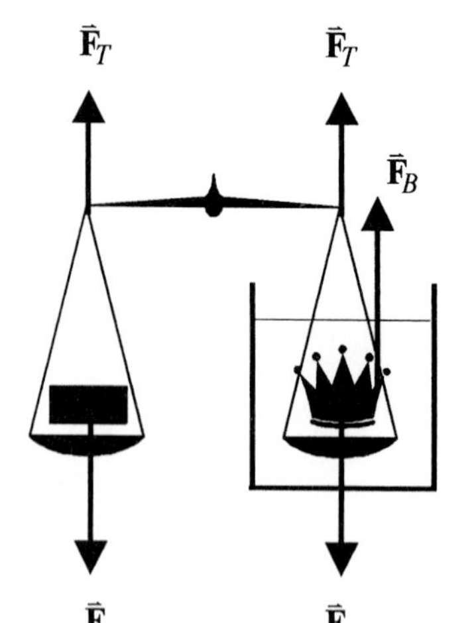

Figure 10.2

EXERCISE 10.12: Archimedes & King Hiero's Crown

King Hiero of Syracuse (third century B.C.) ordered a crown be made of pure gold. Being suspicious of the jeweler, however, the king wondered if the finished crown was truly solid gold. How could the crown's makeup be determined, without cutting into (and destroying) it?

SOLUTION:

Archimedes (287 - 212 B.C.), the great Greek philosopher, was assigned to the case. He suspended the crown, immersed in water, from a balance, as shown in Figure 10.2. Thus, there were three forces acting on the crown—its weight $F_{G,crown}$, the buoyant force F_B from the water, and the tension supplied by the balance cord F_T—and two forces acting on the pan—its weight $F_{G,pan}$ and tension F_T. But by N2, and the fact that the pan is in equilibrium, $F_T = F_{G,pan}$.

* This equation is only true for incompressible fluids. See the next section for a definition of this type of fluid.

Exercise 10.12 Continued...

Now, the buoyant force, according to Archimedes, is equal to the weight of the displaced water: $F_B = F_{G,water}$. Thus, applying Newton's Second Law to the diagram above, we have:

For more on these forces, refer to Sections 7.2.1 and 7.2.10; for more on Newton's Laws, refer to Chapter 6.

$$F_T + F_B = F_{G,crown}$$
$$\Rightarrow F_{G,pan} + F_{G,water} = F_{G,crown}$$

Now, $F_G = mg$, so:

This is in Section 7.2.1.

$$\Rightarrow m_{pan}g + m_{water}g = m_{crown}g$$
$$\Rightarrow m_{pan} + m_{water} = m_{crown}$$

But, $\rho = mV$, or $m = \rho / V$, so:

See Section 11.5.2 for this equation.

$$m_{pan} + \rho_{water}V_{water} = \rho_{crown}V_{crown}$$
$$\Rightarrow \rho_{crown} = \frac{m_{pan}}{V_{crown}} + \rho_{water}\frac{V_{water}}{V_{crown}}$$

Now, the volume of the crown V_{crown} equals the volume of the water displaced by it V_{water}, so we finally get:

$$\therefore \rho_{crown} = \frac{m_{pan}}{V_{water}} + \rho_{water}$$

Using the above formula, Archimedes could calculate the density of the crown by measuring the mass of the objects in the other pan m_{pan}, the volume of water displaced by the crown V_{water}, and the density of water ρ_{water} (a known value). If the calculated value for the crown equaled the known value for the density of gold, then the crown was truly solid gold.

It wasn't. Pity the poor jeweler that thought he could cheat a king!

10.4.2 DYNAMIC FLUIDS

Dynamic fluids flow, but in a highly complex way. It is mathematically simpler to consider the flow of an **ideal fluid**, one that:

Eliminating the Unessential; Section 2.3.

- flows smoothly (is **laminar**);
- doesn't have any internal resistance to flowing (is **non-viscous**);
- maintains a constant density (is **incompressible**);
- doesn't have any eddies (is **irrotational**).

There are many important descriptive properties of ideal fluids; here, we focus on two.

The **continuity equation** tells us that when a fluid travels through regions of different cross-sectional areas, the speed of the flow must be different in the two regions so that the volume of fluid flowing per unit

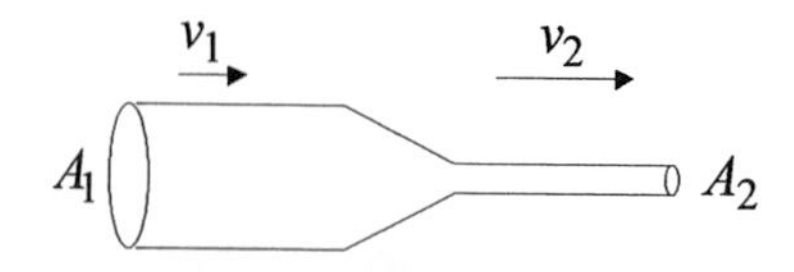

Figure 10.3

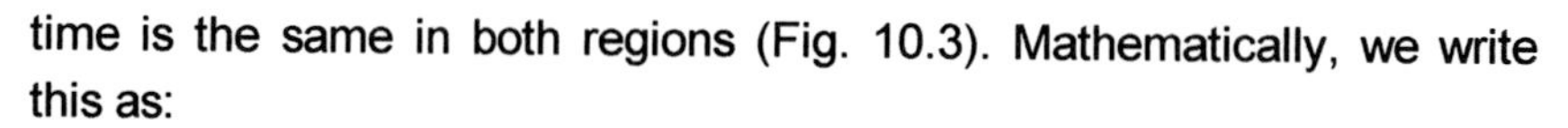

time is the same in both regions (Fig. 10.3). Mathematically, we write this as:

$$A_1 v_1 = A_2 v_2$$

EXAMPLE 10.13: Cholesterol

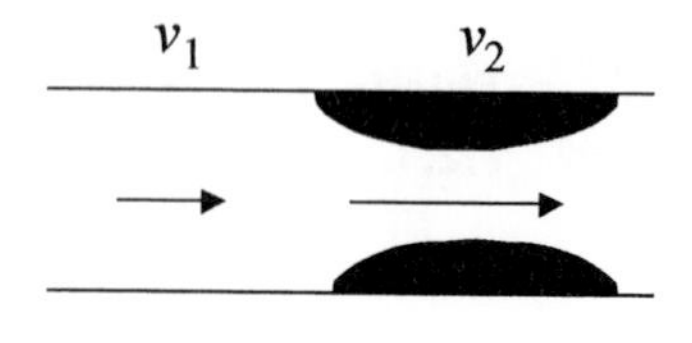

Figure 10.4

Blood moves faster through a cholesterol-blocked region of an artery than through a healthy region (Fig. 10.4). This comes from the continuity equation $A_1 v_1 = A_2 v_2$, or rearranging:

$$v_2 = \left(\frac{A_1}{A_2}\right) v_1$$

Since the area of Region 2 is smaller than the area of Region 1, the fraction A_1 / A_2 is greater than 1. Thus, v_2 must be greater than v_1.

Conservation of energy is in Sections 8.4 and 8.6.

The second important principle for dynamic fluids is **Bernoulli's Law**, a restatement of conservation of energy for fluids. Your textbook or instructor will derive this (very ugly) equation for you:

$$p_1 + \tfrac{1}{2}\rho_1 v_1^2 + \rho g y_1 = p_2 + \tfrac{1}{2}\rho_2 v_2^2 + \rho g y_2$$

where p is pressure, ρ is mass density, v is speed, and y is the height above a convenient reference point; the subscripts indicate regions 1 and 2.

EXERCISE 10.14: Bernoulli's Law

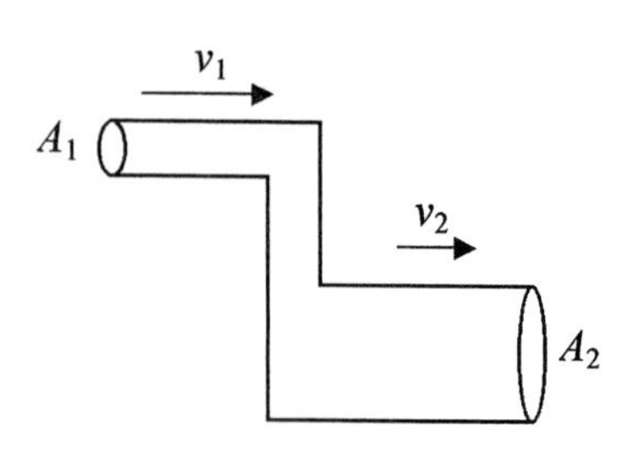

Figure 10.5

Water flows through a pipe of cross-sectional area $4.0\ \text{cm}^2$, with a speed of $5.0\ \text{m/s}$. The pipe then bends downward ten meters, where it now has a cross-sectional area of $8.0\ \text{cm}^2$ (Fig. 10.5). If the pressure at the upper level is $1.5 \times 10^5\ \text{Pa}$, what is the pressure at the lower?

SOLUTION:

First we have to find the speed of the water at the lower level, using the continuity equation, rearranged for us in the last example:

$$v_2 = \left(\frac{A_1}{A_2}\right) v_1 = \left(\frac{4.0\ \text{cm}^2}{8.0\ \text{cm}^2}\right)(5.0\ \text{m/s}) = 2.5\ \text{m/s}$$

Now, we use this figure, plus the above information, in Bernoulli's Law to find the pressure in the second region of the pipe:

$$\begin{aligned} \Rightarrow p_2 &= p_1 + \tfrac{1}{2}\rho(v_1^2 - v_2^2) + \rho g(y_1 - y_2) \\ &= (1.5 \times 10^5\ \text{Pa}) + \tfrac{1}{2}(1000\ \text{kg/m}^3)[(5.0\ \text{m/s})^2 - (2.5\ \text{m/s})^2] \\ &\quad + (1000\ \text{kg/m}^3)(9.8\ \text{m/s})[0 - 10\ \text{m}] \\ &= 6.1 \times 10^4\ \text{Pa} \end{aligned}$$

10.4.3 IDEAL GASES

The above two sections dealt with fluids of any type, gases or liquids. In this section, we are only concerned with gases, the lowest-density fluids.

In particular, here we look at **ideal gases**, those gases with such a low density that their constituent atoms almost never interact. In this special case, the pressure p, volume V, number of moles n, and temperature T (in kelvins!)* of the gas are all related by the **ideal gas law**:

Eliminating the Unessential is in Section 2.3.

Number of moles and temperature are discussed in Section 11.5.2.

$$pV = nRT$$

where R is the **universal gas constant** ($R = 8.31$ J/mole·K).

EXERCISE 10.15: Pressure in a Dorm Room

A typical dorm room with a volume of $50\ \text{m}^3$ contains approximately 420 moles of oxygen. What is the pressure of the oxygen if the room has a temperature of $20°$ C, assuming oxygen is an ideal gas?

SOLUTION:

First, we must convert the temperature to kelvins:

$$T_K = T_C + 273 = 20 + 273 = 293\ \text{K}$$

Now, using the ideal gas law, we can find the oxygen pressure:

$$\begin{aligned} pV &= nRT \\ \Rightarrow p &= \frac{nRT}{V} \\ &= \frac{(420\ \text{mole})(8.31\ \text{J/mole}\cdot\text{K})(293\ \text{K})}{50\ \text{m}^3} \\ &= 2.0 \times 10^4\ \text{Pa} \end{aligned}$$

Ideal gases have many interesting properties. For one thing, the speeds of the atoms in the gas are not random; rather, they follow the **Maxwell Distribution**:

$$P(v) = 4\pi\left(\frac{M}{2\pi RT}\right)e^{-Mv^2/2RT}$$

where M is the molar mass of the gas and T is its temperature. This distribution gives us the *probability* that an atom has speed v, within the range dv around v.

Molar mass is in Section 11.5.2.

From this distribution, we can calculate some important parameters that describe the ideal gas, such as:

* The conversion between Celsius and kelvin is: $T_K = T_C + 273$.

- The average speed of each atom in the gas is $v_{rms} = \sqrt{\dfrac{3RT}{M}}$.
- The average kinetic energy for each atom is $\overline{K} = \frac{3}{2}kT$, where k is the **Boltzmann constant**: $k = 1.38 \times 10^{-23}$ J/K.
- The average distance each atom travels between collisions is the **mean free path** $\overline{L} = V / \sqrt{2}\pi N d^2$, where V is volume of the container holding the gas, N is the number of atoms in the sample, and d is the diameter of each atom.
- The internal energy of a monatomic gas is $E_{\text{int}} = \frac{3}{2}nRT$.

EXERCISE 10.16: Kinetic Energy in a Dorm Room

What is $\overline{K}$ for the oxygen atoms in the previous exercise?

SOLUTION:

$$\overline{K} = \tfrac{3}{2}kT = \tfrac{3}{2}(1.38 \times 10^{-23}\ \text{J/K})(293\ \text{K}) = 6.06 \times 10^{-21}\ \text{J}$$

This seems like a small amount of energy until you realize it is the average energy of just *one* atom!

10.5 THERMODYNAMICS

Energy and energy transfers are discussed in Sections 8.4 – 8.6.

Thermodynamics is the study of temperature and energy transfers in systems associated with changes in temperature. There are three major laws of thermodynamics, numbered zero to two.*

10.5.1 THE ZEROTH LAW

The **Zeroth Law of Thermodynamics** was not developed until after the first and second were discovered. Since it is fundamental to the other two laws, however, it was designated as the "zeroth" law. This law is about *temperature*:

If object *A* has the same temperature as object *B*,
***and* object *B* has the same temperature as object *C*,**
then *A* has the same temperature as *C*.

* Actually, there are four thermodynamic laws, but the fourth one (called the "third law") is strictly for advanced work (or, for chemistry).

Two objects that have the same temperature are said to be in **thermal equilibrium**.

EXAMPLE 10.17: The Zeroth Law

We use the Zeroth Law whenever we use a thermometer.

Say you want to take the temperature of a pot of simmering water (object A). You place a thermometer (object B) in the pot. The thermometer eventually comes to the same temperature as the simmering water; that is, the water and the thermometer become in *thermal equilibrium* with each other. Thus, the thermometer now reads the temperature of the water.

Next, imagine that you take the same thermometer, shake it down, and place it in a *second* pot of simmering water (object C). You will find (eventually) that it comes to the same temperature as the first pot. Therefore, by the Zeroth Law, the water in object A is in thermal equilibrium with the water in object C (even though they aren't touching!).

10.5.2 THE FIRST LAW

The **First Law of Thermodynamics** is just a restatement of the law of conservation of energy for energies flowing in and out of a system by heat or by work.#

For more on conservation of energy and systems, see Sections 8.4 – 8.6.

One way energy flows in and out of systems is via **heat** Q, according to the rule that *energy always flows (by heat) from a high to a low temperature.* When energy flows by heat *into* a system, Q is *positive*.

For more on heat, see Section 8.5.2.

EXAMPLE 10.18: Soup, Part I

Imagine that you get hungry while you are studying. So, you fix yourself a nice big bowl of soup. But your reading is so engrossing that you forget to actually *eat* the soup for 15 minutes. What will happen?

The soup will cool because energy flows out of the soup (the system) into the air (the environment) by heat, so that the internal energy of the soup decreases. This heat flow is evidenced by a decrease in the soup's temperature.

The second way a system can exchange energy with its environment is by changing its volume, and thereby doing work W. If a *system* does work *on* its environment during the volume change, the work W is *positive*.

As we saw in Section 8.5.3, there are many ways that a system can transfer energy. The First Law of Thermodynamics *only* deals with two of these ways: heat and work.

EXAMPLE 10.19: Soup, Part II

Depressed (you really wanted *hot* soup), you decide to reheat your snack. So, you pop a lid on the bowl (you don't want to clean up spatters!), and put the soup in the microwave.

What happens? The lid will swell (and may actually pop off). But why? The atoms trapped in the bowl move faster as they warm up. And the faster they move, the harder they collide against the lid. As the atoms collide with the lid, they do work on it, causing the lid to expand...leaving the atoms more room to move around in.

Internal energy is discussed in Section 8.3.3.

If we combine these two ways of exchanging energy, we have the First Law of Thermodynamics: the *change* in internal energy of a system is equal to the difference between the energy flowing by heat *into* the system and the work done *by* the system:

$$\Delta E_{\text{int}} = Q - W$$

EXERCISE 10.20: Soup, Part III

Assume the change in internal energy in the soup of Example 10.19 was $5\ \text{J}$, such that the microwave supplied $8\ \text{J}$ of energy to it. How much work was done by the atoms in the bowl on the lid?

SOLUTION:

We apply the first law of thermodynamics:

$$\Delta E_{\text{int}} = Q - W$$
$$\Rightarrow W = Q - \Delta E_{\text{int}}$$

Now, the heat is positive because the microwave is *supplying* energy *to* the soup system, so:

$$W = 8 - (+5) \ = +3\ \text{J}$$

The work is positive because the *system* is doing work *on* the lid.

10.5.3 THE SECOND LAW

The **Second Law of Thermodynamics** deals with a new quantity called **entropy**. Entropy S is a measure of the number of available states to a system; or, by another way of looking at things, it is a measure of how *disordered* the system is.

EXAMPLE 10.21: Entropy and You, Part I

How much entropy is in your dorm room? The more disordered a system is, the more entropy it has. So, the messier your room, the more entropy you live with!

The Second Law of Thermodynamics states that in any real world process, the entropy of a system and its environment *must* increase ($\Delta S > 0$). That is, real world situations *always* get more disordered.

EXAMPLE 10.22: Entropy and You, Part II

Can't seem to keep your dorm room clean? Blame it on entropy!

The Second Law also implies that real world processes are **irreversible**; that is, there is no way to "rewind" time and end up at *exactly* the same place you began.

EXAMPLE 10.23: Soup, Part IV

Now that you are finally ready to eat your soup, you sprinkle some crushed crackers on top and stir the whole mess with a spoon. As the crackers mix into the hot soup, you start to wonder if it is possible, assuming you were a really good stirrer, to "unstir" the soup and get the crackers back on top.

The answer is no. Reconstituting the crackers would *decrease* the disorder of the system. That is, the soup + cracker system is *more* disordered when the crackers are mixed into the soup than when the crackers are nicely arranged on top. Since nature always increases disorder, the process of stirring crackers into the soup is *irreversible*.

10.6 CONCLUSION

In this chapter, we discussed three states of matter—solids, liquids, and gases—and their deformation, flow, and reactions to heat. We also took a brief look at the Laws of Thermodynamics. This information will allow us to understand the world in a much more detailed, realistic way than would otherwise be possible if we continued to "Eliminate the Unessential," and viewed the world ideally. And with that, we conclude our look at the topics of introductory physics.

...This darkness of the mind, must needs be scattered not by the rays of the sun...but by the outer view and inner law of nature.
LUCRETIUS
Ancient Roman Poet, 99 - 55 B.C.

PART III: THE ESSENTIAL MATH

Whether we like it or not, mathematics is a *vital* part of physics: without a solid foundation of basic mathematical concepts and techniques, learning college-level physics is much more difficult than it has to be.

Luckily, you have probably already learned most of the math necessary for success in physics. If you need to sharpen up your memory of these concepts a little (or a lot), or merely want to review a topic or two, this part of the book contains all of the mathematics essential (and not so essential) for a typical introductory physics course.

Do not worry about your difficulties in math..., I can assure you mine are still greater.
ALBERT EINSTEIN
American Physicist, 1879 - 1955

This part is organized into six chapters, as follows:

- *Chapter 11: Measurement* examines the proper way to handle measurements, an important component of any scientific class.
- *Chapter 12: Tricks of the Trade* offers mathematical short cuts that physicists use to cut through their heavy numerical workload.
- *Chapter 13: Algebra, Geometry, & Trigonometry* and *Chapter 14: Calculus* review mathematical material you have probably learned in earlier classes.
- *Chapter 15: Vectors* and *Chapter 16: Mathematics For Data Sets* study new kinds of mathematics you probably have not seen before.

How shall you tackle all of these chapters? We recommend the following plan:

- Read through Chapters 11 and 12 *in their entirety* before, or shortly after, your class begins. These brief chapters contain important information that will make your mathematical life *much easier*, so they are well worth the extra time spent reviewing them before you begin.
- Then, leaf through Chapters 13 and 14 to familiarize yourself with their content. If your math memory is fairly sharp, you can use these chapters as *references* to check your recall of small details. On the other hand, if you find your math memory is a little cloudier than you would prefer, you can skim or read through these chapters in more detail as the information becomes applicable to your course.
- Finally, depending on your particular course, Chapters 15 and/or 16 may be optional. Ask your instructor if this material is relevant to your class. If it is (in whole or in part), we recommend that you look through these chapters *before* their enclosed material comes up in the course.

As has been the case throughout this book, Part III purposely eliminates many important details critical to full understanding of the material—such as mathematical proofs and in-depth explanations of certain techniques—preferring instead to give you *just enough* information to get you started.

The downside to this approach, however, is that it is easy to get into the habit of memorizing pat answers to subtle mathematical questions, rather than understanding the deeper issues at the root of these techniques. Therefore, *you* must take it upon yourself to dig into each method, to try to understand *why*—not just how—it works.

However, despite our best efforts to cut through the superfluous information, there is a *lot* of material to cover in the next six chapters. Nevertheless, you should find that the text itself is actually very brief; the bulk of what remains is taken up by worked-out examples. Therefore, you will probably get through Part III—long as it is—without *too* much difficulty. Just take what you need to get started, and leave the rest for another day.

11
MEASUREMENTS

11.1 INTRODUCTION

Many of the numbers seen in a physics class—whether in homework problems, labs, lectures, or reading assignments—will be **measurements** (physical quantities determined directly from nature), even though sometimes they are not explicitly announced as such. Therefore, the dual subjects of how to write, and use, measurements correctly are critical components of your physics course.

Research is the process of going up alleys to see if they are blind.
MARSTON BATES
English Author, 1906 - 1974

Chapter 16 deals with the latter issue; this chapter looks at the former, showing you how to write measurements correctly, briefly, and clearly, using significant figures, rounding, scientific notation, and units.

In Chapter 16.

11.2 SIGNIFICANT FIGURES

As we shall learn in Chapter 16, measurements are inherently error-ridden: there is *no way* to take perfect, error-free measurements. However, as you might expect, some measurements are more accurate and precise than others. Therefore, it is desirable to have a quick and easy way to determine the precision of any given measurement. This method is the technique of significant figures.

Error analysis is discussed in Section 16.2.3.

A **significant figure** is any digit in a measurement that is *exactly known* relative to the error in the measurement. To calculate the number of significant figures in a number, follow these rules:*

1. Non-zero numbers are *always* significant.#
2. Zeros *are* significant when:
   (a) They act as placeholders in a number (*e.g.*: 50.090 has five significant figures);
   (b) They are to the *left* of an *explicit* decimal point (*e.g.*: 7400. has four significant figures) EXCEPT when the zero is the symbolic convention for decimals (*e.g.*: 0.37 has two significant figures).
3. Zeros are *not* significant when:
   (a) They are to the *left* of an *implied* decimal point (*e.g.*: 380 has two significant figures);^
   (b) They are to the *right* of a decimal point, but *before* any significant figures (*e.g.*: 0.000205 has three significant figures).

Come forth into the light of things.
Let Nature be your teacher.
WILLIAM WORDSWORTH
English Poet, 1770 - 1850

EXERCISE 11.1: Significant Figures, Part I

How many significant figures does each quantity have?

(a) 4902.005 (b) 345,000 (c) 0.00640

SOLUTION:

(a) Seven. All of the non-zero digits are significant (Rule 1 above), and, since they are acting as placeholders (Rule 2 (a)), all of the zeros are significant as well.

(b) Three. There is no decimal point printed for this number, so none of the zeros are significant in this case (Rule 3 (a)). *If* there had been a decimal point printed, however, this number would have six significant figures (Rule 2 (b)).

(c) Three. The last zero is significant, since it is a placeholder (Rule 2 (a)), but the two zeros immediately to the right of the decimal point, are not (Rule 3 (b)). The zero to the left of the decimal point is a symbolic convention, and is also not significant (Rule 2 (b)).

---

* The related subjects of significant figures (this section) and rounding (Section 11.4) are, unfortunately, not completely standardized. The rules presented in this book are accepted by a majority of scientists—but not all. Therefore, check with your instructor for the specific rules used in *your* class.

There is one technical exception to these rules. *When a measurement is near a magnitude change, treat the larger number as having the same number of significant figures as the smaller.* For example, both 99 and 101 have two significant figures, even though 99 has two digits and 101 has three. Both numbers have uncertainties of about one part in one hundred, so neither number can be considered to be more accurate than the other.

^ A few texts consider these zeros to be significant.

Moving the decimal point does not change the number of significant figures.

EXERCISE 11.2: Significant Figures, Part II

How many significant figures do each number have?

(a) 4,902,005 (b) 4.902005 (c) 49,020,005,000,000

SOLUTION:

All of these numbers have seven significant figures.

11.3 SCIENTIFIC NOTATION

Measurements in physics cover a large range. For example, physicists routinely measure lengths as small as an atom (about 0.0000000001 m) or as large as the distance to the nearest galaxy (about 10,000,000,000,000,000,000,000 m). In each of these types of measurements, there are many non-significant zeros, making the numbers lengthy and hard to handle.

We want, therefore, to be able to write all measurements briefly, using only significant figures. The method of **scientific notation** enables us to do so. To write a number in scientific notation, follow these steps:

1. Rewrite the number such that the decimal point is immediately to the right of the first significant figure.
2. Count the number of decimal places you had to move or "jump" the decimal point to do Step 1. If you moved the decimal point to the left, the number of jumps is *positive*. If you moved the decimal point to the right, the number of jumps is *negative*.
3. Raise ten to the number of jumps you obtained in Step 2. This power of ten is called the **order of magnitude** or just **magnitude**.
4. Rewrite the number from Step 1, dropping any non-significant zeros. This rewritten number is called the **coefficient**.
5. Write the coefficient and magnitude in the format:

coefficient × magnitude[!]

Don't worry; it's not as hard as it looks!

[!] Some calculators and computers use the notation of "coefficient E number of jumps" to represent the same thing. Thus, 10^9 on a calculator is not 10E9, but 1E9. To use scientific notation on your calculator, make sure the scientific notation function button is set.

EXAMPLE 11.3: Scientific Notation, Part I

Let's practice writing the number 4,260,000,000 in scientific notation.

According to Step 1 in the directions above, we want to rewrite the number such that the decimal point is immediately to the right of the first significant figure. In the number of this example, the first significant figure is the 4, so we rewrite the number as:

$$4.260000000$$

Next (Step 2), we count how many spaces we had to "jump" from the location of the (implied) decimal point of the original number to the new location as written above. Counting from right to left along the rewritten number, we see that we had to jump *nine* spaces to the *left* (check it!).

According to Step 3 in the directions above, this means that we must raise ten to the +9 power. Our magnitude is thus 10^9.

Now (Step 4), we rewrite 4.260000000 again, dropping all non-significant zeros. Thus, our coefficient is 4.26.

Lastly (Step 5), we write the coefficient and magnitude in the correct format:

$$4.26 \times 10^9$$

On a calculator, this number would be displayed as 4.26E9.

EXERCISE 11.4: Scientific Notation, Part II

Rewrite the following numbers in scientific notation:

(a) 35,442 (b) 83 million (c) 0.00000000531

SOLUTION:

(a) $35{,}442 = 3.5442 \times 10^4$

(b) 83 million = $83{,}000{,}000 = 8.3 \times 10^7$

(c) $0.00000000531 = 5.31 \times 10^{-9}$

A great advantage to scientific notation is the ease by which very large or very small numbers can be mathematically manipulated. Use these rules for arithmetic operations in scientific notation:

If you do not rest on the good foundation of nature, you will labor with little honor and less profit.
LEONARDO DA VINCI
Italian Experimentalist, 1452 - 1519

- *Add or subtract in scientific notation normally, using the coefficients, but be sure that all numbers have the same order of magnitude before you begin.* If they don't, rewrite them so that they do. (When the math is complete, you may have to rewrite the final answer to get back into correct scientific notation.)

- *Multiply or divide, in scientific notation, the magnitudes and the coefficients separately.* When the math is complete, recombine the coefficient and magnitude into correct scientific notation.

EXAMPLE 11.5: Adding in Scientific Notation

Let's practice the first rule for arithmetic in scientific notation by adding the numbers 4.86×10^9 and 3.49×10^8 together.

To add numbers in scientific notation, according to the rule above, we must first rewrite the numbers so that they have the same order of magnitude. But which magnitude do we choose: the 10^8 or the 10^9?

The answer is that we get to pick the way that is *easier* for the problem at hand. In this example, let's try doing it both ways, so that we can practice the process twice.

* * *

We start with the slightly easier option, choosing to rewrite the 10^8 figure in terms of a 10^9 magnitude. To help us do this, we look to the rules of exponents in Chapter 13. According to one rule of exponents given there, to *increase* the exponent by one digit, we must move the decimal point in the coefficient one space to the *left*. Thus, our number 3.49×10^8 can be rewritten as 0.349×10^9.

See Section 13.4.6 for more on exponents.

Now that both numbers are written in terms of the 10^9 magnitude, we can proceed normally with the addition:

$$\begin{aligned} 4.86 \times 10^9 + 3.49 \times 10^8 &= 4.86 \times 10^9 + 0.349 \times 10^9 \\ &= (4.86 + 0.349) \times 10^9 \\ &= 5.209 \times 10^9 \end{aligned}$$

This answer is not quite finished; see Example 11.9 for more.

Notice that our answer is already in correct scientific notation.

* * *

Now that we have added our numbers using the 10^9 magnitude, let's try it the other way, using the 10^8 magnitude.

Therefore, we must rewrite 10^9 figure in terms of 10^8 magnitude. Looking to the exponent rules of Chapter 13, we see that to go *decrease* the exponent by one digit, we must move the decimal point in the coefficient to the *right* one position. Therefore, 4.86×10^9 becomes 48.6×10^8. The addition then follows:

$$\begin{aligned} 4.86 \times 10^9 + 3.49 \times 10^8 &= 48.6 \times 10^8 + 3.49 \times 10^8 \\ &= (48.6 + 3.49) \times 10^8 \\ &= 52.09 \times 10^8 \end{aligned}$$

However, this answer is not correctly written in scientific notation. So, we must rewrite it. To get the decimal point in the right position, we must make a "jump" one decimal place to the left, increasing the exponent by $+1$. Our final answer is then:

$$5.209 \times 10^9$$

which is of course the same answer as we obtained before.

EXAMPLE 11.6: Dividing in Scientific Notation

Now, let's practice the second rule for arithmetic using scientific notation by dividing the same two numbers from Example 11.5 above, namely 4.86×10^9 and 3.49×10^8.

According to the division rule given above, we must first isolate the coefficients from the magnitudes: 4.86 and 3.49 are kept separate from 10^9 and 10^8.

Next, we separately do the math on the coefficients and magnitudes. Thus, we divide the magnitudes separately from the coefficients:

$$(4.86) \div (3.49) = 1.392550143$$

$$(10^9) \div (10^8) = 10^{9-8} = 10^1$$

This answer is also not quite finished; see Example 11.9 for more.

Now, we must recombine the coefficient with the magnitude to get the final answer. We do so, and get 1.392550143×10^1.

EXERCISE 11.7: Scientific Notation, Part III

Make the following calculations, and write the results in correct scientific notation.

(a) $(5.27\times10^{-4})+(7.3\times10^{-4})$ (b) $(8.50\times10^{5})-(9.81\times10^{3})$

(c) $(6.203\times10^{2})\times(9.4\times10^{-4})$ (d) $(7.61\times10^{-2})\div(2.78\times10^{-4})$

SOLUTION:

These answers are also not quite finished; see Exercise 11.10 for more.

(a) $$(5.27\times10^{-4})+(7.3\times10^{-4}) = (5.27+7.3)\times10^{-4}$$
$$=12.57\times10^{-4}$$
$$=1.257\times10^{-3}$$

(b) $$(8.50\times10^{5})-(9.81\times10^{3}) = 850\times10^{3} - 9.81\times10^{3}$$
$$=840.19\times10^{3}$$
$$=8.4019\times10^{5}$$

(c) $$(6.203\times10^{2})\times(9.4\times10^{-4}) = [(6.203)(9.4)]\times10^{2+4}$$
$$=58.3082\times10^{6}$$
$$=5.83082\times10^{7}$$

(d) $$(7.61\times10^{-2})\div(2.78\times10^{-4}) = [7.61\div2.78]\times10^{-2--4}$$
$$=2.737410072\times10^{2}$$

Luckily, most calculators today allow you to compute in scientific notation, without having to rewrite coefficients. It is very worthwhile to spend a few minutes with your owner's manual figuring out how to do this.

11.4 ROUNDING

When measurements are combined mathematically, the result can only be as precise as the *least precise* original measurement. That is, the result can only have as many significant figures as the original measurement in the calculation with the *least* number of significant figures.*

I profess to learn and to teach... from the fabric of nature.
WILLIAM HARVEY
English Physician, 1578 - 1657

EXAMPLE 11.8: Determining Significant Figures

For example, recall that in Exercise 11.7 (a) above, the numbers 5.27×10^{-4} and 7.3×10^{-4} were added together to obtain a raw result of 1.257×10^{-3}.

However, according to the rounding rule stated above, the final result of a mathematical operation can only be as precise as the least precise original measurement. Since the number 7.3×10^{-4}, having two significant figures, has the least precision of the two original numbers, our final answer can only have two significant figures as well.

This answer is not quite finished; see Example 11.10 for more.

To ensure the correct level of precision in your calculations, you may need to **round** your answers to the appropriate number of significant figures. *Never round until all calculations are complete*, however, or the rounding itself may introduce errors into your computation.

To round any number, follow these steps.

1. Determine the correct number of significant figures for the measurement.
2. Moving from left to right along the number, count digits until you reach the last significant figure as determined in Step 1.
3. Look at the number to the immediate *right* of that figure.
   (a) If it is in the range 6 - 9, add one to the last significant figure, and delete all numbers to the right of it.
   (b) If it is in the range 0 - 4, leave the last significant figure unchanged, and delete all numbers to the right of it.
   (c) If it is a 5, use the rule (a) above if the last significant figure is odd; otherwise, use the rule (b) above.

Finally, *never round exact numbers* (numbers which are not the result of measurement, like π or e).

* This is the standard rule for rounding. Recent research, however, suggests that the results of multiplication or division of measurements can have *one more* significant figure than the number of significant figures in the least precise original measurement. Because this rule is very new, however, be sure and check with your instructor on which rule is used in your class.

EXAMPLE 11.9: Rounding, Part I

Let's practice rounding using the results of Examples 11.5 and 11.6.

* * *

Because each of these examples use the same original measurements, 4.86×10^9 and 3.49×10^8, we can start by determining the correct number of significant figures for these numbers (Step 1 in the procedure listed above.) Each number has three significant figures. Thus, according to the rule above, our final, rounded answers for both examples will also have three significant figures.

* * *

We first look at the result of Example 11.5, 5.209×10^9. According to Step 2 in our rounding process, we must count digits, from left to right, until we get to the third significant figure. The third significant figure in the number 5.209×10^9 is the 0.

So, according to Step 3 in the directions above, we now examine the number immediately to the right of the third significant figure. This number is the 9. Since it lies in the range 6 - 9, we know that we must add 1 to the 0. Thus, 0 becomes 1. Furthermore, we must delete all remaining digits. Thus, we discard the 9.

Therefore, our correctly rounded result from Example 11.5 is written:

$$5.21 \times 10^9$$

* * *

Now, let's round the result from Example 11.6, which is 1.3925501×10^1. We know that, like in the previous case, we must round this number to three significant figures.

Counting from left to right, we see that the third significant figure in this number is the 9. The number immediately to the right of the 9 is a 2. Since 2 is in the range 0 - 4, we leave the 9 alone, and delete all other numbers. Thus, we discard the 25501.

So, our final answer from Example 11.6 is:

$$1.39 \times 10^1$$

EXERCISE 11.10: Rounding, Part II

Round the results of Exercise 11.7 to the correct number of significant figures.

SOLUTION:

(a) $1.257 \times 10^{-3} \approx 1.2 \times 10^{-3}$ (b) $8.402 \times 10^5 \approx 8.40 \times 10^5$

(c) $5.83082 \times 10^7 \approx 5.8 \times 10^7$ (d) $2.737410072 \times 10^2 \approx 2.74 \times 10^2$

Rounding is especially important when using calculators. Calculators always display ten digits, regardless of the significance of these numbers. Writing down all ten digits is not only *wrong* (unless there are ten significant figures in the problem!), but also time-consuming.

11.5 UNITS

I have taken the course...to express myself in terms of number, weight, and measure...[and] to consider only such causes as have visible foundations in nature.
SIR WILLIAM PETTY
English Natural Philosopher, 1623 - 1685

It is not enough, unfortunately, to write a measurement in correct scientific notation with the correct number of significant figures. A measurement is a quantity of *something*; therefore, a measurement must be *labeled* so everyone knows what something has been measured. Such labels are called **units**.

11.5.1 UNITS & STANDARDS

A unit is a precisely defined amount of a quantity. When attached to a measurement, the unit tells us how many multiples of the fixed amount was measured. The reference for the multiple is called the **standard**. The best standards, those that are **operationally defined**, are measured directly from nature.

EXAMPLE 11.11: Units and Standards of Time

A unit of time is the *second*. Thus, a measurement of time is stated in terms of how many seconds have elapsed since the beginning of an experiment.

The standard for the second is a certain number of molecular vibrations of a particular isotope. Thus, the standard for the second, since it is based on a measurement taken directly from nature, is operationally defined.

There are countless different types of measurable quantities in the natural world. Therefore, the number of possible units and standards is almost unimaginable. Luckily, however, *all measurable quantities can be expressed in terms of only seven* **fundamental quantities**: **length**, **time**, **mass**, **current**, **temperature**, **amount of a substance**, *and* **luminous intensity** (or **luminosity**).*

---

* A very important eighth quantity, the **angle**, is technically not a fundamental quantity because it can be created by finding the ratio of two lengths. However, because this quantity is used so extensively in the introductory physics course—in particular, in association with rotational motion—we will discuss it in this chapter on the same level as the seven (truly) fundamental quantities.

Basic research is what I'm doing when I don't know what I'm doing.
WERNHER VON BRAUN
German Rocket Scientist, 1912 - 1977

All other physical quantities in the universe are related (via an equation) to one or more of the fundamental quantities...*and nothing else*. We say that these other quantities are *derived* from the fundamental quantities; hence we call them **derived quantities**.

This definition of velocity is in Section 5.3.1.

EXAMPLE 11.12: One Derived Quantity

As an example of a derived quantity, consider *velocity*. The formula for velocity, $v = \Delta x / \Delta t$, is written completely in terms of the fundamental quantities of length and time.

Because all physical quantities can be expressed in terms of just seven fundamental quantities, we only need seven units (and their associated seven standards)* to describe them. This saves us from the unpleasant prospect of having different units (and standards) for each and every possible physical quantity.

The units of the derived quantities are derived from the fundamental units by multiplying or dividing the units—though never adding or subtracting them—and, not surprisingly, they are called **derived units**.

Your textbook's appendix may have a longer list.

The set of *all* units used in modern science—both fundamental and derived—is called the **International System of Units**, abbreviated **SI** (from the French), and also known as the **metric system**. Some SI units are listed in Table 11.1.

Table 11.1: SI Units for the Fundamental & Some Derived Quantities

FUNDAM'TL	UNIT	SYMBOL	DERIV'D QUANTITY	UNIT	SYMBOL
Length	meter	m	Force	newton	$N = m{\cdot}kg/s^2$
Time	second	s	Energy	joule	$J = kg{\cdot}m^2/s^2$
Mass	kilogram	kg	Power	watt	$W = J/s$
Current	ampere	A	Pressure	pascal	$Pa = kg/m^2{\cdot}s^2$
Temperature	kelvin	K	Charge	coulomb	$C = A{\cdot}s$
Amount of Substance	mole	mol	Potential Difference	volt	$V = J/C$
Luminosity	candela	cd	Electric Resistance	ohm	$\Omega = V/A$
Angle	radian	rad	Capacitance	farad	$F = C/V$

You should notice two things about this table. First, all units are represented by a single letter or abbreviation that is neither italicized nor written with a period. These symbols are the same in every language. Second, even though many of the units are named after important scientists (for example, the newton), the name of the unit itself is not capitalized—though the abbreviation is.

* Plus one for angle.

EXAMPLE 11.13: One Derived Unit

Again think about velocity. This derived quantity is made up of the fundamental quantities of length—which has the fundamental unit of the meter (m)—and time—with fundamental unit of the second (s).

Thus, the derived unit for velocity is the meter per second (m/s). This expression contains only fundamental units and nothing else.

NOTE 11.14: The Importance of Being Metric

You may feel uncomfortable using the metric system; after all, in the United States, the British unit system is used for everything from buying gasoline (dispensed in gallons, not liters), using a ruler (you probably use the inch side, not the centimeter edge), and keeping track of weight (your scale reads pounds, not kilograms).

Unfortunately, though, the metric system is the unit system of choice for physics, science, and indeed, the entire rest of the world. So, to succeed in your physics class, you *must* use SI.

If you are still one of those who think they can get along just fine with the British system, however, consider this: A recent space probe to Mars crashed simply because someone mistakenly input data into the probe's computer in British—not metric—units.

11.5.2 THE FUNDAMENTAL QUANTITIES

Let's look at some of the fundamental quantities more closely.

LENGTH

Length, usually symbolized by the letters l or d, has two meanings:

- It is a measure of the *straight-line spatial extent* of an object ("how long" an object is). In this sense, length is always a positive number.
- It is also a measure of *how far* an object is (in a straight line) from a fixed reference point. In this meaning, length can have positive *or* negative values.

The SI unit for length is the **meter** (m). The operationally defined standard for the meter is the distance traveled by light in a vacuum during 1/299,792,458 of a second. Table 11.2 below lists approximate values of some lengths.

Table 11.2: Some Order of Magnitude Lengths (in meters)

Distance to the edge of the universe	10^{26}	Height of a tall person	2
Distance to the nearest galaxy	10^{24}	Length of a fly	10^{-3}
Distance to the nearest star	10^{16}	Diameter of a red blood cell	10^{-5}
Distance to the sun	10^{11}	Diameter of a virus	10^{-7}
Diameter of Earth	10^{7}	Diameter of a hydrogen atom	10^{-10}
Height of tallest skyscraper	10^{2}	Diameter of a proton	10^{-15}

TIME

Time *t* has two different meanings:

What then is time? If no one asks me, I know; if I wish to explain it to one that asketh, I know not.
ST. AUGUSTINE
English Archbishop, d. 604

- It is a measure of *when* an event occurs, relative to some starting point. This number can be positive or negative.*
- It is also a measure of the *duration* of an event. In this meaning, time is always positive.

The SI unit for time is the **second** (s). The operationally defined standard for the second is the time it takes cesium-133 to vibrate 9,192,631,770 times. Table 11.3 lists approximate values of some times.

Table 11.3: Some Order of Magnitude Times (in seconds)

Age of the universe	5×10^{17}	Average human reaction time	10^{-1}
Age of the Earth	10^{17}	Period of AM radio	10^{-6}
Existence of humans	10^{14}	Period of molecular rotation	10^{-12}
Age of the pyramids	10^{11}	Period of visible light	10^{-15}
Average life span of a person	10^{9}	Period of nuclear vibration	10^{-20}
Length of one year	10^{7}	Time it takes light to travel across a proton	10^{-23}
Length of one day	10^{5}	Age of universe when laws of physics begin to apply	10^{-43}
Length of one physics lecture	10^{3}		
Length of one heartbeat	1		

Time is God's way of keeping things from happening all at once.
ANONYMOUS

Time has the unusual characteristic that it *always moves forward.* That is, there seems to be an **arrow of time** that points in only one direction —toward the future. This leads to the **principle of causality**, which states that causes *must* always occur earlier than effects.

MASS

Mass, symbolized by the letter m, has many different meanings. Each of these definitions reveals a different aspect of this still mysterious quantity, but each definition is equivalent to the others (**principle of mass equivalence**). Regardless of the definition used, mass is always a positive number.

---

* Negative times are undesirable for many reasons. Therefore, physicists usually "start the clock" on experiments at the beginning, so that all measured times are—*by definition*—positive.

- It is a measure of the quantity of matter possessed by an object.
- It is also a measure of an object's resistance to a change in motion; or, looking at it from another perspective, a determinant of how much an object is accelerated when a force is applied to it.
- It is that which participates in the gravitational interaction.

In Section 10.2.

In Section 6.3.

In Section 7.2.1.

The SI unit for mass is the **kilogram** (kg). The kilogram standard is a platinum-iridium cylinder located in France; all other kilograms in the world are compared to the standard by weighing. Thus, of all of the fundamental quantities, only mass has a standard that is not operationally defined. This is due to the fact that the true nature of mass is not well understood.

Table 11.4 lists approximate values of some masses.

Table 11.4: Some Order of Magnitude Masses (in kilograms)

Galaxy	10^{41}	Apple	1
Sun	10^{30}	Textbook page	10^{-2}
Earth	10^{24}	Raindrop	10^{-6}
Mountain	10^{18}	Red blood cell	10^{-13}
Pyramid	10^{10}	Virus	10^{-19}
Jet	10^{5}	Uranium atom	10^{-25}
Compact car	10^{3}	Proton	10^{-27}
Human	65	Electron	10^{-30}

Mass is related to, *but is not the same thing as*, weight. Weight is the product of mass and g, the **gravitational field**: $F_G = mg$. Since g varies with location, and can even be zero, while the mass of an object is constant, the weight of an object can vary. #

For more on weight, see Section 7.2.1; refer to Sections 7.2.1 and 7.3.1 for more on the gravitational field.

Another quantity related to mass is the **mass density**. Mass density is the ratio of the mass to volume: $\rho = m / V$. This quantity is *constant* for a given liquid or solid. Table 11.5 lists ρ for various materials.

Table 11.5: Some Densities (in 10^{-3} grams per cubic centimeter)^

Hydrogen	0.838	Carbon	2260
Helium	0.166	Nitrogen	1.16
Lithium	534	Oxygen	1.33
Beryllium	1850	Fluorine	1.70
Boron	2340	Neon	0.839

For most places on Earth, g is about 9.8 m/s^2, though the higher in elevation you are, the lower g. For locations outside of Earth, such as the moon, g is very different from 9.8 m/s^2, and g is zero in empty space.

You may wonder why we can measure masses by weighing, since weight varies with location. Luckily, the ratio of mass to weight *at a single location* is constant, so weighing is a good method for calculating mass.

^ Gas densities are applicable at standard pressure (1 atm) and temperature (273 K).

CURRENT

In Section 7.2.2. 

Current, symbolized by the letter i, is moving charge. Charge q is the quantity that participates in the Coulombic force, just like mass is the quantity that participates in the gravitational force. The smallest amount of charge encountered in most introductory physics classes is the magnitude of charge of one electron: $e = 1.60 \times 10^{-16}$ C.* All other measurable charges are multiples of this charge.

Current is the amount of charge through time: $i = \Delta q/\Delta t$. The unit of current is the **ampere** (A); the operationally defined standard of current is one coulomb of charge flowing through in a wire in one second.

TEMPERATURE

Temperature, usually symbolized by the letter T, has two meanings:

It's an experience like no other experience I can describe, the best thing that can happen to a scientist, realizing that something that's happened in his or her mind exactly corresponds to something that happens in nature.
LEO KADANOFF
American Physicist. 1937 -

- It is a measure of the warmth of a body.
- It is also a measure of the average energy of the atoms in the body under measure.

The SI unit for temperature is the **kelvin** (K). The operationally defined standard for the kelvin is 1/273.16 of the temperature of the **triple point** of water, a unique temperature where the gas, liquid, and solid forms of water all coexist.

The lowest possible temperature—called **absolute zero**—is at 0 K. All other temperatures are higher than this (and therefore, positive). Table 11.6 lists some temperatures.

Table 11.6: Some Temperatures (in kelvins)

Absolute zero	0	Lead melts	600
Helium boils	4.15	Gold melts	1340
Dry ice freezes	194	Platinum melts	2050
Water freezes	273	Tungsten melts	3680
Room tempera-	293	Carbon arc	5770
Body temperature	310	Surface of the sun	6270
Water boils	393	Iron welding arc	6290

Two other temperature scales exist; the conversions between the **Celsius**, **Fahrenheit**, and **Kelvin** scales are:

$$T_K = T_C + 273.15 \text{ and } T_C = \tfrac{5}{9}(T_F - 32)$$

Note that while Celsius and Fahrenheit temperatures are notated with the degree symbol (°), Kelvin temperatures are *not*.

AMOUNT OF SUBSTANCE

The **amount of a substance** is given in terms of the number of **moles** n of the substance:

* Some introductory physics courses may include a discussion of **quarks**, the building blocks of protons and neutrons (among other things). Quarks have a charge equal to $+\frac{2}{3}e$ or $-\frac{1}{3}e$.

$$n = \frac{N}{N_A} = \frac{m}{M}$$

where N is the number of molecules in the sample, N_A is **Avogadro's Number** $= 6.02 \times 10^{23}$ per mole, m is the mass of a sample of the material, and M is the material's **molar mass** (the mass of one mole of the substance; it is a number you look up in charts).

Your textbook should have a table of molar masses.

EXAMPLE 11.15: Number of Moles

Just as there are twelve items in a dozen—no matter what the items are—there are 6.02×10^{23} items in a mole. So, a dozen donuts contains twelve donuts; a mole of donuts contains 6.02×10^{23} of them!

ANGLE

The angle, as we have said before, is not a true fundamental quantity. However, because it is used so often in the introductory physics class, we take a closer look at it here.

Angles, usually symbolized by the Greek letters θ or φ, have two meanings:

- They are a measure of the *curved spatial extent* of an object. This is always a positive number.
- They are also a measure of *how far* an object is along a circular arc from a fixed reference point. This can be a positive or negative number.

The SI "unit" for angles is the **radian**,# which is $\frac{1}{2\pi}$ or about 16% of the angle of an entire circle (Fig. 11.1). The "standard" for the radian is the constant ratio of a circle's circumference to its radius: $C/r = 2\pi$.

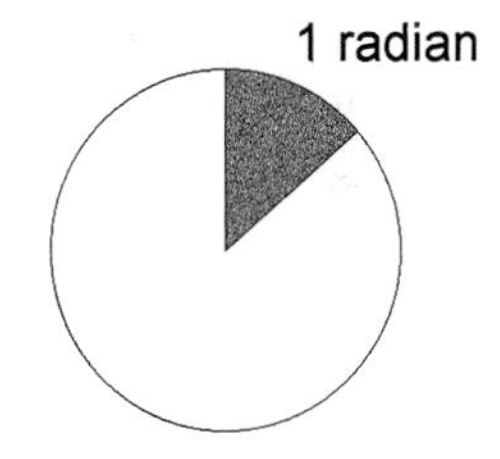

Figure 11.1

11.5.3 THE FUNDAMENTAL CONSTANTS

Throughout the history of physics, it has been shown that certain numbers appear in different theories with intriguing regularity. These numbers, some of which are listed in Table 11.7, are known as the **fundamental constants**.

The "unit" for angles, the radian, is not a true unit, just as the angle is not a true fundamental quantity. It is defined via purely mathematical concepts, and therefore is not measurable from nature in the same way as the units of most other fundamental quantities are.

However, in many ways, the radian is manipulated just like any other unit. Hence, we can consider the radian as a "unit": part true unit, part mathematical construct.

Solid angles have the unit *steradian* (sr).

Table 11.7: Fundamental Constants

CONSTANT	SYM.	VALUE	CONSTANT	SYM.	VALUE
Speed of light (vacuum)	c	3.00×10^{8} m/s	Electron mass	m_e	9.11×10^{-31} kg
Elementary charge	e	1.60×10^{-19} C	Proton mass	m_p	1.67×10^{-27} kg
Gravitation constant	G	6.67×10^{-11} m^3/kg·s^2	Neutron mass	m_n	1.68×10^{-27} kg
Planck constant	h	6.63×10^{-34} J·s	Permittivity of free space	ε_0	8.85×10^{-12} F/m
Avogadro's number	N_A	6.02×10^{23} /mol	Permeability of free space	μ_0	1.26×10^{-6} T·m/A

When the fundamental constants are used in the defining relations of derived quantities, we can *reduce* the number of fundamental quantities needed to describe the universe.

EXAMPLE 11.16: Relating Fundamental Quantities

One familiar fundamental constant is c, the speed of light. This constant can be used to reduce two fundamental quantities—length and time—to one composite quantity called length-time.

This equation is in Section 5.3.1.

Recall that the distance to an object is related to the time of travel via the derived quantity velocity: $v = \Delta x/\Delta t$. We can solve this equation for length:

$$x = v\,t$$

For more on proportionality, see Section 13.4.9.

This equation says that length is proportional to velocity, $x \propto v$, and time, $x \propto t$. But, if velocity were *constant* in this equation, then length would be *only* proportional to time. The speed of light is such a constant, so:

$$x \propto t \text{ only, if } v = c$$

This new equation says that at the speed of light, length and time are *interchangeable*. Length is time; time is length!

* * *

Dimensions are in Section 12.6.

We can emphasize the equivalent nature of length and time by taking a clever (but not obvious) next step. We define a system of units where c is exactly equal to one (without units or dimension). In this system, length and time are not only equal, they have the same dimension: $\mathrm{L} = \mathrm{T}$! So, we have just reduced the seven fundamental quantities to six.

One happy consequence of this special system is that other quantities besides length and time can be related as well, as long as c is defined to be 1 in the relating equation. For instance, Einstein's familiar equation tells us that $E = mc^2$. Because $c \equiv 1$, we can see that energy is intimately related to mass; in fact, in this system, they have the same dimension: $\mathrm{E} = \mathrm{M}$!

11.6 UNIT CONVERSION

The SI units of Table 11.1 are not always appropriate for every problem. Sometimes measurements are very large or very small; in these cases, we want a unit that is better scaled to the size of the number. Other times, measurements are more naturally presented in non-SI unit systems; therefore, it is often necessary to *convert* SI units into different unit systems.

Unit conversion, a method for dealing with both of these situations, is a very simple, but very important, procedure to master. The basis of unit conversion is the **conversion factor**. A conversion factor is a *ratio of units* equal to one.

See your text's appendix for a lengthy list of conversion factors.

EXAMPLE 11.17: Conversion Factors

For example, the conversion factor between minutes and seconds is 1 minute / 60 seconds, because one minute equals sixty seconds.

In the same way, the conversion factor between seconds and minutes is 60 seconds / 1 minute.

Note that these expressions are *not* written 1 / 60 or 60 / 1, for *without the units,* these ratios do not equal one.

To convert units, multiply the original measurement by one or more conversion factors. The conversion factors should be ordered such that units in the numerator cancel out with the *same* unit in the denominator of the next conversion factor.

Where observation is concerned, chance favors only the prepared mind.
LOUIS PASTEUR
French Biologist, 1822 - 1895

EXERCISE 11.18: Unit Conversion, Part I

Calculate the number of seconds in a year using unit conversion.

SOLUTION:

To convert seconds to years, we need several conversion factors: seconds to minutes, minutes to hours, hours to days, and days to years. We know these conversion factors from everyday experience to be:

$$\frac{60 \text{ seconds}}{1 \text{ minute}}, \quad \frac{60 \text{ minutes}}{1 \text{ hour}}, \quad \frac{24 \text{ hours}}{1 \text{ day}}, \text{ and } \frac{365 \text{ days}}{1 \text{ year}}$$

Thus, we can write:

$$\frac{60 \text{ secs}}{1 \text{ min}} \times \frac{60 \text{ mins}}{1 \text{ hour}} \times \frac{24 \text{ hours}}{1 \text{ day}} \times \frac{365 \text{ days}}{1 \text{ year}} = 31{,}536{,}000 \frac{\text{secs}}{\text{year}}$$

Exercise 11.18 Continued...

Notice how the minutes in the denominator of the first term cancel out the minutes in the numerator of the second term; the hours in the denominator of the second term cancel out the hours in the numerator of the third term, and so on. (Also note that we did not have to worry about whether or not a term was plural for the canceling to work.)

(By the way, the above-calculated value is just about $\pi \times 10^7$ s/yr; a value that makes the number of seconds per year easy to remember!)

Conversions are especially straightforward in the metric system, for all conversion factors are fractions or multiples of ten.*

EXERCISE 11.19: Unit Conversion, Part II

Convert 528 cm to kilometers. There are 1000 meters per kilometer, and 100 centimeters per meter.

SOLUTION:

The conversion factors are given in the statement of the problem. We need to rewrite these as ratios:

$$1000 \text{ meters per kilometer} = \frac{1000 \text{ m}}{1 \text{ km}}$$

$$100 \text{ centimeters per meter} = \frac{100 \text{ cm}}{1 \text{ m}}$$

However, we have to be a little bit careful, because we begin the problem with centimeters. Thus, our centimeter ratio must have centimeters in the *denominator* to cancel out the original centimeters in the numerator. In the same way, meters must be in the denominator of the other ratio to cancel out the meters in the numerator of the other conversion factor. Thus, we need to invert our conversion factors:

$$\frac{1 \text{ km}}{1000 \text{ m}} \text{ and } \frac{1 \text{ m}}{100 \text{ cm}}$$

Now we can do the conversion:

$$528 \text{ cm} \times \frac{1 \text{ m}}{100 \text{ cm}} \times \frac{1 \text{ km}}{1000 \text{ m}} = 5.28 \times 10^{-3} \text{ km}$$

* Except for time conversions. There *have* been attempts to "metricize" time conversions—most notably during the French Revolution—but none have caught on.

To simplify matters even further, some factors of ten in the metric system have a label, as listed in Table 11.8. These labels are attached to the beginning of a unit name, multiplying the unit by the associated conversion factor.# All prefixes are pronounced with emphasis on the first syllable.

Table 11.8: Prefixes for SI Units

POWER	PREFIX	SYMBOL	POWER	PREFIX	SYMBOL
10^{24}	yetta	Y	10^{-1}	deci	d
10^{21}	zetta	Z	$\mathbf{10^{-2}}$	**centi**	**c**
10^{18}	exa	E	$\mathbf{10^{-3}}$	**milli**	**m**
10^{15}	peta	P	$\mathbf{10^{-6}}$	**micro**	μ
10^{12}	tera	T	$\mathbf{10^{-9}}$	**nano**	**n**
10^{9}	giga	G	$\mathbf{10^{-12}}$	**pico**	**p**
$\mathbf{10^{6}}$	**mega**	**M**	10^{-15}	femto	f
$\mathbf{10^{3}}$	**kilo**	**k**	10^{-18}	atto	a
10^{2}	hecto	h	10^{-21}	zepto	z
10^{1}	deka	da	10^{-24}	yocto	y

EXAMPLE 11.20: Unit Prefixes, Part I

Since "kilo-" means 1000, a "kilometer" is 1000 meters. This word is pronounced KILL-o-meter, not kill-AH-mi-ter—its more popular version.

EXERCISE 11.21: Unit Prefixes, Part II

Rewrite the following measurements using unit prefixes:

(a) 6000 meters
(b) 0.01 seconds
(c) 53,000,000 radians
(d) 7.45×10^{-11} grams

SOLUTION:

(a) 6 kilometers
(b) 1 centisecond
(c) 53 megaradians
(d) 74.5 picograms

To convert between unit systems, we use the basic "cancel out" multiplication technique discussed above, and an appropriate chart of conversion factors.

Some of the prefixes of Table 11.8 are used more commonly than others. We mark these with bold lettering. By convention, prefixes representing factors of 1000 (10^3) are used most. Thus, you will rarely use the prefixes centi-, deci-, deka-, and hecto-. The use of centi- in "centimeter" is an important exception.

EXERCISE 11.22: Unit Conversion, Part III

Make the following conversions:

(a) Express 2467 meters in terms of millimeters.

(b) Express 843 km/min in terms of m/s.

(c) Express 94 feet in kilometers. There are 3.281 feet in a meter.

SOLUTION:

(a) A millimeter is one thousandth of a meter, so:

$$2467 \text{ m} = 2467 \text{ m} \times \frac{1000 \text{ mm}}{1 \text{ m}} = 2{,}467{,}000 \text{ mm} = 2.467 \times 10^{6} \text{ mm}$$

(b) Using the information given to us in the problem statement:

$$843 \frac{\text{km}}{\text{min}} = 843 \frac{\text{km}}{\text{min}} \times \frac{1000 \text{ m}}{1 \text{ km}} \times \frac{1 \text{ min}}{60 \text{ s}} = 14{,}050 \text{ m/s} \approx 1.41 \times 10^{4} \text{ m/s}$$

(c) Since we know that 3.821 feet $=$ 1 meter, we can do the conversion *as long as* we invert the ratio:

$$94 \text{ ft} = 94 \text{ ft} \times \frac{1 \text{ m}}{3.28 \text{ ft}} \times \frac{1 \text{ km}}{1000 \text{ m}} = 0.028658536 \text{ km} \approx 2.9 \times 10^{-2} \text{ km}$$

Once you understand units, prefixes, and conversions, you should take some time to get a "feel" for the size of different units.

EXAMPLE 11.23: Unit Size, Part I

For example, how long is a meter, really? A meter is just a smidgen longer than a yard (or three feet). Thus, a tall person has a height of about two meters.

How about the length of a centimeter, or a millimeter? A centimeter is about the length of your entire thumbnail. A millimeter, by contrast, is about the length of just the white of your thumbnail (if you keep your nails short).

Try these on for size. What is the mass of a kilogram? A smallish apple has a mass of about a kilogram. What about a gram? A paper-clip has a mass of about a gram.

EXERCISE 11.24: Unit Size, Part II

Spend some time now relating the size of other units to things with personal meaning. How long is a second? How big is an angle of π radians? $\frac{\pi}{3}$ radians? And so on.

11.7 CONCLUSION

As you progress through the course, you will begin to understand the vital importance of measurements to physical understanding. The contents of this chapter are critical for that learning process; unfortunately, however, they are almost always skipped in college-level physics textbooks. Therefore, make sure that you have read and understood this chapter closely before continuing on with your studies.

A few observations and much reasoning lead to error; many observations and a little reasoning lead to truth.
ALEXIS CARREL
Nobel Prize Winner in Medicine, 1873 - 1944

12
TRICKS OF THE TRADE

12.1 INTRODUCTION

Physicists are highly capable mathematicians. But their mathematical proficiency is more than just an ability to manipulate equations or add long sums. It is also an intuitive understanding of the size and meaning of numbers, as well as a capacity for analyzing equations and applying a broad menu of mathematical short cuts. These skills are collectively known as **numeracy**, or mathematical literacy.

To succeed at physics, you too need to be numerate. The mathematical requirements of the course demand it. Luckily, though, numeracy is not an inborn quality, or a special characteristic unique to mathematicians and physicists: it is a *learned* skill.

The purpose of this chapter is to teach you how to increase your numeracy. We look at some mathematical "tricks of the trade" that give physicists such an easy command of numbers. These techniques will help you cope with the high mathematical demands of a typical introductory physics class, as well as provide you with some opportunities to have a little mathematical fun along the way.

It is the highest degree astonishing to see what a large number of general theorems...[Faraday] found...without the help of a single mathematical formula....
HERMANN von HELMHOLTZ
German Scientist, 1821 - 1894

12.2 NUMERACY

One of the most important skills in physics, and in life, is to have a "feel" for the size and meaning of a number. It is very easy (and fun!) to learn this technique: simply *compare* the value of an abstract number to a quantity with which you are already familiar. You may have to do some sort of unit conversion to make this work, however.

For more on unit conversion, see Section 11.6. 

EXAMPLE 12.1: Numeracy, Part I

Let us practice this technique together by answering the question: "How *big* are the numbers million and billion, in terms of something with which we are familiar?"

✷ ✷ ✷

One thing we all have in common—and therefore, something that is intimately familiar to each of us—is breathing. At rest, a person breathes about five times per minute. You can check this by quietly observing your own breathing patterns.

Now, we can use this fact to make the abstract numbers million and billion more concrete and real. For example, you might ask yourself: "How long would it take a resting person to breathe one million times?"

Let's see. If a person breathes five times per minute, it would take:

$$1{,}000{,}000 \times \frac{1\,\text{minute}}{5\,\text{breaths}} = 200{,}000\,\text{minutes}$$

But just how long *is* 200,000 minutes? We convert units to find out:

$$200{,}000\,\text{minutes} \times \frac{1\,\text{hour}}{60\,\text{minutes}} \times \frac{1\,\text{day}}{24\,\text{hours}} \times \frac{1\,\text{month}}{30\,\text{days}} \approx 4.5\,\text{months}$$

200,000 minutes works out to about four and a half months! So, it takes roughly one-third of a year to breathe one million times at rest.

How does this compare to the number billion? Using the same technique as above, we see that it would take:

$$1{,}000{,}000{,}000 \times \frac{1\,\text{minute}}{5\,\text{breaths}} = 200{,}000{,}000\,\text{minutes}$$

such that:

$$200{,}000{,}000\,\text{minutes} \times \frac{1\,\text{hour}}{60\,\text{minutes}} \times \frac{1\,\text{day}}{24\,\text{hours}} \times \frac{1\,\text{year}}{365\,\text{days}} \approx 380\,\text{years}$$

Almost four hundred years!

Thus, while it takes about four months to breathe a million breaths, it takes about four hundred *years* to breathe a billion of them. Whereas a person will take tens of millions of breaths in one lifetime, no person at rest can take a *billion* breaths.

✷ ✷ ✷

To sum up, by comparing the numbers million and billion with something we closely understand—our own breathing—we are better able to understand the magnitudes or "large-ness" of these numbers.

We especially need imagination in science. It is not all math, nor all logic, but it is somewhat beauty and poetry.
MARIA MITCHELL
American Astronomer, 1818 - 1889

EXERCISE 12.2: Numeracy, Part II

(a) The current American budget deficit is around one trillion dollars. If every man, woman, and child in the country contributes ten dollars per year to eliminate this deficit, how long will it take to pay it off, assuming no interest accrues and the debt doesn't otherwise grow? Assume a population of 300 million people in the United States.

(b) Which is bigger, one millionth of the distance from coast-to-coast, or one billionth of the distance around the world?

SOLUTION:

(a) To find the number of years required to pay down the debt, assuming it doesn't grow, we do a short calculation:

$$\$1{,}000{,}000{,}000{,}000 \times \frac{1\text{ person}}{\$10} \times \frac{1\text{ year}}{300{,}000{,}000\text{ people paying}} = 333\text{ years}$$

Therefore, even if the deficit doesn't grow (a *very* unlikely assumption), your grand-children's grandchildren will still face a national debt!

(b) Again we make short calculations. First, we calculate one-millionth of the distance across the country. The width of the United States is about 5000 kilometers (or 3000 miles):

$$5000\text{ km} \times \frac{1}{1{,}000{,}000} = 0.005\text{ km} \times \frac{1000\text{ m}}{\text{km}} = 5\text{ m}$$

Thus, one-millionth of the distance from coast-to-coast is about five meters (or sixteen feet), the width of a large living room.

On the other hand, the circumference of the Earth is about forty thousand kilometers:

$$40{,}000\text{ km} \times \frac{1}{1{,}000{,}000{,}000} = 0.00004\text{ km} \times \frac{1000\text{ m}}{1\text{ km}} \times \frac{100\text{ cm}}{1\text{ m}} = 40\text{ cm}$$

Thus, one billionth of this distance is about forty centimeters (or approximately twenty inches), about two feet.

So, one millionth of the distance from coast-to-coast is longer than one billionth of the distance around the earth. Surprised?

EXERCISE 12.3: Numeracy On Your Own

Spend a few minutes thinking of similar numeracy examples on your own. You should try to get familiar with the sizes of various types of unusual numbers, like a trillion (a one with 12 zeros), a quadrillion (15 zeros), and a googol (100 zeros!); as well as fractions like a thousandth, a billionth, etc.

12.3 QUICK FIGURING

Another important mathematical skill to have in your physics toolbox is quick figuring. **Quick figuring** is the art of making rapid, rough calculations *in your head*. For simple problems, believe it or not, this method can be faster and easier to use than calculators.

Mathematicians use intuition, conjecture, and guesswork all the time....
JOSEPH WARREN
American Engineer, 1926 -

The essence of quick figuring is the fact that some numbers are easier to manipulate than others. For example, it is easier to divide ten by two than it is to divide eleven by two. So, to make a quick figure, replace actual numbers in a problem with simpler, but *similar*, numbers. Then, proceed with the mathematics as usual.

The result of a quick figure is an *estimate*, not an exact value—though, with a little care in your choice of replacement numbers, a quick figure can often come very close to the correct answer.

EXAMPLE 12.4: Quick Figuring, Part I

Let's work an example of quick figuring together by developing a good estimate of the quantity $10{,}455 \times \pi^2$, without using a calculator.

* * *

To begin any quick figure, we replace numbers in a problem with *simpler*, but *approximately equivalent*, values. For example, we can approximate $10{,}455$ as about $10{,}000$, since $10{,}000$ is fairly close to $10{,}455$, but much simpler to work with.

In the same way, we want to replace π^2 by an approximate, similar value. We can think of π, which is equal to $3.1415...$, as being a little bit bigger than 3. Thus, π^2 will be somewhat bigger than 3^2, which is 9. For simplicity, we'll just call it 10.

Thus, the result of our number replacements is:

$$10{,}455 \times \pi^2 \approx 10{,}000 \times 10$$

Now, the next step in quick figuring is to do the math (on the right-hand-side of this equation) in the usual way:

$$10{,}000 \times 10 = 100{,}000$$

Therefore, our quick figure for $10{,}455 \times \pi^2$ is about $100{,}000$. How close did we get? The actual value of the original calculation is $103{,}187$. Thus, the error on this quick figure is about three percent: a very small error, considering how easily we obtained this estimate.

Refer to Section 16.2.3 for more on calculating errors.

* * *

We can get an even better estimate, if we wish, by simply replacing the π^2 by 10, and leaving the $10{,}455$ alone:

$$10{,}455 \times 10 = 104{,}550$$

This quick figure has an error of only *one* percent.

EXERCISE 12.5: Quick Figuring, Part II

Estimate the approximate value of each of the following quantities:

(a) $\sqrt{97} \times 7$ (b) $\dfrac{(0.0452 \times 143)}{47}$ (c) $\dfrac{(73)(855+492)}{(523-346)}$

SOLUTION:

(a) 97 is approximately equal to 100. The square root of 100 is 10. So $\sqrt{97} \times 7 \approx 10 \times 7 = 70$. The actual answer for this calculation is about 68.9, yielding an error of less than two percent.

(b) 0.0452 is approximately equal to 0.05, 143 is approximately equal to 150, and 47 is approximately equal to 50. Now, we can do the math:

$$\frac{(0.0452 \times 143)}{47} \approx \frac{(0.05 \times 150)}{50} = 0.05\left(\frac{150}{50}\right) = 0.05(3) = 0.15$$

Our quick figure yielded a value of 0.15. The actual answer to this problem is about 0.14, an error of about seven percent.

(c)
$$\begin{aligned}\frac{(73)(855+592)}{(503-346)} &\approx \frac{(70)(850+600)}{(500-350)} \\ &= \frac{(70)(1450)}{(150)} \\ &\approx \frac{(70)(1500)}{(150)} \\ &= (70)(10) \\ &= 700\end{aligned}$$

The actual value of this computation is about 673, an error of about four percent. Notice that we quick figured *twice* here!

12.4 ORDER-OF-MAGNITUDE ESTIMATION

A third important mathematical technique is **order-of-magnitude estimation**. There are many names for order-of-magnitude estimation, such as "back-of-the-envelope calculation," "ball-park figuring," and "guesstimation," reflecting the fact that order-of-magnitude estimation is an extremely powerful tool in many areas besides physics.

Life is the art of drawing sufficient conclusions from insufficient premises.
SAMUEL BUTLER
English Poet, 1835 - 1902

An order-of-magnitude estimation starts with a problem for which we desire a rough numerical answer. We may initially have *no idea* what the answer will be, but we press ahead with the estimation anyway.

To do the estimation, we divide up the problem into small, manageable pieces. Then, for each piece, we make an *educated guess* as to its value. Finally, we combine all of the guesses into one answer.

EXAMPLE 12.6: Order-of-Magnitude Estimation

Order-of-magnitude estimation is best learned by example. Let's look at one first used by Enrico Fermi (1901 - 1954), an important twentieth century physicist and enthusiastic proponent of guesstimation.

Fermi once asked: "How many piano tuners are in the United States?" Now, it is important to understand that Fermi was not interested in an *exact* answer to this question, only a rough value. After all, it is likely that Fermi—a particle physicist—probably did not know the true answer to this question himself. But he obtained a rough answer anyway, and so shall we.

* * *

To begin the estimation, we divide the problem up into small, manageable pieces. For example, in order to find out how many piano tuners are in the United States, we need to know (1) how many pianos need to be tuned, and (2) how many pianos one piano tuner can tune alone.

We can break these two pieces down even further. For instance, to get a rough estimate of the number of pianos that need to be tuned (point (1) above), we might first estimate the United States population, then guess at the percentage of the population that owns pianos, and finally, calculate how often the pianos need tuning.

* * *

Scientific notation is in Section 11.3. 

Let's proceed according to that plan. So, we first need a figure for the population of the United States. We could look this fact up in an encyclopedia if we wanted, but we could just as easily pick a ballpark figure. Let's do the latter. Let's estimate that the population of America is around 300 million or 3×10^8 people.

Now we ask ourselves, "How many of those people own a piano?" To answer this rather obscure question, we must *guess* (unless, of course, we have a table of detailed statistics on piano ownership in the US handy). To make a *good* guess, then, we should *stop and think* for a minute: How many of our friends and neighbors own a piano?

From these few seconds of thought, we *guesstimate* that maybe as many as one in five people or as few as one in twenty own a piano. Taking the middle ground, we *choose* a figure of one in ten as our estimate.

Combining our first two guesses, then, we find that in the United States there are about:

$$(3 \times 10^8 \text{ persons}) \times \frac{1 \text{ piano}}{10 \text{ persons}} = 3 \times 10^7 \text{ pianos}$$

Next, we consider how often pianos need tuning. Again, we must make a guess: about once per year seems reasonable. Thus, we have:

$$(3 \times 10^7 \text{ pianos}) \times \frac{1 \text{ tuning}}{1 \text{ year}} = 3 \times 10^7 \frac{\text{tunings}}{\text{year}}$$

Example 12.6 Continued...

We have just answered the first piece of the puzzle; namely, we now have a rough estimate of the number of pianos that need tuning in the United States per year. Now we can move on to our second question: "How many pianos can one tuner tune in a year?"

* * *

To answer this question, we must make a few more reasonable guesses. Let's assume that piano tuners work a regular schedule of five days a week for fifty weeks a year. Let's further assume that a piano tuner can tune two pianos per day, once in the morning and once in the afternoon. Therefore, each tuner can tune about:

$$\frac{2\text{ pianos}}{1\text{ day}}\times\frac{5\text{ days}}{1\text{ week}}\times\frac{50\text{ weeks}}{1\text{ year}}=500\,\frac{\text{pianos}}{\text{year}}$$

This answers the second question: a piano tuner can tune about five hundred pianos per year. Combining this result with the previous one, we can obtain a rough answer to our original question: "How many piano tuners are there in the United States?" If we divide the number of pianos needing tuning per year, by the number of pianos one tuner can tune in a year, we get:

$$\frac{3\times10^7\text{ pianos/year}}{500\text{ pianos/year}}=60{,}000\text{ piano tuners in America}$$

Thus, there are about $60{,}000$ piano tuners in America. We have an answer!!! Hooray! But is this answer correct? And what does it *mean*?

* * *

The amazing thing is that this answer is probably pretty close to the actual value. However, *it doesn't matter* whether the number is *exactly* correct or not, because we—and Fermi—only wanted an *approximate* answer in the first place.

So, if this number is not necessarily correct, what meaning does it have, then? It tells us that there are *probably* around $60{,}000$ piano tuners in the United States. We would therefore be *very* surprised if someone told us there were less than 6000 piano tuners, or more than $600{,}000$. And if someone claimed that there were 6 million piano tuners in the US, we would *know* that they were flat wrong.

This information seems all the more satisfying when we remember that when we started this problem, we had *no idea* how many piano tuners there were in the United States. And yet we obtained a reasonable answer *anyway*. And therein lies the power of order-of-magnitude estimation!

EXERCISE 12.7: Pennies in the Sears Tower

How many pennies can fill the Sears Tower?

SOLUTION:

Our plan of attack is to (1) figure the volume of the tower, and then (2) figure out how many pennies could fill that volume.

* * *

Exercise 12.7 Continued...

First, we need to estimate the volume of the tower. Since the tower is shaped roughly like a rectangular solid, its volume is approximately given by the relation (length) × (width) × (height). To work this equation, therefore, we need guesstimates for the length, width, and height of the tower.

We can assume that the base of the tower takes up about one square city block. We can also assume that each city block is about 250 feet long, or about 80 m. Thus the length and width of the tower are each about seventy meters.

Next, we need to estimate the tower's height. Let's call it 110 stories, each story having a height of about 15 feet, or about 4.5 meters. Thus, the total height of the building is about 500 meters.

Now, using these estimates, we can calculate the tower's volume:

$$(\text{length})\times(\text{width})\times(\text{height}) = (80\,\text{m})(80\text{m})(500\,\text{m}) = 3.2\times10^6\ \text{m}^3$$

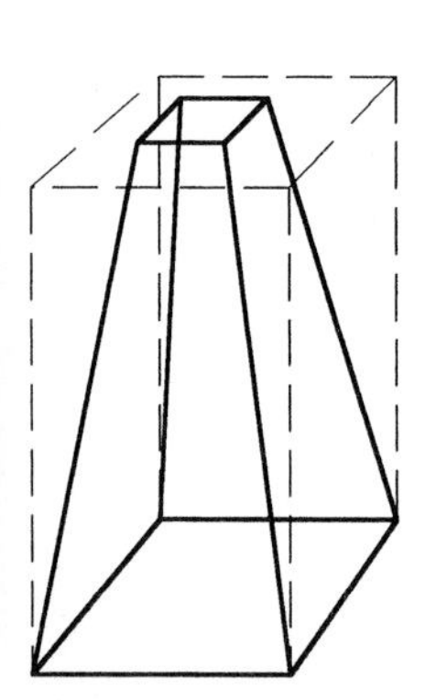

Figure 12.1 (a)
NOT TO SCALE

We can get fancy with this figure, if we'd like. For example, the tower is not shaped like a perfectly rectangular box. Instead, the top is significantly narrower than the bottom (see the schematic in Figure 12.1 (a)). Thus, we might adjust the above figure by taking away, say, 20% of the calculated volume to account for the narrowing. This would yield a new volume of:

$$(0.8)(3.2\times10^6\ \text{m}^3) \approx 2.6\times10^6\ \text{m}^3$$

If we wanted, we could adjust this value even more by taking into account the tower's infrastructure (insulation, elevator shafts, wiring, etc.). In essence, we are subtracting out places in the tower where we could not put pennies. Let's say the infrastructure reduces the tower's volume by another 20%. Repeating the above calculation once more, we get:

$$(0.8)(2.6\times10^6\ \text{m}^3) \approx 2.0\times10^6\ \text{m}^3$$

or about two million cubic meters.

* * *

Okay, we now have a figure for the volume of the Sears Tower. Now we can turn to the second part of the question, "How many pennies can fill this volume?" To answer this, we need to know the volume of one penny.

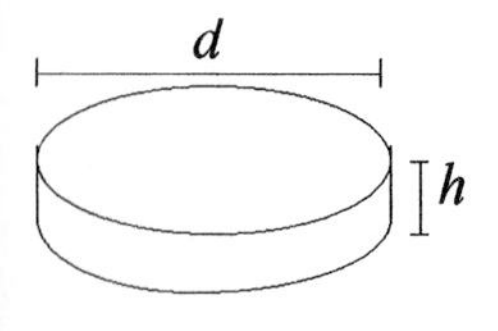

Figure 12.1 (b)
NOT TO SCALE

A penny is shaped like a squat cylinder. See Figure 12.1 (b). The volume of a cylinder is therefore $\frac{\pi}{4}d^2h$, where d is the cylinder's diameter, and h is its height. Now, each penny has a diameter of about a centimeter and a height of about a tenth that. Thus the total volume of penny is about:

$$\tfrac{\pi}{4}d^2t = \tfrac{\pi}{4}(0.01\,\text{m})^2(0.001\,\text{m}) \approx 7.8\times10^{-8}\ \text{m}^3$$

Let's call it $8\times10^{-8}\ \text{m}^3$.

* * *

Finally, to get the number of pennies in the Sears Tower, we need to divide the approximate volume of the tower by the approximate volume of one penny:

$$\frac{2\times10^6\ \text{m}^3}{8\times10^{-8}\ \text{m}^3} = \tfrac{1}{4}\times10^{6--8} = 0.25\times10^{14} = 25\times10^{12}\ \text{pennies}$$

So the Tower could hold about 25 trillion pennies, or a quarter of a trillion dollars worth!

EXERCISE 12.8: Order-of-Magnitude Estimation

You should spend some time thinking about other order-of-magnitude estimations. Try these on for size: How fast do your fingernails grow in miles per hour? How many people are born in the world per day? How many planets in the universe can support life? How many words will you speak in your lifetime? How many bananas are consumed in the US per year? And so on. Think of your own, too!

12.5 SCALING

All science is directed by the idea of approximation.
BERTRAND RUSSELL
English Philosopher, 1872 - 1970

A technique that is similar to, but slightly more involved than, order-of-magnitude estimation is scaling. **Scaling** allows the salient attributes of a problem to be understood even though an exact equation for the problem is lacking.

To use this technique, we choose an appropriate variable in a problem, vary it, and see how changing its scale alters other aspects of the phenomenon. We may have to make some assumptions about the problem, however, in order to complete the scaling.

EXAMPLE 12.9: Could King Kong Really Exist?!?

As with order-of-magnitude estimation, the best way to learn about scaling is by example. Let's consider the case of King Kong: could he exist in the real world?

Before we begin, we must make an assumption about King Kong. From various B-movies, we learn that King Kong was an average ape, normal in every way, until a terrible accident made him very large. From this information, we can *assume* that King Kong is essentially just a very large ape: that is, he has the same bones, musculature, etc., as any other ape. This assumption will be used several times in the discussion below.

* * *

We start the scaling by choosing a scaling variable. To do this, we must take a few minutes and think about the *main* difference between King Kong and other apes. Since we have already assumed that King Kong is merely a very large ape, we know that the key distinction between King Kong and other apes is "size." Thus, we can choose "length" as our scaling variable.

Now, to make the discussion concrete, we want to (arbitrarily) choose a number that describes King Kong's "size" in comparison to other apes. Ideally, this number should be nice, round, and easily manipulated. One number that fits the bill is 100; that is, we can consider King Kong to be 100 times "bigger" than other apes. In other words,

This equation is in Section 11.5.2.

This equation is in Section 11.5.2.

Example 12.9 Continued...

we are *assuming* that every length scale on King Kong is 100 times longer than the corresponding length on an ape.

✱ ✱ ✱

The next step is to start to think a little bit more carefully about apes and "size." What aspects of an ape depend upon *length*? We can be as straightforward, or as fancy, as we'd like with this type of analysis...we'll do a little bit of both.

For example, one aspect of apes that depends upon length is *volume*. The volume of any object is proportional to the third power of its length. Thus, if every length on an ape were increased by a factor of 100, the volume of King Kong would increase by $100^3 = 10^6$ times.

On the other hand, another aspect of apes that depends on length is *mass*. How so? Recall from Chapter 11 that $\rho = m / V$. This implies that $m = \rho V$. So, if ρ is a constant—for example, when comparing the density of King Kong and other apes—then mass is directly proportional to volume: $m \propto V$. So, King Kong's mass is proportional to his volume. But his volume is 10^6 times the ape's, so King Kong has 10^6 times as much mass as an ape.

We also know that mass is proportional to weight. So if King Kong is 10^6 times as massive as an ape, he will be 10^6 times heavier than the ape.

Let's get a little bit fancy now. Let's consider cross-sectional area. Cross-sectional area increases as the square of length. Thus, the cross-sectional area of King Kong is $100^2 = 10^4$ times the cross-sectional area of an ape.

Similarly, the surface area of an object also depends on the square of length. Therefore, the surface area of King Kong is 10^4 that of an ape.

✱ ✱ ✱

This is probably enough information to get a rough idea on the scaling of King Kong. So we'll stop doing math, and start analyzing our results. Let's summarize them:

VARIABLE	FACTOR	VARIABLE	FACTOR
Volume	10^6	Cross-Sectional Area	10^4
Mass	10^6	Surface Area	10^4
Weight	10^6		

What do these results say about King Kong? Well, first of all, we know that King Kong's head weighs a million times more than an ape's head. However, King Kong's neck has a cross-sectional area only ten thousand times larger. Thus, the muscles in King Kong's neck would have to support $10^6 \div 10^4 = 10^2 = 100$ times more weight than the ape's neck muscles. But since King Kong has the *same* musculature as an ape (by assumption), King Kong would not be able to hold up his own head!

Let's try another one. We know that King Kong has ten thousand times more skin (surface area) than an ape. However, this extra skin holds back a *million* times more weight. Thus, King Kong's guts are pressing out with $10^6 \div 10^4 = 10^2 = 100$ times more pressure than in the case of the ape. But an ape's skin will not be able to withstand this pressure, and so King Kong will explode long before he gets to Fay Wray.

We can get even more obscure, if we wish. For example, you may not know that the amount of food an animal eats is proportional to its surface area. Thus, King Kong eats ten thousand times more food than an ape. However, he weighs a *million* times more. Therefore, King Kong could eat only about $1/100$ of the food he needed to sustain his weight. No wonder he's so mad! He's starving!

We could go on with examples like these. But suffice it to say, from the examples we have already seen, King Kong could *not* exist in the real world. *His scale is all wrong.*

EXAMPLE 12.10: Scaling in Real Life

Think about the scaling arguments behind the following problems:

(a) If King Kong, which is just a large ape, can't exist, then why could dinosaurs, which can be thought of as large lizards, exist?

(b) Why is kindling easier to ignite than logs?

(c) If a football player can bench-press his weight, could a tiny football player, only half as tall, do the same?

SOLUTION:

(a) The difference between King Kong and dinosaurs is in our assumption that King Kong is just a large ape. This assumption is *not* true in the case of dinosaurs: dinosaurs are *not* large lizards. Dinosaurs evolved entirely different body types, musculatures, and skeletons to cope with the demands of being very large.

For example, their muscles are stronger with less weight, and their bones are disproportionately larger and lighter, than those of smaller lizards. Thus, dinosaurs could exist whereas King Kong could not, because they evolved adaptations to *compensate* for their very large scale.

(b) Kindling has a smaller cross-sectional area, and less surface area, than logs. Thus, more fire reaches a larger percentage of the wood in a shorter time, and ignition happens faster.

(c) The tiny player would have only $0.5^3 = 0.125$ times the weight of the regular player, but would have arms with cross-sectional area $0.5^2 = 0.25$ times as small. Therefore, the ratio of cross-section to weight, which is 1 for the regular player, is 2 for the tiny player. Thus the tiny footballer could bench-press *more* than his weight (twice as much!), assuming all else was equal. (By the way, this is *not* a good assumption; see part (a) above.)

12.6 THE BEST TRICK: DIMENSIONAL ANALYSIS

This chapter has mostly been concerned with calculations that can be made without the use of explicit formulas. While there are many situations in introductory physics where these techniques are useful, in most circumstances, an equation is required. However, we can apply some of the same sorts of non-equational techniques we have just learned to the analysis of equations. The process of analyzing equations for interesting properties is called **dimensional analysis**.

The equation is the final arbiter.
WERNER HEISENBERG
American Physicist, 1901 - 1976

Every fundamental quantity has an associated **dimension**. For example, "lengthy-ness" is different from "mass-ness," so "length" and "mass" have different dimensions. Dimensions are *not* units; instead, they describe the essential nature of quantities themselves. Therefore, dimensions are *unitless*. However, you can mathematically manipulate dimensions just like units; that is, you can multiply or divide different dimensions, but not add or subtract them.

Dimensions are written inside brackets (see Exercise 12.11 for an example of this): L for length, T for time, M for mass, I for current, τ for temperature, A for amount of a substance, and $\mathcal{L}$ for luminous intensity. In an introductory physics class, you will only need the dimensions of L, T, and M. Angles, numbers, and transcendental functions are **dimensionless**. The dimensions of derived quantities are expressed in terms of the dimensions of the fundamental quantities.

EXERCISE 12.11: Dimensional Analysis, Part I

What is the dimension of velocity?

SOLUTION:

This equation is discussed in Section 5.3.1.

Velocity is related to length and time by $v = \Delta x/\Delta t$. Now, length has dimension L, and time has dimension T. Thus, the dimension of velocity is $[v] = \mathrm{L} / \mathrm{T}$. This can be written more simply as $[v] = \mathrm{LT}^{-1}$.

We can use dimensional analysis to determine the dimension of an unknown quantity, as long as there is an equation relating the unknown quantity to known quantities. This is because *all equations must have the same overall dimension in every term.*

EXERCISE 12.12: Dimensional Analysis, Part II

We can write acceleration in terms of time t, positions x and x_i, and velocity $v_{i,x}$ as:

This equation is discussed in Section 5.4.2.

$$x = x_i + v_{i,x}t + \tfrac{1}{2}a_x t^2$$

What is the dimension of acceleration?

SOLUTION:

Since we know the dimensions of length, time, and velocity, we can determine the dimension of acceleration by plugging the known dimensions into the defining equation above. There is no need to worry about the $\frac{1}{2}$ while doing this, though, for $\frac{1}{2}$, being a number, is dimensionless.

Exercise 12.12 Continued...

So, we plug in the known dimensions:

$$\mathrm{L} = \mathrm{L} + \frac{\mathrm{L}}{\mathrm{T}}\mathrm{T} + [a_x]\mathrm{T}^2$$

Notice that the only term on the left-hand side of the equation, as well as the first two terms on the right-hand side, all have overall dimensions of *length*. Therefore, the third term on the right-hand side of the equation, which contains acceleration, must also have an overall dimension of length.

In order to make this happen, we need to manipulate the dimension in the position marked by "a_x". We need an L in the numerator, and a T^{-2} in the denominator (to cancel out the multiplicative T^2) to do this. That is, "a_x" must have dimensions of LT^{-2}.

Therefore, acceleration has dimensions of length per time squared.

We can also use dimensional analysis to check an equation's accuracy. To do this, we plug the dimension of each quantity into the equation. If one or more terms in the equation now have the incorrect dimension, then the equation *must* be wrong.

Be careful with this technique, though, for dimensionless terms in the equation *cannot* be checked with this method. Therefore, while dimensional analysis can prove an equation wrong, it *cannot* prove an equation correct. However, in most cases, knowing that an equation is dimensionally correct is better than not knowing anything at all.

EXERCISE 12.13: Checking Your Work

Imagine that, as a part of a homework assignment, you are asked to solve the equation in Exercise 12.12 for acceleration. Use dimensional analysis to check that your new equation has the correct dimensions. Can we be absolutely sure that this new equation is correct?

SOLUTION:

First we do the algebra:

$$\tfrac{1}{2}a_x t^2 = (x - x_i) - v_{i,x}t$$

$$\Rightarrow a_x = \frac{2[(x - x_i) - v_{i,x}t]}{t^2}$$

We have thus obtained a conditional equation for acceleration. Now, we want to check and see that this equation is dimensionally correct.

We know that acceleration has the dimensions of LT^{-2} from the previous example. We also know the dimensions of every other term in

Exercise 12.13 Continued...

this equation, as well. So, we can plug this information into our conditional equation:

$$\frac{\mathrm{L}}{\mathrm{T}^2} = \frac{2\left[(\mathrm{L}) - \left(\frac{\mathrm{L}}{\mathrm{T}}\right)(\mathrm{T})\right]}{\mathrm{T}^2}$$

$$\Rightarrow \frac{\mathrm{L}}{\mathrm{T}^2} = \frac{\mathrm{L}}{\mathrm{T}^2}$$

This indeed does work out to have the correct dimensions of LT^{-2}. Our equation is at least not dimensionally wrong.

Even though the dimensional analysis checked out, however, we cannot be absolutely sure that this equation is 100% correct. We only know that it is *dimensionally* correct—which is better than nothing.

A mathematician may say anything he pleases, but a physicist must be at least partially sane.
JOSIAH WILLARD GIBBS
American Physicist, 1839 - 1903

Dimensional analysis is an important tool for physicists on the front line of equation development. Physicists can often determine the rough equational relationship between variables solely on the basis of dimensional analysis.

EXERCISE 12.14: Dimensional Analysis, Part III

A physicist determines that the acceleration of an object moving in a circle depends *only* on the object's velocity v and the radius r of the circle in which the object travels. Determine the functional relationship between a, v and r using dimensional analysis.

SOLUTION:

Since acceleration depends solely on the object's velocity and radius of motion, we know that a must be proportional to $v^x r^y$, where x and y are some possible exponents that we need to determine.

However, we also know the dimensions of acceleration (LT^{-2}), velocity (LT^{-1}), and length (L). Thus, we can determine x and y using dimensional analysis:

$$[a] = \frac{\mathrm{L}}{\mathrm{T}^2} = [v^x r^y] = \left(\frac{\mathrm{L}}{\mathrm{T}}\right)^x \mathrm{L}^y$$

Looking at this equation, we see that we can obtain the necessary T^2 in the denominator of a if x is a 2. But that leaves us with an L^2 in the numerator of the same term...not to mention the L in the multiplicative term! These two terms need to combine somehow to give us just one L in the numerator. We can do this if y is -1 (check it!). Thus we have:

$$[a] = \frac{\mathrm{L}}{\mathrm{T}^2} = [v^2 r^{-1}] = \frac{v^2}{r}$$

Exercise 12.14 Continued...

Now, this formula, obtained through dimensional analysis, may not be completely correct, for there may be some dimensionless terms in the equation that we cannot determine. Luckily, though, in this *real* example, the actual equation *is* $a = v^2 / r$. Thus, in this case, dimensional analysis gave us our equation *exactly*!

See Section 7.2.9 for more on this equation.

12.7 SYMMETRY

We have seen that, today, physical laws are judged as much for their *symmetry* as they are for their ability to predict experimental outcomes. A theory—or anything else, for that matter—is considered **symmetric** if it appears exactly the same following some specific procedure.

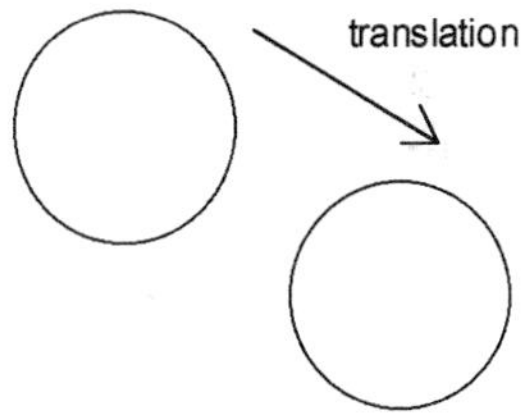

EXAMPLE 12.15: Symmetry and Circles, Part I

For example, a circle has *translational symmetry.* When you move it in a straight line, no matter where you move it, the circle does not change its appearance. See Figure 10.2 (a).

Circles also have *rotational symmetry.* No matter how you rotate a circle—by a large angle or a small one, clockwise or counterclockwise—the circle always appears exactly the same. See Figure 10.2 (b).

translation

Figure 10.2 (a)

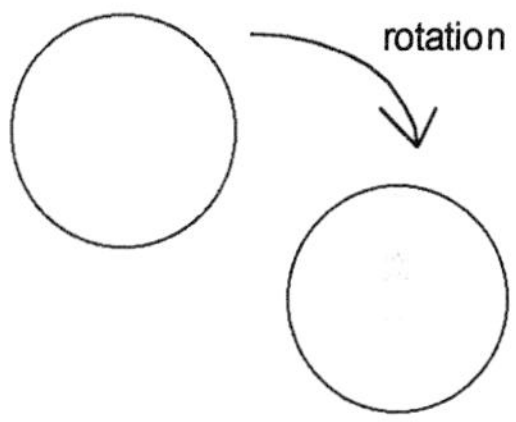

rotation

Figure 10.2 (b)

Symmetry can be **broken**. That is, some procedures can end the symmetry of the theory or the object forever.

EXAMPLE 12.16: Symmetry and Circles, Part II

For example, say that you place a small mark on your circle. Now, when you rotate the circle through any angle other than 2π, the circle will change its appearance. See Figure 10.3. Thus, the circle's rotational symmetry has been *broken* by the small mark.

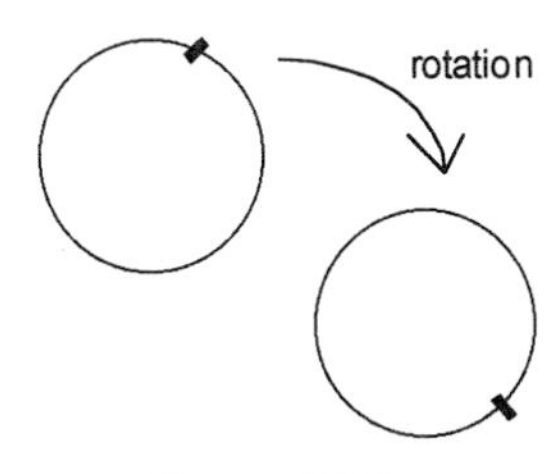

Figure 10.3

Symmetry is a vastly useful guiding principle when evaluating the validity of theories, especially in regards to the conservation laws. It has been shown (**Noether's Principle**) that for every type of symmetry in the universe, there is a corresponding conservation law (and vice versa).

Beauty is the first test:
there is no place in this world for ugly mathematics.
GODFREY HAROLD HARDY
English Mathematician, 1877 - 1947

EXAMPLE 12.17: Symmetry & Conservation Laws

For example, in all cases that have been tested to date, it has been found that the laws of physics do not change when moved along a straight line, so, physical laws have translational symmetry (or, using fancier words, the universe seems to exhibit *homogeneity of space*). By Noether's Principle, there is a corresponding conservation law; in this case, it is the *conservation of linear momentum.*

See Section 8.7 for more on linear momentum.

Similarly, it has been found that the laws of physics don't change when moved around a fixed axis; physical laws therefore seem to obey rotational symmetry (or *isotropy of space*). The corresponding conservation law is *conservation of angular momentum.*

Section 8.7 has more on angular momentum.

Other symmetries and corresponding conservation laws exist. For example, the laws of physics don't change through time. The conservation law corresponding to this *temporal symmetry* (or *homogeneity of time*) is the *conservation of energy.* Likewise, physical laws don't change when we change coordinate systems; this *gauge symmetry* corresponds with *conservation of charge.*

Refer to Sections 8.4 and 8.6 for energy conservation.

* * *

Interestingly, there are some symmetries in nature that we would *expect* to be true, but *aren't.* Noether's Principle tells us, therefore, that there cannot be a corresponding conservation law. For example, we might expect (even though we could never see it happen!) that physical laws would not change even if time ran backwards. However, it has been proven that physical laws *do* change if we go backwards in time. Similarly, although we might expect it to be true, experiments tell us that the laws of physics are *not* the same when reflected in a mirror. Hence, there is no *parity.*

12.8 CONCLUSION

In mathematics I can report no deficience, except that men do not sufficiently understand [its] excellent use.
FRANCIS BACON
English Philosopher, 1561 - 1626

This chapter has introduced some of the "tricks of the trade" that physicists use to make mathematics easier to grasp. Your ability to do physics is greatly enhanced when you use these techniques. Furthermore, most of these techniques can be used in your every day life as well. Therefore, spending some time mastering these techniques before you tackle your physics course will certainly pay you dividends sooner or later.

13
ALGEBRA, GEOMETRY, & TRIGONOMETRY

13.1 INTRODUCTION

This chapter reviews all of the non-calculus mathematics—including algebra, geometry, and trigonometry—you will need (and some you won't!) for a typical introductory physics course.

This chapter is very long; indeed, it is the longest (by far) of the entire book. Therefore, we do not intend that you read this chapter page by page; rather, you should *flip through it* before your class begins to ensure that you are familiar with its material.* Then, you can use this chapter as a *reference* when relevant mathematical topics come up in your class.

As you look through this chapter, you will undoubtedly find subjects that appear unfamiliar to you. If your memory of any of this material is a little fuzzy, it may be comforting to know that *you are not alone*: there are certainly other students in your class who also find math unfamiliar.

But, like it or not, mathematics is an integral part of physics, and it cannot be avoided. It is better to take some time to relearn old math skills now, than to be unable to understand physical concepts later.

So, take a deep breath, and jump in!

[Mathematics] brings together the most diverse phenomenon and discovers hidden conformities which unite them.... It renders [them] present and measurable.

What is more remarkable still, it follows the same method in the study of all phenomenon; it interprets them in the same language, as if to affirm the unity and simplicity of the plan of the universe, and to make still more manifest the immutable order which presides over all natural events.

JEAN BAPTISTE FOURIER
French Mathematician, 1768 - 1830

---

* If you are not familiar with a *majority* of the topics in this chapter, we strongly suggest that you take some time to freshen up your math skills, using this chapter as a guide, as soon as possible. Math deficiencies will hurt your ability to learn physics.

13.2 COORDINATE SYSTEMS

In the scientific method, Section 2.3.1.

The first step in the process of representing reality mathematically, even though it is usually skipped in physics textbooks, is the assignment of a coordinate grid to a problem. This section will teach you to set up your own coordinate systems.

13.2.1 INTRODUCTION TO COORDINATE SYSTEMS

Coordinate systems are important for vectors (Chapter 15) and kinematics (Chapter 5).

A **coordinate system** is a labeled set of **axes** that superimposes a grid of numbers onto a real world situation. In some coordinate systems, the axes intersect at a pre-determined reference point, called the **origin**. However, it is not *necessary* that the axes do so; sometimes, only two of the axes intersect at the origin, and sometimes, none of them do.

> **EXAMPLE 13.1:** A Road Map, Part I
>
> The most familiar kind of coordinate system is the grid on a road map. A good map superimposes a network of lines over an image of real world terrain. The map grid has two labeled axes, representing the "north-south" and "east-west" directions, which intersect at the lower left-hand corner of the map.

The region described by the coordinate system, called the **space**, has a certain number of **dimensions**, or directions that are distinct from each other. For instance, the everyday space we live in has three discrete dimensions: length, width, and depth. A good coordinate grid has one axis for every dimension in the space. These axes are **orthogonal*** to each other, so that each axis defines one and only one dimension.

> **EXAMPLE 13.2:** A Road Map, Part II
>
> The space of a road map is a small area of the *surface* of the world, which has two dimensions. So, a road map has two dimensions, too. The axes of the road map are perpendicular, or orthogonal, to each other, as are the lines on the grid.

Each point in the system is assigned a set of numbers, called the **coordinates**. In a set, there is one coordinate for each dimension.

---

* One way to understand orthogonality is to think of the axes as perpendicular to each other. For example, the direction of "up/down" is perpendicular to the direction of "left/right." The "up/down" direction, therefore, is completely distinct from the "left/right" direction.

Be careful, though! Just because axes are orthogonal *does not necessarily mean* the axes are perpendicular. It's just a way to think about the idea while you are getting used to the new term. For more information on this topic, see any advanced mathematics text.

EXAMPLE 13.3: A Road Map, Part III

Positions on a road map are identified via a system of coordinates. Since there are two dimensions on a road map, each point has two coordinates. Each position's coordinates are (usually) listed on the back or side of the map, and positions are found via labeled axes and grid.

It is important to point out that coordinate grids are *not* the same things as graphs, though the two look very similar (for example, they both have axes): coordinate systems are used to fix positions in space, while graphs are used to plot the relationship between one or more quantities in a system.

We discuss graphs in more detail in Section 16.2.1.

In the following sections, we examine several important kinds of coordinate systems.

13.2.2 THE CARTESIAN COORDINATE SYSTEM

The simplest and most important coordinate system used in introductory physics is the **Cartesian** or **rectangular coordinate system** (Fig. 13.1).

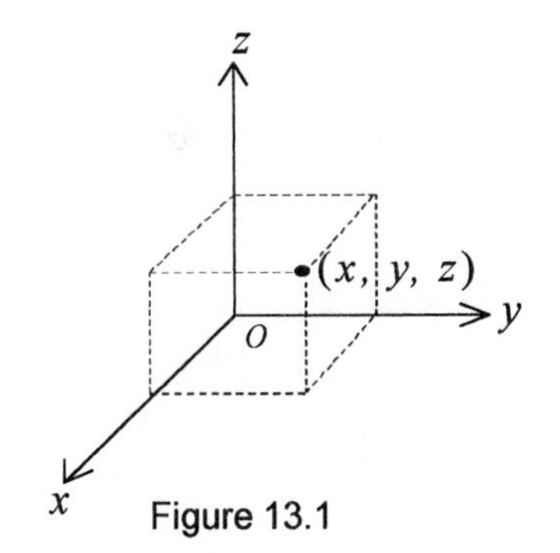

Figure 13.1

In this system, the axes—labeled x, y, and z in three-dimensional space—are mutually perpendicular and intersect at a single origin O. By convention, the sense of Cartesian axes is *right-handed*: that is, the positive axes lie in the same configuration as the thumb, index, and middle fingers of your *right* hand.#

Positions in the system are labeled by the set of numbers (x, y, z), where x is the (straight line) distance to the point from the y-z-plane; y is the distance from the x-z-plane; etc.^ Refer to Figure 13.1 again.

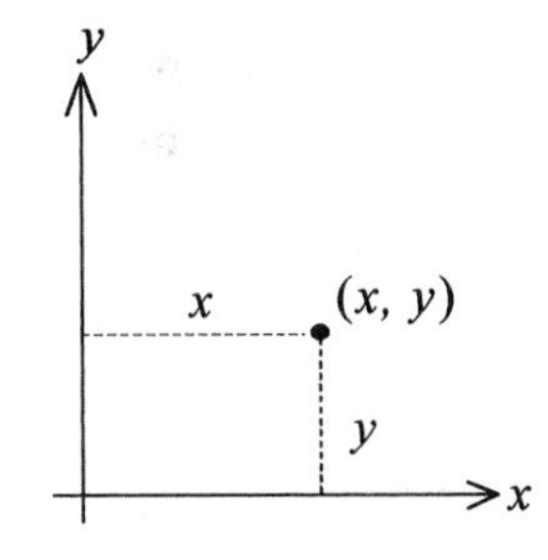

Figure 13.2

In most situations in your introductory physics class, however, you will probably use two-dimensional Cartesian systems, rather than the three-dimensional ones. A two-dimensional Cartesian coordinate system consists of the x- and y-axes only (Fig. 13.2). A point in this system is labeled by (x, y), where x is the distance measured from the y-axis, and y is the distance from the x-axis. We recognize four **quadrants** in the two-dimensional grid, as sketched in Figure 13.3. Note that only the first quadrant was sketched in Figure 13.2.

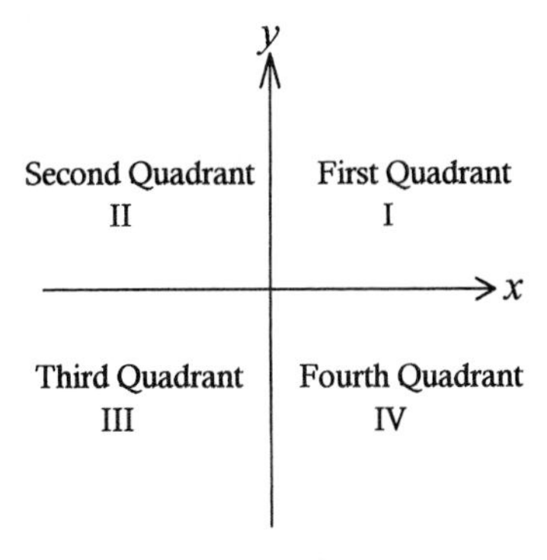

Figure 13.3

EXERCISE 13.4: Cartesian Coordinates

Answer the following questions regarding the two-dimensional Cartesian coordinate system sketched in Figure 13.4 below:

What would a left-handed coordinate system look like? Not the same!

^ Notice that both the axes, *and* the positions on the grid, are labeled x, y, and z. Do not get these labels mixed up.

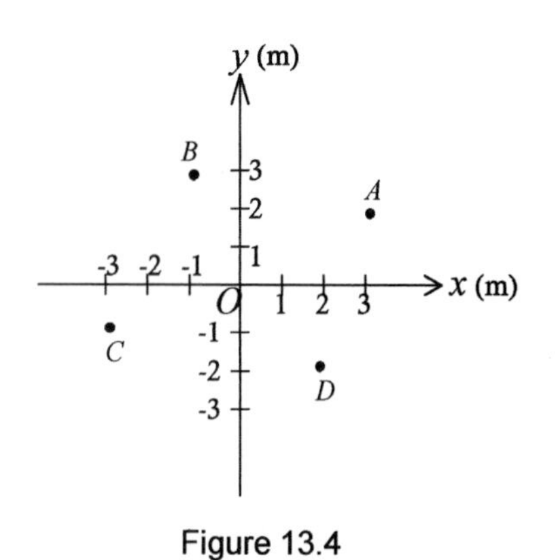

Figure 13.4

Exercise 13.4 Continued...

(a) What is the distance from the x-axis (the y-coordinate) for point A?

(b) How far from the y-axis (the x-coordinate) is point B?

(c) In what quadrant does point C lie?

(d) What are the coordinates of point D?

(e) In which quadrant would the point (-3 m, 2 m) lie?

SOLUTION:

(a) 2 m (b) 1 m (c) III (d) (2 m, -2 m) (e) II

13.2.3 THE POLAR COORDINATE SYSTEM

Despite its simplicity and familiarity, the Cartesian coordinate system is not always the most convenient choice for certain problems.

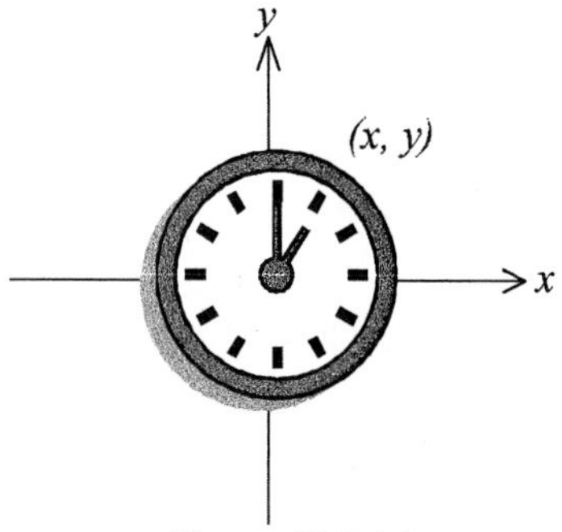

Figure 13.5 (a)

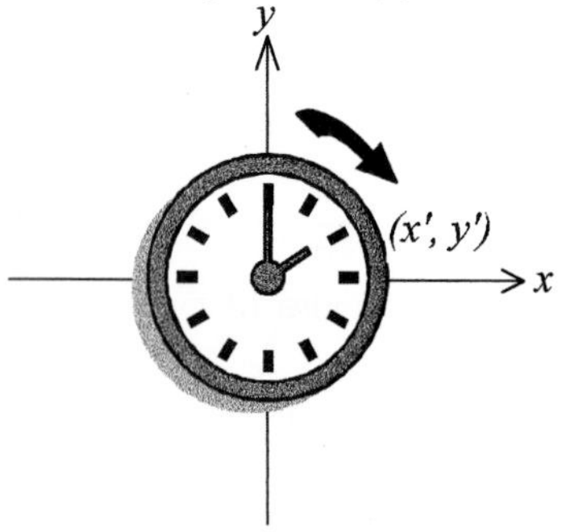

Figure 13.5 (b)

EXAMPLE 13.5: Clocks

To see that this is true, think of the hour hand of a clock as it rotates around the clock's face.

We can describe the (moving) location of the tip of the hand using Cartesian coordinates (x, y), as in Figure 13.5 (a). However, as the hand circles, the location of the tip changes (Fig. 13.5 (b)). Thus, the x and y values for the tip of the hand, in this system, change as well, to (x', y').

Therefore, while it is certainly possible to use Cartesian coordinates in this situation, it is not very *simple* to do so. A better system would take into account the fact that the length of the hour hand does not change through time; only the angle through which it sweeps changes. Such a choice of coordinate system would then only have one changing variable, not two.

For many situations, the **polar** or **circular coordinate system** is a better choice than the Cartesian system. In the polar system, a point P has coordinates r, the **radial distance** from O to P, and θ, the **angle** measured in radians *counter-clockwise* from the x-axis to r (Fig. 13.6) rather than x and y. Thus the polar coordinates of P are (r, θ).*

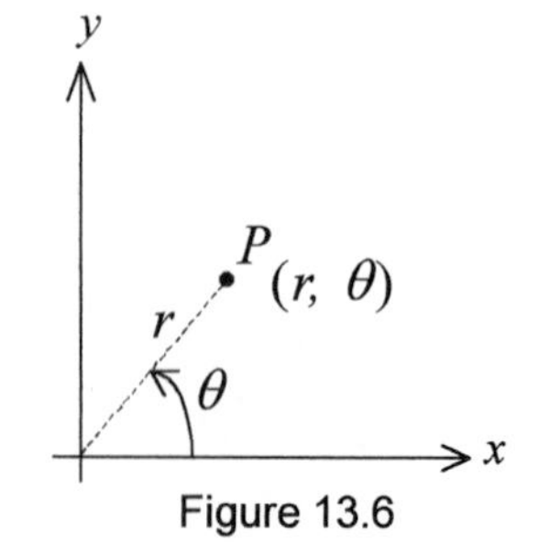

Figure 13.6

* Very unfortunately, the notation for the polar coordinates varies with textbook. Sometimes these coordinates are written (ρ, θ) or (ρ, φ) rather than (r, θ).

We can convert between the Cartesian and polar coordinate system using Figure 13.4 and a little trigonometry:

$$x = r\cos\theta \qquad y = r\sin\theta$$

$$r = \sqrt{x^2 + y^2} \qquad \theta = \tan^{-1}\left(\frac{y}{x}\right)$$

WARNING!

The equations for x, y, r, and θ above *only* apply to a coordinate system defined as in Figure 13.4. If the x-y axes are rotated, or if θ has another definition, the relationships can *change*. Therefore, do *not* memorize these equations, but instead know how to *derive* similar equations for each new case you encounter.

To obtain equations appropriate to your situation, sketch a triangle similar to Figure 13.4. Then apply trigonometry to the sketch. The equations obtained will be similar in form to the ones given above (*i.e.*: they will have sines, cosines, etc.), but may not be *exactly* the same.

Trigonometry is covered in Section 13.6.

WARNING!

When using a calculator to figure the inverse tangent function, be aware that the calculator displays the *smallest* correct angle. For example, the inverse tangent of 1.5 is both 0.98 rads and 4.1 rads, but the calculator only displays 0.98 rads.

The displayed angle may not be the appropriate answer for the problem at hand. Therefore, *always* check your answer for plausibility when using the inverse tangent function. Use the quadrant grid to help. If you find that the larger angle is the correct one for your situation, add π rads to the smaller angle obtained by the calculator.

EXERCISE 13.6: Converting Between Systems

The Cartesian coordinates of two points in the x-y plane are (a) (2.3 m, 4.7 m), and (b) (–2.3 m, –4.7 m). Find the corresponding polar coordinates for these points.

Polar coordinate systems are useful with circular motion and rotation. See Sections 5.4, 5.5, and 6.5 for more.

SOLUTION:

First, we sketch the situation:

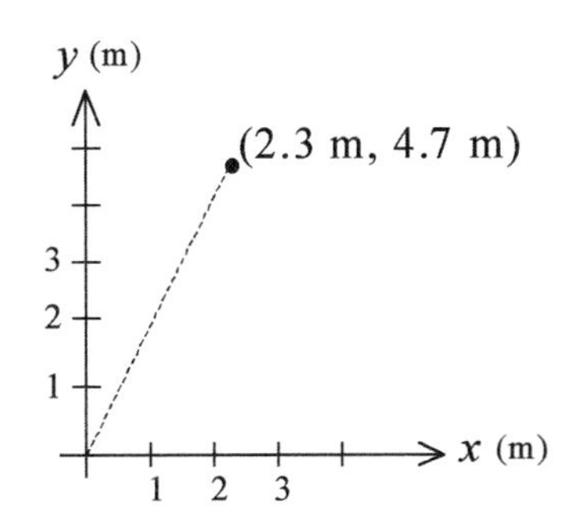

Exercise 13.6 Continued...

Since the axes in this problem are defined in the same way as Figure 13.4, we can use the equations given above verbatim:

$$r = \sqrt{x^2 + y^2} = \sqrt{(2.3\text{ m})^2 + (4.7\text{ m})^2} = \sqrt{27.38\text{ m}^2} \approx 5.2\text{ m}$$

$$\theta = \tan^{-1}\left(\frac{y}{x}\right) = \tan^{-1}\left(\frac{4.7\text{ m}}{2.3\text{ m}}\right) \approx 1.1\text{ rad}$$

(b) First, we sketch the situation:

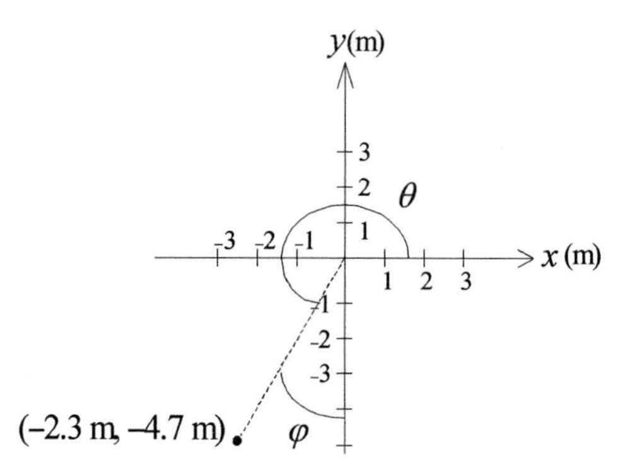

This situation is *not* defined the same way as in Figure 13.4. While r still has the same value, $r = 5.2$ m (check it!), θ has changed.

To analyze this new situation, we must draw a right triangle. One choice for this triangle is sketched below, using the angle φ rather than θ:

φ

5.2 m

(−2.3 m, −4.7 m)

4.7 m

Since we know the length of the hypotenuse, $r = 5.2$ m, and the length of the vertical leg (it is the y coordinate), $y = 4.7$ m, we can write:

$$\cos\varphi = \frac{opp}{hyp} = \frac{4.7\text{ m}}{5.2\text{ m}} = 0.9038$$

$$\Rightarrow \varphi = \cos^{-1}(0.9038) \approx 0.46\text{ rad}$$

Now, from the original sketch at the top of the page, we know that $\theta = \frac{3}{2}\pi - \varphi$. Thus, $\theta = \frac{3}{2}\pi - 0.46 = 4.3$ rads.

13.2.4 THE CYLINDRICAL & SPHERICAL SYSTEMS

Rarely, you might be required to use three-dimensional angular coordinate systems in your introductory physics course. There are two types.

The **cylindrical coordinate system** is used for three-dimensional situations that are symmetrical about one Cartesian axis, like a cylinder (see Fig. 13.7 (a)). The **spherical coordinate system** is for three-dimensional situations that are symmetrical about all three Cartesian axes, like a sphere (refer to Fig.13.7 (b)).

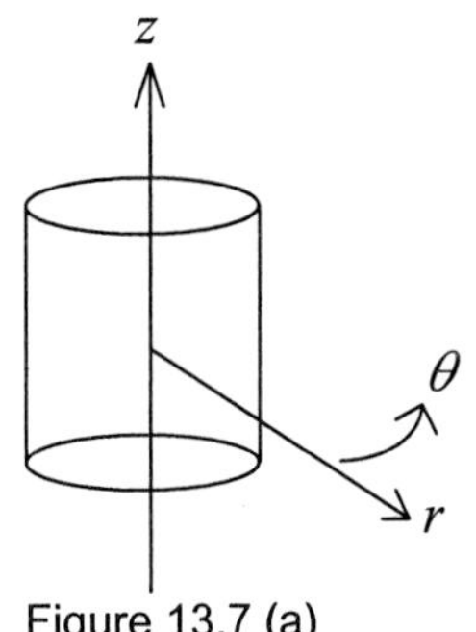

Figure 13.7 (a)

In the cylindrical coordinate system, a point P is labeled by r, θ, and z, written (r, θ, z) (see Fig. 13.8 (a)).* r and θ are defined in the same way as in the polar coordinate system, while z is defined in the same way as in the Cartesian coordinate system. The conversion formulas between cylindrical and Cartesian coordinates should therefore be familiar:

$$x = r\cos\theta \qquad y = r\sin\theta \qquad z = z$$

$$r = \sqrt{x^2 + y^2} \qquad \theta = \tan^{-1}\left(\frac{y}{x}\right) \qquad z = z$$

The warnings of Section 13.2.3 apply here as well.

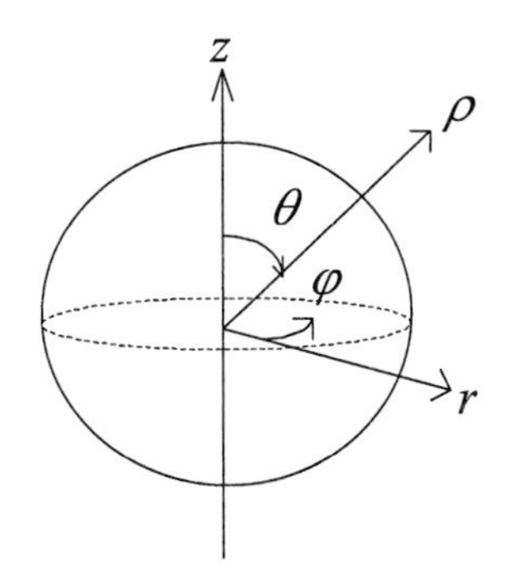

Figure 13.7 (b)

In the spherical coordinate system, a point P is labeled by ρ, φ, and θ, written (ρ, φ, θ) (see Fig. 13.8 (b)).# ρ is the radial distance from O to P on the sphere (and not on the circle through the x-y plane as was the case with r for cylindrical coordinates). φ is defined as θ is in polar coordinates, while θ here is the angle measured from the z-axis to the radius ρ.

Notice that whereas φ can vary from 0 to 2π (a complete circle), θ can only vary from 0 to π (positive z-axis to negative z-axis).

The conversion formulas between spherical and Cartesian coordinates are not quite as obvious as for the cylindrical coordinates. The warnings of Section 13.2.3 apply here too.

$$x = \rho\sin\theta\cos\varphi \qquad y = \rho\sin\theta\sin\varphi \qquad z = \rho\cos\theta$$

$$\rho = \sqrt{x^2 + y^2 + z^2} \qquad \varphi = \tan^{-1}\left(\frac{y}{x}\right) \qquad \theta = \cos^{-1}\left(\frac{z}{\sqrt{x^2 + y^2 + z^2}}\right)$$

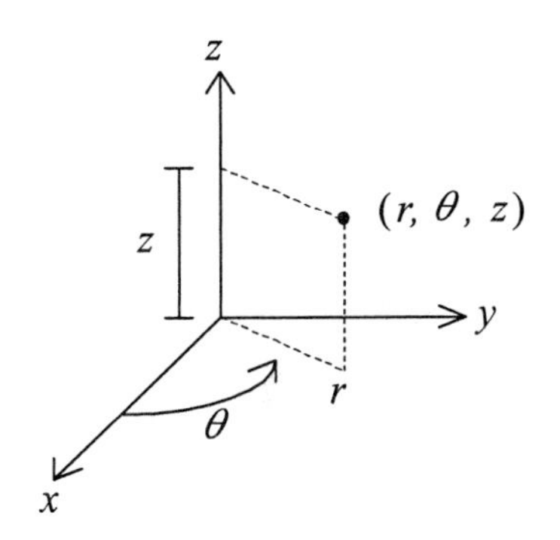

Figure 13.8 (a)

EXERCISE 13.7: Cylindrical & Spherical Coordinates

Convert the Cartesian coordinates $(5\text{ m}, 3\text{ m}, 8\text{ m})$ into cylindrical and spherical coordinates.

SOLUTION:

$x = 5\text{ m}$, $y = 3\text{ m}$, and $z = 8\text{ m}$, so in cylindrical coordinates, we have...

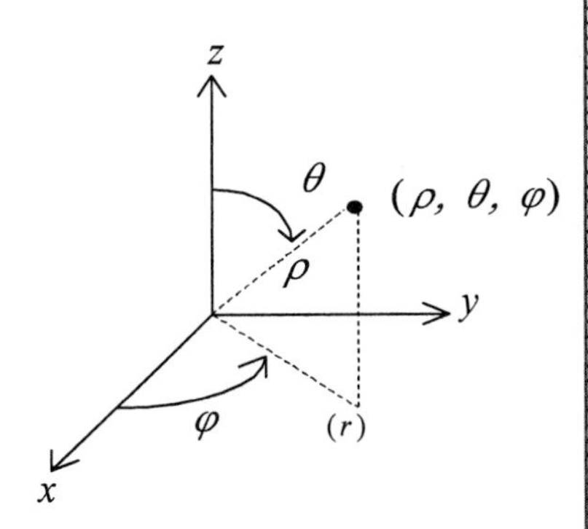

Figure 13.8 (b)

* The symbols for these coordinates also vary with textbook. When the polar coordinates are written (ρ, θ), the cylindrical coordinates are (ρ, θ, z). On the other hand, when the polar coordinates are written (ρ, φ), the cylindrical coordinates are (ρ, φ, z).

These symbols vary as well. When polar coordinates are written (ρ, θ), the spherical coordinates are (r, φ, θ), but when the polar coordinates are (ρ, φ), the spherical coordinates are (r, θ, φ). And as if these confusions weren't enough, in math texts, the convention is to call what we have termed θ the angle φ, and vise versa.

Exercise 13.7 Continued...

$$\rho = \sqrt{x^2 + y^2} = \sqrt{(5\text{ m})^2 + (3\text{ m})^2} = \sqrt{34\text{ m}^2} \approx 5.8\text{ m}$$

$$\theta = \tan^{-1}\left(\frac{y}{x}\right) = \tan^{-1}\left(\frac{3\text{ m}}{5\text{ m}}\right) \approx 0.54\text{ rads}$$

$$z = 8\text{ m}$$

...and in spherical coordinates, we have:

$$\rho = \sqrt{x^2 + y^2 + z^2} = \sqrt{(5\text{ m})^2 + (3\text{ m})^2 + (8\text{ m})^2} = \sqrt{98\text{ m}^2} \approx 9.9\text{ m}$$

$$\varphi = \tan^{-1}\left(\frac{y}{x}\right) = \tan^{-1}\left(\frac{3\text{ m}}{5\text{ m}}\right) \approx 0.54\text{ rads}$$

$$\theta = \cos^{-1}\left(\frac{z}{\sqrt{x^2 + y^2 + z^2}}\right) = \cos^{-1}\left(\frac{8\text{ m}}{\sqrt{98\text{ m}^2}}\right) \approx 0.63\text{ rads}$$

13.2.5 CHOOSING THE SYSTEM

Just as you can pick which *type* of coordinate system is best for your circumstances—Cartesian, polar, etc.—you can also choose the *position of the origin, orientation,* and *dimensionality* of the system. Often a situation presents a natural choice for these, which, if prudently selected, will simplify later calculations. Thus, you always want to take time to think about your choices *before* you define your coordinate system.

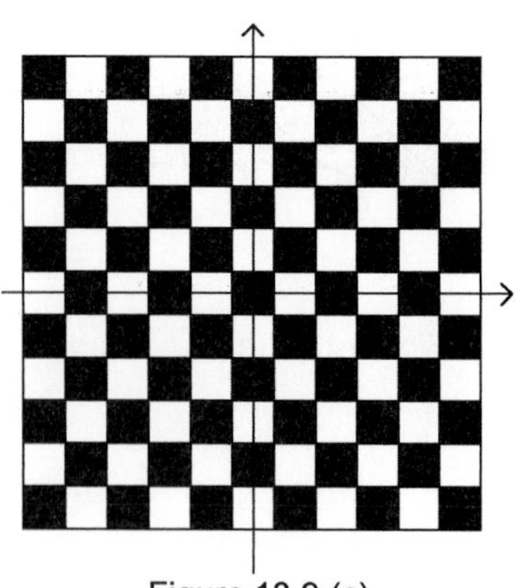

Figure 13.9 (a)

Figure 13.9 (b)

EXAMPLE 13.8: Choosing the Origin

What are some logical locations for the origin of a Cartesian system on a checkerboard? When would each choice be appropriate?

SOLUTION:

While there are an infinite number of possible locations for an origin on a checkerboard, there are really only two *good* choices.

First, the origin could be placed at the *center* of the board, so that one-quarter of the board resides in each of the four Cartesian quadrants (Fig. 13.9 (a)). This choice of origin location is good for problems that are symmetrical about a center point.

Second, the origin could be located at the lower left hand *corner* of the board, so that the entire board is located in a single, positive quadrant (Fig. 13.9 (b)). This choice is good for problems when all coordinates must be of the same sign.

EXAMPLE 13.9: Choosing the Orientation

What would be the best orientation of a Cartesian system for motion on a hill? With the x-axis parallel to the slope! See Figures 13.10. In this way, all of the motion (that is, sliding down the hill) takes place in only one dimension, the x direction.

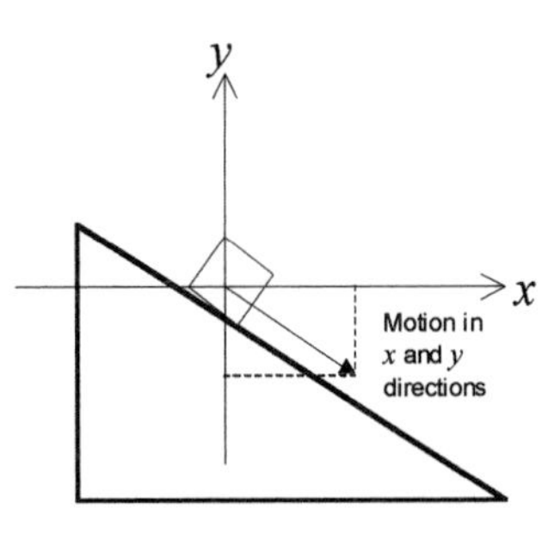

Figure 13.10 (a)

EXERCISE 13.10: Choosing the Dimensions

For each of the following motions, determine the number of coordinate system dimensions needed:

(a) A flying airplane;

(b) A horizontally-thrown ball;

(c) A rock dropped from a tall building.

SOLUTION:

(a) Since the plane can fly up or down, left or right, or forward and back, we must use three dimensions to describe this motion.

(b) The ball will initially travel horizontally because of the throw, but it will also eventually drop (in the vertical direction). Thus, we need a two-dimensional coordinate system to describe this motion.

(c) The rock will drop straight down. This is one-dimensional motion, requiring a one-dimensional coordinate system.

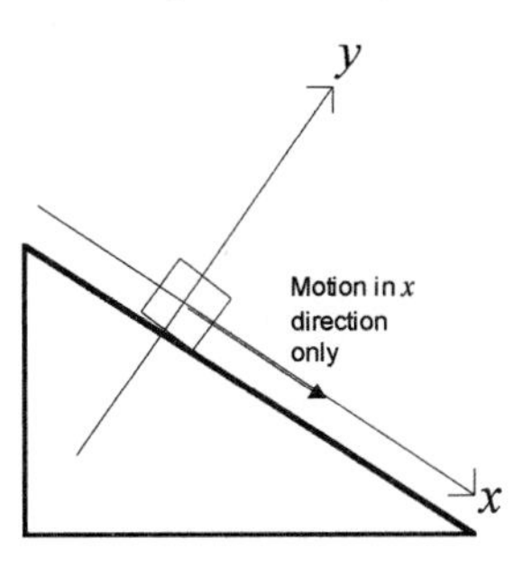

Figure 13.10 (b)

The reason that you can pick and choose different coordinate systems at will—including the location of origin, grid orientation, and number of dimensions—is that coordinate systems are *not real.* They are just mathematical constructs that make life a little more ordered for the physicist. Thus, *the choice of coordinate system does not change the solution to the problem; it only facilitates the solving of it.*

Choosing the type and location of a coordinate system may seem unimportant, yet these simple techniques are vastly powerful. The proper choice of coordinate system often determines whether a problem is trivially easy or impossibly hard to solve. It is therefore well worth the time spent looking for the appropriate coordinate system before you begin each problem.

13.3 NUMBERS

Numbers are an inescapable aspect of mathematics. This section will remind you of some of their important properties.

Truth...is inherent in the nature of number, and inbred in it.
PHILOLAUS
Ancient Greek Philosopher, c. 475 B.C.

13.3.1 REAL NUMBERS

Real numbers are conventional, everyday, non-imaginary numbers. There are three types: integers, rationals, and irrationals.

- **Integers** are real numbers representable without decimal points, such as 5 or 918,745,612.

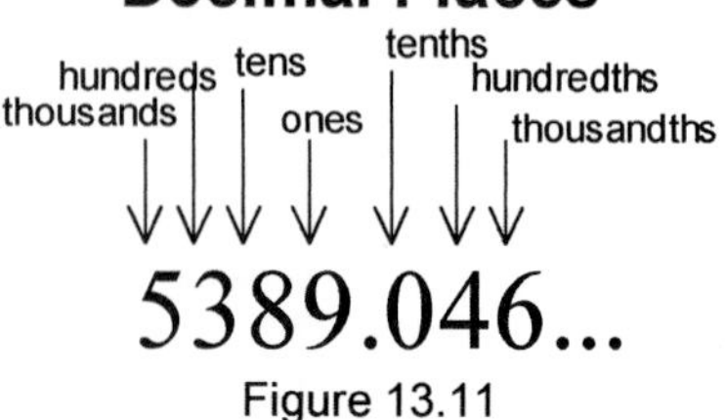

Figure 13.11

I could never make out what those damned dots meant.
RANDOLF CHURCHILL, in regards to decimal points
English Author, 1911 - 1968

Significant figures are in Section 11.2. 

- **Rational numbers** are real numbers composed of the ratio of two integers, such as $\frac{367}{89}$. Because of their fractional nature, rational numbers can be also written in decimal notation. To convert between rationals and decimals, simply divide the numerator by the denominator. Figure 13.11 illustrates the place names for each decimal position. There are two types of rational numbers:

  1. **Repeating rational numbers** never end, having a series of digits that repeat infinitely, like 5.927927927927.... We can write repeating rationals more compactly by installing a bar over the repeating series, such as $5.\overline{927}$.
  2. **Terminating rational numbers** eventually end, as in 54.346. We can write terminating rational numbers in the same way as repeating rationals by inserting an infinite number of zeros after the significant digits, as in 54.3460000000000....

- **Irrational numbers** are unlike rationals in that their decimal notations neither terminate nor repeat, nor can they be expressed in terms of a ratio of integers. They are simply a series of random, unending numbers. The most famous irrational number is pi, $\pi = 3.14159...$, but there are countless other irrational numbers such as $\sqrt{2} = 1.4142...$ and $e = 2.7182...$. See Table 13.1 for a list of common irrational numbers.

Table 13.1: Some Common Irrational Numbers

$\log 2 = 0.301029995...$	$\sqrt{2} = 1.414213562...$	$e = 2.718281828...$
$\ln 2 = 0.69314718...$	$\sqrt{3} = 1.732050808...$	$\pi = 3.141592654...$
$\ln 10 = 2.302585093...$	$\frac{1}{\sqrt{3}} = 0.577350269...$	$\frac{\sqrt{3}}{2} = 0.866025403...$
	$\frac{1}{\sqrt{2}} = 0.707106781...$	

Number is the within of all things.
PYTHAGORAS (Att.)
Greek Mathematician, c. 580 - 500 B.C.

A **prime number** is any number that can only be evenly divided by itself and one. For example, 2, 3, and 5 are prime numbers.

13.3.2 SIGNED NUMBERS & ABSOLUTE VALUE

We distinguish values that are "more than" and "less than" zero with **signed numbers**. Numbers that are greater than zero, like $+4$ (or just 4), are called **positive**; numbers that are less than zero, like -4, are called **negative**.

We can visualize the relationships between positive and negative numbers on a **number line** like the one in Figure 13.12. Notice the convention in which positive numbers are placed to the *right* of zero, and negative numbers to the *left* of zero. Two number lines that intersect perpendicularly at zero make up a Cartesian coordinate system. See Figure 13.13.

The **absolute value function**, symbolized by $|\,|$, makes any number positive. Thus, the absolute value of -3, $|-3|$, is 3, but so is the absolute value of $+3$, $|+3| = 3$.

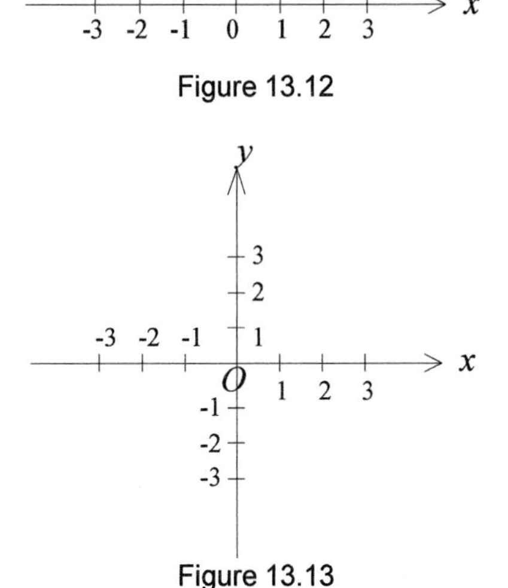

Figure 13.12

Figure 13.13

13.3.3 IMAGINARY & COMPLEX NUMBERS

Imaginary numbers are real numbers multiplied by i, where $i = \sqrt{-1}$.[*] For example, while 5 is a real number, $5i$ is imaginary. Note that "imaginary" does not mean "pretend" here! These numbers have meaning to the physicist just like real numbers do.

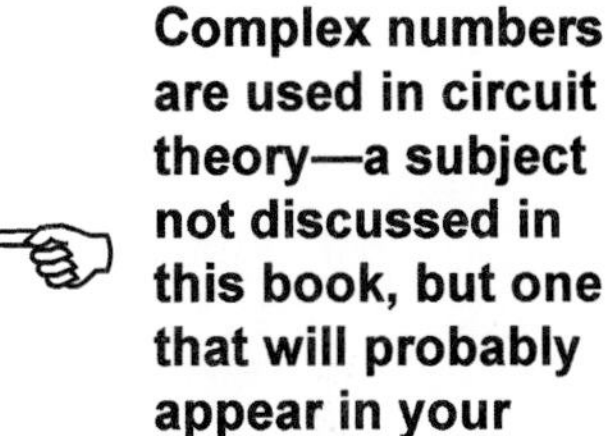
Complex numbers are used in circuit theory—a subject not discussed in this book, but one that will probably appear in your class.

Complex numbers are composites of real and imaginary numbers. They have the form $z = x + iy$, where x and y are both real, but z is complex. We term x the **real part** of z, written $x = \mathrm{Re}(z)$, and y the **imaginary part** of z, written $y = \mathrm{Im}(z)$.[#] We define the **conjugate** of the complex number $z = x + iy$ as $\bar{z} = x - iy$.

We manipulate complex numbers according to the following rules:

Addition/Subtraction:

$$(x_1 + iy_1) \pm (x_2 + iy_2) = (x_1 + x_2) \pm i(y_1 + y_2)$$

Multiplication:[^]

$$(x_1 + iy_1)(x_2 + iy_2) = (x_1 x_2 - y_1 y_2) + i(x_1 y_2 + x_2 y_1)$$

Division:[!]

$$\frac{x_1 + iy_1}{x_2 + iy_2} = \frac{x_1 + iy_1}{x_2 + iy_2} \cdot \frac{x_2 - iy_2}{x_2 - iy_2} = \frac{(x_1 x_2 + y_1 y_2) + i(y_1 x_2 - x_1 y_2)}{x_2^{\,2} + y_2^{\,2}}$$

---

[*] Some texts use j rather than i.

[#] Notice that y is called the "imaginary part of z" even though it is a real number!

[^] Note that this law implies that the product of a complex number with its conjugate is:

$$(x + yi)(x - yi) = x^2 + y^2$$

[!] To divide one complex number by another, as we have done here, you must multiply both the numerator and the denominator by the conjugate of the denominator. Then you multiply the two fractions according to the multiplication rules, obtaining the answer given.

EXERCISE 13.11: Complex Numbers

Consider two complex numbers: $z_1 = 5 + 3i$, and $z_2 = 9 - 4i$.

(a) What is the real part of z_1?

(b) What is the imaginary part of z_2?

(c) What is the conjugate of z_1?

(d) What is the sum of z_1 and z_2?

(e) What is the product of z_1 and z_2?

(f) What is the quotient of z_1 and z_2?

SOLUTION:

(a) 5 (b) −4 (c) $\bar{z}_1 = 5 - 3i$

(d) $z_1 + z_2 = (5+3i) + (9-4i) = (5+9) + (3-4)i = 14 - i$

(e)
$$\begin{aligned} z_1 z_2 &= (5+3i)(9-4i) \\ &= [(5)(9) - (3)(-4)] + [(5)(-4) + (3)(9)]i \\ &= 57 + 7i \end{aligned}$$

(f)
$$\begin{aligned} \frac{z_1}{z_2} &= \frac{(5+3i)}{(9-4i)} \\ &= \frac{[(5)(9) + (3)(-4)] + [(3)(9) - (5)(-4)]i}{9^2 + (-4)^2} \\ &= \frac{33 - 47i}{97} \\ &\approx 0.34 - 0.48i \end{aligned}$$

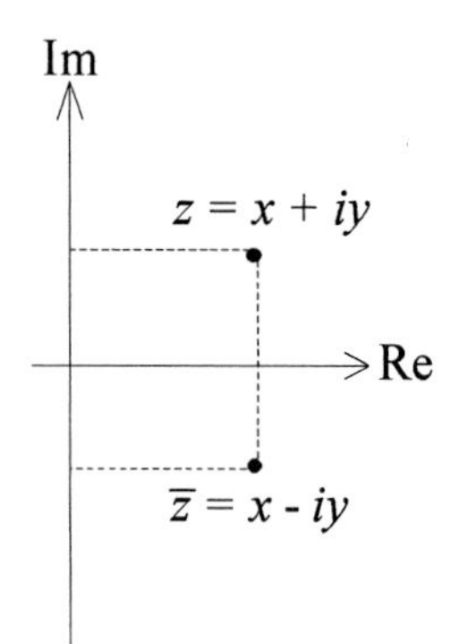

Figure 13.14

PLOTTING COMPLEX NUMBERS

We often plot complex numbers on a two-dimensional Cartesian coordinate system so that we can visualize the situation (Fig. 13.14). The real part of z becomes the x coordinate, and the imaginary part of z becomes the y coordinate. In this case, we call the x-axis the **real axis**, and the y-axis the **imaginary axis**.

Cartesian quadrants are in Section 13.2.2.

EXERCISE 13.12: Plotting Complex Numbers

Plot z_1 and z_2 from Exercise 13.11 above. In which quadrant is $\bar{z}_2$?

SOLUTION:

We plot these in the sketch below:

Exercise 13.12 Continued...

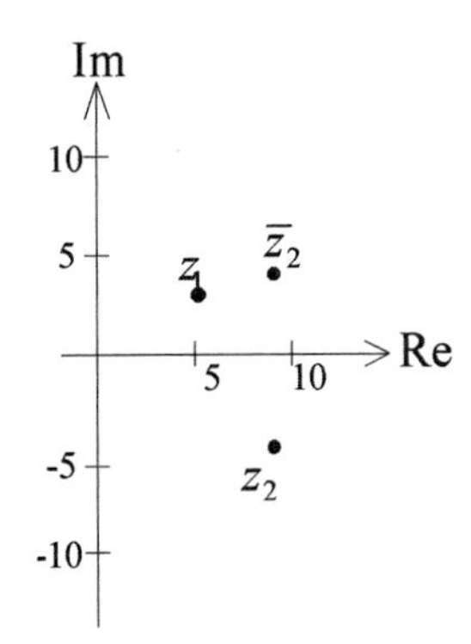

The conjugate of z_2 is in quadrant I.

MAGNITUDE & ARGUMENT

Just as we could write Cartesian coordinates (x, y) in terms of polar coordinates (r, θ), we can write the real and imaginary parts of z in polar coordinates as well (Fig. 13.15). When used for complex numbers, however, (r, θ) are called the **magnitude** and **argument** of z, rather than the radial distance and the angle.

Referring to Figure 13.15, and using some trigonometry, we see that $z = r\cos\theta + ir\sin\theta = r(\cos\theta + i\sin\theta) = x + iy$. Thus, the conversions between (x, y) and (r, θ) are the same as in Section 13.2.3, with the warnings therein still applicable:

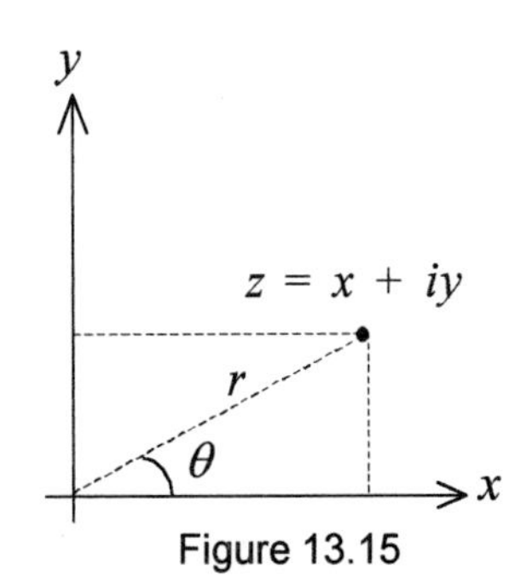

Figure 13.15

$$x = r\cos\theta \qquad r = \sqrt{x^2 + y^2}$$

$$y = r\sin\theta \qquad \theta = \tan^{-1}\left(\frac{y}{x}\right)$$

EXERCISE 13.13: Converting Between (x,y) & (r,θ)

What is the magnitude of z_1 in Exercise 13.11? What is the argument of z_2 in that exercise?

SOLUTION:

The magnitude of z_1 is $r = \sqrt{x^2 + y^2} = \sqrt{5^2 + 3^2} = \sqrt{34} \approx 5.8$

The argument of z_2 is $\theta = \tan^{-1}\left(\frac{y}{x}\right) = \tan^{-1}\left(\frac{-4}{9}\right) \approx -0.42$ rads

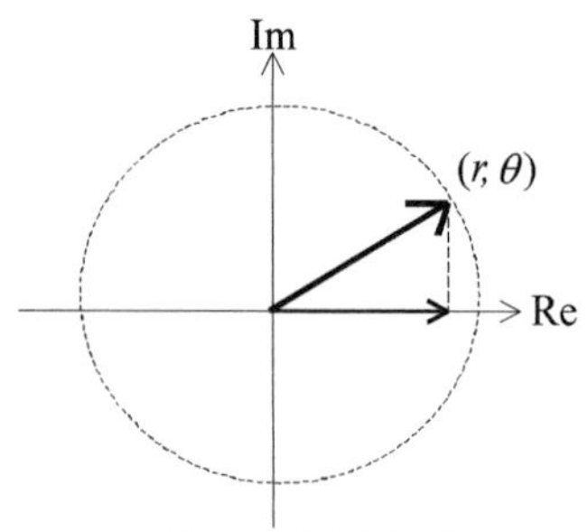

Figure 13.16 (a)

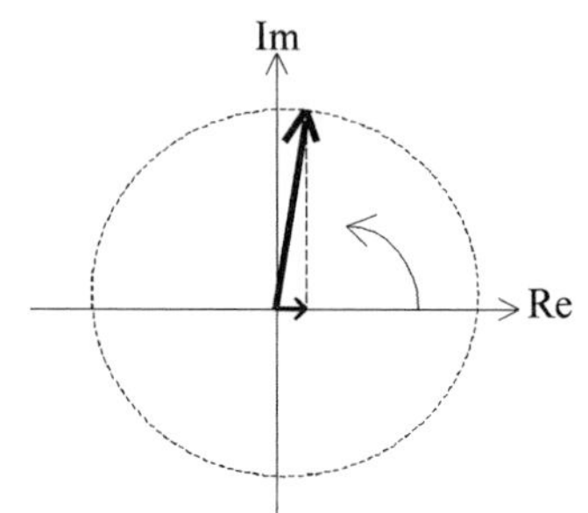

Figure 13.16 (b)

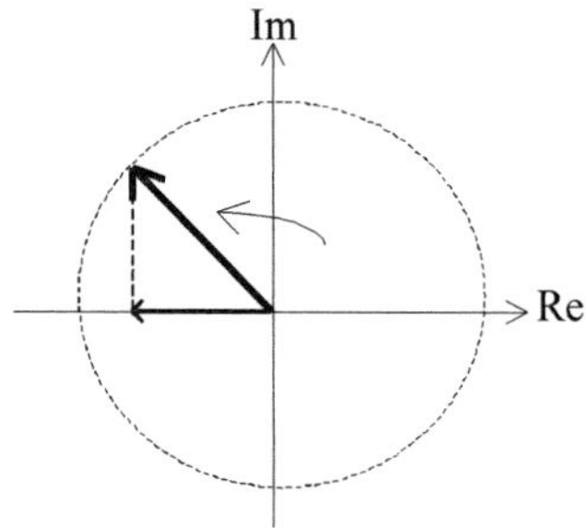

Figure 13.16 (c)

PHASORS

An arrow that is drawn with its base at the origin and its head on the imaginary coordinate $(r,\ \theta)$ is called a **phasor** (Fig. 13.16 (a)). Because the body of a phasor passes through both real and imaginary space on the coordinate grid, the phasor is considered to have both real and imaginary parts.

Phasors are most useful when the coordinate $(r,\ \theta)$ changes with time. If the phasor traces out a circle, the projection of the phasor on the x (or y) axis oscillates from a maximum value, to zero, to a negative maximum, to zero, and back. See Figures 13.16 (a) - (c) for some examples of phasor motion. Thus, the *circular* motion of a phasor on the *complex plane* is directly linked to the *linear* motion of a point on the *real* (or *imaginary*) line. Phasors, therefore, are especially useful for analyzing rotational motion (in Chapters 5, 6, and 8).

13.4 ALGEBRA

Algebra is the art of manipulating equations. Physicists use algebra far more than any other mathematics, so the importance of mastering this subject cannot be overstated.

Algebra is the intellectual instrument which has been created for rendering clear the quantitative aspect of the world.
ALFRED NORTH WHITEHEAD
English Mathematician, 1861 - 1941

13.4.1 ALGEBRAIC SYMBOLISM

Algebraic equations use symbols to represent quantities that are unknown or that may vary.

There are two types of algebraic symbols used: alphabetic and mathematical. The alphabetic symbols most often used in math and physics come from the English and Greek alphabets. Table 13.2 lists the Greek alphabet for reference.

Table 13.2: The Greek Alphabet

Alpha	A	α	Nu	N	ν
Beta	B	β	Xi	Ξ	ξ
Gamma	Γ	γ	Omicron	O	o
Delta	Δ	δ	Pi	Π	π
Epsilon	E	ε	Rho	P	ρ
Zeta	Z	ζ	Sigma	Σ	σ
Eta	H	η	Tau	T	τ
Theta	Θ	θ	Upsilon	Y	υ
Iota	I	ι	Phi	Φ	ϕ
Kappa	K	κ	Chi	X	χ
Lambda	Λ	λ	Psi	ϑ	φ
Mu	M	μ	Omega	O	o

CAUTION 13.14: Physical Quantities & Symbols

The alphabetic symbols for various physical quantities have been chosen in different ways.

Some symbols have the same first letter (or letters) as the name of the quantity. For example, the quantity of "time" has the symbol t. Likewise, the quantity "torque" has the symbol τ, which is the Greek letter t.

On the other hand, some symbols have, over time, come to indicate a *group* of similar quantities. For instance, the symbols x, y, and z often represent (straight line) *length* or *distance*. The Greek letters θ, φ, and ϕ often represent *angles*. The symbols a, b, c, and k, or α and β often denote arbitrary *constants*.

Sometimes it is not immediately clear why certain symbols were picked to represent a quantity. For instance, p represents "momentum." Some effort is made to prevent using the same symbol twice in any one area of physics, but unfortunately this goal is not always met.

Time is in Section 11.5.2.

Torque is in Section 6.52.

Length and angles are in Section 11.5.2.

Momentum is in Section 8.7.

Science is a redirection of the bewildering diversity of unique events to manageable uniformity within one of a number of symbol systems.
ALDOUS LEONARD HUXLEY
English Novelist, 1894 - 1963

Because there are not enough symbols to give every physical quantity a unique label,* you need to take care to keep the symbolic definitions of various physical quantities straight. Context will often make the correct usage apparent.

NOTE 13.15: Keeping Them Straight

A good way to keep track of all of the symbols that you will encounter in your class is to create a running list of physical quantities and their symbols as you progress through the class. Alternatively, some textbooks have a list of quantities and their symbols in the Appendix: mark this page and refer to it throughout the term.

In some cases, alphabetic symbols represent quantities that can *vary* in value; hence, we refer to these as **variables**. When there is more than one value for a variable, the symbol is labeled with a subscript to keep the variables straight.

Nature is written in signs and symbols.
JOHN GREENLEAF WHITTIER
English Author, 1807 - 1892

EXAMPLE 13.16: Variable Notation

For example, the values of momentum at two different times would be written as p_1 and p_2, or, alternatively, as p_i and p_f.

* Furthermore, unfortunately, the same quantity might have different symbols in different fields. See the footnotes on symbolism in polar coordinates, for example.

Algebra also requires an understanding of *mathematical* symbols as well as alphabetic symbols. Table 13.3 below lists some mathematical symbols common in an introductory physics course.

Uniformity in nomenclature... is of such vast and paramount importance...as to outweigh every consideration of technical convenience or custom.
SIR JOHN HERSCHEL
English Astronomer, 1792 - 1871

Table 13.3: Some Mathematical Symbols and Their Meaning

SYMBOL	MEANING	SYMBOL	MEANING
$=$	Is equal to	$\Rightarrow$	Implies
$\approx$	Is approximately equal to	$\therefore$	Therefore
$\sim$	Is of the same order as	$\pm$	Plus or minus
$\propto$	Is proportional to	$\overline{x}$	Average of x
$\equiv$	Is defined to be	Δ	Change
$\neq$	Is unequal to	$!$	Factorial
$>$	Is greater than	Σ	Sum
$>>$	Is much greater than	Π	Product
$\geq$	Is greater than or equal to	∞	Infinity
$<$	Is less than	d/dx	Derivative
$<<$	Is much less than	$\partial/\partial x$	Partial derivative
$\leq$	Is less than or equal to	$\int$	Integral

EXAMPLE 13.17: Taking a Closer Look

Let's look at the meaning of a few of the more commonly used mathematical symbols in Table 13.3.

- The equals sign = really indicates three things.
  - A relationship between a cause and an effect: *cause = effect.* For example, Newton's Second Law, $\vec{\mathbf{F}}_{net} = m\vec{\mathbf{a}}$, relates a cause (a net force) to its effect (acceleration).
  - A relationship between two time periods: *before = after.* For example, in momentum problems, when momentum is conserved before and after a collision, we write $p_1 = p_2$.
  - The ability to say the same thing in two different ways, one on either side of the equation. For example, when $A = A$; or, when it is found through algebra that $x = 2$.
- The definition symbol $\equiv$ is used when defining a quantity in terms of other, known quantities. For example, when defining velocity in terms of distance and time, we write $v \equiv \Delta x/\Delta t$.
- The proportionality symbol $\propto$ is used when two quantities are proportional. For example, since we know that $x = v\,t$, we could write $x \propto t$, which says that x is proportional to t.
- The imply symbol $\Rightarrow$ is often used when deriving algebraic equations. For example, we could write $v = x/t \Rightarrow x = v\,t$.

Newton's Second Law is covered in Section 6.3.

Momentum is discussed in Section 8.7.

This equation is in Section 5.3.1.

For more on proportionality, see Section 13.4.9.

Example 13.17 Continued...

- The symbol, Δ, known as **delta notation**, is defined as a change; for example, in the case of a change in position x from position x_1 to position x_2: $\Delta x = x_2 - x_1$. Sometimes the lower case delta symbol, δ, is used to indicate very small values of Δx.
- The summation symbol Σ allows us to write the sum $f_1 + f_2 + \ldots + f_n$ more compactly as Σf_i.

In an algebraic equation, the alphabetic symbols are related via mathematical symbols and numbers to create "mathematical sentences." Mathematical sentences are in many ways like regular sentences, and you should strive to read them as such. To start this process, try spelling each element of the equation in words, either out loud or in your head.

EXAMPLE 13.18: One Equation

For example, the algebraic equation $5x + 3y = 2$ relates the variables x and y via the mathematical symbols $+$ and $=$, and the numbers 5, 3, and 2. You might "spell out" this equation in words as "Five of something added to three of something else equals two."

The fundamental task of algebra is to solve equations for one variable in terms of the remaining (known) variable(s) and numbers. The variable for which we solve is called the **independent variable**, and all other variables in the equation are **dependent variables**. With some equations, completing this algebraic task is not difficult at all. With other equations, the number of possible solving methods gives an indication of how difficult solving can be.

For more on independent and dependent variables, refer to Sections 13.4.12 and 16.2.1.

It is very important for learning physics that you become as proficient at solving equations as possible. You should strive for accuracy as well as speed in your solving, though accuracy should be your first concern. Speed will come with time and diligent practice.

13.4.2 SOLVING ALGEBRAIC EQUATIONS

The general procedure for solving an algebraic equation is called **collecting like terms**. To use this technique, "collect" all terms in the equation containing the independent variable to one side of the equals sign, and "collect" all *other* terms to the other side. You may have to add, subtract, multiply, divide, or otherwise manipulate the equation to complete this task, *but remember that every operation performed on one side of the equation must also be performed on the other.*

CAUTION 13.19: Think with Algebra!

"Collecting like terms" is a classic example of the kind of pat rule discussed in the introduction to this part. The rule works as a good mnemonic device, but examine it too closely, and it seems hollow and meaningless. In fact, if you apply the rule mechanically and without thought, you may get a wrong answer! Therefore, you need to *think* about the meaning underlying this rule—and rules like it—before you blindly apply it.

EXERCISE 13.20: Solving Algebraic Equations

Solve for x in the following equations:

(a) $5x + 7 = 42$ (b) $ax - b = cx + d$

SOLUTION:

(a) To solve this problem, we need to collect like terms. That is, we want to *isolate* x on the left-hand side of the equation by putting everything else (in this case, the 7) on the right-hand side. Therefore, we need to subtract 7 from both sides of the equation:

$$5x + 7 = 42$$
$$5x + 7 - 7 = 42 - 7$$
$$5x = 35$$

To find our value of x, we now divide by 5 on both sides:

$$x = 7$$

(b) Collect the x terms on the left-hand side of the equation, and the constants on the right-hand side, by adding and subtracting:

$$ax - b = cx + d$$
$$ax - b + b - cx = cx - cx + d + b$$
$$ax - cx = d + b$$

Factoring is covered in Section 13.4.10. 

Now, to solve for x, we factor out the x on the left-hand side, and divide by the factored term:

$$(a - c)x = d + b$$
$$x = \frac{b+d}{a-c}$$

When an equation is an inequality, containing the symbols $<$ or $>$, solving can be slightly more complicated. You can still do any mathematical operation you wish (as long as you do it to both sides), EXCEPT that *if you multiply or divide the equation by a negative number, you must reverse the inequality.** The result will not be a unique number, but instead will be a *range* of numbers satisfying the inequality.

EXERCISE 13.21: Solving Inequalities

Solve for x in the following equations:

(a) $2x - 3 < 4x + 7$ (b) $-3x + 2 \geq -x - 6$

SOLUTION:

(a) Isolate x and solve just as if this were a regular equation:

$$\begin{aligned} 2x - 3 &< 4x + 7 \\ 2x - 4x &< 7 + 3 \\ -2x &< 10 \\ \Rightarrow \quad x &> -5 \end{aligned}$$

Note that to get our final answer for x, we had to divide the next to last equation by a negative two. But the negative sign *reverses* the inequality, so that our answer is $x > -5$. This equation says that x can be any number greater than -5—for example, -4, 0, or 129.

(b) Solve normally:

$$\begin{aligned} -3x + 2 &\geq -x - 6 \\ -3x + x &\geq -6 - 2 \\ -2x &\geq -8 \end{aligned}$$

Since the negative sign *reverses* the inequality, our answer is $x \leq 4$ (x is greater than four).

13.4.3 FRACTIONS

A **fraction** (also known as a **rational function**) is a ratio of two numbers, written a/b or $\frac{a}{b}$. In this notation, a is called the **numerator**, and b is termed the **denominator**.

When the magnitude of a is larger than the magnitude of b, the entire fraction is greater than one; but when the magnitude of b is larger than the magnitude of a, the entire fraction is less than one.

Add, subtract, multiply, and divide fractions with the following rules:

The man who discredits the supreme certainty of mathematics is feeding on confusion....
LEONARDO DA VINCI
Italian Experimentalist, 1452 - 1519

* An additional, more advanced rule about inequalities is that if you invert both sides of the inequality (via division), you must also reverse the inequality.

Addition/Subtraction: $\frac{a}{b} \pm \frac{c}{d} = \frac{ad \pm bc}{bd}$ *

Multiplication: $\left(\frac{a}{b}\right)\left(\frac{c}{d}\right) = \frac{ac}{bd}$

Division: $\left(\frac{a}{b}\right) \div \left(\frac{c}{d}\right) = \left(\frac{a}{b}\right)\left(\frac{d}{c}\right) = \frac{ad}{bc}$

WARNING!

The first rule does NOT say that $\frac{1}{a} + \frac{1}{b} = \frac{1}{a+b}$. Instead, the rule for adding fractions is more complicated, *i.e.*, $\frac{1}{a} + \frac{1}{b} = \frac{a+b}{ab}$.

EXAMPLE 13.22: Look Out!

For example, $\frac{1}{2} + \frac{1}{2} \neq \frac{1}{4}$. Think, "A half dollar plus a half dollar does not equal a quarter." Rather, $\frac{1}{2} + \frac{1}{2} = \frac{2+2}{2(2)} = \frac{4}{4} = 1$. It equals a dollar!

EXERCISE 13.23: Fun with Fractions

Solve for x in the following equations:

(a) $a = \frac{1}{1+x}$

(b) $\frac{1}{5} = \frac{7-x}{2x}$

SOLUTION:

(a) To solve this equation, the $(1 + x)$ term should be in the numerator. To get this, we multiply both sides of the equation by $(1 + x)$:

$$a = \frac{1}{1+x}$$

$$a(1+x) = 1 \cdot \frac{(1+x)}{(1+x)}$$

$$\Rightarrow a(1+x) = 1$$

Now, we divide both sides of the equation by a to isolate the x:

* This method is known as *finding a common denominator*.

Exercise 13.23 Continued...

$$1 + x = \frac{1}{a}$$

Finally, we solve for x:

$$x = \frac{1}{a} - 1$$

To simplify the answer (optional), we could find a common denominator. We write the 1 on the right-hand side as a fraction:

$$x = \frac{1}{a} - \frac{1}{1}$$

Then, we apply the addition formula:

$$x = \frac{1(1) - a(1)}{a(1)} = \frac{1-a}{a}$$

(b) **Cross-multiply** each side by the denominator of the other to get rid of the fractions.

$$\frac{1}{5} = \frac{7-x}{2x}$$
$$1\,(2x) = 5\,(7-x)$$
$$2x = 35 - 5x$$

Then collect terms and solve for x:

$$2x + 5x = 35$$
$$7x = 35$$
$$x = 5$$

We can check this answer simply by plugging 5 back into our equation for x:

$$\frac{1}{5} \stackrel{?}{=} \frac{7-(5)}{2(5)} = \frac{2}{10} = \frac{1}{5} \quad \checkmark$$

Learn how and why to check your work in Section 4.4.1.

13.4.4 THE POINT-SLOPE EQUATION

One simple, but important, algebraic equation is the **point-slope equation** $y = mx + b$, which describes the straight line of Figure 13.17. The point-slope equation is very valuable in introductory physics and is often used in laboratory analysis.

The **y-axis intercept**, b, in this equation is the position where the line crosses (intercepts) the y-axis. The **slope** m of the line, on the other hand, is a measure of the line's "slantiness":

$$m = \frac{rise}{run} = \frac{y_2 - y_1}{x_2 - x_1} \equiv \frac{\Delta y}{\Delta x}$$

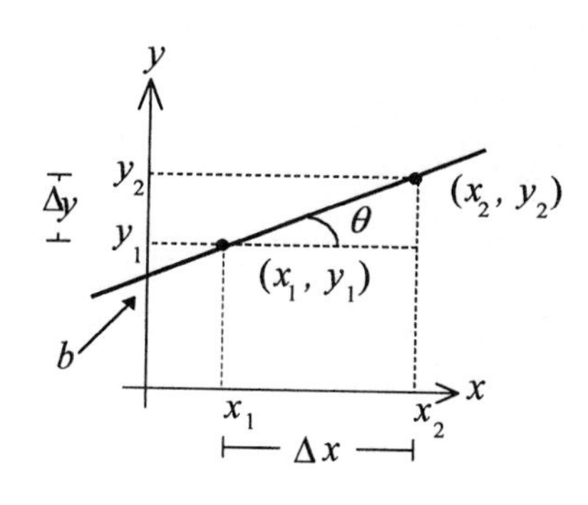

Figure 13.17

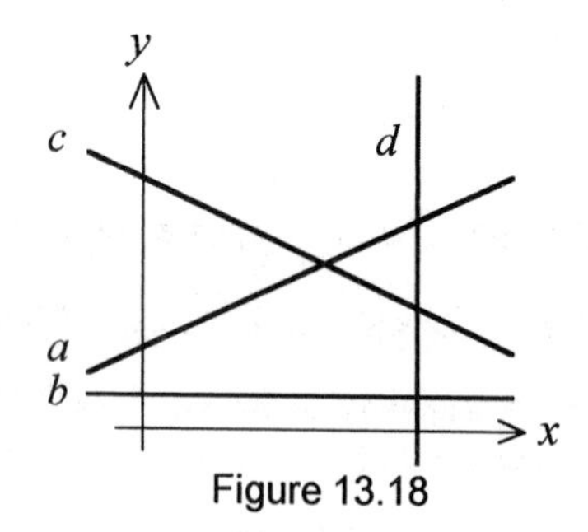

Figure 13.18

(a) Positive slope;
(b) Zero slope;
(c) Negative slope;
(d) Undefined slope

In many cases, especially in laboratory situations, the slope has a dimension; *i.e.*, x and y have different units, so their quotient (the slope) has a derived unit. However, in the special case that x and y have the *same* units, the slope is dimensionless (because the units cancel out in the division).

In the particular case where x and y have both the same units *and* the same scale, the slope is equal to $m = \tan\theta$, where θ is the angle between the line in question and horizontal (refer to Figure 13.17 again). So, slopes can be positive (θ acute), zero (horizontal line), negative (θ obtuse), or undefined (vertical line); Figure 13.18 shows examples of each of these.

Two lines are perpendicular if the product of their slopes is -1.

EXERCISE 13.24: Graphing the Point-Slope Equation

Plot the following linear equations:

(a) $y = x + 3$ (b) $y = -3x - 4$

SOLUTION:

Graphs are covered in Section 16.2.1. 

(a) The slope of this equation is 1, as can be seen by matching the equation $y = x + 3$ to $y = mx + b$. Therefore, the line should travel to the right one tick for every tick traveled upwards. The y-intercept, $b = +3$, is the point at which the line will cross the y-axis. Thus the graph will look like:

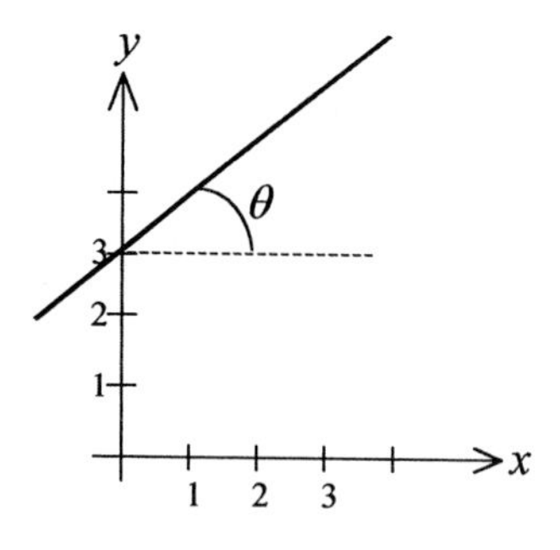

(b) Here, the slope is -3, and the y-intercept is $y = -4$. The graph is:

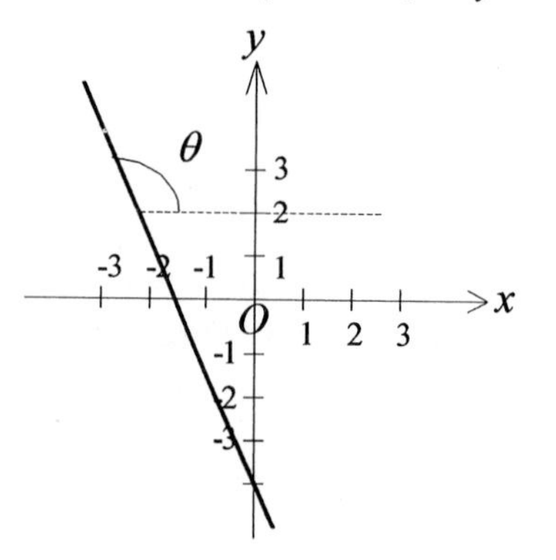

EXERCISE 13.25: Finding Slopes and Angles

Calculate the slopes of the lines that pass through the following points. What is the smallest angle subtended by a horizontal line intersecting these lines? Each of the axes in this case has the same units and scales. These lines are sketched in Figures 13.19.

(a) $(4, 5)$ and $(7, 8)$ (b) $(-3, 0)$ and $(2, -4)$

SOLUTION:

(a) $m = \frac{y_2 - y_1}{x_2 - x_1} = \frac{8-5}{7-4} = \frac{3}{3} = 1$ and $\theta = \tan^{-1} m = \tan^{-1} 1 = \frac{\pi}{4}$ rad

(b) $m = \frac{y_2 - y_1}{x_2 - x_1} = \frac{-4-0}{2--3} = \frac{-4}{5}$ and $\theta = \tan^{-1} m = \tan^{-1} \frac{-4}{5} \approx -0.67$ rad

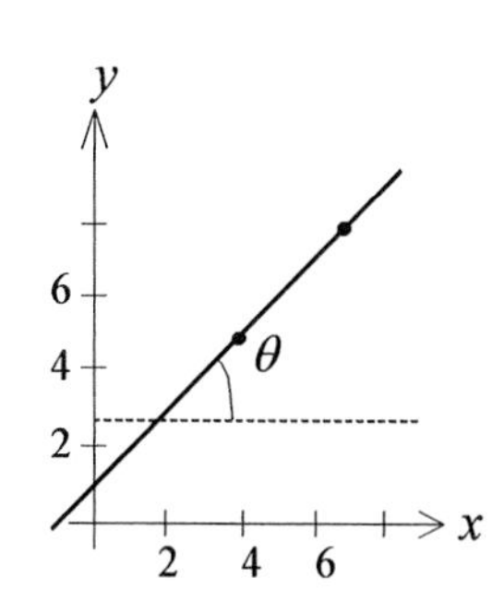

Figure 13.19 (a)

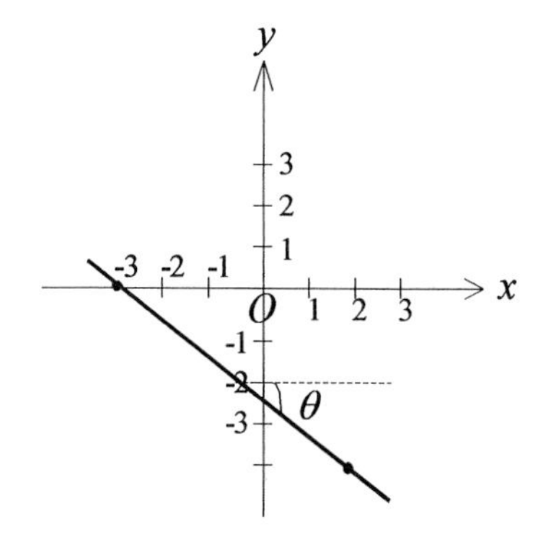

Figure 13.19 (b)

13.4.5 SIMULTANEOUS LINEAR EQUATIONS

See Section 13.4.6 for more on exponents.

A **linear equation** is any equation without exponents. In general, a linear equation can have any number of variables.

However, equations with more than one variable are only solvable if there is the same number of pieces of information (usually equations) as variables.* The multiple equations required for solution are called **simultaneous equations**.

EXAMPLE 13.26: The Point-Slope Equation

For example, the point-slope equation $y = mx + b$ is one type of linear equation because neither the x nor the y is raised to a power. This equation, having two variables, requires two simultaneous equations to obtain unique solutions for both x and y.

We can solve simultaneous linear equations graphically or algebraically. The techniques differ depending on whether there are two, or more, variables in the equations.

* Another criterion is that *all equations must be **independent** of each other*, that is, there must be no way in which any two (or more) of the equations can be combined to create another equation already in the set. Virtually all of the equations you will use in an introductory physics class will meet this requirement automatically, so you will probably never need this criterion.

SOLVING TWO-VARIABLE SIMULTANEOUS LINEAR EQUATIONS

To solve two-variable simultaneous equations *graphically* (something you will not need to do often in an introductory physics class, and, when you do, you can usually use a graphing calculator), follow these steps:

1. Write both equations in the form of the point-slope equation; that is, solve equation for y in terms of x.
2. Sketch each line on the same coordinate grid.
3. The solution is the single point where the lines intersect.

EXERCISE 13.27: Solving Graphically

Solve the following simultaneous equations graphically.

(a) $8x + 4y = 3$
$3x - y = 5$

(b) $47 + x = 12y$
$34 + 21x = -2y$

SOLUTION:

(a) First we solve both equations for y, to get the point-slope form:

$$8x + 4y = 3 \Rightarrow y = \tfrac{1}{4}(3 - 8x) = -2x + \tfrac{3}{4}$$
$$3x - y = 5 \Rightarrow y = 3x - 5$$

Then we plot each line on the same grid:

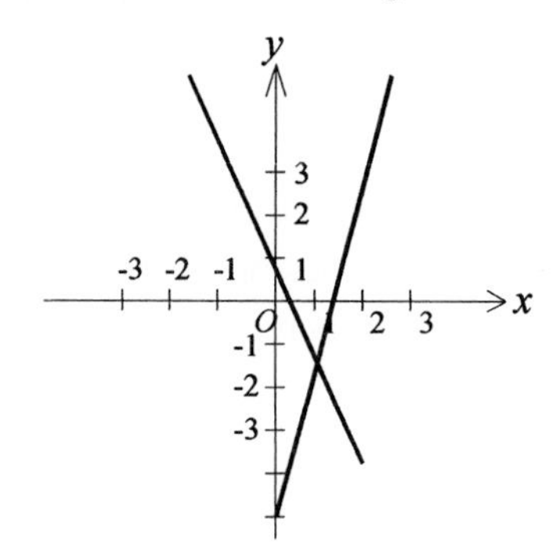

The intersection of the lines is approximately $(1.2, -1.6)$.

(b) Solve both equations for y, and plot together:

$$47 + x = 12y \Rightarrow y = \tfrac{1}{12}(47 + x) = \tfrac{1}{12}x + 3.9$$
$$34 + 21x = -2y \Rightarrow y = -\tfrac{1}{2}(34 + 21x) = -10.5x - 16$$

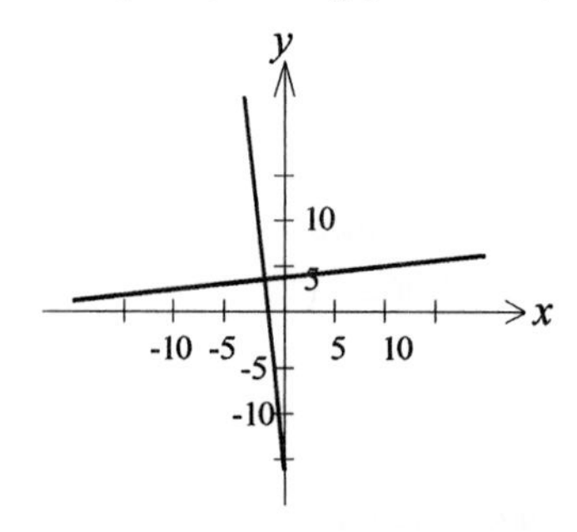

The intersection of the lines is approximately $(3.8, -2.0)$.

To solve two-variable simultaneous linear equations *algebraically*, follow one of these two procedures.

Neither admits nor hopes for any principles in Physics other than those which are in Geometry or in abstract Mathematics, because thus all the phenomena of nature are explained, and some demonstration of them can be given.
RENE DESCARTES
French Mathematician, 1596 - 1650

Procedure 1

1. Solve both equations in the form of the point-slope equation.
2. Set the equations equal to each other.
3. Solve the new composite equation for x.
4. Plug this value for x into the easier of the two original equations, and solve for y.

EXERCISE 13.28: Solving Algebraically, Part I

Solve the simultaneous equations of Exercise 13.27 algebraically using Procedure 1.

SOLUTION:

(a) First we solve both equations for y, to get the point-slope form:

$$8x + 4y = 3 \Rightarrow y = \tfrac{1}{4}(3 - 8x)$$
$$x - y = 5 \Rightarrow y = 3x - 5$$

Then we set the two equations equal to each other:

$$y = \tfrac{1}{4}(3 - 8x) = 3x - 5$$

We solve the new composite equation for x:

$$\begin{aligned} \tfrac{1}{4}(3 - 8x) &= 3x - 5 \\ \Rightarrow 3 - 8x &= 4(3x - 5) \\ &= 12x - 20 \\ \Rightarrow 20x &= 23 \\ x &= \tfrac{23}{20} \approx 1.2 \end{aligned}$$

Now, we plug this value for x into the second equation to get y:

$$y = 3(\tfrac{23}{20}) - 5 = \tfrac{69-100}{20} = -\tfrac{31}{20} \approx -1.6$$

Notice that these answers are identical to those we obtained in the problem above by graphing.

(b) Solve both equations for y.

$$47 + x = 12y \Rightarrow y = \tfrac{1}{12}(47 + x)$$
$$34 + 21x = -2y \Rightarrow y = -\tfrac{1}{2}(34 + 21x)$$

Set the equations equal to each other and solve for x:

$$\begin{aligned} y = \tfrac{1}{12}(47 + x) &= -\tfrac{1}{2}(34 + 21x) \\ \Rightarrow 47 + x &= -6\,(34 + 21x) \\ &= -204 - 126x \\ \Rightarrow -127\,x &= 251 \\ x &= -\tfrac{251}{127} \approx -2.0 \end{aligned}$$

Now plug x into the first equation to get y:

$$y = \tfrac{1}{12}(47 - 2.0) \approx 3.8$$

Procedure 2

1. Examine both equations. Look for terms, which if the two equations were to be added or subtracted, will cancel out. You may need to multiply one or more of the equations by some factor to make this work.
2. Add or subtract the equations, obtaining a brand new equation.
3. Solve the new composite equation for one variable.
4. Plug this value into one of the two original point-slope equations, whichever is easier, and solve for the other variable.

EXERCISE 13.29: Solving Algebraically, Part II

Solve the simultaneous equations of Exercise 13.27 algebraically using Procedure 2.

SOLUTION:

(a) According to Step 1 above, we want to examine our equation for terms that might cancel. In this problem, we see that there is a + sign in front of the y term in the first equation, and a – sign in front of the y term in the second. Thus, if we add the two equations, we can get rid of the y term *if* we also multiply the second equation by a 4. So let's do that. We multiply the second equation by 4:

$$4(3x - y = 5) \Rightarrow 12x - 4y = 20$$

Now we add together the first equation and the new second equation (Step 2):

$$\begin{array}{r} 8x + 4y = 3 \\ \underline{12x - 4y = 20} \\ 20x + 0y = 23 \end{array}$$

Thus, we now have one equation in terms of one unknown:

$$20x = 23$$
$$x = \tfrac{23}{20} \approx 1.2$$

Plug this value for x into the second, original equation to get y:

$$y = 3(\tfrac{23}{20}) - 5 = \tfrac{69-100}{20} = -\tfrac{31}{20} \approx -1.6$$

(b) We look at the two equations for a while, and realize that if we multiply the second equation by 6, the y term will drop out when the two equations are added:

$$6(34 + 21x = -2y) \Rightarrow 204 + 126x = -12y$$

So we add:

$$\begin{array}{r} 47 + x = 12y \\ \underline{204 + 126x = -12y} \\ 251 + 127x = 0y \end{array}$$

This implies that:

$$127x = -251$$
$$x = -\tfrac{251}{127} \approx -2.0$$
$$\Rightarrow y = \tfrac{1}{12}(47 - 2.0) \approx 3.8$$

Most calculators today can easily solve two-variable simultaneous equations *and* plot the result. So, it is well worth your time to read the instruction manual to find out how. But make sure you can solve these types of equations without your calculator as well: solving two-variable simultaneous equations is an important skill to have in your physics toolbox!

Mathematicians are all a bit crazy.
PAUL ERDOS
American Mathematician, 1913 - 1996

SOLVING MULTI-VARIABLE SIMULTANEOUS LINEAR EQUATIONS

It can be very difficult to solve *multi*-variable (more than two) equations *graphically* without the use of a computer, so we will not discuss it here.

To solve multi-variable simultaneous linear equations *algebraically*, however, there are two methods. Neither method works in all possible situations; but when taken together, these techniques cover most cases you will encounter.

Procedure 1

1. Solve the least-complex equation(s) for the variable(s) of your choice. This might result in an immediate answer for one variable.
2. Solve the remaining equations for the remaining variables, one equation for each variable.
3. Pick the simplest equation from Step 2. Solve this equation totally in terms of one variable, using other equations from Step 2, and any answers obtained in Step 1, as needed.
4. Repeat Step 3, continually picking *new* equations, until you can write one equation totally in terms of known values. Solve this equation for a number.
5. Now plug this number into another equation to get another number, and so on, until you have numbers for all the variables.

Be careful with this method. Achieving a unique solution for all variables becomes increasingly difficult with each additional equation in the set. The best route is to eliminate as many variables as possible *before* you begin substituting equations. Also, only use each equation once; using an equation more than once will result in garbage solutions.

EXERCISE 13.30: Multi-Variable Equations, Part I

Solve the following three simultaneous equations using Procedure 1:

$$5x + 3y - 6z = 10$$
$$3x - 4y - 3z = 9$$
$$7x - 4z = 2$$

SOLUTION:

Notice that the last equation contains just two variables, x and z. Thus, using this equation, we can write z totally in terms of x, thereby eliminating one variable immediately (Step 1):

Exercise 13.30 Continued...

$$7x - 4z = 2 \Rightarrow z = \tfrac{1}{4}(-2 + 7x) = \tfrac{7}{4}x - \tfrac{1}{2}$$

Now, we can solve the first equation for x, and the second equation for y (Step 2):

$$5x + 3y - 6z = 10 \Rightarrow x = \tfrac{1}{5}(10 + 6z - 3y) = 2 - \tfrac{3}{5}y + \tfrac{6}{5}z$$
$$3x - 4y - 3z = 9 \Rightarrow y = \tfrac{1}{4}(-9 - 3z + 3x) = -\tfrac{9}{4} + \tfrac{3}{4}x - \tfrac{3}{4}z$$

Plug the equation for z into the equation for y to solve for y in terms of x alone:

$$\begin{aligned} y &= -\tfrac{9}{4} + \tfrac{3}{4}x - \tfrac{3}{4}(\tfrac{7}{4}x - \tfrac{1}{2}) \\ &= (-\tfrac{9}{4} + \tfrac{3}{8}) + (\tfrac{3}{4} - \tfrac{21}{16})x \\ &= \tfrac{1}{16}[(-36+6) + (12-21)x] \\ &= \tfrac{-1}{16}[30 + 9x] \end{aligned}$$

Finally, we plug this equation into the equation for x, along with the equation for z, and solve for x.

$$\begin{aligned} x &= 2 - \tfrac{3}{5}[\tfrac{-1}{16}(30+9x)] + \tfrac{6}{5}[\tfrac{7}{4}x - \tfrac{1}{2}] \\ &= [2 + \tfrac{90}{80} - \tfrac{6}{10}] + [\tfrac{27}{80} + \tfrac{42}{20}]x \\ &= \tfrac{1}{80}[(160 + 90 - 48) + (27 + 168)x] \\ &= \tfrac{1}{80}[202 + 222x] \\ &\Rightarrow 80x = 202 + 222x \\ 222x - 80x &= -202 \\ x &\approx -1.4 \end{aligned}$$

Once we have x, we can plug it into the other two equations to get values for y and z:

$$y = \tfrac{1}{16}[-30 - 9x] = \tfrac{1}{16}[-30 - 9(-1.4)] \approx -1.1$$
$$z = \tfrac{7}{4}x - \tfrac{1}{2} = \tfrac{7}{4}(-1.4) - \tfrac{1}{2} \approx -3.0$$

Notice that we used each equation only *once* in each process.

Procedure 2

1. Examine all equations. Look for terms, which if the equations were to be added or subtracted, would cancel out. You may need to multiply one of the equations by a factor to make this work. (Sometimes you may not find a relationship between *all* of the equations. If this happens, solve some of the equations by pairs according to Procedure 2 of the last section.)
2. Add or subtract the equations, obtaining a brand new equation.
3. Solve the new composite equation for the one variable.
4. Plug this value into one of the remaining original equations, or into the new equation, whichever is easier, and solve again.
5. Repeat as often as necessary to solve for all variables.

EXERCISE 13.31: Multi-Variable Equations, Part II

Solve the three simultaneous equations from Exercise 13.30 using Procedure 2.

SOLUTION:

We solve this problem by multiplying the first two equations by a factor so that the y terms drop out completely (the factors are 4 and 3, respectively):

$$4(5x + 3y - 6z = 10) \Rightarrow 20x + 12y - 24z = 40$$
$$3(3x - 4y - 3z = 9) \Rightarrow 9x - 12y - 9z = 27$$

Now we add the new equations together:

$$\begin{array}{r} 20x + 12y - 24z = 40 \\ 9x - 12y - 9z = 27 \\ \hline 29x + 0y - 33z = 67 \end{array}$$

Combining this new equation with the third original equation, we now have two two-variable simultaneous equations. These can be solved by substitution. Recall that $z = \frac{7}{4}x - \frac{1}{2}$, so:

$$29x - 33(\tfrac{7}{4}x - \tfrac{1}{2}) = 67$$
$$\Rightarrow 29x - \tfrac{231}{4}x + \tfrac{33}{2} = 67$$
$$\Rightarrow 116x - 231x = 268 - 66$$
$$\Rightarrow 115x = -202$$
$$\Rightarrow x \approx -1.7$$

Thus, $z = \frac{7}{4}x - \frac{1}{2} = \frac{7}{4}(-1.7) - \frac{1}{2} \approx -3.5$, and:

$$3(-1.7) - 4y - 3(-3.5) = 9$$
$$\Rightarrow -5.1 + 10.5 - 9 = 4y$$
$$y \approx -1.0$$

OTHER METHODS

There are several more advanced algebraic methods for solving multi-variable simultaneous linear equations. Although it is not necessary to know these methods to learn introductory physics, they can help in the clutch. Refer to any linear algebra text for more information on these and other advanced algebraic methods.

As we have said before, however, *most calculators can solve these equations at the push of a button.* It is very worthwhile to read your owner's manual to learn how to do this, for it will save you a lot of time in solving, and reduce your chance of error.

Although mathematical concepts and operations are formulated to represent aspects of the physical world, mathematics is not to be identified with the physical world. However, it tells us a good deal about that world, if we are careful to apply it and interpret it properly.
MORRIS KLINE
American Mathematician, 1908 -

For example, the kinematic equations are non-linear. See Section 5.4.2.

13.4.6 EXPONENTS

Many equations in physics are **non-linear**: one or more variables have **exponents** or **powers**. Mathematical manipulation of exponents is straightforward with the following rules:

Addition: $a^x + a^x = 2a^x$,
but $a^x + b^x$ and $a^x + a^y$ do not reduce.

Base Multiplication: $x^a x^b = x^{a+b}$

Exponent Multiplication: $(x^a)^b = x^{ab}$

Exponent Division: $\dfrac{x^a}{x^b} = x^{a-b}$

Roots: $x^{1/a} = \sqrt[a]{x}$ (e.g.: $x^{1/2} = \sqrt{x}$).

Finally, note that $x^0 = 1$ regardless of the value of x.

CAUTION 13.32: Think with Exponents!

Notice that the fourth rule implies that $x^{-n} = 1/x^n$. Also note that the second and fifth rules imply that $\sqrt[n]{a}\sqrt[n]{b} = \sqrt[n]{ab}$.

EXERCISE 13.33: Figuring Exponents

Use your calculator to figure the following:

(a) 243^0 (b) 5638^1 (c) $5^3 \times 5^2$

(d) $3^3 \div 3^2$ (e) $(4^5)^2$ (f) $3^{\frac{1}{2}}$

SOLUTION:

(a) $243^0 = 1$

(b) $5638^1 = 5638$

(c) $5^3 \times 5^2 = 5^{3+2} = 5^5 = 3125$

(d) $3^3 \div 3^2 = 3^{3-2} = 3^1 = 3$

(e) $(4^5)^2 = 4^{5(2)} = 4^{10} = 1{,}048{,}576$

(f) $3^{\frac{1}{2}} = \sqrt{3} = 1.732...$

13.4.7 POLYNOMIALS

A **polynomial** is any expression in which one or more variables may have an exponent.* The **order** or **degree** of a polynomial is the value of the largest exponent in the expression. For instance, the point-slope equation $y = mx + b$ is a first order-polynomial (as well as being a linear equation!) because the x and the y are only raised to the first power.

EXAMPLE 13.34: A Polynomial, Part I

The polynomial $y = ax^4 + bx^3 + cx^2 + dx + e$ is of the fourth order, because the highest exponent in the equation is a 4.

The **Fundamental Theorem of Algebra** says that every polynomial of degree n has n **roots**, or solutions found when the equation is set equal to zero. These solutions may not be all different, however! The roots can be real or complex numbers.

EXAMPLE 13.35: A Polynomial, Part II

Since the polynomial in the previous example is of the fourth order, there will be four solutions to the equation. However, it is possible that some or all of the solutions will have the same value, and that some or all of the solutions will be complex.

13.4.8 THE QUADRATIC EQUATION

One important quadratic equation in physics is the trajectory equation for projectile motion. See your textbook for more.

The **quadratic equation** $ax^2 + bx + c = 0$ is a second-order polynomial with two solutions:

$$x_1 = \frac{-b + \sqrt{b^2 - 4ac}}{2a} \text{ and } x_2 = \frac{-b - \sqrt{b^2 - 4ac}}{2a}$$

Although the quadratic equation yields two roots, often only *one* answer is acceptable. Always check your result for plausibility!

If $b^2 \geq 4ac$, the root is a real number, and if $b^2 < 4ac$, it is imaginary.

* With no transcendental terms like sinusoids or logarithms in the equation.

Time is in Section 11.5.2. 

EXERCISE 13.36: Solving Quadratics

Solve for the elapsed time t in the equation $7t^2 + 5t - 3 = 0$, t in seconds. Elapsed time is a quantity that is always positive.

SOLUTION:

We use the quadratic formula, where $a = 7\ \text{s}^{-2}$, $b = 5\ \text{s}^{-1}$, and $c = -3\ \text{s}$:

$$t = \frac{-5 \pm \sqrt{25 - 4(7)(-3)}}{2(7)} = \tfrac{1}{14}(-5 \pm \sqrt{109})$$

$$\Rightarrow\ t \approx 0.39\ \text{s} \ \text{ or } \ t \approx -1.1\ \text{s}$$

However, because elapsed time can never be negative, the second result *must* be incorrect. We therefore discard it. So, the final answer is $t = 0.39$ s only.

13.4.9 CONSTANTS & PROPORTIONALITIES

When dealing with equations, it is sometimes desirable to consider how certain terms in the equation behave in relation, or *proportion,* to others.

A term is **directly proportional** to another if the ratio of the two is constant. A term is **inversely proportional** to another if the product of the two is constant.

EXAMPLE 13.37: Proportions, Part I

Recall that $v = \Delta x / \Delta t$. From this equation, we can say that velocity is directly proportional to displacement (an increase in displacement increases speed), but we can also say that velocity is inversely proportional to time interval (an increase in time interval decreases speed). That is, we can write $v \propto \Delta x$ and $v \propto 1 / \Delta t$.

A great feature of proportionalities is that they can always be turned into equalities with the help of a **multiplicative constant**.

EXAMPLE 13.38: Proportions, Part II

For example, if we knew that a variable x was directly proportional to another quantity y, so that $x \propto y$; *and* we also knew that there were no other variables in this relationship, then we could create an equation relating x and y by introducing a multiplicative constant.

Let's call the constant in this case k. Thus, our relation $x \propto y$ would become the equation $x = ky$.

There are three main types of constants in physics:

- **Fundamental constants**, like e and c;

In Section 11.5.3.

- **Numerical constants**, like 2 and π;
- **Coefficients**, like μ, the coefficient of friction. Coefficients are measured experimentally, and vary with material, conditions, etc. Therefore, they are not technically *constants* like the fundamental and numerical constants above. However, they are used in much the same way as the other constants, to unite proportionalities.

In Section 7.2.6.

13.4.10 FACTORING

One way to simplify linear equations or polynomials is to **factor** out common terms. Recall these formulas for factoring polynomials:

Constant factor: $ax + ay + az = a(x + y + z)$

Perfect square: $a^2 + 2ab + b^2 = (a+b)^2$

Difference of squares: $a^2 - b^2 = (a+b)(a-b)$

The first rule above can be used with any order polynomial. Rules 2 and 3 are used with second-order non-quadratic polynomials. Notice, however, that Rules 2 and 3 can only be used if the equation has the correct form given. Be careful: this is *not* always the case.

We can, however, *make* second-order polynomials perfect squares using the method known as **completing the square**. To complete the square, follow these steps:

1. Check that the coefficient of the second-order term is a one. If so, proceed to Step 2. If not, divide all terms in the entire equation by that coefficient.
2. Collect all first- and second-order terms inside one set of parentheses. Factor these terms so that you have one first-order coefficient, and one second-order coefficient. Collect all constants outside the parentheses.
3. Halve the first-order coefficient, then square it.
4. Add the number obtained in Step 3 to the terms inside the parentheses. Subtract the same number from the constants outside the parentheses.
5. You have just forced a perfect square among the quadratic terms in the parentheses, so factor them as such.
6. Continue with other algebra as needed.

EXERCISE 13.39: Fun with Factoring

Solve these equations for x using factoring:

(a) $4x + 2 = 6$ (b) $x^2 + 6x + 9 = 0$ (c) $9x^2 - 4 = 0$ (d) $x^2 - 4x - 12 = 0$

SOLUTION:

(a) Factor a 2 from both sides of the equation:

$$2\,(2x + 1) = 2\,(3)$$

Cancel the 2 and solve:

$$\begin{aligned} 2x + 1 &= 3 \\ 2x &= 2 \\ x &= 1 \end{aligned}$$

(b) This equation is a perfect square:

$$\begin{aligned} x^2 + 6x + 9 &= 0 \\ (x + 3)^2 &= 0 \\ \Rightarrow x + 3 &= 0 \text{ and } x + 3 = 0 \\ \therefore x &= -3, -3 \end{aligned}$$

(c) This equation is a difference of squares:

$$\begin{aligned} 9x^2 - 4 &= 0 \\ (3x - 2)(3x + 2) &= 0 \\ \Rightarrow 3x - 2 &= 0 \text{ and } 3x + 2 = 0 \\ \therefore x &= \frac{2}{3}, -\frac{2}{3} \end{aligned}$$

(d) To solve this equation, we must complete the square. First, we collect the first- and second-order terms, the x and x^2, inside parentheses. Then, the constants go outside the parentheses:

$$\begin{aligned} x^2 - 4x - 12 &= 0 \\ (x^2 - 4x) - 12 &= 0 \end{aligned}$$

Looking at this equation, we see that the coefficient attached to the first-order term is a 4. By Step 3, we halve the 4 to 2, and then square the 2 to a 4. Next, we add the 4 inside the parenthesis *and* subtract it outside the parenthesis:

$$\begin{aligned} (x^2 - 4x + 4) - 12 - 4 &= 0 \\ \Rightarrow (x^2 - 4x + 4) - 16 &= 0 \end{aligned}$$

Now, we have a perfect square in side the parenthesis, which we know how to solve from part (c) above:

$$\begin{aligned} (x - 2)^2 &= 16 \\ x - 2 &= \pm 4 \\ x &= \pm 4 + 2 \\ x &= 6 \text{ or } x = -2 \end{aligned}$$

Notice that when we took the square root of both sides, we had to account for *both* positive and negative solutions.

13.4.11 "UNFACTORING": POWER SERIES EXPANSION

Sometimes you are faced with a factored equation that needs to be "un-factored" or **expanded**. To do so, choose the appropriate formula from Table 13.4 below and plug into the suitable expression.

Table 13.4: Power Series Expansions

$(a+b)^n = a^n + \frac{n}{1!}a^{n-1}b + \frac{n(n-1)}{2!}a^{n-2}b^2 + \cdots$	$\tan x = x + \frac{x^3}{3} + \frac{x^5}{15} + \cdots$
$e^x = 1 + \frac{x^1}{1!} + \frac{x^2}{2!} + \frac{x^3}{3!} + \cdots$	$\sin^{-1} x = x + \frac{1}{6}x^3 + \frac{3}{40}x^5 + \cdots$
$\ln(1 \pm x) = \pm x - \frac{1}{2}x^2 \pm x^3 + \cdots$	$\cos^{-1} x = \frac{\pi}{2} - \sin^{-1} x$
$\sin x = x - \frac{x^3}{3!} + \frac{x^5}{5!} - \cdots$	$\tan^{-1} x = x - \frac{x^3}{3} + \frac{x^5}{5} - \cdots$
$\cos x = 1 - \frac{x^2}{2!} + \frac{x^4}{4!} - \cdots$	

These formulas relate the factored form of an equation to a polynomial, usually (but not always!) with an infinite number of terms. A polynomial with an infinite number of terms is called a **power series**, hence the name **power series expansion** for this technique.

Note that many formulas in this table contain the symbol "!," called the **factorial**. To use a factorial, multiply the number or variable being factorialized by a series of numbers, each number exactly one less than the one before, until you get to one.

The factorial of zero, by definition, is one: $0! \equiv 1$.

EXAMPLE 13.40: Factorials

For example, the factorial of 3, $3!$, is $3 \times 2 \times 1 = 6$. The factorial of 8 is $8! = 8 \times 7 \times 6 \times 5 \times 4 \times 3 \times 2 \times 1 = 40{,}320$. And the factorial of n is $n! = n\,(n-1)(n-2) \ldots. (1)$.

THE BINOMIAL THEOREM

The most important formula from Table 13.4, the **Binomial Theorem**, does not always have an infinite number of terms. The formula presented in that table can be rewritten in a more compact fashion, in terms of a sum:

$$(a \pm b)^n = \sum_{k=0}^{n} \left(\frac{n!}{(n-k)!k!} \right) a^{n-k} b^k \equiv \sum \binom{n}{k} a^{n-k} b^k$$

EXERCISE 13.41: Using the Binomial Theorem

Expand the term $(x + 5)^3$ using the Binomial Theorem.

SOLUTION:

To expand this term, we apply the Binomial Theorem formula:

$$(x+5)^3 = \sum_{k=0}^{3}\left(\frac{3!}{(3-k)!k!}\right)a^{3-k}b^k$$

$$= \left[\left(\frac{3!}{(3-0)!0!}\right)x^{3-0}5^0\right] + \left[\left(\frac{3!}{(3-1)!1!}\right)x^{3-1}5^1\right] + \left[\left(\frac{3!}{(3-2)!2!}\right)x^{3-2}5^2\right] + \left[\left(\frac{3!}{(3-3)!3!}\right)x^{3-3}5^3\right]$$

$$= \left[\left(\frac{3\cdot2\cdot1}{(3\cdot2\cdot1)\cdot(1)}\right)(1)x^3\right] + \left[\left(\frac{3\cdot2\cdot1}{(2\cdot1)(1)}\right)(5)x^2\right] + \left[\left(\frac{3\cdot2\cdot1}{(1)(2\cdot1)}\right)(5^2)x\right] + \left[\left(\frac{3\cdot2\cdot1}{(1)(3\cdot2\cdot1)}\right)(5^3)(1)\right]$$

$$= x^3 + 15x^2 + 75x + 125$$

1
1 1
1 2 1
1 3 3 1
1 4 6 4 1
⋮

Figure 13.20

The Binomial Theorem is graphically expressed in **Pascal's Triangle** (Fig. 13.20). Each line in this infinite triangle is created by summing the numbers to the immediate left and right above it.

EXAMPLE 13.42: Pascal's Triangle

For example, the left-hand 4 in the fifth line of Pascal's triangle, as seen in Figure 13.20, is created by the summation of the 1 and 3 in the fourth line immediately above it. And the 6 to the right of the 4 is created by adding the two 3's right above it.

We use Pascal's Triangle as a mnemonic to help us "unfactor" equations like $(a \pm b)^n$. To do so, we use the numbers in the $(n + 1)$th line as coefficients for each term in the polynomial expansion. Each term is created by multiplying a and b together, such that the exponent on a continually decreases from n to 0, while the exponent on b continually increases from 0 to n.

EXAMPLE 13.43: Using Pascal's Triangle, Part I

For instance, in the equation $(a+b)^n$, let $n = 3$. The expansion will then have four terms: a^3b^0, a^2b^1, a^1b^2, and a^0b^3. The coefficients for each of these terms will be taken from the *fourth* line of Pascal's Triangle. According to Figure 13.20, that line consists of the numbers 1 3 3 1. Thus, our final expansion for $(a + b)^3$ is:

$$(a+b)^3 = 1a^3b^0 + 3a^2b^1 + 3a^1b^2 + 1a^0b^3 = a^3 + 3ab(a+b) + b^3$$

EXERCISE 13.44: Using Pascal's Triangle, Part II

"Unfactor" the term from Exercise 13.41 using Pascal's Triangle.

SOLUTION:

$$(x+5)^3 = (1)(x^3)(5^0) + (3)(x^2)(5^1) + (3)(x^1)(5^2) + (1)(x^0)(5^3)$$
$$= x^3 + 15x^2 + 75x + 125$$

Notice that we obtained the same answer as in Exercise 13.41, but with much less work. Thus, Pascal's Triangle is a good method to use when n is small (less than five or so). Otherwise, the Binomial Theorem is the way to go.

POWER SERIES EXPANSIONS

The remaining power series expansions of Table 13.4 have an infinite number of terms. Thus, if x is large, these series will diverge and the equation approaches infinity.

However, if x is small ($x \ll 1$), these series will converge upon a unique, non-infinite number. Therefore, when x is small, it is usually not necessary to calculate all of the terms in the expansion: since x *has* to be small for the series to converge, often only the first one or two terms in the series are all that is necessary to obtain an approximate value for the series.

EXERCISE 13.45: Power Series Expansions

Assuming that $x \ll 1$, write out the following power series expansions:

(a) $(1+x)^{-4}$ (b) $\cos 5x$

SOLUTION:

(a) $$(1+x)^{-4} = 1 + \frac{-4}{1}x + \frac{-4(-5)}{2\cdot 1}x^2 + \frac{-4(-5)(-6)}{3\cdot 2\cdot 1}x^3 + \cdots \approx 1 - 4x + 10x^2$$

(b) $$\cos 5x = 1 - \frac{(5x)^2}{2!} + \frac{(5x)^4}{4!} - \frac{(5x)^6}{6!} + \cdots \approx 1 - \tfrac{25}{2}x^2 + \tfrac{625}{24}x^4$$

If you have an expression that does not look like one of the power series expansions in Table 13.4, try a little algebra and see if you can't force the appropriate form.

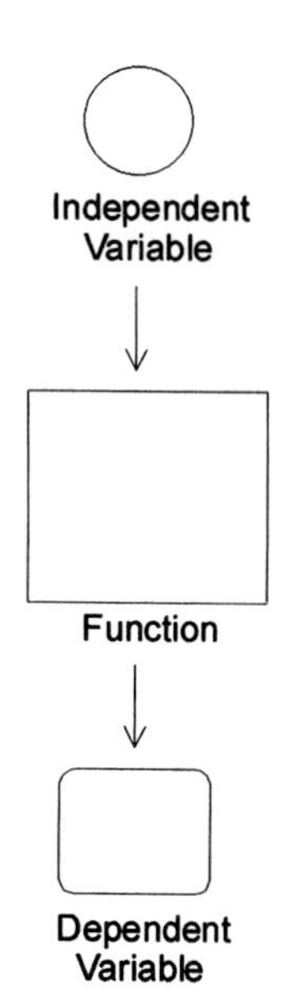

Figure 13.21

13.4.12 FUNCTIONS

A **function** is a mathematical relationship between independent and dependent variables. A function is like a computer in that it takes input—the **independent variable**(s)—and then returns output—the **dependent variable**. See Figure 13.21.

EXAMPLE 13.46: Functions, Part I

To help you understand the nature of functions, consider the polynomial equation $y = ax^4 + bx^3 + cx^2 + dx + e$.

To use this equation, we select a value for x, plug it in, and get a value for y. So, the formula $ax^4 + bx^3 + cx^2 + dx + e$ is like a computer that takes different values for x, and spits out unique values of y. The polynomial equation, therefore, is a *function* that relates the independent variable, x, and the dependent variable, y, in such a way that no matter what the value of x, we can always find the value of y.

Note that the value for y *completely depends* upon the value of x. We can, if it suits us, therefore think of y as merely being *dependent* on x, rather than a separate entity of its own.

Functions come in many forms. However, all functions share the property that the value of the dependent variable is completely determined by the value(s) of the independent variable(s) in the equation; hence their names. To indicate this subservient relationship, we often write the dependent variable in function notation as $y = f(x)$.* If there is more than one independent variable in the function, this notation easily expands to $f(x,y)$ or $f(x, y, z)$.

EXAMPLE 13.47: Functions, Part II

We can emphasize the functional nature of the polynomial $y = ax^4 + bx^3 + cx^2 + dx + e$ in Example 13.46 above by rewriting the equation as $f(x) = ax^4 + bx^3 + cx^2 + dx + e$ instead.

Because the definition of a function—a mathematical relationship between independent and dependent variables—is so basic, just about any equation qualifies as a function. Some functions are therefore useless at best, and problematic at worst. For this reason, most physicists only use functions that meet the following common-sense criteria. Well-defined functions are:

---

*$f(x)$ is read as "the function of the variable x" or "f of x".

- **Single-valued**: For every value of the independent variable, there is one *and only one* value of the dependent variable (Figure 13.22 (a) shows a non-single-valued function).
- **Continuous**: There are no gaps or kinks in the function (or in derivatives of the function). See Figure 13.22 (b) for examples of non-continuous functions.

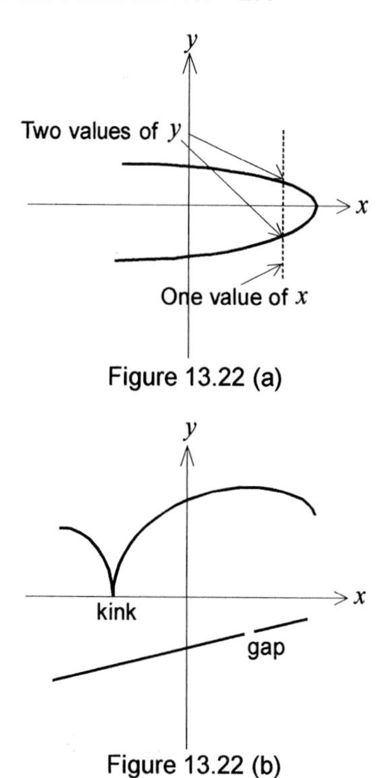

Figure 13.22 (a)

Figure 13.22 (b)

EXERCISE 13.48: Well Defined Functions

Are the functions sketched below well defined?

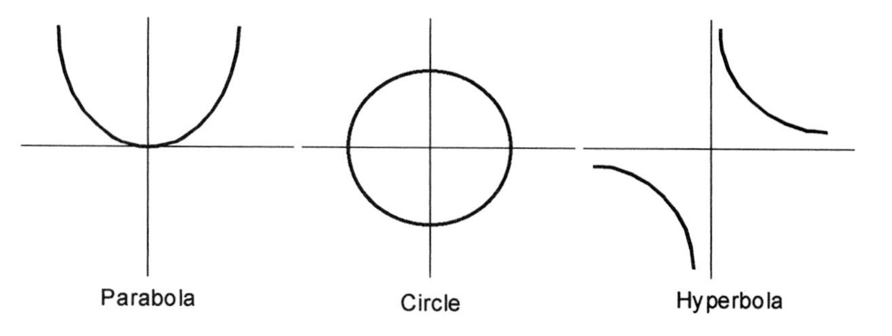

SOLUTION:

The parabola and the hyperbola both meet the criteria of single-valuedness, for a vertical line through the x-axis passes through the function only once. On the other hand, a vertical line passes through the circle twice, so a circle is not a single-valued function:

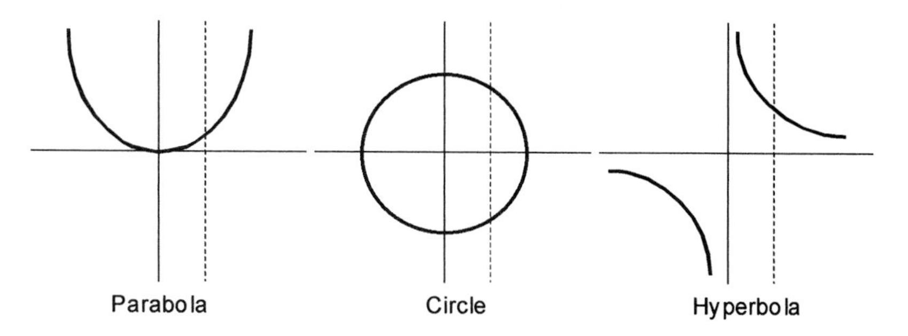

On the other hand, the parabola and the circle are both continuous functions, for they have no gaps or kinks. The hyperbola clearly has a gap at $x = 0$, however, so it is not continuous. (If we only consider one arm of the hyperbola, though, the condition of continuity is met.)

Thus, only the parabola can be considered a well-defined function, since it is the only one of the three that is *both* continuous and single-valued.

Many simple functions appear again and again in introductory physics. You should be able to recognize the following eight functions, as well as their graphs, by sight.

1. The **linear function** $f(x) = mx + b$ is a straight line (Fig. 13.23 (a)).
2. The **quadratic function** $f(x) = ax^2 + bx + c$ is a parabola (Fig. 13.23 (b)).
3. The **cubic function** $f(x) = ax^3 + bx^2 + cx + d$ is roughly in the shape of an "S" (Fig. 13.23 (c)).
4. The **quartic function** $f(x) = ax^4 + bx^3 + cx^2 + dx + e$ has the rough shape of a "M" or a "W" (Fig. 13.23 (d)).
5. The **inverse function** $f(x) = k/x$ is hyperbolic (Fig. 13.23 (e)).*
6. The **inverse squared function** $f(x) = k/x^2$ drops toward zero faster than the inverse function (Fig. 13.23 (f)).
7. The **exponential functions** $f(x) = e^{\pm x}$ also drop toward zero (Fig. 13.23 (g)).
8. The **trigonometric functions** $f(x) = \sin x$, $\cos x$, and $\tan x$ oscillate (Figs. 13.23 (h), (i), and (j), respectively).

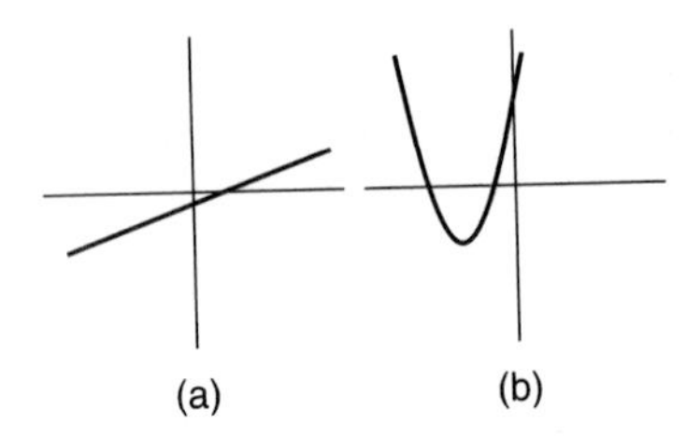

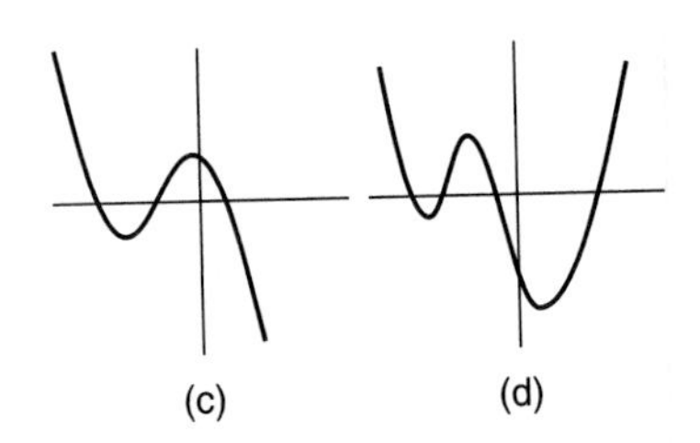

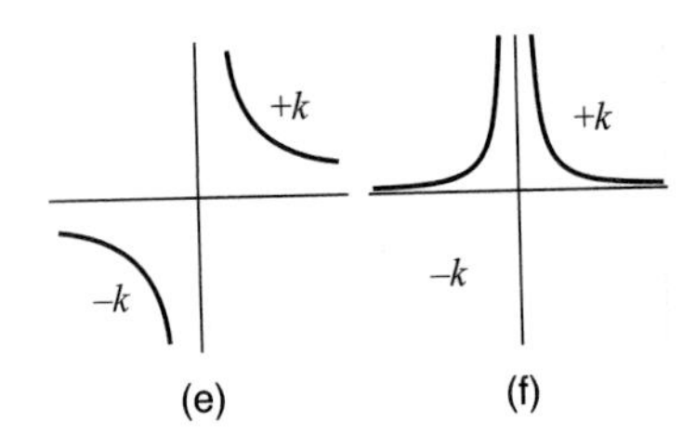

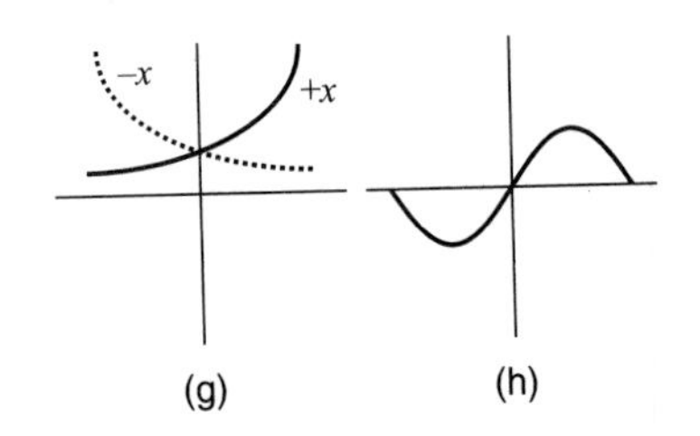

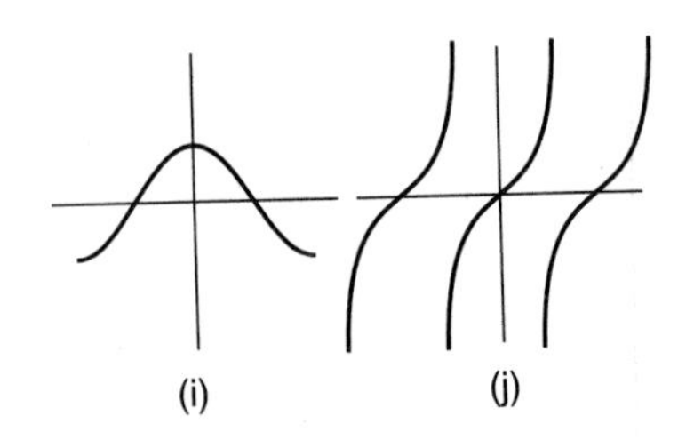

Figure 13.23

EXERCISE 13.49: Some Important Functions

Which of the above functions are single-valued? Which are continuous? Is it a problem that some of these functions are not well defined?

SOLUTION:

All of the functions in the above list are single-valued (when oriented as sketched, that is). Most of the functions are continuous as well, except for the inverse function, the inverse squared function, and the tangent function.

Each of these three functions can be *made* well defined, however, by limiting the effective area of the function. For example, the inverse function and the inverse squared function are continuous in the regions $x > 0$ and $x < 0$, and the tangent function is continuous in the region $-\frac{\pi}{2} < x < \frac{\pi}{2}$.

As long as we work in the realm in which these three functions are continuous, then, they are well defined and there is no problem using them for physical analysis.

* Notice that as x increases in the inverse function, $f(x)$ decreases, but the product $x \cdot f(x)$ remains constant.

13.4.13 PARAMETRIC EQUATIONS

Until now, we have been discussing functions of the form $y = f(x)$; that is, equations where the value of y depends in a direct way upon the value of x. There are many occasions, however, when it is desirable to separate y from x, making both quantities dependent upon a *third* variable t, called the **parameter**. Then the function $f(x)$ can be represented as *two* functions $x(t)$ and $y(t)$, called the **parametric equations**.#

Parametric equations are used with kinematics, Section 5.4.2.

Parametric equations may seem to complicate matters, since, after all, one equation is being replaced by two. However, in physics, parametric equations—in which the parameter t represents time, and x and y represent positions in two-dimensional space—allow us to develop equations with much more ease than would otherwise be the case.

To convert parametric equations $x(t)$ and $y(t)$ to the familiar equation $y = f(x)$, solve $x(t)$ for t in terms of x, then plug the new equation into $y(t)$, yielding $f(x)$.

EXERCISE 13.50: Parametric Equations

The parametric equations for x and y in a gravitational field as a function of time are:

$$x(t) = x_i + v_{i,x}t$$

$$y(t) = y_i + v_{i,y}t - \tfrac{1}{2}gt^2$$

where x_i, y_i, $v_{i,x}$, $v_{i,y}$, and g are constants. Find $y = f(x)$.

These equations can be derived from the equations of constant acceleration in Section 5.4.2. Your textbook will show the derivation.

SOLUTION:

First, we solve $x(t)$ for t as a function of x:

$$t = \frac{x - x_i}{v_{i,x}}$$

Now, we plug this equation into $y(t)$ to find $y = f(x)$:

$$y = y_i + v_{i,y}\left(\frac{x - x_i}{v_{i,x}}\right) - \tfrac{1}{2}g\left(\frac{x - x_i}{v_{i,x}}\right)^2$$

$$\Rightarrow \Delta y = \left(\frac{v_{i,y}}{v_{i.x}}\right)\Delta x - \left(\frac{g}{2v_{i,x}^2}\right)\Delta x^2$$

Because this equation depends on Δx^2, Δy is parabolic.

Polar coordinates can also be parametrized: $\rho(t)$ and $\theta(t)$.

13.4.14 LOGARITHMS

Logarithms are often used with wave intensities.

This rule is in Section 9.4.1 on exponents.

Given the xth-order polynomial equation $y = a^x$, we can define a related quantity $x \equiv \log_a y$, where x is the termed the **logarithm** of y and a is called the **base number**. Logarithms, like trigonometric functions, are **transcendental functions**, because they are non-algebraic.

From the definition of the logarithm, and from the fact that $x^0 = 1$, it follows that $\log_a 0 = 1$. The following rules also apply to logarithms:

Addition "is" Multiplication: $\log a + \log b = \log ab$

Subtraction "is" Division: $\log a - \log b = \log\left(\frac{a}{b}\right)$

Bring Powers Down: $\log(a^b) = b\log a$

Switching Bases: $(\log_a b)(\log_b c) = \log_a c$

EXERCISE 13.51: Logarithm Laws

Rewrite the expression $\log\left(\frac{ab^c}{d^e f}\right)$ using logarithm laws.

SOLUTION:

$$\log\left(\frac{ab^c}{d^e f}\right) = \log(ab^c) - \log(d^e f) \quad \text{(second rule)}$$

$$= \log a + \log b^c - \log d^e - \log f \quad \text{(first rule)}$$

$$= \log a + c\log b - e\log d - \log f \quad \text{(third rule)}$$

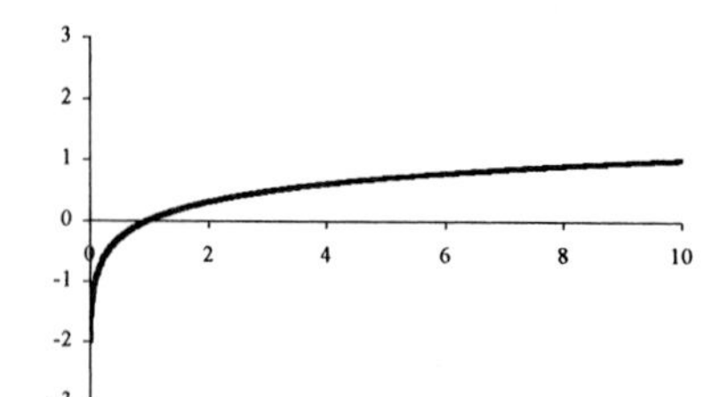

Figure 13.24 (a)

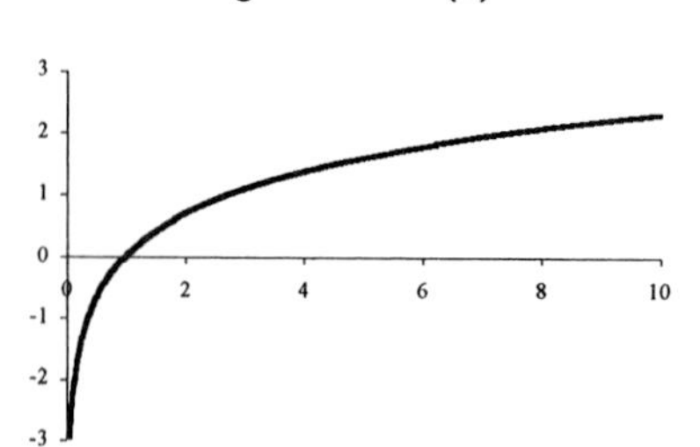

Figure 13.24 (b)

Although any base number a is theoretically possible for logarithms, the two most common ones are **base 10** ($a = 10$) and **base *e***, ($a = e$, where $e = 2.71828...$). Base e logarithms, called **natural logarithms**, are usually written $\ln x$ rather than $\log_e x$.

Refer to Figure 13.24 (a) for a sketch of the base 10 logarithm, and Figure 13.24 (b) for a sketch of the natural logarithm, each on standard grids. In both cases, you should note that negative logs are *not* defined.

EXERCISE 13.52: The Natural Logarithm

What is $\ln e$?

SOLUTION:

$\ln e = 1$, because $e^1 = e$.

On many occasions in your physics class, you will need to plot logarithms. There are two basic methods, depending upon the type of graphing grid you choose: a regular grid—one in which the tick-marks are equally spaced (Fig. 13.25 (a))—or a logarithmic grid—a grid in which the tick-marks are spaced unequally (larger spaces for numbers beginning in 1's and 2's, smaller spaces for numbers beginning in 8's and 9's).

There are two types of the latter kind of graph papers: "log-linear" (or "semi-log"), in which only the y-axis is logarithmic (Fig. 13.25 (b)), and "log-log," in which both axes are logarithmic (Fig. 13.25 (c)).

To plot logarithms, choose Method A if you have regular graph paper, and Method B if you have either kind of logarithmic paper.

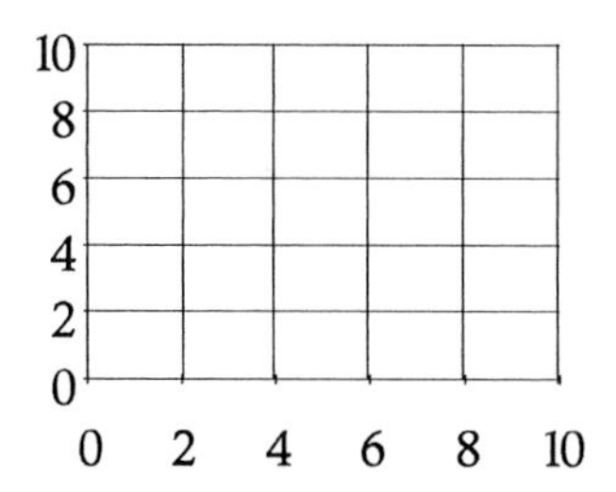

Figure 13.25 (a)
Use for Log-Linear or Log-Log

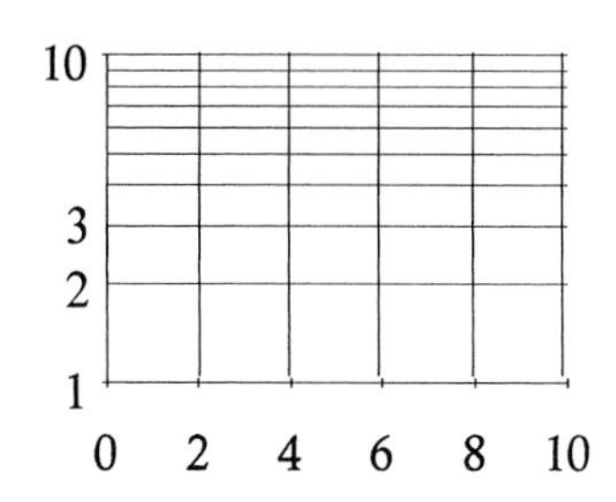

Figure 13.25 (b)
Use for Log-Linear Only

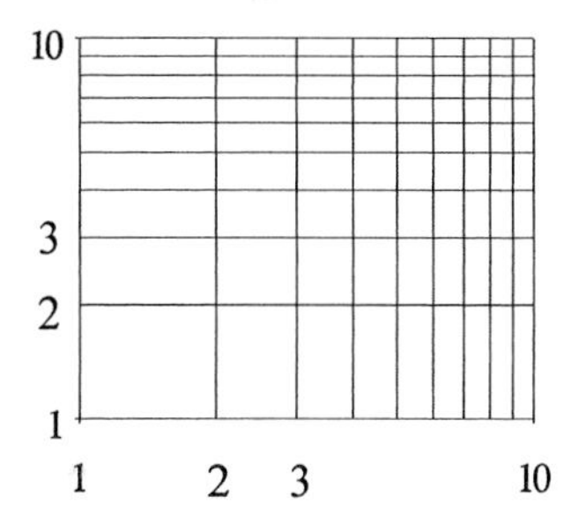

Figure 13.25 (c)
Use for Log-Log Only

Method A

For use with a regular graphing grid

1. Create a table. In the first and second columns of the table, list the x and y values.
2. *Skip this step if you plotting an exponential function.* For each value of x, calculate $\log x$. Input these values into the third column of your table.
3. For each value of y, calculate $\log y$. Input these values into the fourth column of the table (third column if you skipped Step 2).
4. Plot the values of the third and fourth (or first and third columns if you skipped Step 2) on your graph.
5. The graph will be a straight line.

Method B

For use with a logarithmic grid

1. Plot each value of x and y on the grid.
2. The graph will be a straight line.

Notice that Method B consists of far fewer steps than Method A. Thus, using logarithmic graph paper is a much easier way to plot logarithms, *once you get a hang of the strange format of the grid.* So, it is well worth your time to learn how to use log paper.

EXERCISE 13.53: Plotting Logarithms

The size of an exponentially growing bacterial colony is measured once per minute. The results are tabulated below. Plot these measurements on a regular grid and a log-linear grid. Plot the second graph using both methods.

Time (min)	Amount (l)	Time (min)	Amount (l)
1	0.015	5	0.25
2	0.030	6	0.50
3	0.060	7	1.0
4	0.12		

Exercise 13.53 Continued...

SOLUTION:

A normal plot of this data on a regular grid, without using Method A, looks like:

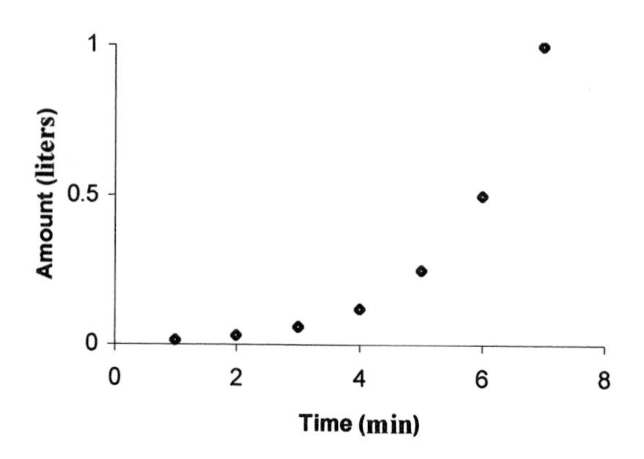

Now, we want to plot the same data, on the same kind of graph paper, using Method A above. This method asks us to begin by creating a table with the x and y values of the experiment (Step 1). This step has already been done for us above, where time is x and amount of bacteria is y.

We can further skip Step 2 because bacterial growth is described by an exponential function (according to the problem statement). So, Step 3 asks us to add a third column to our table, in which we list values of $\log y$. We write our new table, then, as follows:

x (min)	y (l)	$\log y$	x (min)	y (l)	$\log y$
1	0.015	−1.8	5	0.25	−0.60
2	0.030	−1.5	6	0.50	−0.30
3	0.060	−1.2	7	1.0	0
4	0.12	−0.90			

As Step 4 indicates, we now can plot the first and third columns of this table on a regular grid:

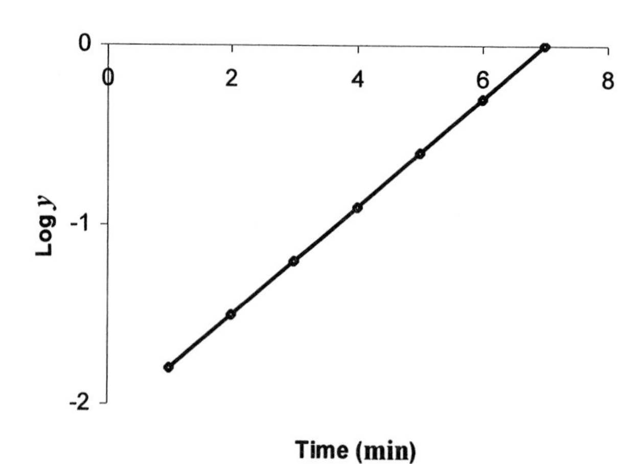

The result is a straight line, as predicted! We could, if we wished, analyze this line using the techniques of Section 16.2.2. Finally, we now want to plot our growth data using Method B. To do so, we use the original data table, but plot it on a log-linear grid:

Exercise 13.53 Continued...

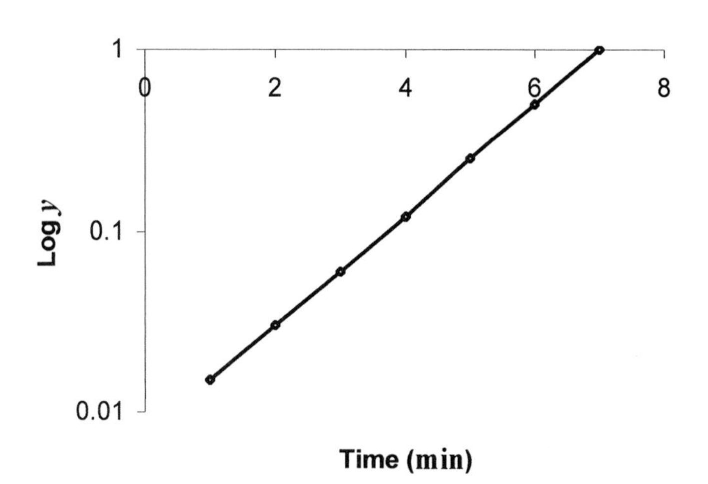

This line is the same as the one drawn above, but without all of the work!

13.4.15 EXPONENTIAL FUNCTIONS

The inverse of the natural logarithm is $y = e^x$. This function, together with the negative exponent $y = e^{-x}$, make up a class of special functions called the **exponential functions**, also written $\exp(x)$ and $\exp(-x)$ (Figs. 13.26 (a) and (b)).

Exponential functions are especially useful with radioactivity.

EXERCISE 13.54: The Exponential Function

What is $e^{\ln x}$?

SOLUTION:

Let $y = e^{\ln x}$. To find out what y is, we must take the natural logarithm of both sides of the equation:

$$y = e^{\ln x}$$
$$\ln y = \ln(e^{\ln x})$$

Since we know from Exercise 13.52 that $\ln e = 1$, this equation is:

$$\ln y = \ln x \Rightarrow y = x$$

Thus $e^{\ln x}$ is just x. Because of their reciprocal relationship, $\ln x$ and e^x are called **inverse functions**.

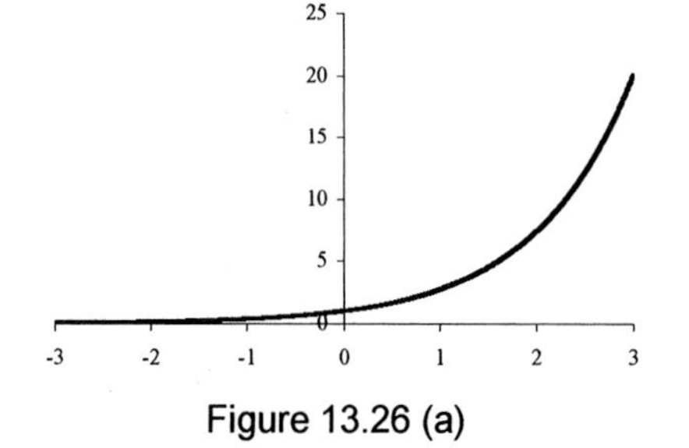

Figure 13.26 (a)

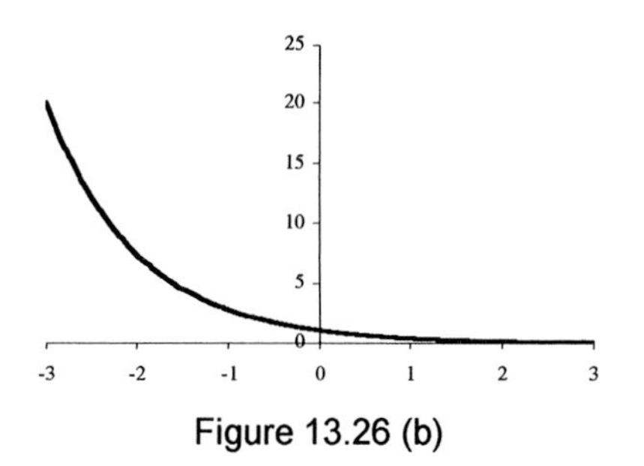

Figure 13.26 (b)

Any phenomenon that increases like the positive exponential function e^x is said to have **exponential growth**. An exponentially growing population that begins with $N(0)$ members and grows at a rate R will have $N(t)$ members a time t later, where:

$$N(t) = N(0)\, e^{Rt};\quad N(t) > N(0)$$

Assuming the growth rate R remains constant, an exponentially growing population increases by the same factor in the same amount of time, *regardless of the initial population size.* A measure of this growth factor is the **doubling time** $\tau = \ln 2/R$, the time required to double the initial population, $N(t) = 2\, N(0)$.

EXERCISE 13.55: Exponential Growth

Currently, the world's population is about six billion people. It is growing by 1.7% per year.

(a) Calculate the time required to double the world's population.

(b) What will the population be in five years, assuming it continues to grow at this rate?

SOLUTION:

(a) $\tau = \dfrac{\ln 2}{R} = \dfrac{0.693...}{0.017} \approx 41 \text{ years}$

(b) $N(4) = N(t)\, e^{Rt} = (6 \text{ billion})\, e^{(0.017)(5)} \approx 6.5 \text{ billion people}$

Any phenomenon which decreases like the negative exponential function e^{-x} is said to exhibit **exponential decay**. An exponentially decaying population that begins with $N(0)$ members and decays at a rate R will have $N(t)$ members a time t later, where:

$$N(t) = N(0)\, e^{-Rt};\quad N(t) < N(0)$$

Assuming the decay rate R remains constant, an exponentially decaying population decreases by the same factor in the same amount of time, *regardless of the initial population size.* A measure of this decay factor is the **half-life** $\tau = \ln 2/R$, the time required to halve the initial population, $N(t) = \frac{1}{2} N(0)$.

EXERCISE 13.56: Radioactive Half-Life

A radioactive sample loses 28% of its material in 42 s. What is the half-life of the sample?

Exercise 13.56 Continued...

SOLUTION:

Percent loss is the ratio of $N(t)$ with respect to $N(0)$, or $N(t) / N(0) = 0.28$. Thus we need to solve the above equation for τ in terms of this ratio:

$$N(t) = N(0)e^{-t/\tau}$$
$$\Rightarrow \frac{N(t)}{N(0)} = e^{-t/\tau}$$
$$\Rightarrow \ln\left(\frac{N(t)}{N(0)}\right) = -\frac{t}{\tau}$$
$$\therefore \tau = -\frac{t}{\ln\left(\frac{N(t)}{N(0)}\right)}$$

Now we plug in the appropriate numbers:

$$\tau = -\frac{42\,\text{s}}{\ln(0.28)} = 33\,\text{s}$$

13.5 GEOMETRY

Geometry is the study of regular shapes in Cartesian coordinates.

13.5.1 DISTANCE

The straight-line distance d (Fig. 13.27 (a)) between two points with coordinates (x_1, y_1) and (x_2, y_2) is:

$$d = \sqrt{(x_2 - x_1)^2 + (y_2 - y_1)^2} \equiv \sqrt{\Delta x^2 + \Delta y^2}$$

The **arc length** s (Fig. 13.27 (b)) between two points on the rim of a circle of radius r separated by angle θ (in radians) is $s = r\theta$.

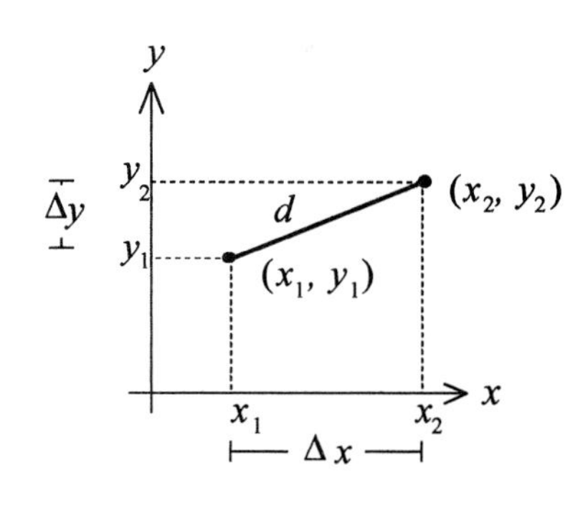

Figure 13.27 (a)

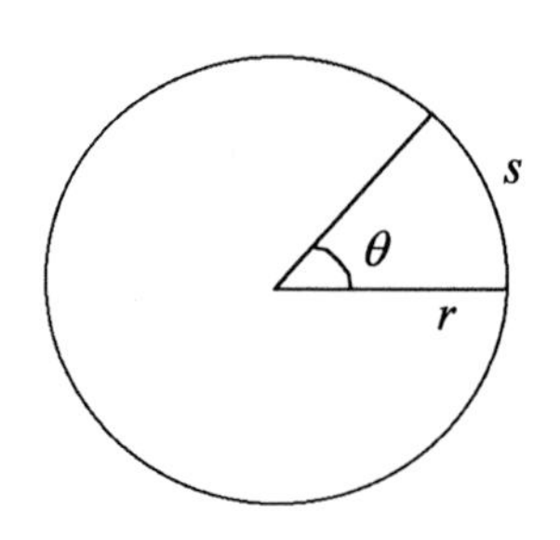

Figure 13.27 (b)

EXERCISE 13.57: Distance

(a) Find the straight-line distance between the points (5 m, 2 m) and (6 m, –8 m).

(b) Find the arc-length subtended by an angle of $\frac{\pi}{6}$ rads on the rim of a circle of radius 5 cm.

Exercise 13.57 Continued...

SOLUTION:

(a) $d = \sqrt{(x_2 - x_1)^2 + (y_2 - y_1)^2} = \sqrt{(6\text{m} - 5\text{m})^2 + (-8\text{m} - 2\text{m})^2} = \sqrt{101\text{m}^2} \approx 10\text{m}$

(b) $s = r\theta = (5\text{ cm})(0.52\text{ rads}) \approx 2.6\text{ cm}$

13.5.2 PERIMETERS, AREAS, & VOLUMES

God eternally geometrizes.
PLATO
Greek Philosopher, 428 - 348 B.C.

Two-dimensional objects, such as circles and squares, are characterized in terms of perimeters and cross-sectional areas. The **perimeter** is the length of the object's rim; the perimeter of a circle is its **circumference**. The **cross-sectional area** is the two-dimensional expanse of the object's face. Figure 13.28 below lists perimeters and cross-sectional areas of six simple two-dimensional objects:

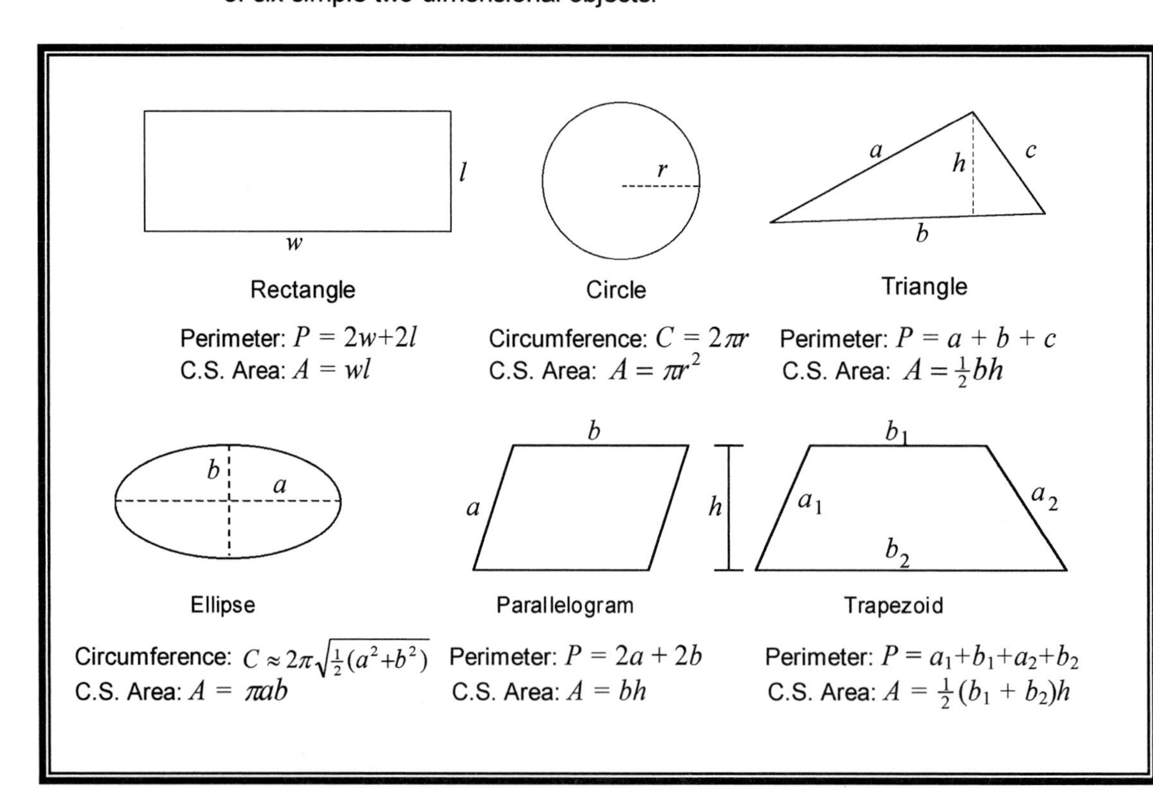

Figure 13.28

Mighty is geometry; joined with art, resistless.
EURIPEDES
Greek Dramatist, 484 - 406 B.C.

Three-dimensional objects, like spheres and cubes, are characterized by surface area and volume. The **surface area** is the area of the object's surface (only). The **volume** is a measure of the space occupied by the object. Figure 13.29 below lists surface areas and volumes of five simple three-dimensional shapes:

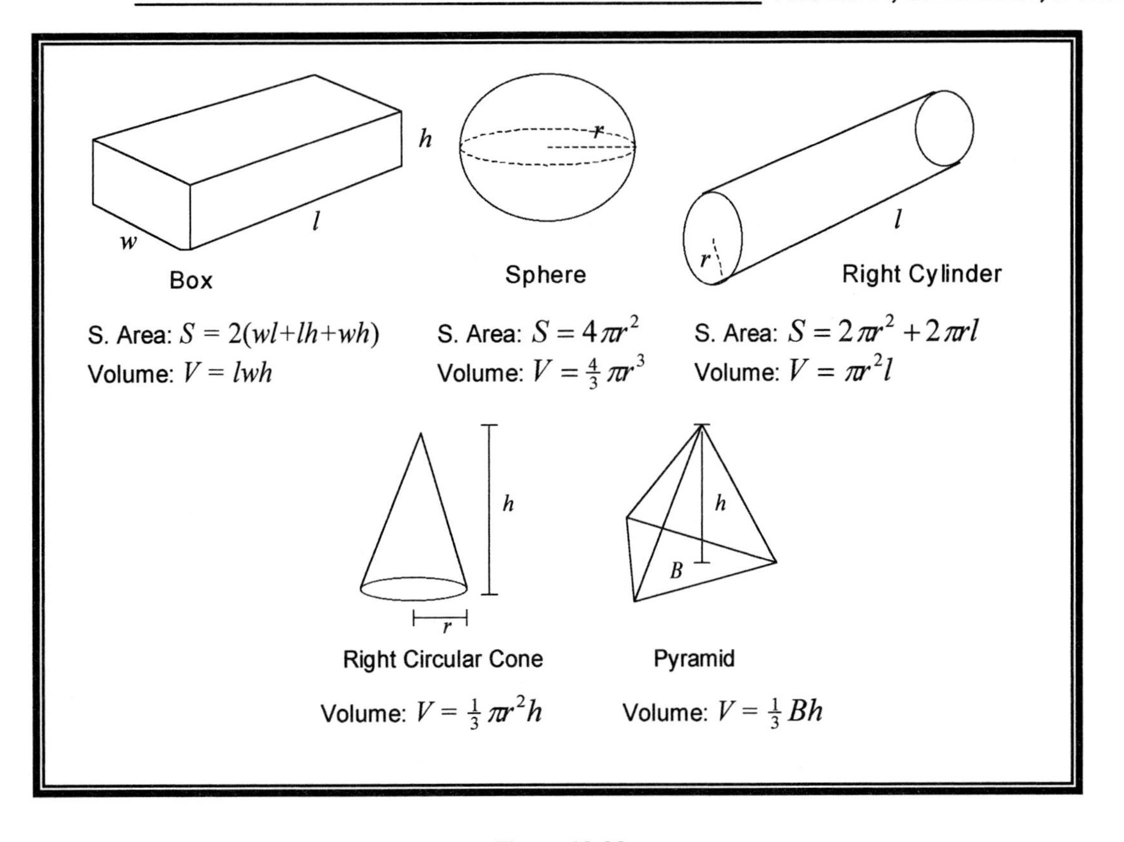

Figure 13.29

EXERCISE 13.58: Try 'Em Out!

(a) Calculate the perimeter of a square of side 5 m.

(b) What is the cross-sectional area of a triangle of height 0.5 m and base 1.5 m?

(c) Find the surface area of a sphere of radius 40 cm.

(d) Figure the volume of a right cylinder of radius 12 cm and length 20 cm.

SOLUTION:

(a) $P = 2w + 2l = 2(5\text{ m}) + 2(5\text{ m}) = 20\text{ m}$

(b) $A = \frac{1}{2}bh = \frac{1}{2}(1.5\text{ m})(0.5\text{ m}) = 0.375\text{ m}^2$

(c) $S = 4\pi r^2 = 4\pi(40\text{ cm})^2 \approx 20{,}000\text{ cm}^2$

(d) $V = \pi r^2 l = \pi(12\text{ cm})^2(20\text{ cm}) \approx 9000\text{ cm}^3$

Where there is matter, there is geometry.
JOHANNES KEPLER
Italian Astronomer, 1571 - 1630

13.5.3 CONIC SECTIONS

Circle
Ellipse
Hyperbola
Parabola

Figure 13.30

All of the important regular planar figures—such as circles, ellipses, parabolas, and hyperbolas—are created when a plane intersects a cone (Fig. 13.30). These curves, however, also have geometric definitions and algebraic equations. In this section, we define each of the conic sections and provide their associated equation. To make the equations useful, however, you must *adapt* them to your particular situation.

EXAMPLE 13.59: Conic Sections

For example, the equation given below for the parabola describes a curve that opens along the positive y-axis. To make the equation useful for a parabola opening along the positive x-axis, you must interchange the x and y in that equation; to make it open along the negative y-axis, you must put a negative sign in front of the y, and so on. You may also have to do some algebra to make your equation fit the form given.

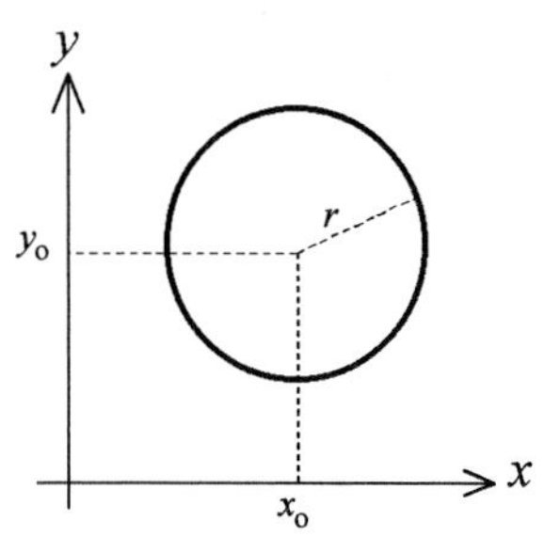

Figure 13.31 (a)

The simplest conic section is the **circle** (Fig. 13.31 (a)), defined as the set of all points lying a fixed distance from a single center point (x_o, y_o). Circles are symmetric about all axes. The equation of a circle with radius r is:

$$(x-x_o)^2+(y-y_o)^2=r^2$$

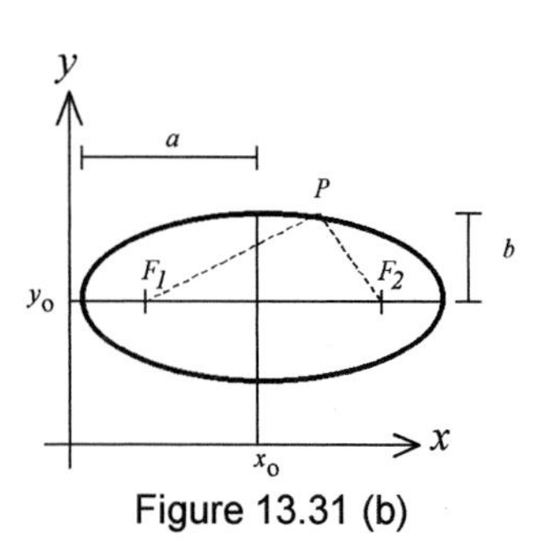

Figure 13.31 (b)

An **ellipse** (Fig. 13.31 (b)) is like a flattened circle but with two "centers" or **foci**. It is the set of all points such that the sum of the distances from each focus to any point on the curve P is constant: $\overline{F_1P}+\overline{F_2P}=k$.

All ellipses are symmetric about the line through $\overline{F_1F_2}$ connecting each far edge of the ellipse, called the **major axis**, as well as the perpendicular bisector of that line, known as the **minor axis**. The equation of an ellipse centered at (x_o, y_o) with major axis $2a$ and minor axis $2b$ is:

$$\frac{(x-x_o)^2}{a^2}+\frac{(y-y_o)^2}{b^2}=1$$

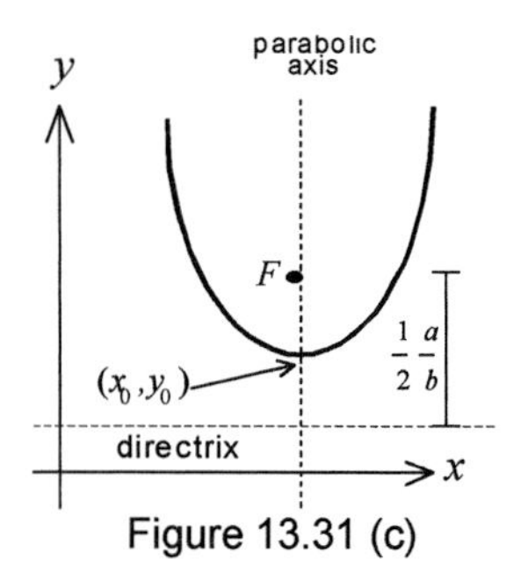

Figure 13.31 (c)

Another important conic section is the **parabola** (Fig. 13.31 (c)). A parabola is all points equally distant from a fixed point, called the **focus**, and a fixed line, called the **directrix**. The distance from the focus to the directrix is $\frac{1}{2}\frac{a}{b}$.

Parabolas are symmetric about a line perpendicular to the directrix, passing through the focus, called the **parabolic axis**. The equation of a parabola with **vertex** at (x_o, y_o) is:

$$(y-y_o)=\frac{a}{b}(x-x_o)^2$$

A **rectangular hyperbola** (Fig. 13.31 (d)) is the set of all points such that the difference of the distances to each focus to any point on the curve P is constant: $\overline{F_1P} - \overline{F_2P} = k$.

Hyperbolas are symmetric about the line $\overline{F_1F_2}$, the **transverse axis**, as well as the perpendicular bisector of that line, the **conjugate axis**. The equation of a **rectangular hyperbola** centered at (x_o, y_o) is:

$$\frac{(x-x_o)^2}{a^2} - \frac{(y-y_o)^2}{b^2} = 1$$

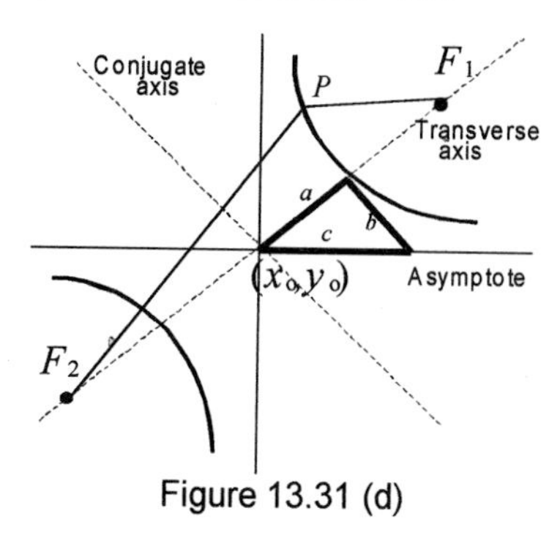

Figure 13.31 (d)

The **eccentricity** of the ellipse, parabola, and hyperbola is $e = \sqrt{(b/a)^2 + 1}$.

EXERCISE 13.60: Plotting a Circle

Plot the circle $x^2 + y^2 - 4x + 6y = 12$.

SOLUTION:

First, we must put this equation into the general form given above, $(x-x_o)^2 + (y-y_o)^2 = r^2$, by completing the square:

$$x^2 + y^2 - 4x + 6y = 12$$
$$(x^2 - 4x) + (y^2 + 6y) = 12$$
$$(x^2 - 4x + 4) + (y^2 + 6x + 9) = 12 + 4 + 9$$
$$(x-2)^2 + (y+3)^2 = 25 = 5^2$$

Completing the square is in Section 13.4.10.

Now we can plot the circle by comparing this result with the defining equation above. From that comparison, we see that $x_o = 2$, $y_o = -3$, and $r = 5$:

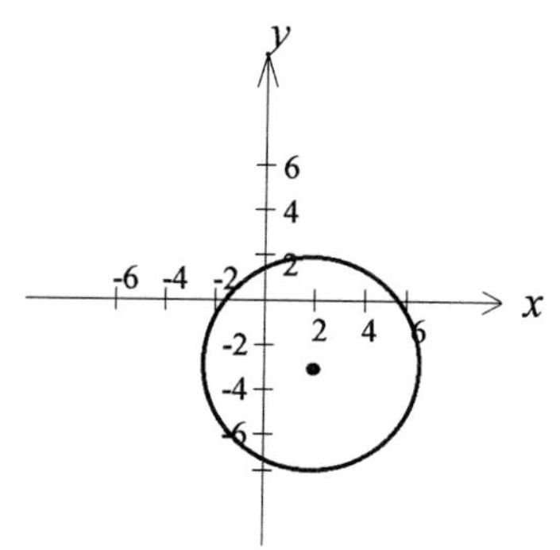

There is nothing strange in the circle being the origin of any and every marvel.
ARISTOTLE
Greek Philosopher, 384 - 322 B.C.

EXERCISE 13.61: Plotting a Parabola

Plot the parabola $x = 70y - y^2$.

SOLUTION:

Again, we first need to put the equation into standard form by completing the square:

Geometry will draw the soul toward truth....
PLATO
Greek Philosopher, 428 - 348 B.C.

Exercise 13.61 Continued...

$$\begin{aligned} x &= 70y - y^2 \\ &= (-1225 + 70y - y^2) + 1225 \\ x - 1225 &= -(y - 35)^2 \end{aligned}$$

This equation tells us that the parabola is centered at $(1225, 35)$. Because the x and y are interchanged from the defining equation, however, we also know this parabola opens parallel to the x-axis rather than the y-axis. The negative sign in front of the y term further indicates the parabola opens in the negative direction.

Now we can plot; notice the scales chosen for each axis.

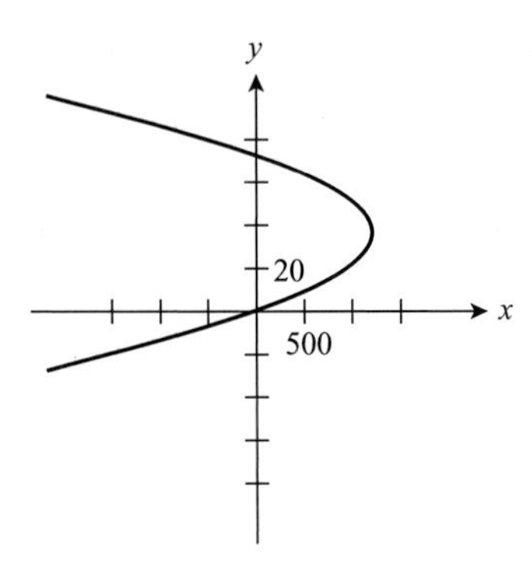

13.6 TRIGONOMETRY

Trigonometry is the mathematics of angles and triangles. It is used throughout introductory physics, so a good working knowledge of the subject is invaluable. If you remember nothing else about this section, though, remember this: *it is always best to measure angles in radians, not degrees.*

13.6.1 ANGLES & TRIANGLES

Right

Acute

Obtuse

Figure 13.32

ANGLES

There are three types of angles (Fig. 13.32). A **right angle** has $\frac{\pi}{2}$ radians. An **acute angle** is an angle of less than $\frac{\pi}{2}$ radians, while an **obtuse angle** is an angle greater than $\frac{\pi}{2}$ radians.

Any two angles that *sum* to $\frac{\pi}{2}$ rads are termed **complementary**; any two angles that sum to π rads are called **supplementary**.

By convention, angles are considered positive if they are measured counter-clockwise from the positive x-axis, negative if they are measured clockwise. See Figure 13.33.

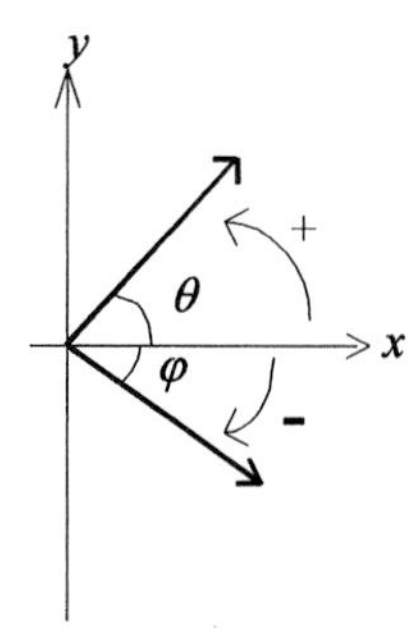

Figure 13.33

Two angles are equal of any of the following is true:

1. Their sides are parallel (Fig. 13.34 (a)).
2. Their sides are mutually perpendicular (Fig. 13.34 (b)).
3. They have the same complement (Fig. 13.34 (c)).
4. They are **vertical angles** (Fig. 13.34 (d)).

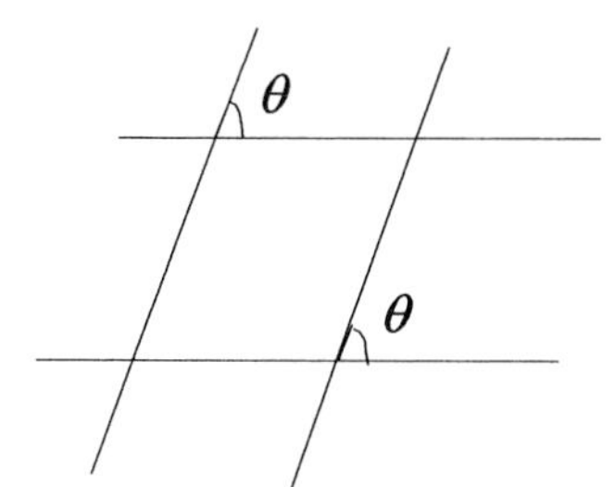

Figure 13.34 (a)

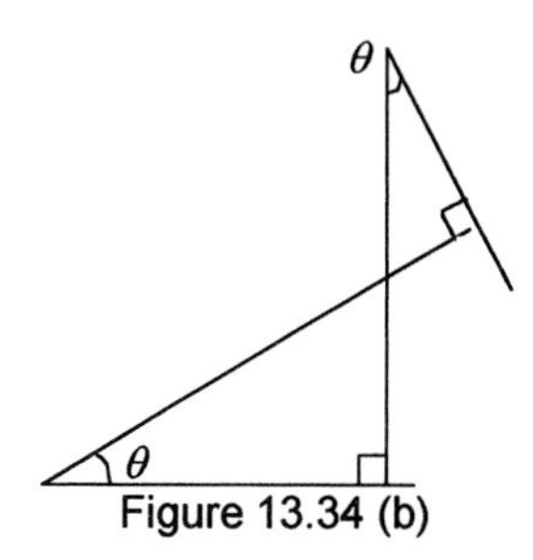

Figure 13.34 (b)

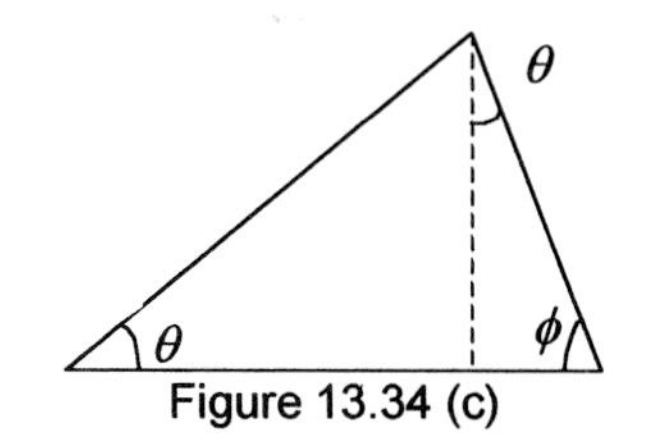

Figure 13.34 (c)

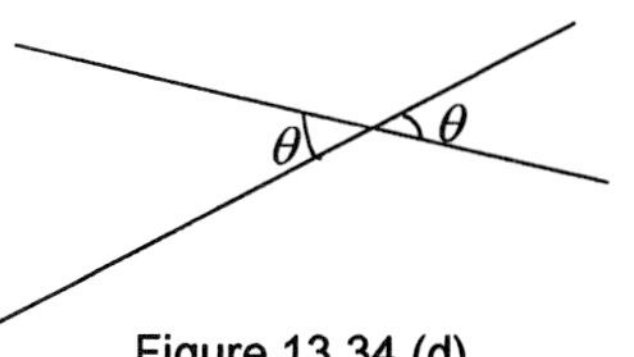

Figure 13.34 (d)

EXERCISE 13.62: Find the Equal Angles

In the diagram below, which angles are equal to θ?

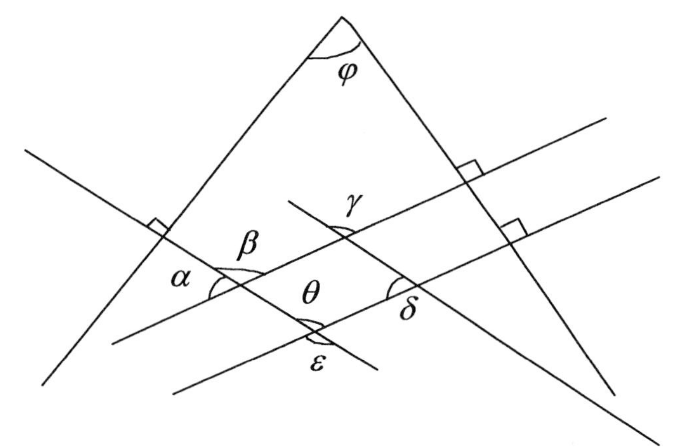

SOLUTION:

β, γ, ε, and δ are equal to θ. α and φ are supplementary to θ.

TRIANGLES

A **triangle** is a geometrical object consisting of three lines joined with three angles. *The sum of the three angles in a triangle is always equal to π radians.*

An **isosceles triangle** has two sides of equal length; the angles opposing these sides are also equal. An **equilateral triangle** has three equal-length sides; thus, all angles are equal and $\frac{\pi}{3}$ rads each.

Any two triangles are **similar** if any two of their angles are equal, any two of their sides are equal, or one side and one angle are equal. The corresponding sides of similar triangles are proportional; the areas of similar triangles are proportional to the square of the sides (Fig. 13.35).

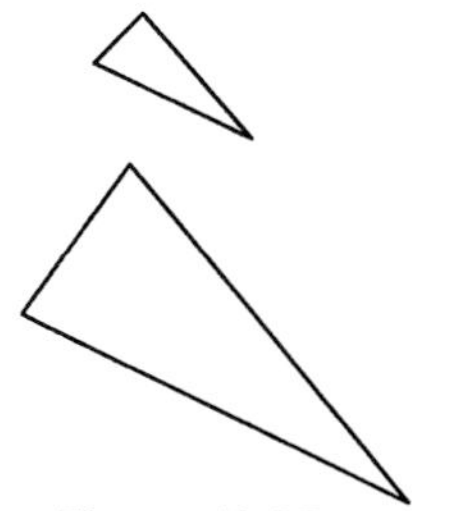
Figure 13.35

13.6.2 RIGHT TRIANGLES & THE TRIGONOMETRIC FUNCTIONS

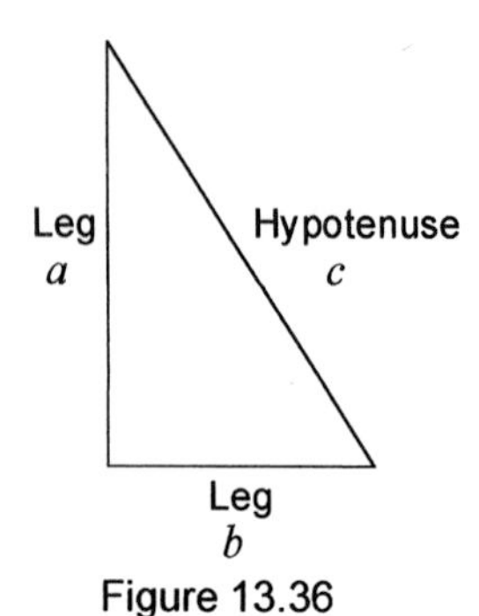

Figure 13.36

RIGHT TRIANGLES

A **right triangle** has one angle that is right. The other two angles in a right triangle are necessarily acute and complementary (think about it!).

The side of the triangle directly across from the right angle is called the **hypotenuse** c, and the other two sides are triangle's **legs** a and b. The **Pythagorean theorem** relates the legs and hypotenuse (Fig. 13.36) of a right triangle as:

$$a^2 + b^2 = c^2$$

Figure 13.37 below shows four special right triangles.

Figure 13.37

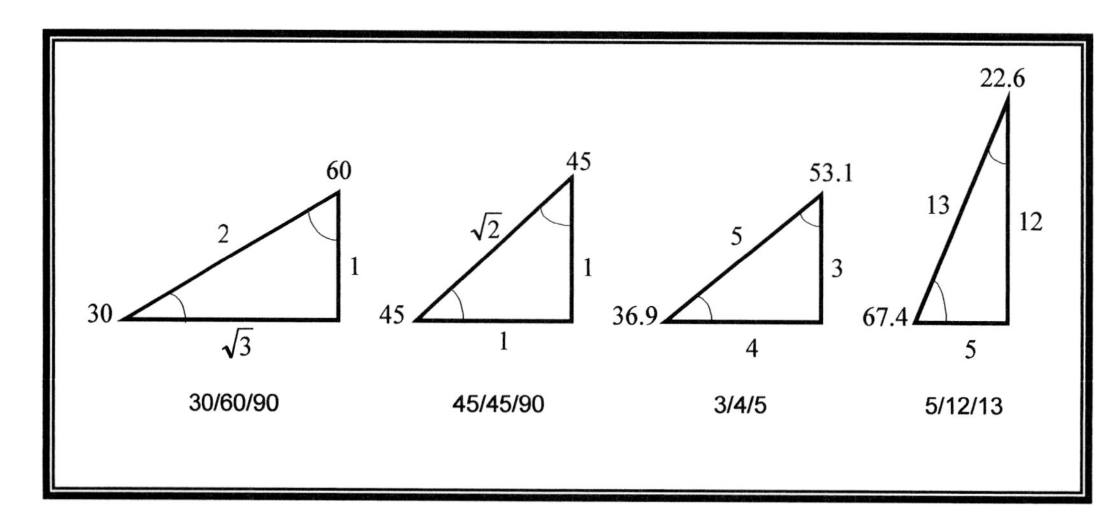

The trig functions are very useful for vector analysis (in Chapter 15).

TRIGONOMETRIC FUNCTIONS

The **trigonometric functions sine**, **cosine**, **tangent**, **cosecant**, **secant**, and **cotangent**—abbreviated as **sin**, **cos**, **tan**, **csc**, **sec**, and **cot**, respectively—relate the angles, or **arguments**, and sides of a right triangle as follows:

$$\sin\theta = \frac{opposite}{hypotenuse} = \frac{a}{c} \qquad \csc\theta = \frac{1}{\sin\theta}$$

$$\cos\theta = \frac{adjacent}{hypotenuse} = \frac{b}{c} \qquad \sec\theta = \frac{1}{\cos\theta}$$

$$\tan\theta = \frac{opposite}{adjacent} = \frac{a}{b} = \frac{\sin\theta}{\cos\theta} \qquad \cot\theta = \frac{1}{\tan\theta} = \frac{\cos\theta}{\sin\theta}$$

Luckily, you probably won't need to know all of these; *you usually only need sines, cosines, and tangents.* One easy way to remember the defining relationships of these functions is with the mnemonic:

SOH CAH TOA

pronounced "sew-kah-toe-ah". This mnemonic stands for "sine is the opposite side divided by the *h*ypotenuse; cosine is the *a*djacent side divided by the *h*ypotenuse; and *t*angent is the *o*pposite side divided by the *a*djacent side."

EXERCISE 13.63: A Little Trigonometry

In Figure 13.38, calculate the length of the missing side, a, and the two angles θ_1 and θ_2.

SOLUTION:

$$a = \sqrt{c^2 - b^2} = \sqrt{3^2 - 1^2} \approx 2.8$$

$$\theta_1 = \sin^{-1}\frac{b}{c} = \sin^{-1}\frac{1}{3} = 0.33 \text{ rads}$$

$$\theta_2 = \tfrac{\pi}{2} - \theta_1 = 1.2 \text{ rads}$$

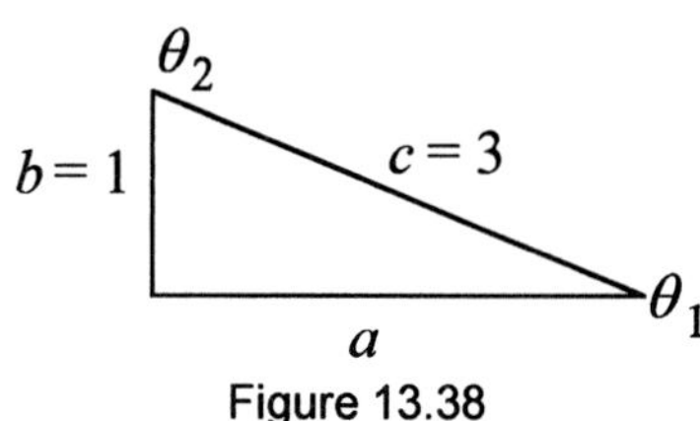

Figure 13.38

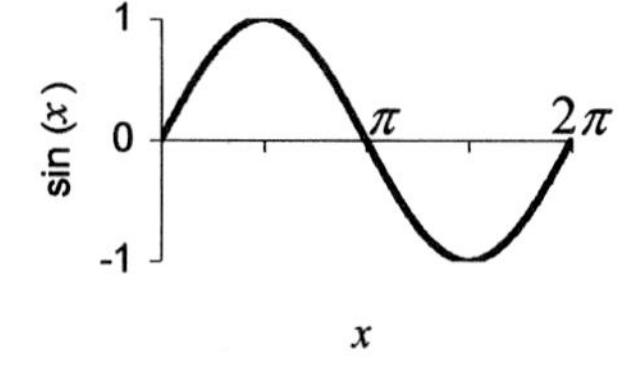

Figure 13.39 (a)

Figures 13.39 (a) - (c) plot the sine, cosine, and tangent functions from 0 to 2π. In the four regions $0 < \theta < \frac{\pi}{2}$, $\frac{\pi}{2} < \theta < \pi$, $\pi < \theta < \frac{3\pi}{2}$, and $\frac{3\pi}{2} < \theta < 2\pi$—corresponding to the four Cartesian quadrants I, II, III, and IV respectively—each function has a particular sign. Figure 13.40 shows the specific signs for each quadrant for each trig function. To remember these quadrants, try this mnemonic:

All are positive in the *first* quadrant;
Sine is positive in the *second* quadrant;
Tangent is positive in the *third* quadrant;
Cosine is positive in the *fourth* quadrant.

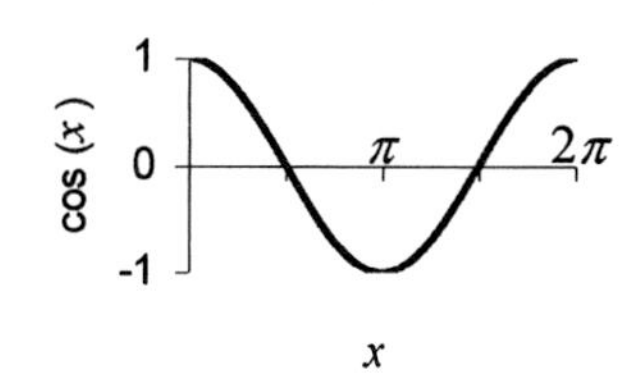

Figure 13.39 (b)

EXERCISE 13.64: Fun with Trigonometric Functions

In what quadrants is the cosine positive? The sine?

SOLUTION:

From Figure 13.40, we see that the cosine is positive in quadrants I and IV, while the sine is positive in quadrants I and II.

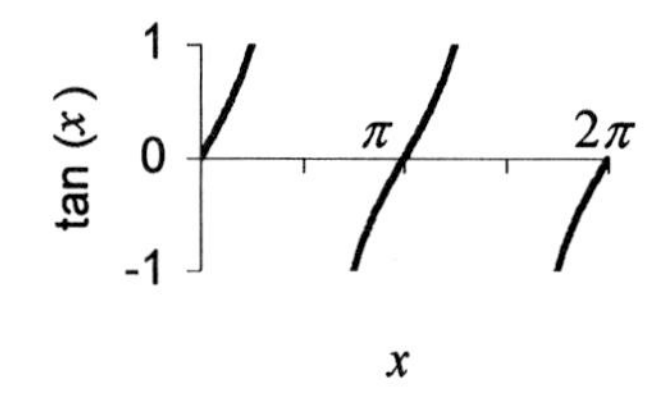

Figure 13.39 (c)

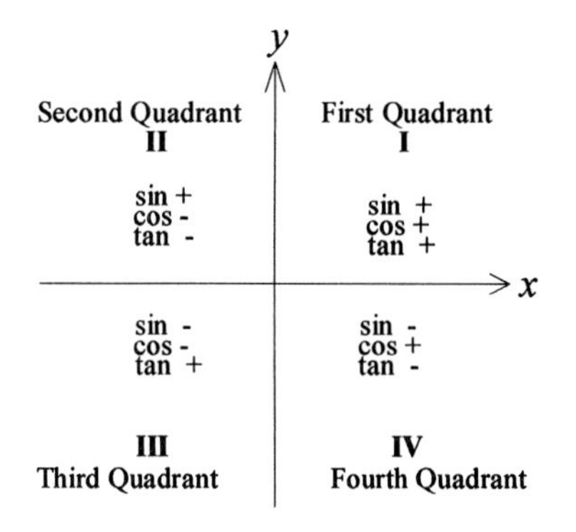

Figure 13.40

THE UNIT CIRCLE

The trigonometric functions are *very* important in physics because they are applicable to many situations besides just right triangles. Therefore, it is helpful to have a memory aide that helps you calculate trigonometric

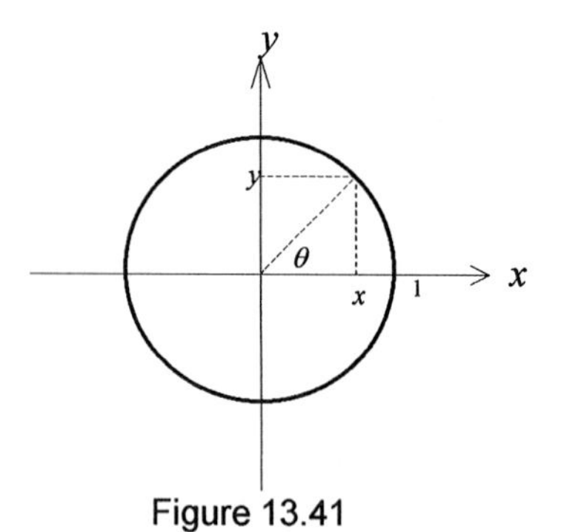

Figure 13.41

functions for any angle, without the use of a right triangle. This memory aide is called the **unit circle**.

The unit circle is simply a circle that is centered on the origin and has a radius of one unit (see Figure 13.41). Because the radius is equal to one, all of the ratios for the two main trigonometric functions, sine and cosine, are equal to their numerators:

$$\sin\theta = \frac{opp}{hyp} = \frac{y}{1} = y$$

$$\cos\theta = \frac{adj}{hyp} = \frac{x}{1} = x$$

Notice that the point (x, y) on the unit circle is equal to $(\cos\theta, \sin\theta)$, because $x = \sin\theta$, and $y = \cos\theta$. We can generalize this result to *any* angle θ, even one greater than $\frac{\pi}{2}$ radians.

TRIGONOMETRIC IDENTITIES

Extrapolation of Figures 13.39 (a) - (c) into the negative x region verifies the following trigonometric identities:

$$\sin(-\theta) = -\sin\theta$$
$$\cos(-\theta) = \cos\theta$$
$$\tan(-\theta) = -\tan\theta$$

Odd functions have the property that negative arguments produce *negative* results with the same magnitude, so the trigonometric functions *sine and tangent are odd functions.* **Even functions** have the property that negative arguments produce *positive* results with the same value, so *the cosine function is an even function.*

Table 13.6 lists values of the sine, cosine, and tangent functions at some important angles. Use this table to verify the following trigonometric identities:

$$\sin\theta = \cos(\tfrac{\pi}{2} - \theta)$$
$$\cos\theta = \sin(\tfrac{\pi}{2} - \theta)$$
$$\cot\theta = \tan(\tfrac{\pi}{2} - \theta)$$

Table 13.6: Values of the Trigonometric Functions at Selected Angles

FUNCTION	0	$\frac{\pi}{6}$	$\frac{\pi}{4}$	$\frac{\pi}{3}$	$\frac{\pi}{2}$
$\sin\theta$	0	$\frac{1}{2}$	$\frac{1}{\sqrt{2}}$	$\frac{\sqrt{3}}{2}$	1
$\cos\theta$	1	$\frac{\sqrt{3}}{2}$	$\frac{1}{\sqrt{2}}$	$\frac{1}{2}$	0
$\tan\theta$	0	$\frac{\sqrt{3}}{3}$	1	$\sqrt{3}$	1

EXERCISE 13.65: More Fun with Trig Functions

(a) $\sin(\frac{\pi}{6})$ is equal to the cosine of what?

(b) What is the tangent of $\frac{\pi}{2}$? $\frac{\pi}{3}$?

SOLUTION:

(a) From $\sin\theta = \cos(\frac{\pi}{2} - \theta)$, we know:

$$\sin(\tfrac{\pi}{6}) = \cos(\tfrac{\pi}{2} - \tfrac{\pi}{6}) = \cos(\tfrac{\pi}{3}) = 0.5$$

(b) From Table 13.6, we know $\tan(\frac{\pi}{2}) = 1$, while $\tan(\frac{\pi}{3}) = 0.808$.

In Mathematicks he was greater
Than Tycho Brahe or Esra Pater.
For he by Geometrick scale
Could take the size of Pots of Ale;
Resolve by Sines and Tangents straight,
If Bread or Butter wanted weight;
And wisely tell what hour o' th' day
The clock doth strike, by Algebra.
SAMUEL BUTLER, *Hudibras*
English Poet, 1835 - 1908

If the argument of a trigonometric function is small ($\theta << 1$ radian), *and* in radians, then the following *important* approximations hold:

$$\sin\theta \approx \theta \qquad \cos\theta \approx 1 \qquad \tan\theta \approx \theta$$

Physics Majors: These relations can be proved using the techniques of Section 13.4.11.

EXERCISE 13.66: Small Angle Approximation

What is the small angle approximation for $\dfrac{8\cos\theta}{\sin^3\theta}$?

SOLUTION:

$$\frac{8\cos\theta}{\sin^3\theta} \approx \frac{8(1)}{\theta^3} = \frac{8}{\theta^3}$$

Many physical equations contain trigonometric functions, often in complex configurations. Use the following trigonometric identities to simplify such equations. Do *not* try to memorize these equations; simply look them up when needed.

Pythagorean Relations: $\begin{Bmatrix} \sin^2\theta + \cos^2\theta = 1 \\ \tan^2\theta + 1 = \sec^2\theta \\ 1 + \cot^2\theta = \csc^2\theta \end{Bmatrix}$

Half-Angle Formulas: $\begin{Bmatrix} \sin^2\dfrac{\theta}{2} = \frac{1}{2}(1-\cos\theta) \\ \cos^2\dfrac{\theta}{2} = \frac{1}{2}(1+\cos\theta) \\ \tan^2\dfrac{\theta}{2} = \dfrac{1-\cos\theta}{1+\cos\theta} \end{Bmatrix}$

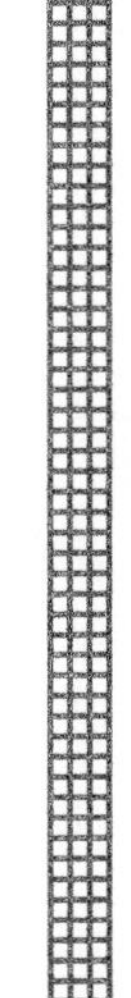

Double Angle Formulas:

$$\left\{\begin{array}{l} \sin 2\theta = 2\sin\theta\cos\theta \\ \cos 2\theta = \cos^2\theta - \sin^2\theta = 2\cos^2\theta - 1 = 1 - 2\sin^2\theta \\ \tan 2\theta = \dfrac{2\tan\theta}{1-\tan^2\theta} \end{array}\right\}$$

Angle Add Formulas:

$$\left\{\begin{array}{l} \sin(\alpha\pm\beta) = \sin\alpha\cos\beta \pm \sin\beta\cos\alpha \\ \cos(\alpha\pm\beta) = \cos\alpha\cos\beta \mp \sin\alpha\sin\beta \\ \tan(\alpha\pm\beta) = \dfrac{\tan\alpha \pm \tan\beta}{1\mp\tan\alpha\tan\beta} \end{array}\right\}$$

Addition Formulas:

$$\left\{\begin{array}{l} \sin\alpha \pm \sin\beta = 2\sin[\tfrac{1}{2}(\alpha\pm\beta)]\cos[\tfrac{1}{2}(\alpha\mp\beta)] \\ \cos\alpha + \cos\beta = 2\cos[\tfrac{1}{2}(\alpha+\beta)]\cos[\tfrac{1}{2}(\alpha-\beta)] \\ \cos\alpha - \cos\beta = 2\sin[\tfrac{1}{2}(\alpha+\beta)]\sin[\tfrac{1}{2}(\alpha-\beta)] \end{array}\right\}$$

EXERCISE 13.67: Factoring Trigonometric Functions

Simplify the following ugly expression: $\dfrac{\cos 2\theta + 2\sin^2\theta}{(\sin^2\frac{\theta}{2})(\cos^2\frac{\theta}{2})}$.

SOLUTION:

$$\frac{\cos 2\theta + 2\sin^2\theta}{(\sin^2\frac{\theta}{2})(\cos^2\frac{\theta}{2})} = \frac{(\cos^2\theta - \sin^2\theta) + 2\sin^2\theta}{\frac{1}{2}(1-\cos\theta)\frac{1}{2}(1+\cos\theta)} = \frac{\cos^2\theta + \sin^2\theta}{\frac{1}{4}(1-\cos^2\theta)} = \frac{4}{\sin^2\theta}$$

The trigonometric functions are related to the exponential function by **Euler's Formula:**

$$(\cos\theta + i\sin\theta) = e^{i\theta}$$

Rewriting the above equation yields equations for the trigonometric functions in terms of exponentials:

$$\sin x = \frac{e^{ix} - e^{-ix}}{2i} \text{ and } \cos x = \frac{e^{ix} + e^{-ix}}{2}$$

Notice that the arguments of the exponential functions are imaginary, while the arguments of the trigonometric functions are not.

INVERSE TRIGONOMETRIC FUNCTIONS

Whereas the trigonometric functions return a number given an angle, the **inverse trigonometric functions**—$\sin^{-1}x$, $\cos^{-1}x$, $\tan^{-1}x$, $\csc^{-1}x$, $\sec^{-1}x$, **and** $\cot^{-1}x$—return an angle given a number (Fig. 13.42 (a) - (c)). Thus, you need the inverse trig functions if you want to calculate trigonometric functions "backward" from the argument to a side of a triangle.

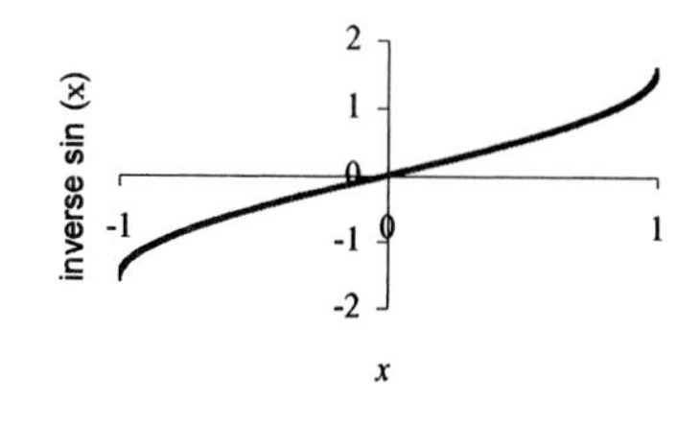

Figure 13.42 (a)

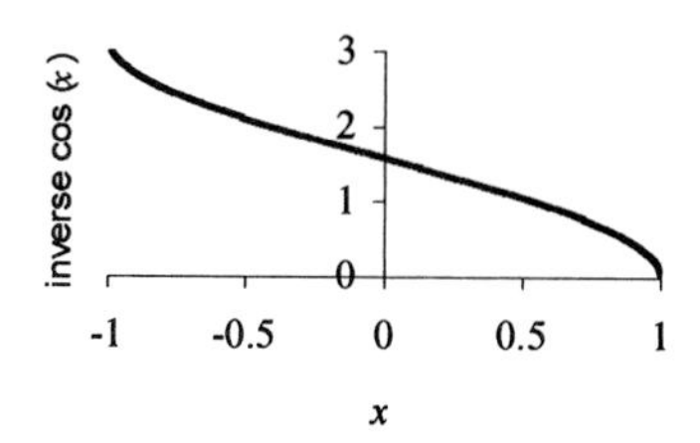

Figure 13.42 (b)

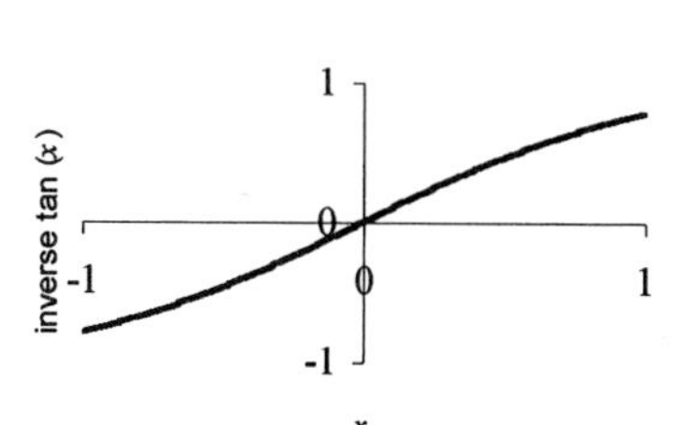

Figure 13.42 (c)

WARNING!

Be careful with notation for inverse trigonometric functions, for $\sin^{-1}x \neq \frac{1}{\sin x}$, nor does $\sin^{-1}x \neq \csc x$! Also be careful about the domains for these functions. The inverse sine, cosine, secant, cosecant, and cotangent are only defined if $-1 < x < 1$. The inverse tangent, however, is defined for all x.

WARNING!

When using a calculator to figure the inverse tangent function, be aware that the calculator displays the *smallest* correct angle. For example, the inverse tangent of 1.5 is both 0.98 rads and 4.1 rads, but the calculator only displays 0.98 rads.

The displayed angle may not be the appropriate answer for the problem at hand. Therefore, *always* check your answer for plausibility when using the inverse tangent function. Use the quadrant grid to help. If you find that the larger angle is the correct one for your situation, add π rads to the smaller angle obtained by the calculator.

EXERCISE 13.68: Inverse Trigonometric Functions

Find each inverse trigonometric function:

(a) $\sin^{-1}(0.5)$ (b) $\cos^{-1}(\frac{\sqrt{2}}{2})$ (c) $\tan^{-1}(\sqrt{3})$

SOLUTION:

(a) $\sin^{-1}(0.5) = \frac{\pi}{6}$ (b) $\cos^{-1}(\frac{\sqrt{2}}{2}) = \frac{\pi}{4}$ (c) $\tan^{-1}(\sqrt{3}) = \frac{\pi}{3}$ *or* $-\frac{\pi}{3}$

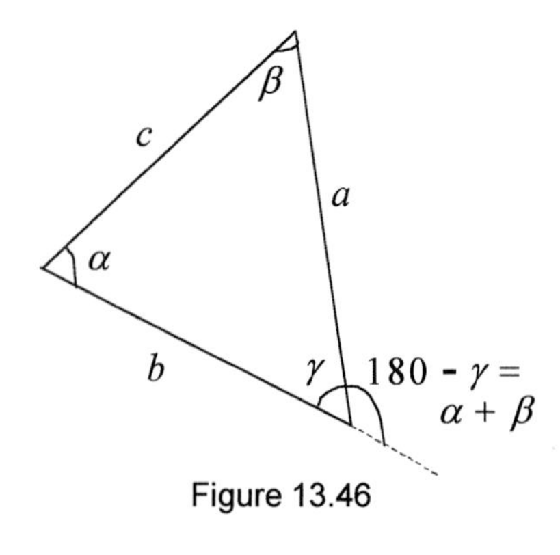

Figure 13.46

13.6.3 NON-RIGHT TRIANGLES

The definitions of trigonometric functions given in the previous section are based on right triangles. Analyze all non-right triangles (Fig. 13.46) with the following laws:

Law of Sines:
$$\frac{a}{\sin\alpha} = \frac{b}{\sin\beta} = \frac{c}{\sin\gamma}$$

Law of Cosines:
$$\begin{Bmatrix} a^2 = b^2 + c^2 - 2bc\cos\alpha \\ b^2 = a^2 + c^2 - 2ac\cos\beta \\ c^2 = a^2 + b^2 - 2ab\cos\gamma \end{Bmatrix}$$

Law of Sides:
$$\begin{Bmatrix} a = b\cos\gamma + c\cos\beta \\ b = a\cos\gamma + c\cos\alpha \\ c = a\cos\beta + b\cos\alpha \end{Bmatrix}$$

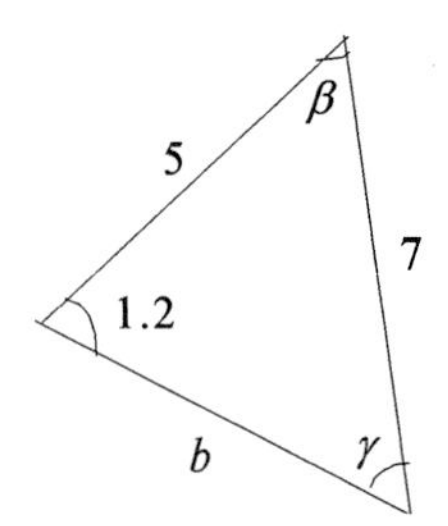

Figure 13.47

EXERCISE 13.69: Non-Right Triangles

Refer to Figure 13.47 to answer the following questions:

(a) Using the Law of Sines, find γ.

(b) Using the Law of Sides, find b.

(c) Using the Law of Cosines, find β.

SOLUTION:

(a) To solve this problem, we apply the Law of Sines, using the first and third terms only:

$$\frac{a}{\sin\alpha} = \frac{c}{\sin\gamma}$$

$$\Rightarrow \gamma = \sin^{-1}\left(\frac{c}{a}\sin\alpha\right) = \sin^{-1}\left(\frac{5}{7}\sin(1.2)\right) \approx 0.73\,\text{rad}$$

(b) Here, we apply the Law of Sides:

$$b = a\cos\gamma + c\cos\alpha = 7\cos(0.73) + 5\cos(1.2) \approx 7.0$$

(c) Now, we use the Law of Cosines:

$$b^2 = a^2 + c^2 - 2ac\cos\beta$$

$$\Rightarrow \beta = \cos^{-1}\left(\frac{a^2 + c^2 - b^2}{2ac}\right) = \cos^{-1}\left(\frac{7^2 + 5^2 - 7^2}{2(7)(5)}\right) = 1.2\,\text{rad}$$

Thus the triangle is isosceles, and the figure is clearly not to scale.

13.7 CONCLUSION

I was at first frightened...when I saw such mathematical force made to bear on the subject [of physics], and then wondered to see that the subject stood it so well.

MICHAEL FARADAY
English Physicist, 1791 - 1867

Michael Faraday, a nineteenth-century physicist, was probably the last scientist to analyze physical problems without using any math. Certainly, you will not have this luxury in your physics course, so you will need the techniques of this chapter.

Although the chapter is frightfully long, remember that most of it should be *review*: once you see it, it should jog memories from high school-level courses. Just keep returning to this chapter as you need it throughout your physics course.

14
CALCULUS

14.1 INTRODUCTION

Calculus is the mathematics of change.

Many quantities in physics change—changing is across distance, time, or otherwise—so calculus is an inescapable aspect of most introductory (college-level) physics courses.

While this fact may concern you—after all, calculus can be challenging for some students—the truth is that physics is actually *harder* to learn without calculus than with it. Therefore, it is worth the extra effort to understand calculus *before* you get too far along in your physics class.

Since 'tis Nature's Law to change,
Constancy alone is strange.
JOHN WILMOT
Second Earl of Rochester, 1647 - 1680

This chapter provides a short survey of important calculus principles that you will need to succeed in a typical introductory physics course, as well as some advanced material you will probably not need right away.

Calculus comes in two flavors: **differentiation** and **integration**. These two processes are inverse and complementary, just like addition and subtraction. And like addition and subtraction, you need to know both parts in order to solve problems. We begin with differentiation.

14.2 DIFFERENTIAL CALCULUS

In this section, we define and learn to calculate derivatives.

14.2.1 WHAT IS A DERIVATIVE?

When a quantity changes, we often want to know the *rate* at which the change occurred. A quantity's (instantaneous) rate of change is its **derivative**.

In Section 5.3.1.

> **EXAMPLE 14.1:** Position and Velocity, Part I
>
> In Chapter 5, we introduced the physical quantity of *position*, a quantity that can change with time. The *rate* at which this happens, as we also learned in that chapter, is *velocity*. That is, velocity is the *derivative* of position (with respect to time).

For more on slopes, refer to Section 13.4.4.

As it turns out, a derivative—the rate of change of a function—is the same thing as a slope. In particular, the derivative is the slope of the line *tangent* to the function. A **tangent line** touches a function exactly once, and is parallel to the function at that point (Fig. 14.1).*

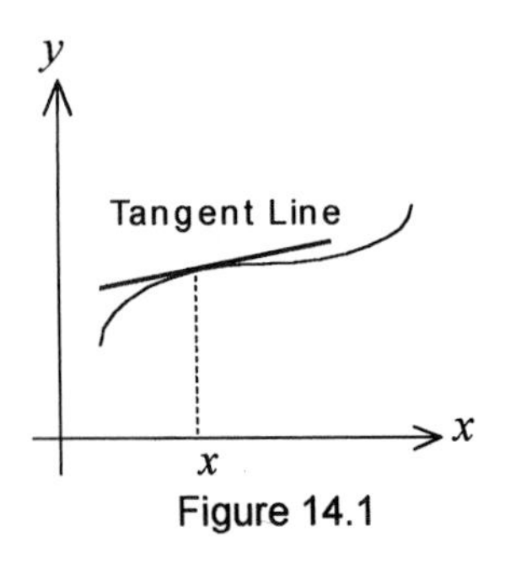

Figure 14.1

> **DEFINITION 14.2:** Derivatives
>
> Let's take a quick look at how derivatives are defined. Recall from Section 13.4.4 that the slope of a line is given by the equation:
>
> $$m = \frac{\Delta y}{\Delta x} = \frac{(y_2 - y_1)}{(x_2 - x_1)}$$
>
> To use this equation, we need two points on a line that intersects the function, like (x_1, y_1) and (x_2, y_2) in Figure 14.2 (a).
>
> The closer the two points (x_1, y_1) and (x_2, y_2) are to each other, the closer the slope of the line connecting them is to the slope of the line tangent to the curve at x (Figs. 14.2 (b) and (c)). In fact, we can imagine a case (Eliminating the Unessential) in which the two points are *so* close to each other as to be right on top of one other (Fig. 14.2 (d)). In this ideal case, the value of m above *is* the slope of the tangent line.
>
> In mathematical jargon, the process of moving (x_1, y_1) and (x_2, y_2) closer and closer together (and thereby getting a better and better approximation to the slope of the tangent line) is called taking the **limit**; that is, we have just taken the limit of the slope equation when $\Delta x = x_2 - x_1$ gets very small (or "goes to zero"):

Eliminating the Unessential is in Section 2.3.

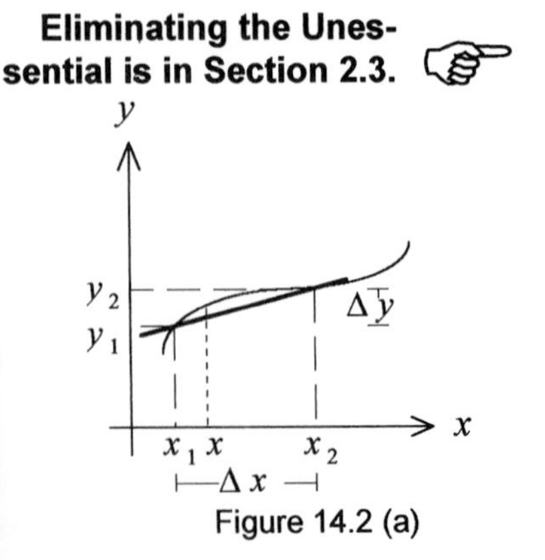

Figure 14.2 (a)

* In the special case that the function is itself a line, the tangent line is identical to the function, and therefore can be considered to touch the function everywhere.

Definition 14.2 Continued...

$$m = \lim_{\Delta x \to 0} \frac{\Delta y}{\Delta x}$$

We *define* this limit to be the derivative of the function $y = f(x)$:

$$\lim_{\Delta x \to 0} \frac{\Delta y}{\Delta x} \equiv \frac{dy}{dx} = \frac{df}{dx}$$

Therefore, the derivative is the slope of the line tangent to the function.

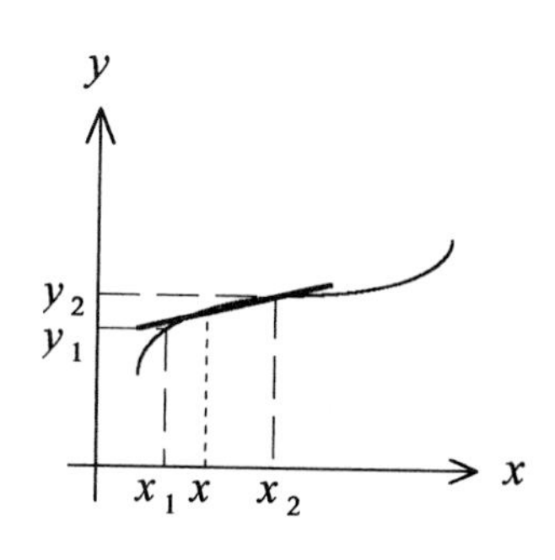

Figure 14.2 (b)

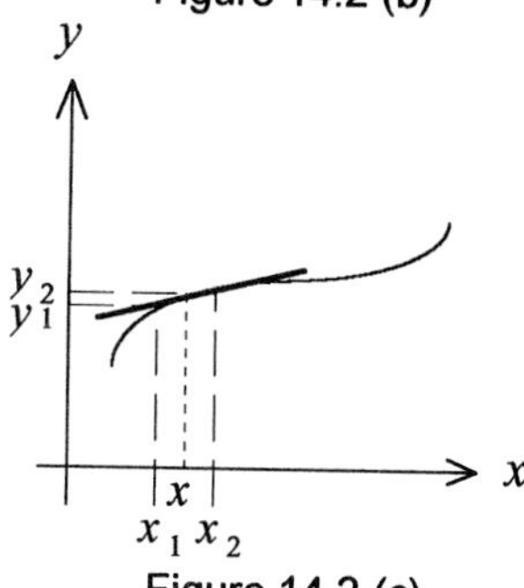

Figure 14.2 (c)

Derivatives are written either as a ratio[#] df/dx, or with a prime $f'(x)$. This notation is spoken as "the derivative of f with respect to x," "dee-f, dee-x," or "f prime of x."

EXAMPLE 14.3: Position and Velocity, Part II

Because velocity is the derivative of position (with respect to time), we can write:

$$v = \frac{dx}{dt} \text{ or } v = x'(t)$$

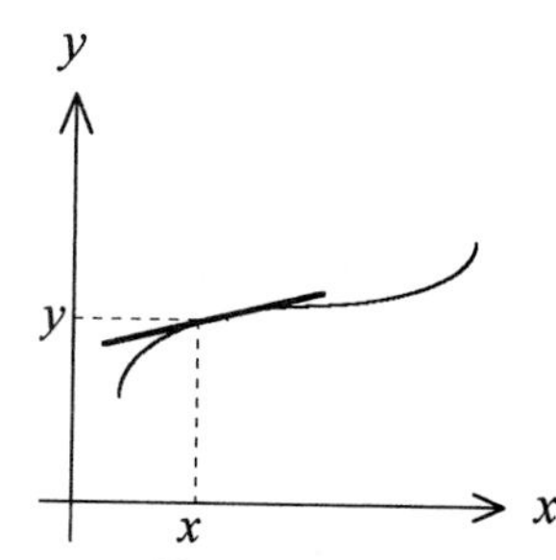

Figure 14.2 (d)

EXERCISE 14.4: Poorly Defined Functions

The discussion in Box 14.2 above only applies when the function $f(x)$ is well defined. Why must functions be well defined in calculus?

SOLUTION:

As we saw in Section 13.4.12, a poorly defined function has a gap, a kink, or more than one value of y for each value of x (Fig. 14.3 (a) and (b)). The problem with a poorly defined function, as far as calculus is concerned, is that at each of these locations on the curve, it is impossible to draw a single tangent line. Thus, the derivative—which depends on the existence of a unique tangent line for its very definition—can't exist at these places.

So, all functions in calculus must be well defined to ensure that one and only one derivative is defined for each point on the curve. Luckily, in introductory physics, all functions you will encounter will be well defined; therefore, you should have no trouble taking derivatives.

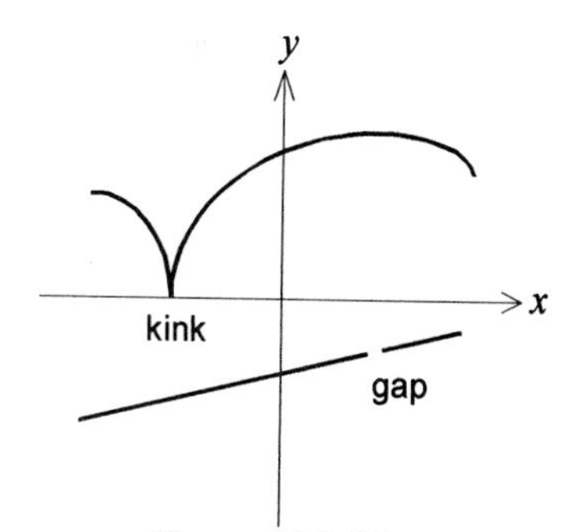

Figure 14.3 (a)

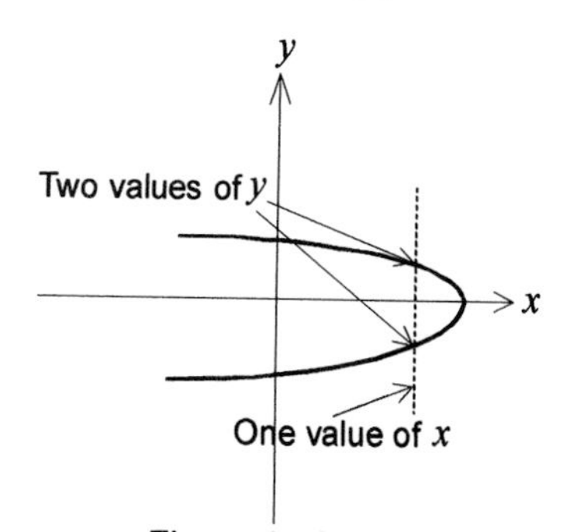

Figure 14.3 (b)

[#] Despite the use of ratio notation in the derivative, df/dx is *not* a true ratio. Sometimes the notation is used *as if* it were a ratio, but keep in mind that df/dx actually constitutes a single symbol.

14.2.2 CALCULATING ELEMENTARY DERIVATIVES

The definition of derivatives notwithstanding, to calculate the derivative of a (single-term) function *with respect to the variable that is in the function*, you do *not* need slopes or tangent lines. You only need to consult a short list like the one in Table 14.1. (Note: In this table, a and n are constants).

Table 14.1: Some Elementary Derivatives

FUNCTION	DERIVATIVE	FUNCTION	DERIVATIVE
$y = a$	$y' = 0$	$y = \csc x$	$y' = -\cot x \csc x$
$y = ax^n$	$y' = nax^{n-1}$	$y = \sec x$	$y' = \tan x \sec x$
$y = e^{ax}$	$y' = ae^{ax}$	$y = \cot x$	$y' = -a \csc^2 ax$
$y = \ln ax$	$y' = 1/x$	$y = \sin^{-1} x$	$y' = \frac{a}{\sqrt{1-(ax)^2}}$
$y = \sin ax$	$y' = a \cos ax$	$y = \cos^{-1} x$	$y' = \frac{-a}{\sqrt{1-(ax)^2}}$
$y = \cos ax$	$y' = -a \sin ax$	$y = \tan^{-1} x$	$y' = \frac{a}{1+(ax)^2}$
$y = \tan ax$	$y' = a \sec^2 ax$		

However, it is important that you do not become dependent on such a table. You should have some of the simpler derivatives *memorized:*

Exponents: Section 13.4.6. **Exponential Function: Section 13.4.15.**

Natural Logs: Section 13.4.14; Inverse Function: Section 13.4.12. **Trigonometric Functions: Section 13.6.2.**

- The derivative of a constant is zero.
- The derivative of x^n is nx^{n-1} (the **power rule**).
- The derivative of e^x is itself, e^x.
- The derivative of $\ln x$ is the inverse function, $1/x$.
- The derivative of $\sin x$ is $\cos x$.
- The derivative of $\cos x$ is $-\sin x$.
- The derivative of $\tan x$ is $\sec^2 x$.

EXERCISE 14.5: Elementary Derivatives

Calculate derivatives of the following functions using Table 14.1 and your memory.

(a) $y = x^5$ (b) $y = \sin 3x$ (c) $y = 4$ (d) $y = \sin^{-1}(9x)$

SOLUTION:

(a) Using the power rule on $y = x^5$, we obtain $y' = 5x^4$.

Exercise 14.5 Continued...

(b) Referring to Table 14.1, we see that $y' = 3\cos 3x$.

(c) Four is a constant, so $y' = 0$.

(d) Again, referring to Table 14.1, we get $y' = \dfrac{9}{\sqrt{1-(9x)^2}}$.

Often a function has more than one term, or is otherwise more complicated than the situations seen in Table 14.1. The following rules help with these kinds of functions:

- If a function $f(x)$ is made up of the *sum* of other functions $g(x)$ and $h(x)$, such that $f(x) = g(x) + h(x)$, then the derivative $f'(x)$ is the sum of the derivatives of $g(x)$ and $h(x)$:

$$f'(x) = g'(x) + h'(x)$$

- If a function $f(x)$ is made up of the *product* of other functions, such that $f(x) = g(x)\ h(x)$, then the derivative $f'(x)$ is found from the **product rule**: the derivative $f'(x)$ is the derivative of g times h, plus the derivative of h times g, or:

$$f'(x) = g'(x)h(x) + g(x)h'(x)$$

- If a function $f(x)$ is made up of the *quotient* of other functions, such that $f(x) = g(x)/h(x)$, the derivative $f'(x)$ is calculated using the **quotient rule**: derivative of g times h, minus the derivative of h times g, over h squared, or:

$$f'(x) = \frac{g'(x)h(x) - g(x)h'(x)}{[h(x)]^2}$$

- If a function is a complicated polynomial $f(x) = g(x)^r$, the derivative $f'(x)$ is calculated using the **generalized product rule**: derivative of the polynomial by the product rule, multiplied by the derivative of g.

$$f'(x) = r\left[g(x)^{r-1}\right]g'(x)$$

EXERCISE 14.6: More Complex Derivatives

Calculate derivatives for the following functions:

(a) $f(x) = 5x^4 + 2x^3 + 7x + 5$

(b) $f(x) = (3x^3)(2e^{4x})$

(c) $f(x) = \dfrac{5x}{1+x^3}$

(d) $f(x) = (1 + 3x^2)^3$

Exercise 14.6 Continued...

SOLUTION:

(a) According to the first rule above, we know that we need to take the derivative of each term in this polynomial by itself, via the power rule:

$$f'(x) = \tfrac{d}{dx}(5x^4) + \tfrac{d}{dx}(2x^2) + \tfrac{d}{dx}(7x) + \tfrac{d}{dx}(5)$$
$$= 20x^3 + 6x^2 + 7$$

(b) This problem requires the product rule. We have two multiplicative terms, $3x^3$ and $2e^{4x}$. Comparing these to the equation $f(x) = g(x)\cdot h(x)$, we see that $g(x) = 3x^3$ and $h(x) = 2e^{4x}$. Thus, we can use the product rule $f'(x) = g'(x)h(x) + g(x)h'(x)$ (and the power rule) as:

$$f'(x) = \left[\tfrac{d}{dx}(3x^3)\right](2e^{4x}) + (3x^3)\left[\tfrac{d}{dx}(2e^{4x})\right]$$
$$= (9x^2)(2e^{4x}) + (3x^3)(8e^{4x})$$
$$= 18x^2e^{4x} + 24x^3e^{4x}$$
$$= 6x^2e^{4x}(3 + 4x)$$

In the last step, we factored to make the final answer simpler.

(c) This is a quotient rule problem, where $g(x) = 5x$ and $h(x) = 1 + x^3$. To solve, we apply the quotient rule as follows:

$$f'(x) = \frac{\left[\tfrac{d}{dx}(5x)\right](1+x^3) - (5x)\left[\tfrac{d}{dx}(1+x^3)\right]}{\left[(1+x^3)\right]^2}$$
$$= \frac{(5)(1+x^3) - (5x)(3x^2)}{(1+x^3)^2}$$
$$= \frac{5 + 5x^3 - 15x^3}{(1+x^3)^2}$$
$$= \frac{5(1-2x^3)}{(1+x^3)^2}$$

(d) Here, we use the generalized product rule, where $g(x) = 1 + 3x^2$:

$$f'(x) = 3(1+3x^2)^2\left[\tfrac{d}{dx}(1+3x^2)\right]$$
$$= 3(1+3x^2)(6x)$$
$$= 18x(1+3x^2)$$

If y is a function of the variable t, $y = f(t)$, while t is itself a function of the variable x, $t = f(x)$, the derivative of y with respect to x (and <u>not</u> t) is calculated using the **chain rule**:

$$\frac{dy}{dx} = \frac{dy}{dt}\frac{dt}{dx}$$

EXERCISE 14.7: Using the Chain Rule

Calculate the derivative of $y = x^2$ with respect to t.

SOLUTION:

To find dy/dt, we must rearrange the chain rule:

$$\frac{dy}{dx} = \frac{dy}{dt}\frac{dt}{dx} \Rightarrow \frac{dy}{dt} = \frac{dy}{dx}\frac{dx}{dt}$$

Therefore:

$$\frac{dy}{dt} = (2x)\frac{dx}{dt}$$

You can take derivatives of derivatives as well. The **second derivative** of $f(x)$ is just the derivative of the first derivative, written $d^2f\,/\,dx^2$ or $f''(x)$.

Second derivatives are a measure of how much curvature, or **concavity**, a function has. Figure 14.4 (a) illustrates a function with a positive second derivative (**concave up**), while Figure 14.4 (b) depicts a function with a negative second derivative (**concave down**).

Third, fourth, and higher derivatives can also be found by continuing to take derivatives of derivatives. However, they are very rarely used in an introductory physics course, and we will not discuss them here.

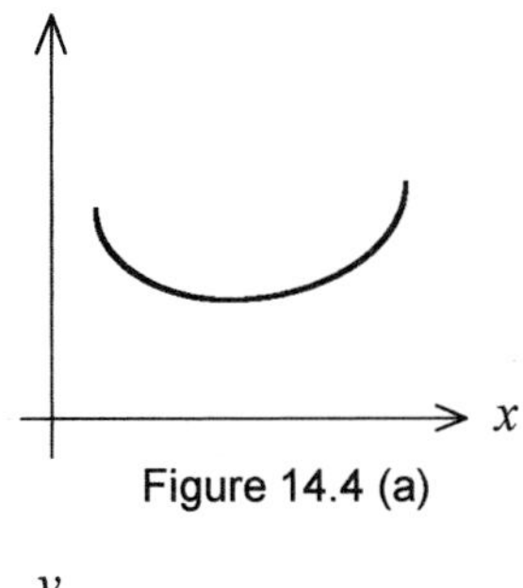

Figure 14.4 (a)

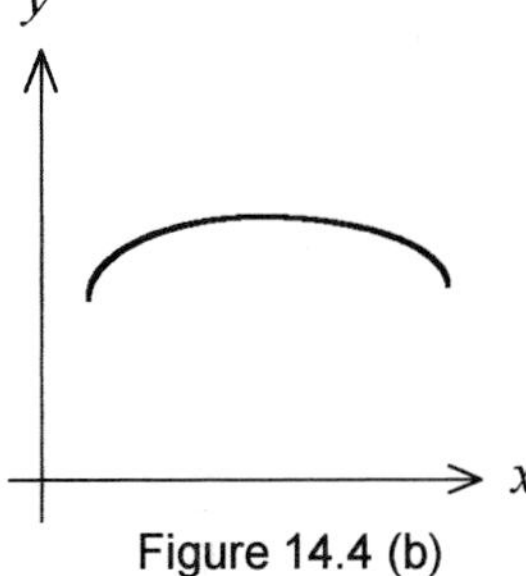

Figure 14.4 (b)

EXERCISE 14.8: Second Derivatives

Find the second derivative of the function in Exercise 14.6 (a).

SOLUTION:

In Exercise 14.6 (a), we found that $f'(x) = 20x^3 + 6x^2 + 7$. To find the second derivative of this function, we must take the derivative of $f'(x)$:

$$f''(x) = \tfrac{d}{dx}f'(x) = \tfrac{d}{dx}(20x^3 + 6x^2 + 7) = 60x^2 + 12x$$

14.2.3 ADVANCED DERIVATION TECHNIQUES

The techniques discussed in the last section will solve most derivatives used in an introductory physics class. However, in a few cases, different methods are needed. The following two techniques will help.

IMPLICIT DERIVATION

Sometimes, you will encounter equations that are not nicely arranged with y on one side of the equation and x on the other; rather, x and y are all jumbled up together.

Often such equations can be made to fit the familiar form $y = f(x)$ with a little algebra and patience—so that the derivative can be taken in the normal way—but other times, it is impossible to do so. To find the derivative of the latter kind of equation, we use **implicit derivation**.

To implicitly derive an equation, we take the derivative of each term in the expression *separately* using the chain rule. Those terms in the equation that are functions of x only will have normal derivatives, but those terms that have y in them will have the term dy/dx in the derivative. Once the entire expression has been derived, we then use algebra to solve for dy/dx.

EXERCISE 14.9: Implicit Derivation

Implicitly find the derivative of y with respect to x if $xy^3 + x^3y = 2$.

SOLUTION:

This equation cannot be solved for y in terms of x (try it!), so we must use implicit derivation to find its derivative.

We start by taking the derivative of each term in the equation separately, according to our regular derivation rules:

First term xy^3: $\frac{d}{dx}(xy^3) = y^3 + x(3y^2)\frac{dy}{dx}$

Second term x^2y: $\frac{d}{dx}(x^3y) = 3x^2y + x^3\frac{dy}{dx}$

Third term 2: $\frac{d}{dx}(2) = 0$

Now, we put the derived terms back together in the same order as in the original equation (and factor the result):

$$\left[y^3 + 3xy^2\frac{dy}{dx}\right] + \left[3x^2y + x^3\frac{dy}{dx}\right] = 0$$

$$\Rightarrow y^3 + 3x^2y + \frac{dy}{dx}(3xy^2 + x^3) = 0$$

Using a little algebra (which we skipped), we now solve for dy/dx:

$$\therefore \frac{dy}{dx} = -\frac{3x^2y + y^3}{3xy^2 + x^3}$$

Notice that the final answer has y in it!

Implicit derivation is often used to solve **related rate problems**. In these kinds of problems, an equation relates two or more variables in such a way that a relationship between the rates of change of these variables can be determined via implicit derivation. The derivatives in related rate problems are usually taken with respect to a third variable (time t) rather than x or y.

Related rate problems are important for relative motion. **See Section 5.3.3 for more.**

EXERCISE 14.10: Related Rates

A particle moves along the circle $x^2 + y^2 = 169$. At a certain time t, the particle is located at $(5 \text{ cm}, 12 \text{ cm})$. If the rate at which the particle's x position changes with time is 100 cm/s, at what rate does the particle's y position change?

SOLUTION:

This is a related rate problem. The equation $x^2 + y^2 = 169$ relates the variables x and y. But, through implicit derivation, it also relates the *rate of change* of x (the derivative of x with respect to t), and the *rate of change* of y (the derivative of y with respect to t). Therefore, we do an implicit derivation of the original equation:

$$\frac{d}{dt}[x^2 + y^2 = 169]$$
$$\Rightarrow \; 2x\frac{dx}{dt} + 2y\frac{dy}{dt} = 0$$

Because we want the rate at which the y position changes, we solve the above expression for dy/dt:

$$\frac{dy}{dt} = -\frac{x}{y}\frac{dx}{dt}$$

Plugging in our known values, we get:

$$\frac{dy}{dt} = -\frac{5 \text{ cm}}{12 \text{ cm}}(100 \text{ cm/s}) \approx -42 \text{ cm/s}$$

The negative sign means the y position is *decreasing*.

PARTIAL DERIVATIVES

Some functions, like $z = f(x, y)$ or $\zeta = f(x, y, z)$, depend on multiple variables. Such functions describe three-dimensional (or higher) curves. To take the derivative of multiple variable functions, we use **partial derivation**.

To take the partial derivative of the arbitrary function $\zeta = f(x, y\ldots)$, follow these steps:

1. Pretend that all variables in the function, except for the first variable x, are constant.

2. Take the derivative of the function (in the normal way) with respect to x. Call this derivative $\partial f / \partial x$.*
3. Now pretend that all variables in the function, except for the *second* variable y, are constant.
4. Take the derivative of the function again, but with respect to y. Call this derivative $\partial f / \partial y$.
5. Repeat this process with all variables in the function.
6. The differential is the sum of the partial derivatives found in Steps 1 through 5, multiplied by the **differentials** dx, dy...:

$$df = \frac{\partial f}{\partial x}dx + \frac{\partial f}{\partial y}dy + \cdots$$

EXERCISE 14.11: Partial Derivatives

What is the differential of $f(x,y) = \dfrac{5xy}{x^2+y^2}$?

SOLUTION:

To take the differential of $f(x, y)$, we first pretend that all variables in the equation, except for x, are constant. In this case, only the variable y must remain constant at first. Therefore, we take the derivative of f with respect to x, while holding y constant:

$$\frac{\partial f}{\partial x} = \frac{(5y)(x^2+y^2)-(5xy)(2x)}{(x^2+y^2)^2}$$
$$= \frac{5y[x^2+y^2-2x^2]}{(x^2+y^2)^2}$$
$$= \frac{5y[y^2-x^2]}{(x^2+y^2)^2}$$

Next, we reverse the process: we take the derivative of f with respect to y, while holding x constant:

$$\frac{\partial f}{\partial y} = \frac{(5x)(x^2+y^2)-(5xy)(2y)}{(x^2+y^2)^2}$$
$$= \frac{5x[x^2+y^2-2y^2]}{(x^2+y^2)^2}$$
$$= \frac{5x[x^2-y^2]}{(x^2+y^2)^2}$$

The differential of $f(x, y)$ is then the sum of these derivatives, multiplied by the differentials dx and dy:

$$df = \frac{5y[y^2-x^2]}{(x^2+y^2)^2}dx + \frac{5x[x^2-y^2]}{(x^2+y^2)^2}dy$$

* Notice that the ∂ signifying partial differentiation is different from the d that we have been using up until now. This different notation is deliberate so that we will never confuse partial differentiation with ordinary differentiation.

14.2.4 APPLICATIONS OF DERIVATIVES

Now that we know how to calculate derivatives, we can take a closer look at how derivatives are used to evaluate physical situations.

CALCULATING EXTREMA

Every function (that is not a line) will have high points (**maxima**) and low points (**minima**). See Figures 14.5 (a) - (b). These special places on the curve are called **critical points**. *All critical points have the characteristic that the derivative is zero there*: $dy / dx = 0$; that is, the tangent line is horizontal to the function at the critical point. This is **Fermat's Theorem.**

We can use Fermat's Theorem to locate maxima and minima— collectively called **extrema**—on the function. This is an important technique in many areas of physics, but is especially useful for potential energy analyses.

To find the extrema of a function $f(x)$, follow these steps:

1. Calculate the first derivative of $f(x)$.
2. Set the first derivative equal to zero, and solve the equation. The values of x found in this way are the critical points.#
3. Calculate the second derivative of $f(x)$.
4. Plug each critical point of Step 2 into the second derivative. For *each* number that results, if it is:

   - Less than zero, then $f(x)$ has a maximum at the corresponding critical point. Proceed to Step 7.
   - Greater than zero, then $f(x)$has a minimum at the corresponding critical point. Proceed to Step 7.
   - Equal to zero, further testing is required. Proceed to Step 5.

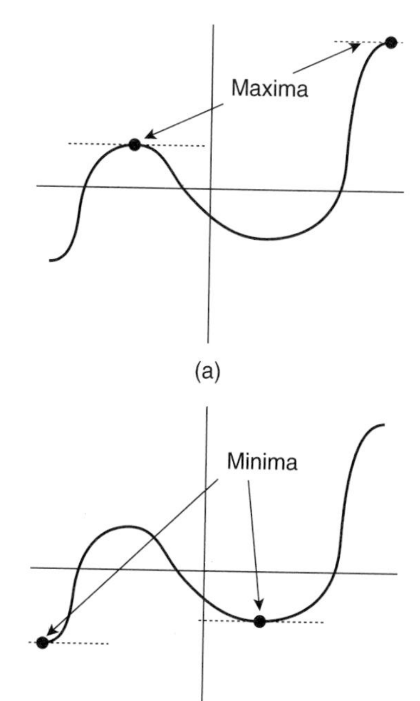

5. Pick one number that is greater than, and one number that is less than, the critical point. Your numbers should be easy to manipulate, and close to the test point (so that the value of another critical point is not inadvertently tested).
6. Plug each of these two numbers into the second derivative. If:

   - Both numbers are less than zero, then $f(x)$ has a maximum at the corresponding critical point. Proceed to Step 7.
   - Both numbers are greater than zero, then $f(x)$ has a minimum at the corresponding critical point. Proceed to Step 7.
   - One number is less than zero and the other number is greater than zero, then $f(x)$ has an inflection point there. An **inflection** or **stationary point** is a place where the curve changes concavity (Figure 14.5 (c)). Ignore this point.

7. Plug the extrema points found in Steps 4 through 6 into the *original* equation to calculate the y values for each extrema.

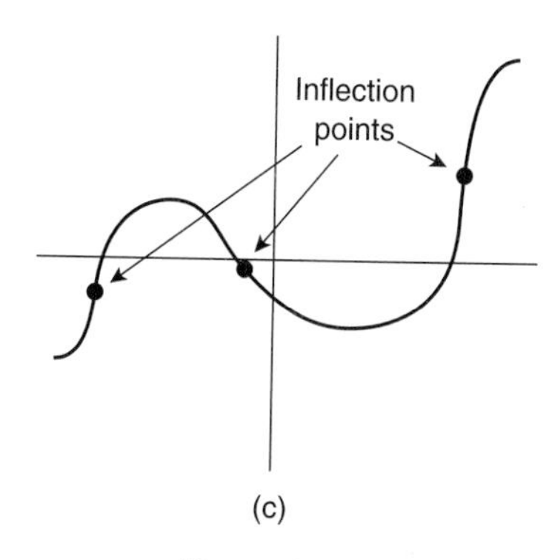

Figure 14.5

In some cases, the first derivative will have places where it is undefined; for example, where a solution for x will cause a denominator to be zero. These values of x, if any, should also be considered in the set of critical points.

The true aspect of delight, the exaltation, the sense of being more than man, which is the touchstone of the highest excellence, is to be found in mathematics as surely as poetry.
BERTRAND RUSSELL
English Mathematician, 1872 - 1970

8. *Optional: Use only if the function is defined only in a certain region* $[x_1, x_2]$. Plug the endpoints of the region into the *original* equation to calculate their corresponding y values. The y values will determine whether the endpoints are maxima or minima.
9. *Optional: Use only if there is more than one maxima or minima defined on a function.* The maximum that is the highest is denoted the **absolute maximum**, while all others are **local maxima**. The lowest minimum is called the **absolute minima**, and all others are **local minima**.

EXERCISE 14.12: Finding Extrema

Find all extrema for the function $y = x^3 + x^2$ in the region $[1, -1]$.

SOLUTION:

We start by calculating the first derivative of the function (Step 1):

$$y' = 3x^2 + 2x = x(3x+2)$$

Notice that we did a little factoring of the result at the end. Now, we set this equation equal to zero, and solve for all x (Step 2):

$$y' = x(3x+2) = 0$$

Factoring is in Section 13.4.10. 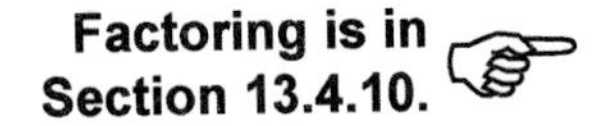

From our techniques of Chapter 13, we know that the solutions to this equation are $x = 0$ and $-\frac{2}{3}$. These numbers are thus our critical points. Next, we calculate the second derivative of y (Step 3):

$$y'' = 6x + 2 = 2(3x+1)$$

Again, we factored at the end to make the expression nicer to work with. According to Step 4, we plug in our critical points $(0, -\frac{2}{3})$ into this second derivative equation. Let's chart our results:

Critical Point	y''	$<, >, =$ to 0?	Test Result
0	2	$>$	Minimum
$-\frac{2}{3}$	–2	$<$	Maximum

Thus, from the second derivative test, we know that there is a maximum at $x = 0$, a minimum at $x = -\frac{2}{3}$, and no inflection points. We can therefore skip to Step 7, which tells us to plug each of the critical points into the *original* equation $y = x^3 + x^2$. Thus, our extrema for this function are located at:

Minimum: $(0, 0)$
Maximum: $(-\frac{2}{3}, 0.15)$

According to Step 8, now, we must plug the endpoints $(-1, 1)$ into the original equation as well, to see if these are maxima or minima. The results are $(-1, 0)$ and $(1, 2)$: by definition, $(-1, 0)$ is a min, and $(1, 2)$ is a max.

By Step 9, we know that the point $(1, 2)$ is an absolute max, because it has a bigger y value than $(-\frac{2}{3}, 0.15)$ [which in turn is a local max]. In just the same way, we see that $(-1, 0)$ and $(0, 0)$ are both local mins because they have *equal* y values.

EXTREMA AND EQUILIBRIA

At extrema, the function is said to be in **equilibrium**. There are many kinds of equilibriums. We can visualize some of these by imagining that the function is a roller coaster (Figs. 14.6 (a) and (b)).

Imagine that the roller-coaster cart is in a minimum of the function (Fig. 14.7 (a)). If it is given a *little* push (or **perturbation**), the cart will always return to the minimum (after rocking around for a while). We can say, therefore, that at a minimum, the function is **stable** to perturbations. Thus, minima are also called **stable equilibriums**.

Now imagine the cart is at a maximum (Fig. 14.7 (b)). If the cart is pushed, even just the littlest bit, the cart will roll away from the maximum and never return. Thus, at a maximum, we say that the function is **unstable** to perturbations. As a result, maxima are also called **unstable equilibriums**.

A last type of equilibrium occurs on horizontal straight lines. Here there are neither maxima nor minima. Thus, when pushed, the cart will not return to the same point; rather, it will continue along at the same level. This is the case of **neutral equilibrium** (Fig. 14.7 (c)).

Figure 14.6 (a): A Function

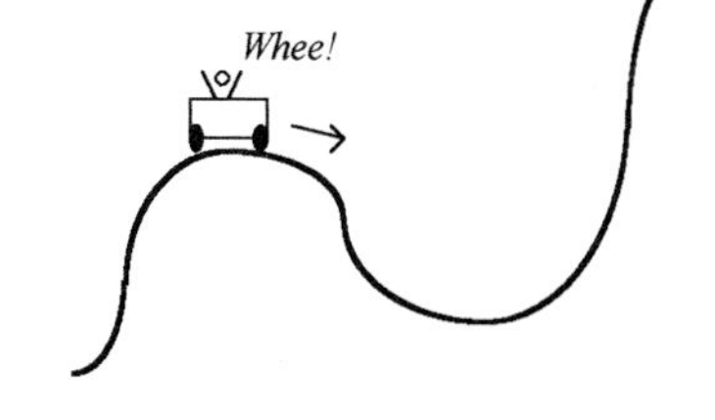

Figure 14.6 (b): A Roller Coaster

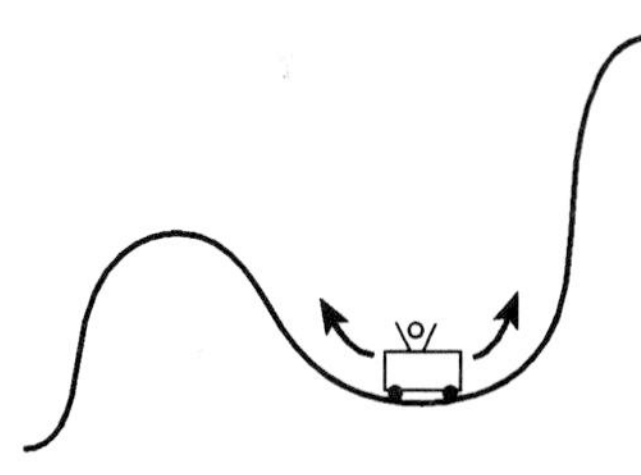

Figure 14.7 (a)

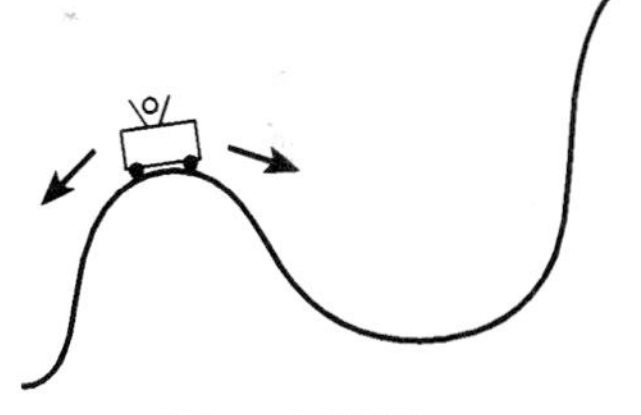

Figure 14.7 (b)

Figure 14.7 (c)

EXERCISE 14.13: Equilibrium

Label the equilibrium points of the graph in Figure 14.8 below as stable, unstable, or neutral.

SOLUTION:

Particles at x_1 are in stable equilibrium. Particles at x_2 are in unstable equilibrium. And particles at x_3 are in neutral equilibrium.

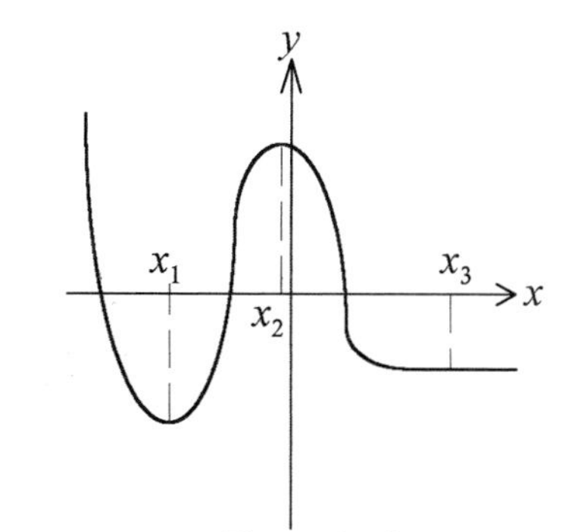

Figure 14.8

SKETCHING QUICK GRAPHS USING EXTREMA

We can use the extrema to roughly sketch a function by hand. To do so, we first complete Steps 1 through 9 in the section CALCULATING EXTREMA above. Then we continue with the following steps. Please note that your calculator should be able to do this kind of graphing very quickly; check your owner's manual to find out how.

10. Make a list of all critical points and endpoints obtained in the procedure above.
11. For each *interval* on the x-axis between the critical points and endpoints—including the intervals between positive and negative infinity and the last critical points on either side of the graph—choose an easily manipulated test value
12. Plug each number from Step 11 into the first derivative. If the result is:

- Positive, the slope of the graph will increase in that interval.
- Negative, the slope of the graph will decrease in that interval.

13. Begin a table. In the first column, list the intervals from Step 11. In the second column, list your test points from Step 11. In the third column, list the test results from Step 12.
14. Set the second derivative equal to zero. Solve this equation for x.*
15. Repeat Step 11 for the numbers of Step 14.
16. Plug each number from Step 15 into the second derivative. If the result is:
    - Positive, the graph will be concave up in that interval.
    - Negative, the graph will be concave down in that interval.
17. Begin a new table. In the first column, list the intervals from Step 15. In the second column, list your test points from Step 15. In the third column, list the test results from Step 16.
18. Set the *original* equation equal to zero. Solve this equation for all x. These are the x-axis intercepts—where the graph will cross the x-axis.
19. Set x in the *original* equation equal to zero. Solve for y. This number is the y-axis intercept—where the graph crosses the y-axis.
20. Now, begin the plot. Draw a grid that is well scaled for the intervals you have determined in Steps 11 and 15.
21. On this grid, plot the critical points and endpoints.
22. Using the first table as a guide, roughly sketch in the slope of each first derivative interval on the graph.
23. Using the second table as a guide, roughly sketch in the concavity of each second derivative interval on the graph.
24. Finally, connect all plotted points, using a smooth line that roughly mimics the slope and concavity sketches.

I [love] math for its own sake...because it allows for no hypocrisy, and no vagueness.
STENDHAL
French Novelist, 1783 - 1842

EXERCISE 14.14: A Quick Graph

Make a quick sketch of the function $y = x^3 + x^2$ from Exercise 14.12 in the region $[1, -1]$.

SOLUTION:

Summarizing our results from that exercise, we know that (Step 10) our critical points and endpoints for this problem are:

Critical Points (x):	$-\frac{2}{3}$ and 0
Endpoints:	1 and -1

Now, Step 11 asks us to break the x region into intervals. That is, our future graph will stretch from -1 to 1 in the x direction, but will be broken into three parts by the critical points $-\frac{2}{3}$ and 0. Hence, our intervals are $(-1, -\frac{2}{3})$, $(-\frac{2}{3}, 0)$ and $(0, 1)$.

* Including any undefined points.

Exercise 14.14 Continued...

Step 11 also asks us to choose test points for each interval. We choose the points -0.75, -0.5, and 0.5 because (1) each one is in a different interval, and (2) they are easy to use. (Check this statement. Then, think about this question: Why can't we choose the even *easier* numbers $1, -1$ or 0?)

Next, according to Step 12, we must plug each of these test points into the first derivative equation, obtained in Exercise 14.12 above. The results are tabulated below (Step 13):

Interval	Test Point	Test Result
$(-1, -\frac{2}{3})$	-0.75	Positive, Increasing
$(-\frac{2}{3}, 0)$	-0.5	Negative, Decreasing
$(0, 1)$	0.5	Positive, Increasing

Now, for Step 14, we set the second derivative equal to zero and solve for all x:

$$y'' = 6x + 2 = 2(3x + 1) = 0$$
$$\Rightarrow x = -\tfrac{1}{3}$$

This point defines two *new* intervals $(-1, -\frac{1}{3})$ and $(-\frac{1}{3}, 1)$. According to Step 15, we need to pick test points for *these* intervals. We choose -0.5 and 0 for the same kinds of reasons as listed above.

Now, in Step 16, we are instructed to plug each of these new test points into the *second* derivative equation. The results are tabulated below (Step 17):

Interval	Test Point	Test Result
$(-1, -\frac{1}{3})$	-0.75	Negative, Decreasing
$(-\frac{1}{3}, 1)$	0	Positive, Increasing

Next, according to Step 18, we set the *original* equation equal to zero and solve for the x-axis intercepts:

$$y = x^3 + x^2 = x^2(x + 1) = 0$$
$$\Rightarrow x = 0, -1$$

So, the graph will cross the x-axis at $(0, 0)$ and $(-1, 0)$. Step 19 then tells us to set x equal to zero in the original equation, and solve for y-axis intercepts:

$$y = 0^3 + 0^2 = 0$$

Therefore, the graph will cross the y-axis at $(0, 0)$—a fact we already knew from the last step, but no matter.

Finally, according to Steps 20 and 21, we can start the plot by drawing an appropriately scaled grid and sketching in the critical points and endpoints. We found the associated y-values in Exercise 14.12 above. We can therefore summarize these points as:

Critical Points: $(-\frac{2}{3}, 0.15)$ and $(0, 0)$
Endpoints: $(-1, 0)$ and $(1, 2)$

Now we can sketch:

Exercise 14.14 Continued...

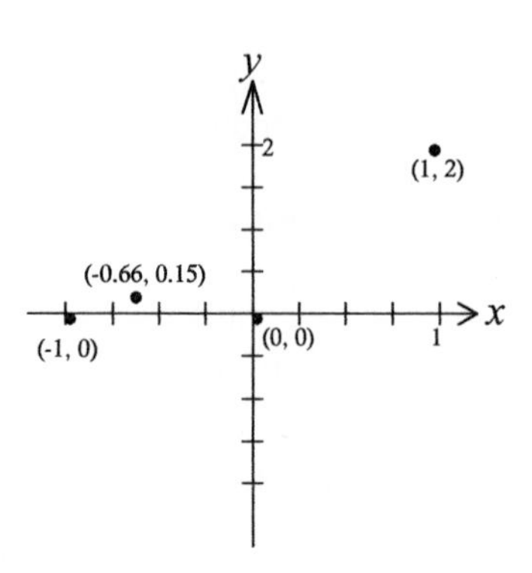

Using the two tables as our guides, we sketch onto this grid the first derivative intervals and the increasing/decreasing information (Step 22), as well as the second derivative intervals and concavity information (Step 23):

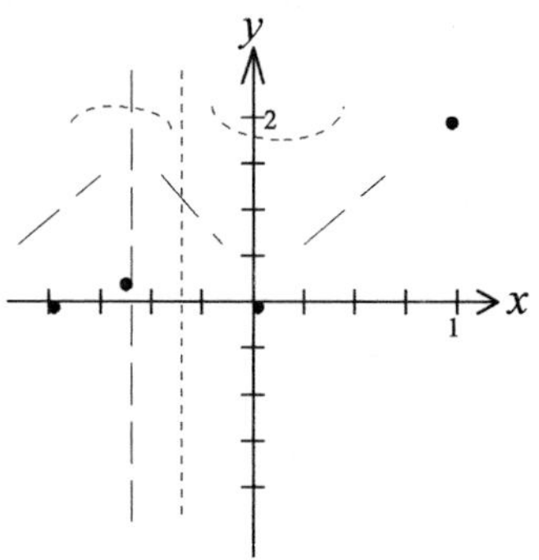

Then, we "connect the dots" with a smooth line that mimics the shapes of the first and second derivative sketches (Step 24):

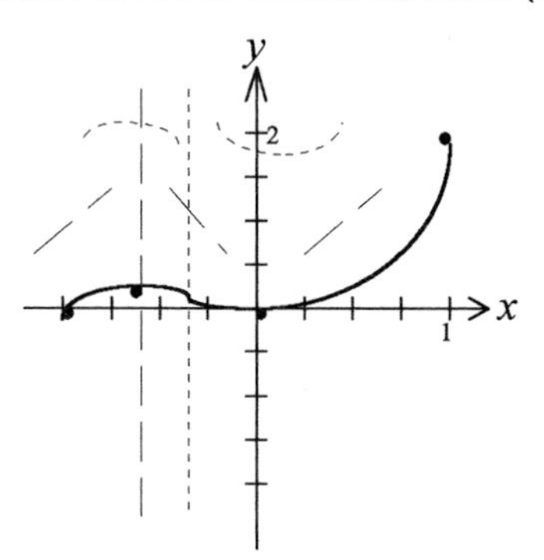

Notice how well our "quick" sketch matches the actual curve:

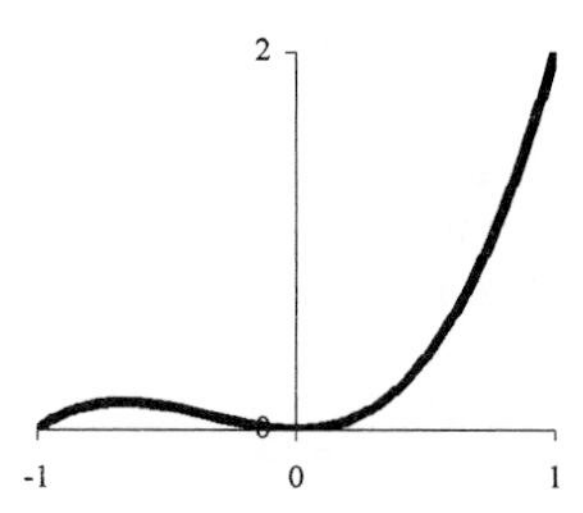

14.3 INTEGRAL CALCULUS

In this section, we define and learn to calculate integrals.

14.3.1 WHAT IS AN INTEGRAL?

When a quantity changes, we often want to know the *net* effect of the change. The total effect of a quantity's change is its **anti-derivative**, calculated via an **integral**.

> **EXAMPLE 14.15:** Position and Velocity, Part III
>
> If velocity v is the derivative of position x (Example 14.1), then position is the anti-derivative of velocity.

Calculating an integral is nothing more than figuring an area. For an arbitrarily shaped function, this is done by slicing up the area into tiny pieces of known shape, and then *adding up* the areas of the pieces.

> **DEFINITION 14.16:** Integrals
>
> Let's quickly examine the definition of anti-derivatives. Recall from Chapter 13 that the area of a rectangle with base b and height h is:
>
> $$A = bh$$
>
> We can apply this equation to the area under an arbitrary function $y = f(x)$ (Fig. 14.9 (a)) by slicing this area up into thin rectangles (Fig. 14.9 (b)). Each rectangle has base $b = \Delta x$ and height $h = y = f(x)$, and so has area $A = f(x) \cdot \Delta x$.
>
> The total area under the function, then, is the sum of all of the rectangular areas:
>
> $$\sum f(x)\Delta x$$
>
> Notice, however, that if we draw narrower rectangles, we can fit many more of them under the curve. By this method, we can get a better approximation of the area under the curve (Fig. 14.9 (c)). In fact, we can imagine a case in which the rectangles are *so* narrow as to be invisible (Fig. 14.9 (d)).
>
> In this ideal case, the sum of the areas of the rectangles *is* the area under the curve. In other words, we have taken the limit of the above equation as Δx goes to zero, which we *define* to be the integral:
>
> $$\lim_{\Delta x \to 0} \sum f(x)\Delta x \equiv \int f(x)dx$$

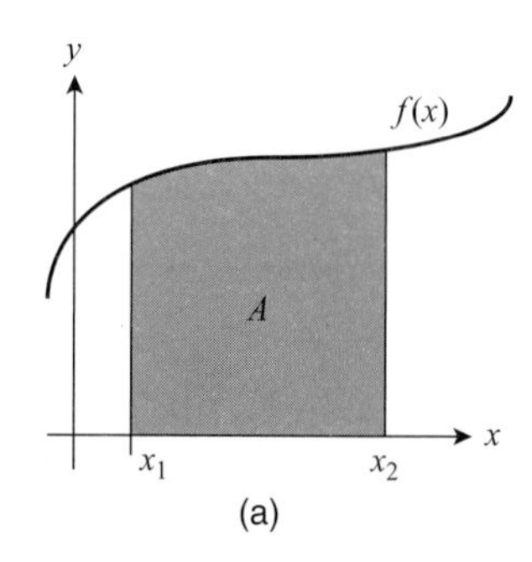

(a)

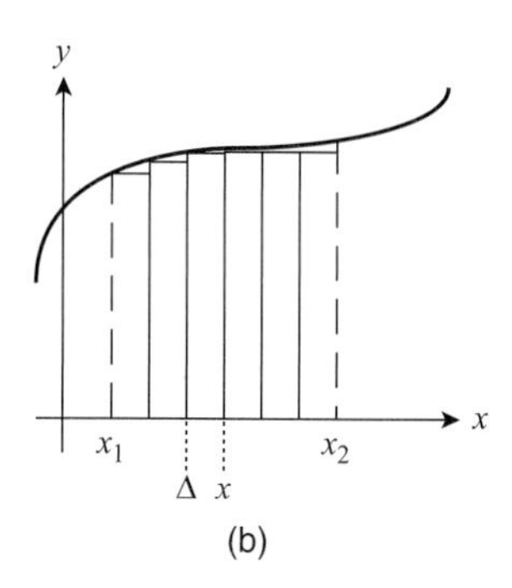

(b)

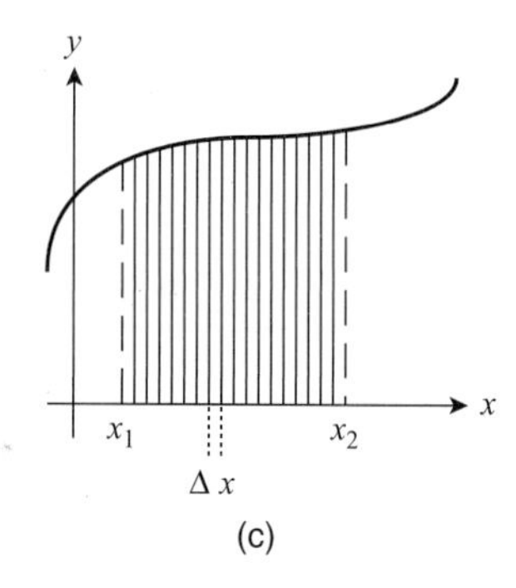

(c)

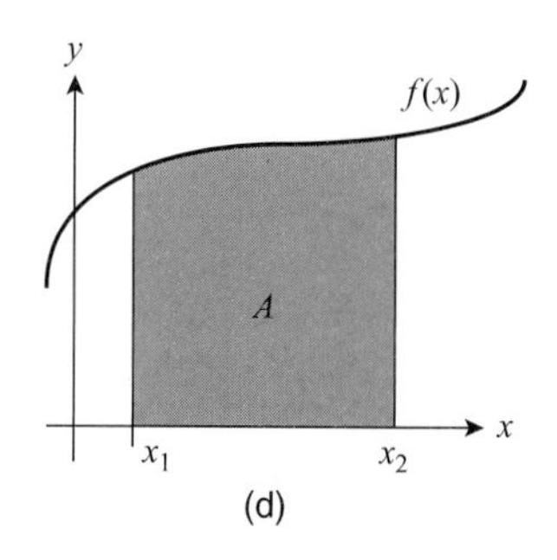

(d)

Figure 14.9

The integral notation $\int f(x)dx$ is read "the integral of $f(x)$ over dx." $f(x)$, here, is called the **integrand** and dx is known as the **differential**.

EXAMPLE 14.17: Position and Velocity, Part IV

Since velocity v is the derivative of position x, $v = dx/dt$, we can calculate position from velocity as an integral:

$$x = \int vdt$$

The result of this integral is the *position* of the object as it moves through time.

14.3.2 CALCULATING ELEMENTARY INTEGRALS

More on indefinite integrals on page 320.

Like derivatives, calculating elementary integrals is mostly a matter of consulting an appropriate list (like the one in Table 14.2) and choosing an integral that matches your problem.

Table 14.2: Some Elementary Integrals

INTEGRAL	INTEGRAL	INTEGRAL
$\int x^n dx = \frac{x^{n+1}}{n+1}$	$\int \frac{dx}{\sqrt{x^2 \pm a^2}} = \ln(x + \sqrt{x^2 \pm a^2})$	$\int \ln axdx = x\ln ax - x$
$\int (a+bx)^n dx = \frac{(a+bx)^{n+1}}{b(n+1)}$	$\int \frac{xdx}{\sqrt{x^2 \pm a^2}} = \sqrt{x^2 \pm a^2}$	$\int \sin axdx = -\frac{1}{a}\cos x$
$\int \frac{dx}{x} = \ln x$	$\int \frac{dx}{a^2+x^2} = \frac{1}{a}\tan^{-1}\frac{x}{a}$	$\int \cos axdx = \frac{1}{a}\sin x$
$\int \frac{dx}{a+bx} = \frac{1}{b}\ln(a+bx)$	$\int \frac{dx}{(x^2 \pm a^2)^{3/2}} = \frac{1}{a^2}\frac{\pm x}{\sqrt{x^2 \pm a^2}}$	$\int \tan xdx = \ln\lvert\sec x\rvert$
$\int \frac{dx}{(a+bx)^2} = -\frac{1}{b(a+bx)}$	$\int \frac{xdx}{(x^2+a^2)^{3/2}} = -\frac{1}{\sqrt{x^2+a^2}}$	$\int \sec axdx = \frac{1}{a}\ln(\sec ax + \tan ax)$
$\int \frac{dx}{(a+bx^2)^n} = -\frac{1}{2b(n-1)(a+bx^2)^{n-1}}; n \neq 1$	$\int \frac{dx}{\sqrt{a^2-x^2}} = \sin^{-1}\frac{x}{a}$	$\int \csc axdx = \frac{1}{a}\ln(\csc ax - \cot ax)$
$\int \frac{xdx}{a+bx^2} = \frac{1}{2b}\ln(a+bx)$	$\int e^{\pm ax}dx = \frac{1}{a}e^{\pm ax}$	$\int \cot axdx = \frac{1}{a}\ln(\sin ax)$
$\int \frac{dx}{(a+bx^2)} = \begin{pmatrix} \frac{1}{\sqrt{ab}}\tan^{-1}\left(\frac{\sqrt{b}}{a}x\right); ab>0 \\ \frac{1}{2\sqrt{ab}}\ln\left(\frac{a-x\sqrt{\lvert ab\rvert}}{a+x\sqrt{\lvert ab\rvert}}\right); ab<0 \end{pmatrix}$	$\int xe^{ax}dx = \frac{e^{ax}}{a^2}(ax-1)$	

However, you should memorize the following important integrals so that you are not completely dependent upon this or other tables:*

- The integral of 0 is an arbitrary constant, C.
- The integral of x^n is x^{n+1}/n (the **power rule**).
- The integral of e^x, is itself, e^x.
- The integral of the inverse function, $\frac{1}{x}$, is $\ln x$.
- The integral of $\sin x$ is $-\cos x$.
- The integral of $\cos x$ is $\sin x$.
- The integral of $\tan x$ is $\ln|\sec x|$.

We said that calculating an integral is *mostly* a matter of looking up an integral in a table; unfortunately, there is one more complication. The value of an integral depends on the location of the endpoints x_1 and x_2 (Fig. 14.10); that is, the closer x_1 and x_2 are to each other, the smaller the value for the anti-derivative. Therefore, you must always *evaluate* the integral at its endpoints to obtain a final solution to the problem.

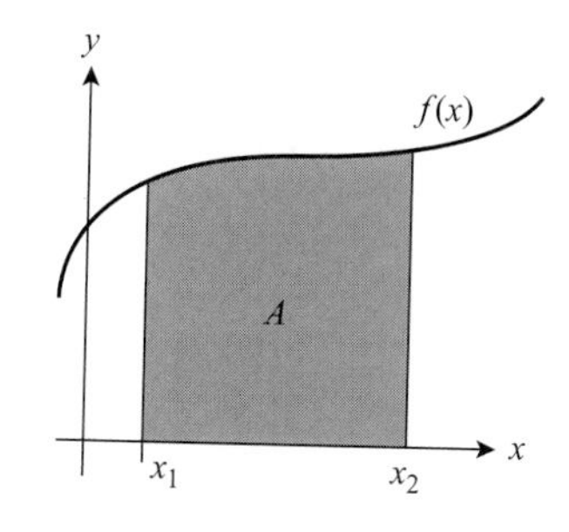

Figure 14.10

DEFINITE INTEGRALS

Definite integrals explicitly specify x_1 and x_2, so that the solution to the integral is *exact*. Such integrals are calculated via the **Fundamental Theorem of Calculus**:

$$\int_{x_1}^{x_2} f(x)dx \equiv F(x)\Big|_{x_1}^{x_2} = F(x_2) - F(x_1)$$

where $F(x)$ is the anti-derivative of $f(x)$. This formula states that to find the definite integral of $f(x)$, separately plug x_1 and x_2 into the anti-derivative $F(x)$, and subtract the resulting two values.

Some interesting definite integrals are listed in Table 14.3.

Table 14.3: Some Definite Integrals

INTEGRAL	INTEGRAL
$\int_0^\infty x^n e^{-ax} dx = \frac{n!}{a^{n+1}}$	$\int_0^\infty x^{2n} e^{-ax^2} dx = \frac{(2n-1)!}{2^{n+1}a^n}\sqrt{\frac{\pi}{a}}$
$\int_0^\infty \frac{dx}{(1+e^{ax})} = \frac{\ln 2}{a}$	$\int_0^\infty \frac{\sin ax dx}{x} = \frac{\pi}{2}$
$\int_0^\infty x^n e^{-ax} dx = \frac{n!}{a^{n+1}}$	

Factorials are in Section 13.4.11.

* A really good list of integrals can be *extensive*. If you are planning to major in physics (or math, for that matter), you might want to consider purchasing a book of integral tables. At the very least, you should know where these books are located in your library. Some good candidates are the *Handbook of Chemistry and Physics* (CRC Press, Inc.) or *Tables of Integrals and Other Mathematical Data* by Dwight (Macmillan).

Many calculators today can do integrals without the need for integral tables. Look in your owner's manual for more information.

EXERCISE 14.18: Definite Integrals

Calculate the definite integrals of the following functions:

(a) $\int_0^2 x^{3/2}dx$ (b) $\int_1^4 (5x^2+7)dx$ (c) $\int_2^5 \frac{dx}{x}$

SOLUTION:

(a) To calculate a definite integral, we use a table of integrals to find the anti-derivative:

$$\int_0^2 x^{\frac{3}{2}}dx = \frac{x^{\frac{5}{2}}}{\frac{5}{2}}$$

But, we *evaluate* this function at the points 2 and 0 according to the Fundamental Theorem of Calculus:

$$\int_0^2 x^{\frac{3}{2}}dx = \left.\frac{x^{\frac{5}{2}}}{\frac{5}{2}}\right|_0^2$$

That is:

$$\int_0^2 x^{\frac{3}{2}}dx = \tfrac{2}{5}\left[(2)^{5/2} - (0)^{5/2}\right] = \tfrac{2}{5}\left[4\sqrt{2} - 0\right] = \tfrac{8\sqrt{2}}{5}$$

(b) First we do the integral, and then we evaluate it:

$$\int_1^4 (5x^2+7)dx = \left.\left(\frac{5x^3}{3}+7x\right)\right|_1^4$$
$$= [\tfrac{5}{3}(4)^3 + 7(4)] - [\tfrac{5}{3}(1)^3 + 7(1)]$$
$$= 126$$

(c) $\int_2^5 \frac{dx}{x} = \ln x\Big|_2^5 = 1.609 - 0.6931 = 0.916$

INDEFINITE INTEGRALS

Indefinite integrals—like those in Table 14.2 above—do not explicitly specify the endpoints x_1 and x_2; therefore, the solutions to such integrals are *uncertain* because it is not known where the area begins (x_1), or where it ends (x_2).

However, the uncertainty in indefinite integrals is limited to an additive constant.* That is, the solution to an indefinite integral is a function,

* The source of the constant of integration is in the inverse relationship between integration and derivation. When we take the derivative of the anti-derivative $F(x)$, $F'(x) = f(x)$, any constants in $F(x)$ will disappear. Therefore, an infinite number of functions $F(x)$—which differ *only* by a constant C—will have *exactly the same* derivative $f(x)$.

given by a table of integrals, and some constant, which we call C, the **constant of integration**. The correct format for the solution of an indefinite integral is therefore:

$$\text{function} + C$$

EXERCISE 14.19: Indefinite Integrals

Calculate the indefinite integrals of the following functions:

(a) $y = x^5$ (b) $y = \sin 3x$ (c) $y = 4$

SOLUTION:

(a) Using the power rule above, and writing the equation in correct format with the constant of integration, we have:

$$\int x^5 dx = \tfrac{1}{6}x^6 + C$$

(b) Referring to Table 14.2, and writing in the correct format:

$$\int \sin 3x dx = -\tfrac{1}{3}\cos 3x + C$$

(c) If we think of 4 as $4x^0$ (recalling that $x^0 = 1$), then by the power rule, we have $4x^{0+1} = 4x$, or:

$$\int 4dx = 4x + C$$

Sometimes there is enough information in a problem, in the form of **initial** or **boundary conditions**, to figure C exactly.

EXERCISE 14.20: Boundary Conditions

Calculate the value of the indefinite integral for part (c) in Exercise 14.19 above, if the boundary condition for the problem is $f(0) = 2$.

SOLUTION:

We know from the boundary condition that the anti-derivative has a value of 2 when x equals 0. We plug this information into the equation in part (c) above, and get:

$$4(0) + C = 2$$

So, $C = 2$. We can therefore write our final solution as:

$$\int 4dx = 4x + 2$$

14.3.3 REWRITING INTEGRANDS

Mathematics, rightly viewed, possesses not only truth, but supreme beauty—a beauty cold and austere like that of sculpture.
BERTRAND RUSSELL
English Mathematician, 1872 - 1970

Some integrals have more complex integrands than can be listed in integral tables. However, many of these integrands can be *rewritten*, so that the tables can still be used for solving. The following two techniques allow you to rewrite such integrals.

CHANGING VARIABLES

The first technique involves **changing variables** in the integrand in order to match the variables seen in the integral table. To change variables, follow these steps:

Implicit derivation is in Section 14.2.3.

1. Carefully choose a term in the integrand that, if replaced, will simplify the problem. This step takes some practice to master, but in general, you should pick the most complicated term.
2. Set the complex expression chosen in Step 1 equal to u.
3. Implicitly derive u with respect to the original variable. This will result in a new differential du in terms of the old differential dx.
4. Solve the equation from Step 3 for dx in terms of du.
5. Plug in the values of u and dx into the original integral.
6. *Optional: Use only if the integral is definite.* Plug each limit of integration into the formula developed in Step 2. This will result in new limits of integration for the new variable. Write the new limits in the rewritten integral.
7. Solve the rewritten integral with an integral table.
8. *Optional: Use only if you need the final answer to be in terms of the original variable.* Replace u with the equation from Step 2.

EXERCISE 14.21: Changing Variables

Solve the indefinite integral $\int (1+5x)^2\,dx$ by changing variables.

SOLUTION:

This integral is not in any integral table. However, the related integral $\int u^2 du$ is. We therefore choose $1 + 5x$ to be our u (Step 1). Thus, $u = 1 + 5x$ (Step 2).

Now, we implicitly derive this expression with respect to the original variable, x (Step 3). Doing so, we get:

$$du = 5dx$$

Next, we solve this equation for dx in terms of du (Step 4):

$$dx = \tfrac{1}{5}du$$

Now, we plug in u and dx into the integral, yielding (Step 5 & 7):

$$\int (1+5x)^2\,dx = \int u^2 (\tfrac{1}{5}du) = \tfrac{1}{5}(\tfrac{1}{3}u^3 + C) = \tfrac{1}{15}u^3 + C$$

Exercise 14.21 Continued...

We can skip Step 6 because this is an indefinite integral. If we finally replace u with $1 + 5x$ (Step 8), we obtain our final answer:

$$\tfrac{1}{15}(1+5x)^3+C$$

PERFECT DIFFERENTIALS

The second technique for rewriting integrals, called the **method of perfect differentials**, is very similar to the first. Therefore, we use the same method as written out above, but with the following changes:

- In Step 1 above, pick the term in the integrand that, when implicitly derived, will yield the second term in the equation.
- This eliminates the need to do Step 4.

EXERCISE 14.22: Perfect Differentials

Solve the indefinite integral $\int \sin x \cos x dx$ using the method of perfect differentials.

SOLUTION:

There are two terms in this integrand, $\sin x$ and $\cos x$, so we know that we must choose one of these as our variable u. The natural derivative of $\sin x$ is $\cos x$, so $\sin x$ is the best choice to be u. We therefore set $u = \sin x$.

Now, by implicit derivation, it follows that:

$$du = \cos x \, dx$$

Replacing u and du into the integrand, we have:

$$\begin{aligned}\int \sin x \cos x dx &= \int u du \\ &= \tfrac{1}{2}u^2+C \\ &= \tfrac{1}{2}(\sin x)^2+C\end{aligned}$$

where we have replaced u with x in our final answer.

14.3.4 ADVANCED INTEGRATION TECHNIQUES

The techniques discussed in the last section will solve most integrals in the introductory physics class. However, in some cases, more advanced methods are needed for solving. The following two techniques will help.

INTEGRATION BY PARTS

One method for solving thorny integrals is **integration by parts**. The goal of this technique is to rewrite the integrand $u\,dv$ in terms of $v\,du$, where v is the integral of the differential dv, and du is the differential of u. Once this is done, the following relation solves the integral:

$$\int u dv = uv - \int v du$$

The key to this method is to spend a little time picking the right u and dv. A good rule of thumb is to always try the more complicated term in the integral as dv first.

EXERCISE 14.23: Integration by Parts

Calculate the indefinite integral $\int xe^x dx$.

SOLUTION:

There are two terms in the integrand, x and e^x. We consider e^x to be more "complicated" than x, so we therefore set $dv = e^x dx$. That leaves u to be x: $u = x$. Now, we integrate dv to get v...

$$v = e^x$$

...and we differentiate u to get du...

$$du = dx$$

...so that the only thing left to do is to plug these new variables into the equation given above:

$$\int u dv = uv - \int v du$$

$$\int xe^x dx = xe^x - \int e^x dx$$

Now, we still have an integral left, but since we know that the integral of the exponential function is just the exponential function, it is easy to take care of. Therefore, our final answer is:

$$\int xe^x dx = xe^x - e^x + C$$

MULTIPLE INTEGRALS

Multiple integrals are important for calculating Gauss' Law, Section 7.3.2.

Many integrals in physics are taken over an area dA or a volume dV, rather than a length dx. These kinds of integrals are called **multiple integrals**.

To solve a multiple integral, the differential dA or dV needs to be *replaced* by a product of linear differentials. Sometimes this will already be done for you, sometimes not. If not, you will need to do the replacing yourself. Your choice of replacement will depend upon the coordinate system you are working in; see Table 14.4 below.

Note that in every case (except for polar coordinates), there is more than one replacement choice for the area differential. The correct one will be determined by the problem.

Table 14.4: Linear Differentials

COORDINATES	AREA (dA)	VOLUME (dV)
CARTESIAN	$dx\,dy$ $dy\,dz$ $dx\,dz$	$dx\,dy\,dz$
POLAR	$r\,dr\,d\theta$	N/A
CYLINDRICAL	$rdrd\theta$ $drdz$ $rd\theta dz$	$r\,dr\,d\theta\,dz$
SPHERICAL	$\rho\sin\varphi drd\theta$ $\rho d\rho d\varphi$ $\rho^2\sin\varphi d\theta d\varphi = d\Omega$	$\rho^2\sin\varphi d\rho d\varphi d\theta$

After you have substituted the differentials, or if they are already in place, integrate each differential as if it were a single integral, holding the other variable(s) constant. Work from the innermost differential outwards. You may want to rearrange the order of the differentials to make the integration easier.*

EXERCISE 14.24: Multiple Integrals

Calculate the multiple integrals below, each of which already has an appropriate choice of differential.

(a) $\iint (xy + x^2 + 2y)dxdy$ (b) $\int_0^{\pi/2}\int_0^5 (25 - r^2)rdrd\theta$

SOLUTION:

(a) Since dx is the innermost differential, we integrate over x first. To do so, we pretend that y is a constant, and can be considered integrate normally:

$$\iint (xy + x^2 + 2y)dxdy = \int (\tfrac{1}{2}x^2 y + \tfrac{1}{3}x^3 + 2xy + C_1)dy$$

Now, we integrate over y, holding x constant:

* Often the only indication that you have a multiple integral is the differential dA or dV; sometimes there is only one integral sign $\int$ explicitly written, rather than two $\iint$ or three $\iiint$

Exercise 14.24 Continued...

$$\int(\tfrac{1}{2}x^2y+\tfrac{1}{3}x^3+2xy+C_1)dy=\tfrac{1}{4}x^2y^2+\tfrac{1}{3}x^3y+xy^2+C_1y+C_2$$

Note that we integrated C_1 just like any other constant, and that we added another constant C_2 because of the second integration over y.

(b) This time we must integrate in polar coordinates. This should not faze us too much, because r and θ can be considered constants just like x and y. We begin by integrating over r while holding θ constant:

$$\int_0^{\pi/2}\int_0^5(25-r^2)rdrd\theta=\int_0^{\pi/2}\int_0^5(25r-r^3)drd\theta=\int_0^{\pi/2}(\tfrac{25}{2}r^2-\tfrac{1}{4}r^4)\Big|_0^5d\theta$$

Now we plug in the r values for the definite integral.

$$\Rightarrow\int_0^{\pi/2}[\tfrac{25}{2}(5)^2-\tfrac{1}{4}(5)^4-(0-0)]d\theta=\int_0^{\pi/2}\tfrac{625}{4}d\theta$$

Finally, we integrate over θ:

$$\int_0^{\pi/2}\tfrac{625}{4}d\theta=\tfrac{625}{4}\theta\Big|_0^{\pi/2}=\tfrac{625\pi}{8}$$

14.4 LINEAR DIFFERENTIAL EQUATIONS

The twin arts of calculus are combined in differential equations.

14.4.1 INTRODUCTION TO DIFFERENTIAL EQUATIONS

A **differential equation** is an equation containing one or more derivatives. The derivatives can be ordinary or partial, but they are usually only of the first- or second-order. Unlike other, more familiar equations, the solution of a differential equation is not a set of *numbers* x that satisfy a function $y = f(x)$, but is instead a set of *functions* $y = f(x)$ that satisfy the differential equation.

EXAMPLE 14.25: A Differential Equation, Part I

For example, one important second-order, ordinary differential equation is $y'' = cy$. It has two solutions: the functions $\cos \omega t$ and $\sin \omega t$, where $c = -\omega^2$.

A **linear** differential equation has no derivatives with exponents. In this special but very important class of differential equations, *the sum of any two solutions to the equation is also a solution.* This property, known as **linearity** or the **superposition principle**, means that if y_1 and y_2 are solutions to a linear differential equation, so is $C_1y_1 + C_2y_2$, where C_1 and C_2 are arbitrary constants.

EXAMPLE 14.26: A Differential Equation, Part II

The differential equation of the previous example, $y'' = cy$, is linear. Therefore, a third solution to this equation is the sum of the first two solutions, namely $C_1 \sin \omega t + C_2 \cos \omega t$.

To verify that a function is a solution to a differential equation, we plug the function *and* its derivatives into the differential equation. Using algebra, we then show that the left side of the equation equals the right side. In order to do this, we often have to solve for C_1 and C_2.

EXERCISE 14.27: Verifying a Differential Equation

Verify that the equation $C_1 \sin \omega t + C_2 \cos \omega t$ is a solution to the differential equation $y'' = cy$, where C_1, C_2, ω and c are arbitrary constants, and the derivative is taken with respect to time.

SOLUTION:

The first thing we do is take the first and second derivatives of our test solution, $C_1 \sin \omega t + C_2 \cos \omega t$, with respect to time:

First derivative: $\omega C_1 \cos \omega t - \omega C_2 \sin \omega t$

Second derivative: $-\omega^2 C_1 \sin \omega t - \omega^2 C_2 \cos \omega t$

Next, we plug the solution and its second derivative into the given differential equation $y'' = cy$:

$$\Rightarrow (-\omega^2 C_1 \sin \omega t - \omega^2 C_2 \cos \omega t) = c(C_1 \sin \omega t + C_2 \cos \omega t)$$

If we factor this equation by $\sin \omega t$ and $\cos \omega t$, we get:

$$\sin \omega t(-\omega^2 C_1 - cC_1) + \cos \omega t(-\omega^2 C_2 - cC_2) = 0$$

For an arbitrary ωt, this equation can only be true *if* the quantities in parentheses are both zero. (That is, any ωt that makes $\sin \omega t$ zero will not make $\cos \omega t$ as well; therefore, the parentheses terms *must* be zero for the equation to be verified). Therefore, we have:

$$C_1(c+\omega^2) = 0 \text{ and } C_2(c+\omega^2) = 0.$$

Notice that in both cases, the equations are true only if $c = -\omega^2$.

14.4.2 SOLVING DIFFERENTIAL EQUATIONS

[Mathematics is] a total disburdening of the mind.
ERNST MACH
Austrian Physicist, 1838 - 1916

To solve a differential equation, in principle you must integrate. In practice, solving differential equations often involves tricky (and not obvious) mathematics unrelated to integration. Furthermore, each type of differential equation requires a different solving method.

We will therefore not attempt to cover the solution of all possible differential equations in this book. Instead, we will cover only the equations that might be seen in an introductory physics class. If you need more information, consult any differential equation textbook.

FIRST ORDER ORDINARY

The simplest type differential equation consists only of a first-order ordinary derivative in the y term. It has the form:

$$y' = \frac{dy}{dx} = \frac{g(x)}{h(y)}$$

Such an equation is solved by **separating variables**: that is, by moving all terms with x's in them (including the differential dx) to one side of the equation, and all terms with y's in it (including the differential dy) to the other side of the equation:

$$g(x)dx = h(y)dy$$

To solve, integrate both sides of the equation normally but separately.

EXERCISE 14.28: Separation of Variables, Part I

Solve the first-order ordinary differential equation $\frac{dy}{dx} = \frac{1}{4y^2}$.

SOLUTION:

First, we put all of the y terms on one side of the equation, and all of the x terms on the other:

$$4y^2 dy = dx$$

Now, we integrate each side normally but separately:

$$\int 4y^2 dy = \int dx$$
$$\tfrac{4}{3}y^3 = x + C$$

Notice that we only need *one* constant of integration here (rather than one for each side of the equation). This is because C is an *arbitrary* number; therefore, it can account for the constant from both sides of the equation. Also notice that while there were no x's in the original differential equation, there *is* an x in the solution. This is because the derivative dy/dx included an x.

Finally, we solve this equation for y in terms of x:

$$y = \sqrt[3]{\tfrac{3}{4}x} + C$$

Another important first order ordinary differential equation has the form:

$$y' = cy$$

The general solution of this differential equation is the exponential function $y = ke^{\pm x}$, where k is an arbitrary constant.

EXERCISE 14.29: Separation of Variables, Part II

Show that the solution of $y' = cy$ is the exponential function $y = ke^{x}$ by using separation of variables.

SOLUTION:

First, we separate variables...

$$\frac{dy}{dx} = cy$$

$$\Rightarrow \frac{dy}{y} = cdx$$

...then we integrate:

$$\int \frac{dy}{y} = \int cdx$$

$$\Rightarrow \ln y = cx + C$$

We want our final answer to have the form $y = f(x)$. Recalling the rules of logarithms from Section 13.4.14, we know that if we raise every term in the equation to the e^{x} power, we will get rid of the natural logarithm on the y term:

In Section 13.4.14.

$$e^{\ln y} = y = e^{cx+C}$$

Using the logarithm law that adding "is" multiplying, we know that:

$$\Rightarrow y = e^{cx} e^{C}$$

But e^{C} is just an arbitrary constant (that is different from C itself). Therefore, we *define* e^{C} as k, so that we have:

$$y = ke^{cx}$$

SECOND ORDER ORDINARY

Another simple type of differential equation has a first and second order ordinary derivative in y:

$$y'' + by' + c = 0$$

Its solution is:

$$y = C_1 y_1 + C_2 y_2 = C_1 e^{k_1 x} + C_2 e^{k_2 x}$$

where C_1 and C_2 are arbitrary constants, and:

$$k_1 = \tfrac{1}{2}\left(-b + \sqrt{b^2 - 4c}\right) \text{ and } k_2 = \tfrac{1}{2}\left(-b - \sqrt{b^2 - 4c}\right)^*$$

Depending upon the value of $b^2 - 4c$, k_1 and k_2 can be:

(1) Both real ($b^2 - 4c > 0$);
(2) Both imaginary ($b^2 - 4c < 0$); or
(3) Equal ($b^2 - 4c = 0$).

If k_1 and k_2 are both imaginary, then the solution to the equation can be written in the above way *or* as:

$$y = e^{\beta x}\left[C_1 \cos \gamma x + C_2 \sin \gamma x\right]$$

where $\beta = -\tfrac{1}{2}b$ and $\gamma = \tfrac{1}{2}\sqrt{4c - b^2}$. If k_1 and k_2 are equal, then the solution to the equation is instead:

$$y = C_1 e^{-\frac{1}{2}bx} + C_2 x e^{-\frac{1}{2}bx}$$

EXERCISE 14.30: Second-Order Differential

Solve the differential equation $y'' + 3y' - 10y = 0$.

SOLUTION:

To see whether our k values are real, imaginary, or equal, we first calculate $b^2 - 4c$, where b is 3 and c is -10:

$$b^2 - 4c = 9 - 4(-10) = 49$$

Since this number is greater than zero, the solutions to the differential equation are real. So, now we can calculate the k's:

$$\begin{aligned} k_1 &= \tfrac{1}{2}\left(-b + \sqrt{b^2 - 4c}\right) \\ &= \tfrac{1}{2}(-3 + \sqrt{9 - 4(-10)}) \\ &= \tfrac{1}{2}(-3 + \sqrt{49}) \\ &= \tfrac{1}{2}(4) \\ &= 2 \end{aligned} \qquad \begin{aligned} k_2 &= \tfrac{1}{2}\left(-b - \sqrt{b^2 - 4c}\right) \\ &= \tfrac{1}{2}(-3 - \sqrt{9 - 4(-10)}) \\ &= \tfrac{1}{2}(-3 - 7) \\ &= \tfrac{1}{2}(-10) \\ &= -5 \end{aligned}$$

Now that we have the k's, we can plug into the main equation:

$$\Rightarrow \; y = C_1 e^{k_1 x} + C_2 e^{k_2 x} = C_1 e^{2x} + C_2 e^{-5x}$$

More on simple harmonic motion in Section 9.3.

The case where k_1 and k_2 are both imaginary is a very important one in physics. It corresponds to naturally occurring, regular (simple harmonic) oscillations, such as those exhibited by pendulums.

* That these solutions look a lot like the solutions for the quadratic equation (Section 13.4.8) is not a coincidence; see any differential equation text for an explanation.

EXAMPLE 14.31: Simple Harmonic Motion

This is the case of second-order, ordinary differential equation in Example 14.25, $y'' = cy$, where $c = -\omega^2$.

PARTIAL DIFFERENTIAL EQUATIONS

Knowing how to solve partial differential equations is not necessary for an introductory physics class, but you should know that they exist. You might see at least one example of a second-order partial linear differential equation in your course: the **wave equation**:

For more on the wave equation, go to Section 7.3.2.

$$\frac{\partial^2 y}{\partial x^2} = \frac{1}{v^2}\frac{\partial^2 y}{\partial t^2}$$

The solution to this equation is an oscillatory function of both x and t:

$$y = y_o \sin(kx \pm \omega t)$$

where y_o is a constant that indicates the height, or **amplitude**, of the wave; k and ω are constants that are related to the **frequency** of the wave.

14.5 CONCLUSION

To live happily, one must be willing to accept change.
LATIN PROVERB

...And to live happily in a physics course, you must accept the importance of calculus. Make sure that you feel very comfortable with the different aspects of calculus, as described in this chapter. Begin this process as early as possible. The time spent doing this will pay off later, when physics gets more complex.

15 VECTORS

15.1 INTRODUCTION

All of the measurable physical quantities an introductory physics class can be classed into just two broad categories: those representable by *one* number—like time t—and those representable by *more than one* number—such as two-dimensional position (x, y). We call former kinds of quantities **scalars**, while the latter are known as **vectors**.*

Vectors were invented because by using them the laws of physics can be stated in simple forms.
DONALD G. IVEY
American Engineer, 1935 -

Scalars are just ordinary numbers, manipulated via the familiar math of Chapters 11 through 14. Vectors, by contrast, are *new* mathematical objects with their own special mathematics. So, the subject of this chapter is vectors and vector math.

Because they are distinct mathematical objects, *scalars and vectors can never equal each other.* To keep them separate, then, we use different notation for each: scalars are represented by single symbols written in italics (for example, time t), while vectors are represented by set notation (x, y), boldface symbols ($\mathbf{r} = (x, y)$), or symbols topped with half-arrows ($\vec{\mathbf{r}}$).

Most textbooks use the boldface notation; however, when writing vectors by hand, you should use the arrow. So, to get you in the habit of writing vectors in this way, we use the half-arrow notation in this book.

* Be careful! The individual numbers making up vectors are themselves scalars. It is the *set* of numbers—taken as a whole—that is the vector.

15.2 A GRAPHICAL INTERPRETATION OF VECTORS

The easiest way to understand vectors is to *visualize* them as arrows.

Figure 15.1

15.2.1 VISUALIZING VECTORS

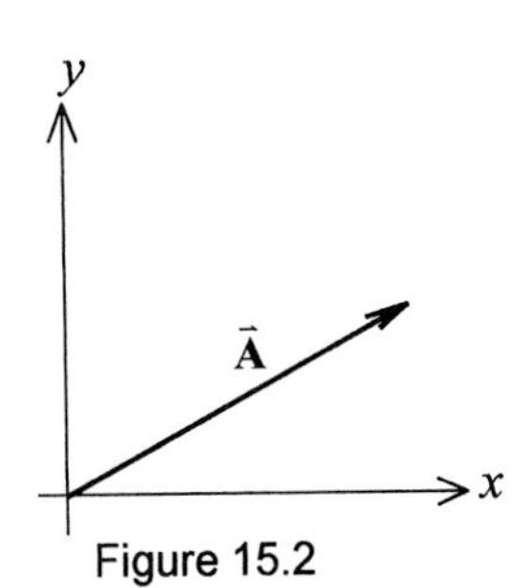

Figure 15.2

Figure 15.1 shows a two-dimensional, arbitrary vector $\vec{\mathbf{A}}$, drawn as an arrow. The length of the arrow (as measured by a ruler) is proportional to the **magnitude** of the vector, symbolized by A or $|A|$. The angle the arrow subtends from a horizontal line (as measured by a protractor) is the **direction** of the vector, symbolized by θ. *The magnitude of a vector is always a positive number,* but the vector's direction can be either positive or negative (depending on whether the angle is measured counterclockwise or clockwise, respectively, from the line).

Vectors can be visualized on coordinate axes, as in Figure 15.2. But the vector itself is *independent* of coordinate system. That is, a vector's magnitude and direction *never change*, even if the coordinate system changes. Therefore, because the position of the vector relative to the coordinate system is irrelevant, *as long as you do not change the vector's magnitude or direction, you may move a vector around at will*—or equivalently, *you may put the vector in different coordinate systems.*

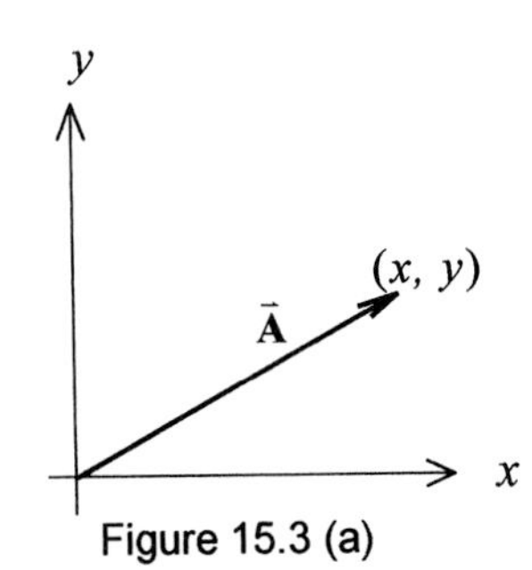

Figure 15.3 (a)

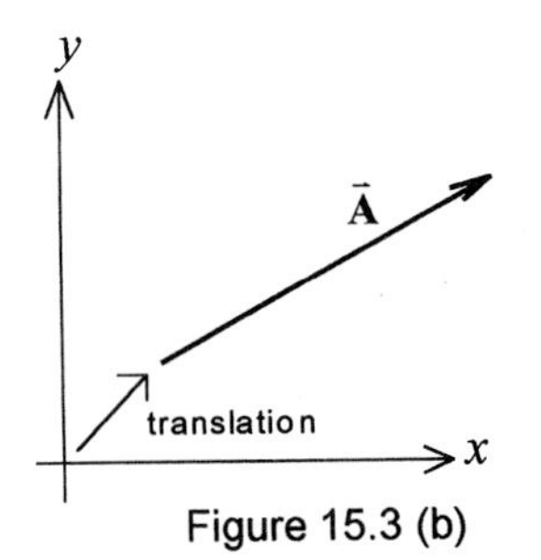

Figure 15.3 (b)

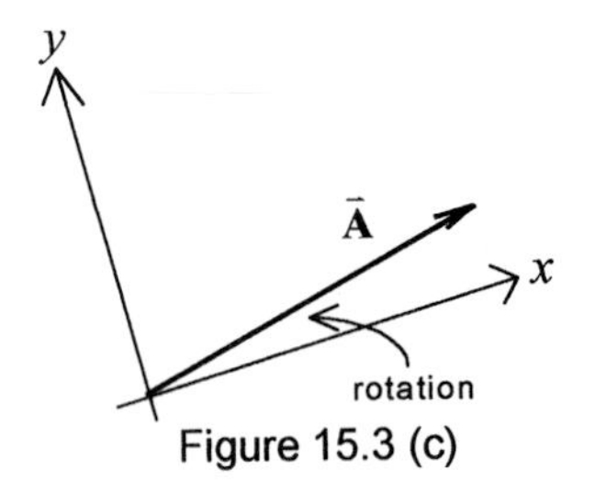

Figure 15.3 (c)

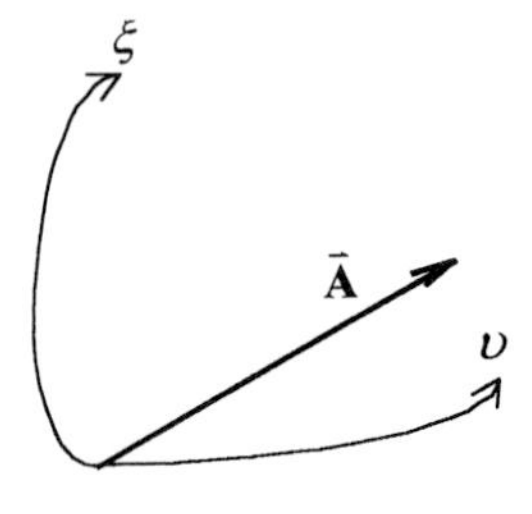

Figure 15.3 (d)

EXAMPLE 15.1: Coordinate Systems and Vectors

Consider an arbitrary vector $\vec{\mathbf{A}}$. We can place this vector in any number of coordinate systems.

For example, we could choose an ordinary Cartesian system, such that the base of the vector is at the origin, and the tip of the vector is at the point (x, y) (Fig. 15.3 (a)). This is the simplest and best choice of coordinate system, since the set description of the vector's position (x, y) parallels the actual location of the tip of the vector in the grid.

However, it is not *necessary* that we choose this kind of system. We could instead choose a coordinate system that is *translated* away from the vector (Fig. 15.3 (b)). Or, we could choose a system that is *rotated* around the vector (Fig. 15.3 (c)). We could even put the vector in some crazy non-Cartesian coordinate system (Fig. 15.3 (d)). No matter what coordinate system we choose—even none at all!—however, the vector's magnitude and direction will remain the same.

Once we understand that vectors are just arrows, the following vector definitions are straightforward:

- Any two vectors with the *same* magnitude *and* direction are **equal** (Fig. 15.4 (a)).

- A vector $-\vec{A}$ is **negative** if it has the same magnitude but opposite direction as $\vec{A}$ (Fig. 15.4 (b)).
- Vectors that point in the same direction are said to be **parallel** (Fig. 15.4 (c)); vectors that point in opposite directions are **anti-parallel** (Fig. 15.4 (d)).
- Vectors that point **into** the page are symbolized by ⊗ (think of the feathers at the back of an arrow, viewed just as the vector disappears *into* the page); vectors that point **out of** the page are symbolized by ⊙ (think of the very tip of the arrow's point as the vector emerges *out of* the page).

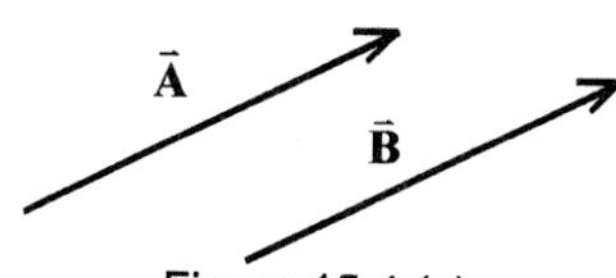

Figure 15.4 (a)

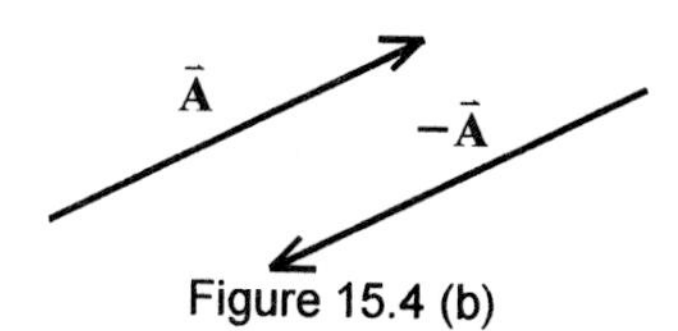

Figure 15.4 (b)

Figure 15.4 (c)

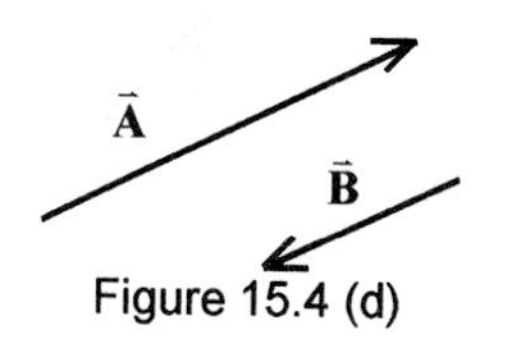

Figure 15.4 (d)

15.2.2 ADDING & SUBTRACTING VECTORS GRAPHICALLY

We can use the convenient, graphical nature of vectors to help us make simple vector sums. There are two ways to do this. The first method—good for adding two vectors at a time, like those in Figure 15.5 (a)—is the **parallelogram** or **tail-to-tail method**:

1. Place the tails of vectors $\vec{A}$ and $\vec{B}$ together, keeping the original direction and magnitudes of the vectors unchanged, such that two sides of a parallelogram are created (Fig. 15.5 (b)).
2. Sketch in the other two sides of the parallelogram (Fig. 15.5 (c)).
3. Draw a vector, known as the **resultant vector** or **resultant**, from the intersection of the vector tails to the intersection of the two new sides (Fig. 15.5 (d)).
4. Measure the length of the resultant vector with a ruler to obtain its magnitude, and measure the angle the resultant makes with the x-axis with a protractor to find its direction.

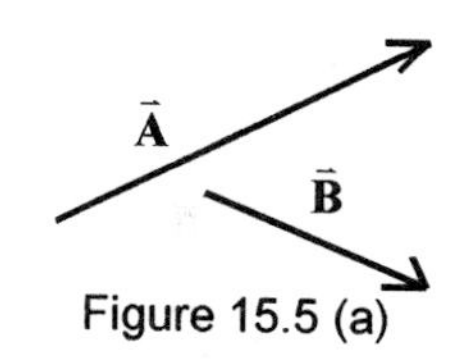

Figure 15.5 (a)

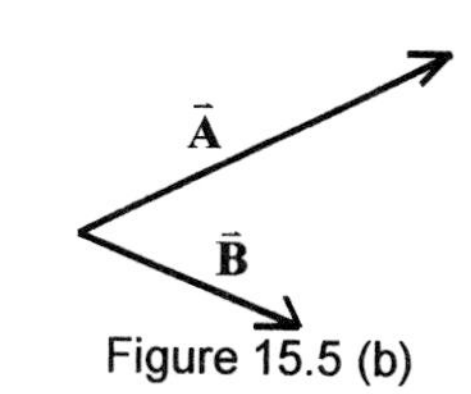

Figure 15.5 (b)

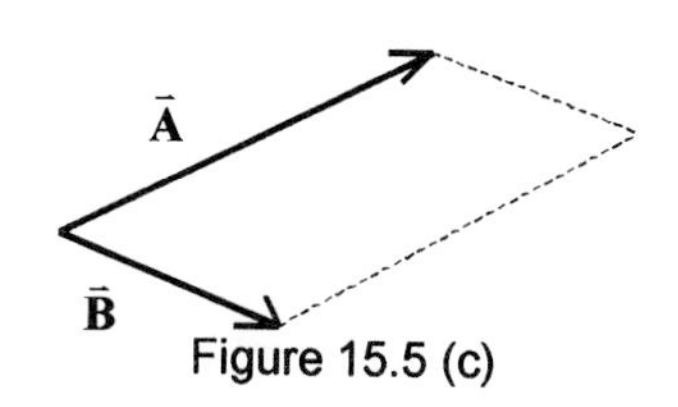

Figure 15.5 (c)

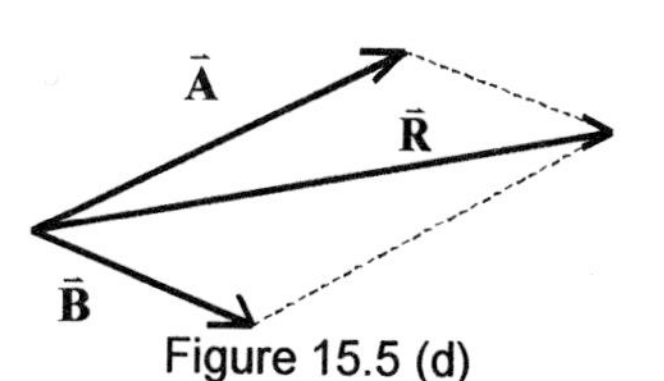

Figure 15.5 (d)

EXERCISE 15.2: Adding Vectors Graphically, Part I

Vectors $\vec{A}$ and $\vec{B}$ are sketched below. Add these two vectors using the parallelogram method.

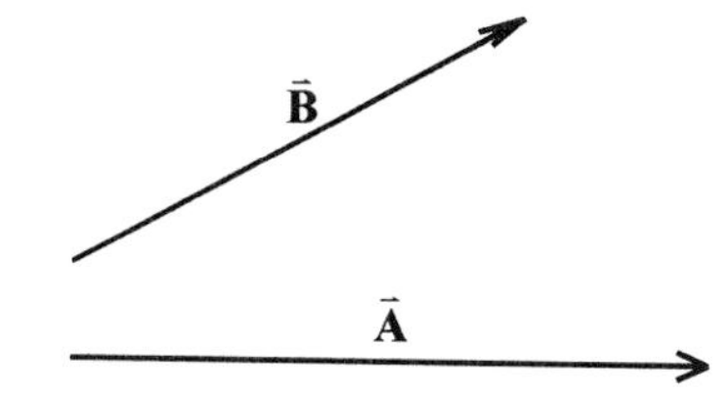

SOLUTION:

First, we place the two vectors tail-to-tail, creating a parallelogram (Steps 1 and 2):

Exercise 15.2 Continued...

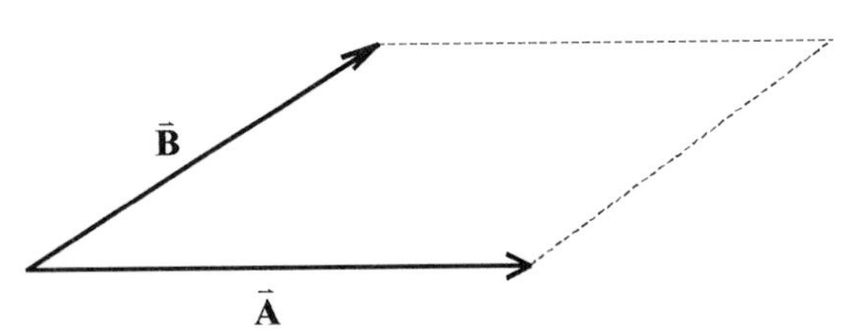

Now, we draw the resultant vector $\vec{\mathbf{R}}$ (Step 3):

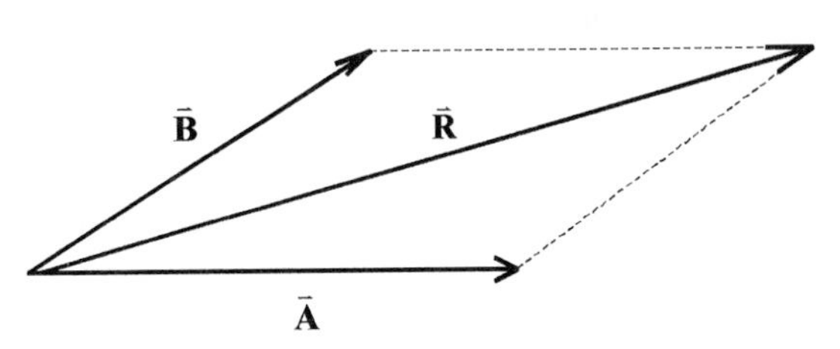

The magnitude and direction of the resultant is then measured with a ruler and a protractor (Step 4).

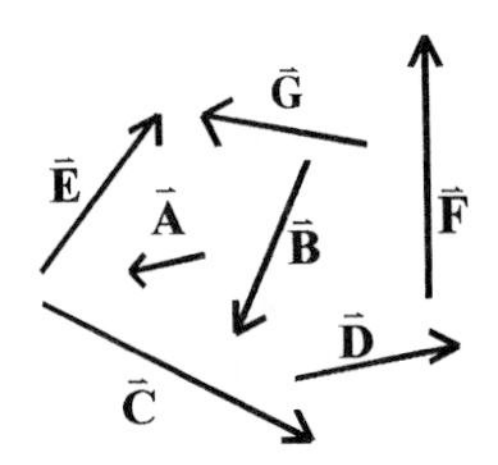

Figure 15.6 (a)

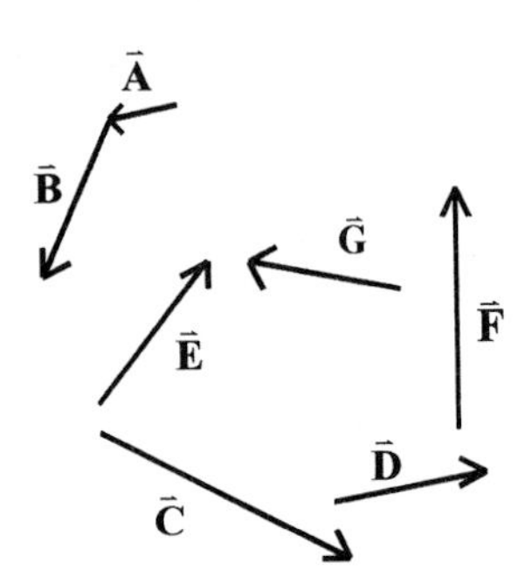

Figure 15.6 (b)

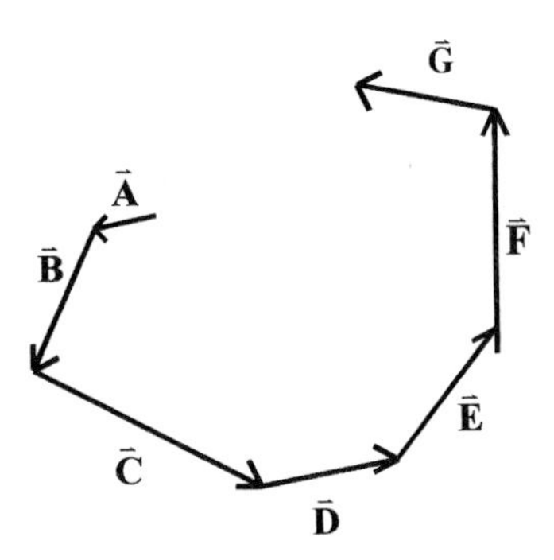

Figure 15.6 (c)

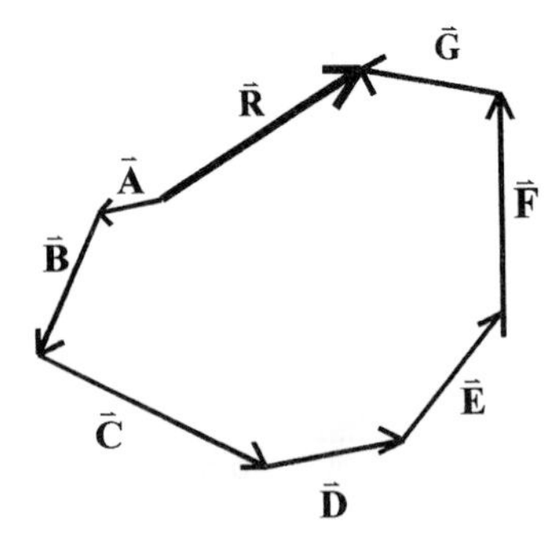

Figure 15.6 (d)

The second method for adding vectors graphically is the **tip-to-tail method**. This method works for adding two *or more* vectors, like those in Figure 15.6 (a).

1. Place the tail of the second vector at the tip of the first, keeping the direction and magnitudes of the vectors unchanged (Fig. 15.6 (b)).
2. Repeat Step 1 for each additional vector (Fig. 15.6 (c)).
3. Draw the resultant vector from the *tail* of the first vector to the *tip* of the last (Fig. 15.6 (d)).
4. Measure the length of the resultant vector with a ruler to obtain its magnitude, and measure the angle the resultant makes with the x-axis with a protractor to find its direction.

EXERCISE 15.3: Adding Vectors Graphically, Part II

Add the vectors $\vec{\mathbf{A}}$ and $\vec{\mathbf{B}}$ from Exercise 15.2 using the tip-to-tail method.

SOLUTION:

First, we place the two vectors tip-to-tail (Step 1):

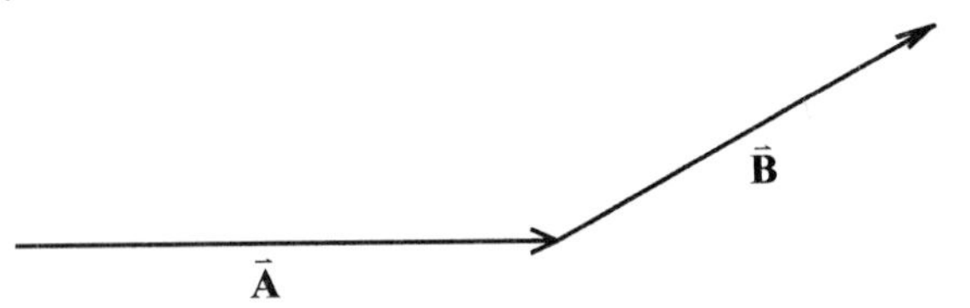

Exercise 15.3 Continued...

We can skip Step 2 because there are only two vectors. Then, (Step 3) we draw the resultant vector $\vec{\mathbf{R}}$ from the base of $\vec{\mathbf{A}}$ to the tip of $\vec{\mathbf{B}}$:

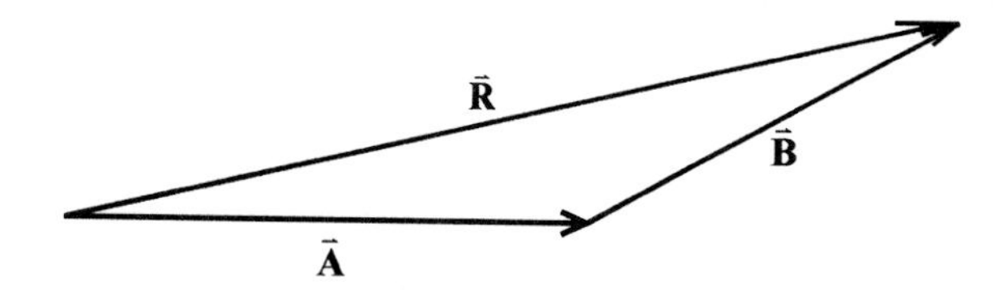

We now measure the length and angle with ruler and protractor (Step 4). Notice that this $\vec{\mathbf{R}}$ is identical to that of the previous exercise.

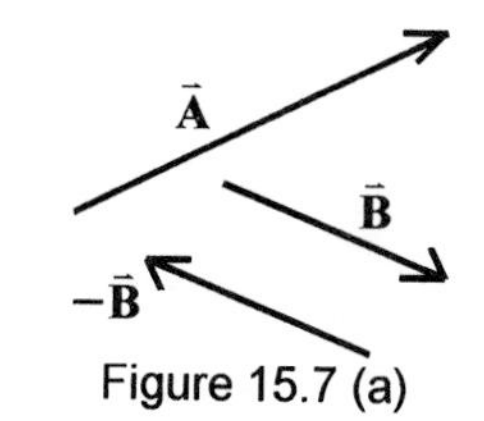

Figure 15.7 (a)

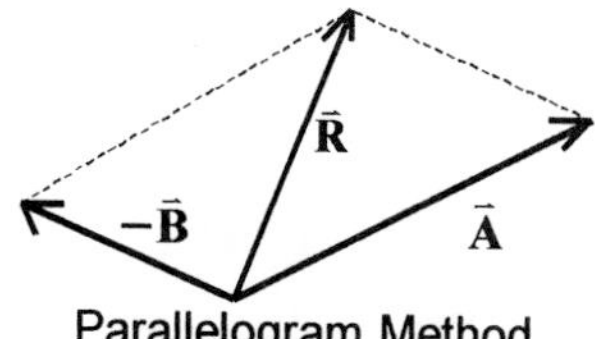

Parallelogram Method
Figure 15.7 (b)

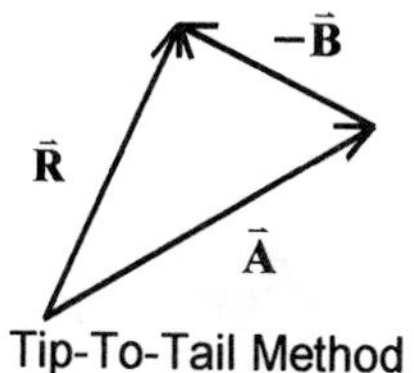

Tip-To-Tail Method
Figure 15.7 (c)

Vector subtraction is identical to vector addition, except that negative vectors are used: *i.e.*, $\vec{\mathbf{A}} - \vec{\mathbf{B}} = \vec{\mathbf{A}} + (-\vec{\mathbf{B}})$ (Fig. 15.7 (a) through (c)).

EXERCISE 15.4: Subtracting Vectors Graphically

Subtract the vectors $\vec{\mathbf{A}}$ and $\vec{\mathbf{B}}$ from Exercise 15.2 to find $\vec{\mathbf{A}} - \vec{\mathbf{B}}$.

SOLUTION:

First, we have to reverse the direction of $\vec{\mathbf{B}}$ to get $-\vec{\mathbf{B}}$:

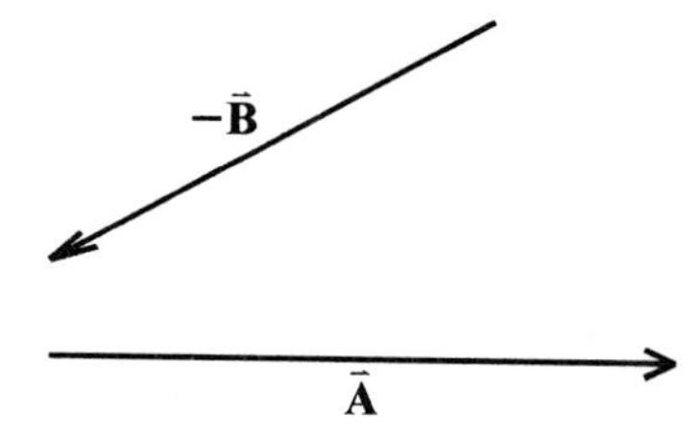

Next, we can find the sum $\vec{\mathbf{A}} + -(\vec{\mathbf{B}})$ by either of the two methods. Let's do it both ways. First, the parallelogram method:

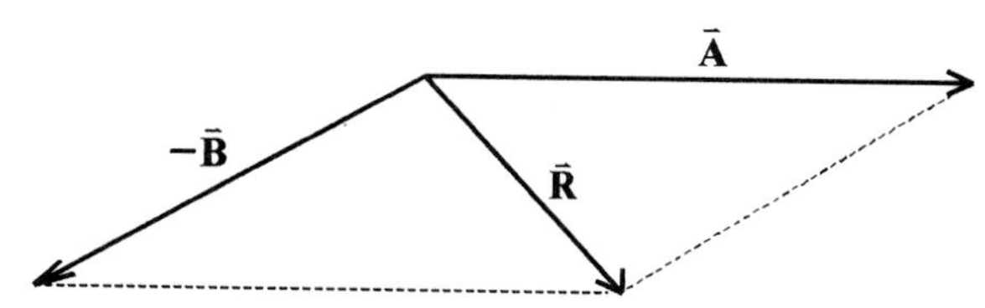

Now, the tip-to-tail method:

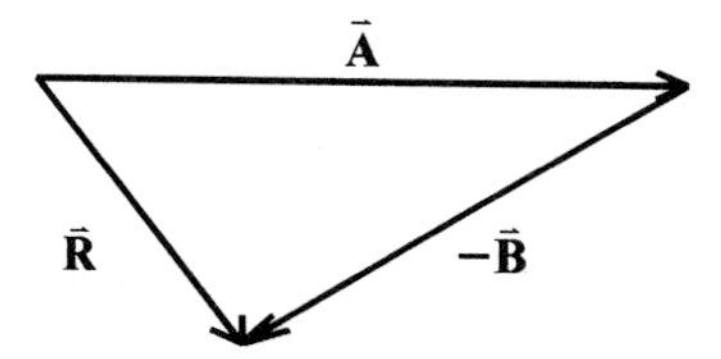

Again, notice that the $\vec{\mathbf{R}}$'s are identical to each other.

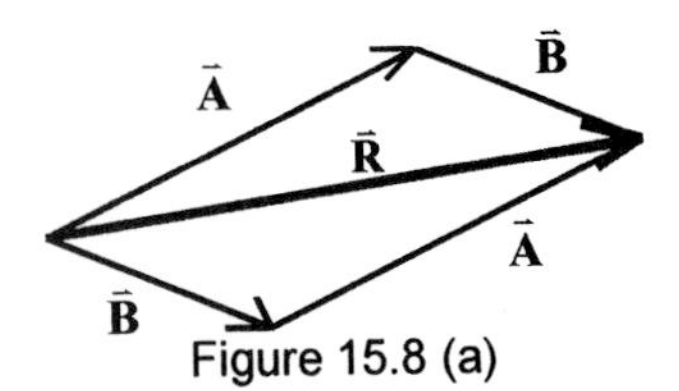

Figure 15.8 (a)

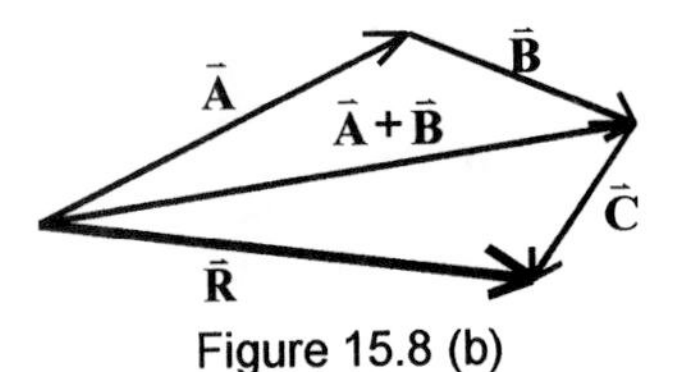

Figure 15.8 (b)

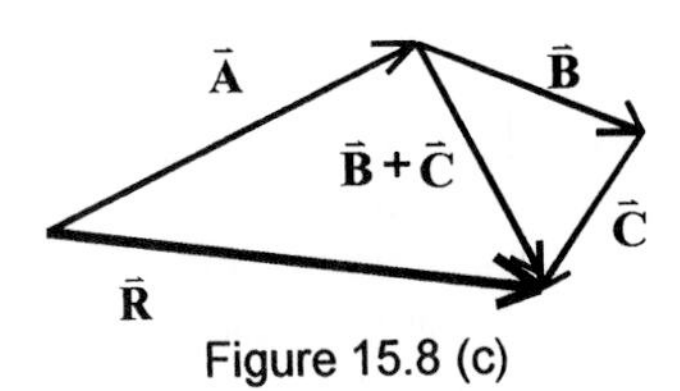

Figure 15.8 (c)

15.2.3 ORDER OF VECTOR ADDITION

Two straightforward laws for vector addition can be proved using the parallelogram and tip-to-tail methods. The first is the **commutative law of addition**. Both branches of the parallelogram in Figure 15.8 (a) give the same additive result, proving:

$$\vec{A}+\vec{B}=\vec{B}+\vec{A}$$

The second is the **associative law of addition**. Figures 15.8 (b) and (c) show that combining $\vec{A}$ and $\vec{B}$, followed by $\vec{C}$, gives the same resultant vector as combining $\vec{B}$ and $\vec{C}$, followed by $\vec{A}$, proving:

$$(\vec{A}+\vec{B})+\vec{C}=\vec{A}+(\vec{B}+\vec{C})$$

These two laws of vector addition, taken together, tell us that *the order in which vectors are added is irrelevant.* Therefore, *it always acceptable to rearrange the order of vector addition* if it will help you solve a problem more easily.

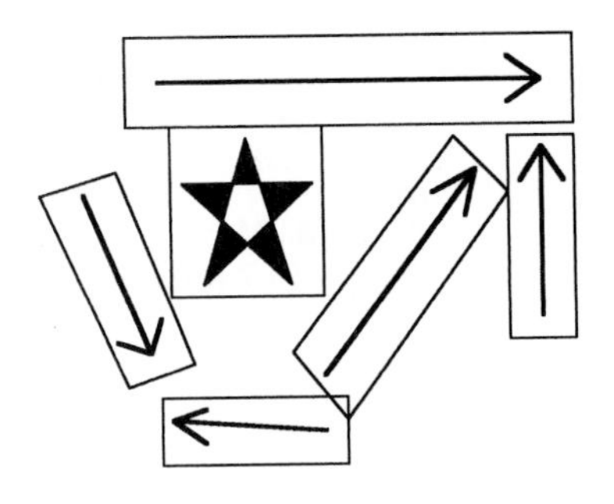
Figure 15.9

EXERCISE 15.5: Adding Vectors: A Treasure Map!

A treasure hunter, after many long years of searching in libraries, antique shops, and museums, has collected six treasure map fragments, as sketched in Figure 15.9. The treasure hunter knows that she is supposed to begin her search at the star, the real world location of which she has discovered in her researches. Furthermore, she knows that these six fragments, in the exact orientations and lengths sketched, make up the entire map. What she doesn't know is which way the pieces should be ordered so that she locates the treasure. Can you help her?

SOLUTION:

The answer is: it doesn't matter in which way the pieces are ordered! As long as she begins at the star, the treasure hunter can take the pieces in *any* order and get to the treasure. This is because the map arrows are vectors, and, as long as their directions and magnitudes are not changed, vectors can be added in any order:

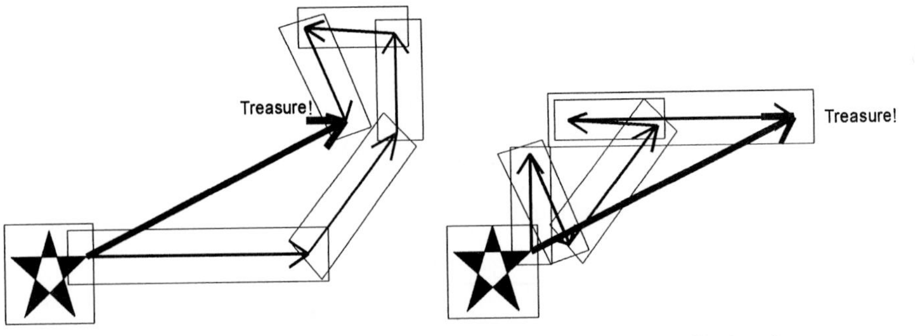

Vectors can help our treasure hunter in another way. If she does a little graphical vector addition while she is still in the library, there is no need for her to actually take each step indicated on the map. Instead, she merely needs to go straight from the star to the final destination (along the resultant vector)!

15.3 VECTOR COMPONENTS

Even though graphical techniques for quickly adding and subtracting vectors are immensely useful and easy to understand, they cannot help us *multiply* vectors—something we will need to do often in introductory physics. So, a non-graphical description of vectors is also needed. Such a technique is the method of components.

15.3.1 COMPONENTS

Vector components are sets of numbers that describe a vector *within a certain coordinate system.* Different coordinate systems yield different components for the same vector. Each component in the set is the **projection** of the vector on a different coordinate axis.

Components are most easily understood in Cartesian coordinate systems, but they are useful in other systems as well.

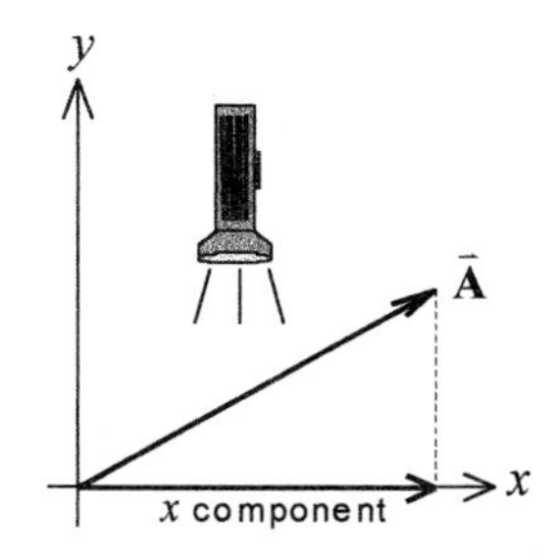

Figure 15.10 (a)

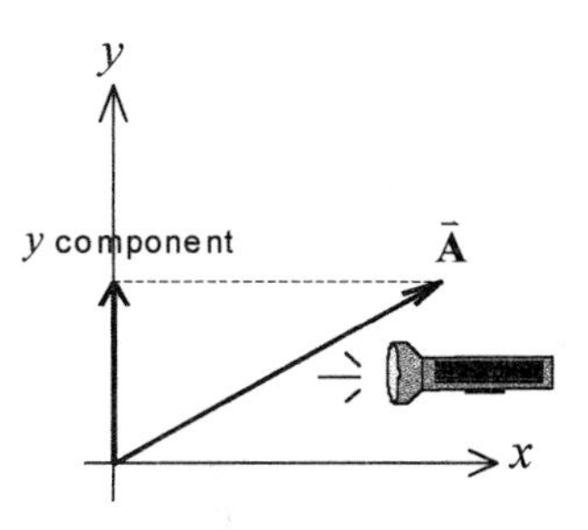

Figure 15.10 (b)

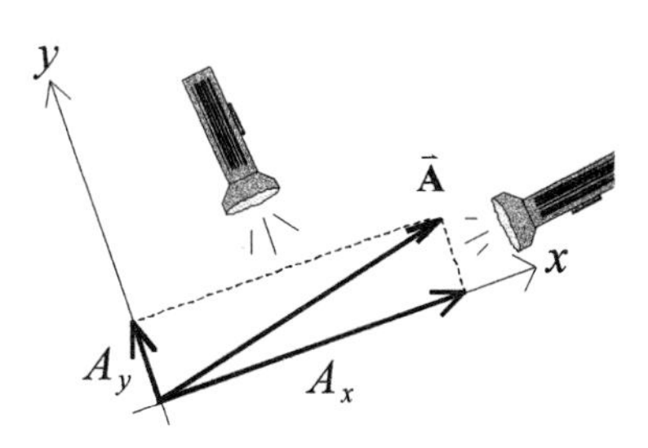

Figure 15.10 (c)

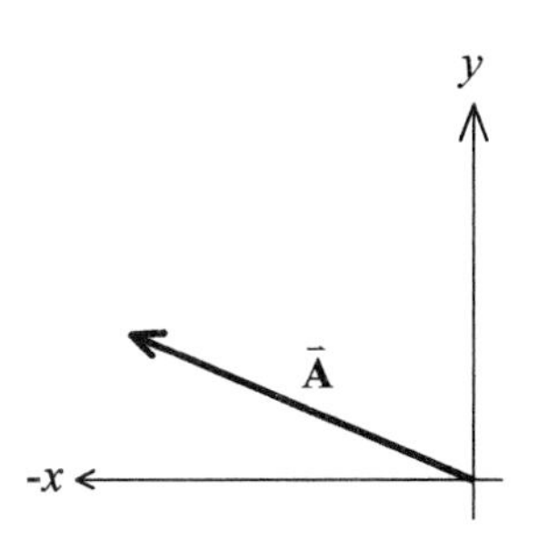

Figure 15.10 (d)

EXAMPLE 15.6: Understanding Vector Components

To visualize vector components, we might imagine a light shining down on a vector from above the x-axis. (Fig. 15.10 (a)). The "shadow" of the vector on the x-axis is the *x-component* A_x of the vector.

In the same way, a light shining on the vector from the right of the y-axis creates a "shadow" on the y-axis that is the *y-component* A_y of the vector (Fig. 15.10 (b)).

And if the coordinate system changes, the vector components also change, *even though* the vector itself remains the same (Fig.15.10 (c)).

How do we find components on vectors in other quadrants? For example, how would we find the y-component of $\vec{\mathbf{A}}$ if that vector were in the second quadrant (Fig. 15.10 (d))? Hint: Try moving the flashlights.

EXERCISE 15.7: Drawing Vector Components

Sketch components of the vectors in Example 15.1, Figures (a) - (c).

SOLUTION:

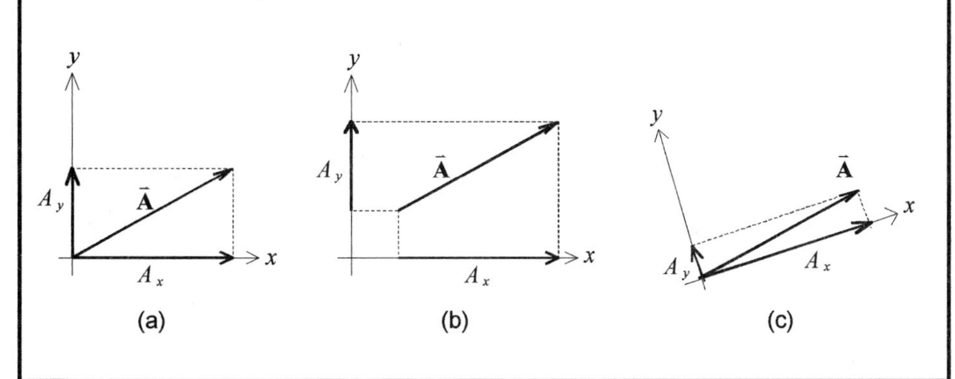

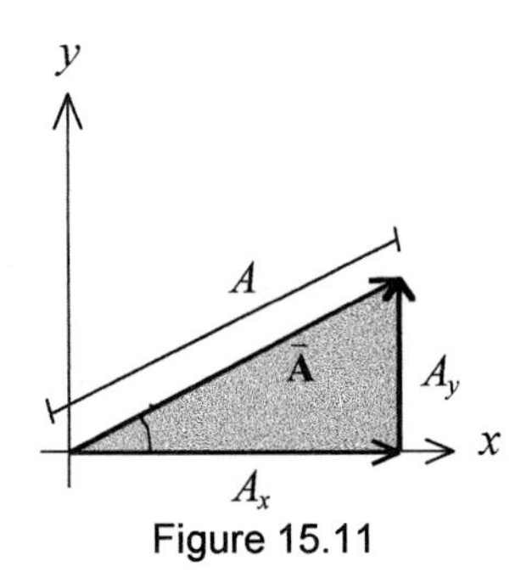

Figure 15.11

Quad.:	I	II	III	IV
A_x	+	–	–	+
A_y	+	+	–	–

Figure 15.12

15.3.2 RESOLVING THE VECTOR

The mathematical process of calculating *numerical* values for the components is called **resolving the vector.**

We begin with an arbitrary vector $\vec{\mathbf{A}}$ and its vector components (in Cartesian coordinates), A_x and A_y, as sketched in Figure 15.11.* Using trigonometry (Section 13.6), the magnitudes of the components can be calculated in terms of the magnitude and direction of the vector:

$$A_x = A\cos\theta \qquad A_y = A\sin\theta$$

Notice that depending on the value of θ, A_x and A_y can be positive, negative, or zero. The correct signs for each quadrant are given in Figure 15.12.

Rewriting the above equations yields relations for the magnitude and direction of the vector in terms of the components:

$$A = \sqrt{A_x^2 + A_y^2} \qquad \theta = \tan^{-1}\left(\frac{A_y}{A_x}\right)$$

WARNING!

The equations for A_x, A_y, A, and θ above *only* apply to a coordinate system defined as in Figure 13.4. If the x-y axes are rotated, or if θ has another definition, the relationships can *change*. Therefore, do *not* memorize these equations, but instead know how to *derive* similar equations for each new case you encounter.

To obtain equations appropriate to your situation, sketch a triangle similar to Figure 13.4. Then apply trigonometry to the sketch. The equations obtained will be similar in form to the ones given above (*i.e.*: they will have sines, cosines, etc.), but may not be *exactly* the same.

WARNING!

When using a calculator to figure the inverse tangent function, be aware that the calculator displays the *smallest* correct angle. For example, the inverse tangent of 1.5 is *both* 0.98 rads and 4.1 rads, but the calculator only displays 0.98 rads.

The displayed angle may not be the appropriate answer for the problem at hand. Therefore, *always* check your answer for plausibility when using the inverse tangent function. Use the quadrant grid to help. If you find that the larger angle is the correct one for your situation, add π rads to the smaller angle obtained by the calculator.

* Notice that the y-component has moved to the right in this figure! Why can we do this?

EXERCISE 15.8: Vector Components, Part I

A certain vector $\vec{\mathbf{A}}$ has magnitude $A = 10$ m and direction $\theta = \frac{\pi}{6}$ rads. Find the components A_x and A_y. This vector is sketched in Figure 15.13.

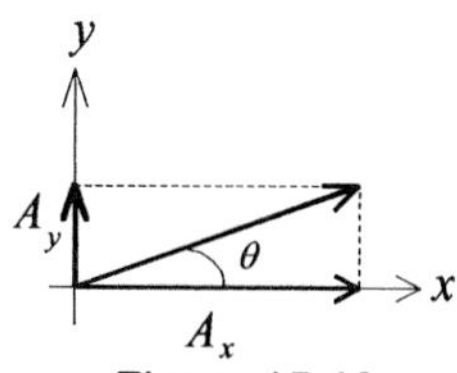

Figure 15.13

SOLUTION:

Figure 15.13 has the same orientation as Figure 15.11 above, so we can use the above formulas verbatim:

$$A_x = A\cos\theta = (10\text{ m})\cos\tfrac{\pi}{6} \approx 8.7\text{ m}$$
$$A_y = A\sin\theta = (10\text{ m})\sin\tfrac{\pi}{6} = 5\text{ m}$$

EXERCISE 15.9: Vector Components, Part II

Find the components A_x and A_y of a vector $\vec{\mathbf{A}}$ with magnitude $A = 10$ m and direction $\theta = \frac{7\pi}{6}$ rads, as sketched in Figure 15.14.

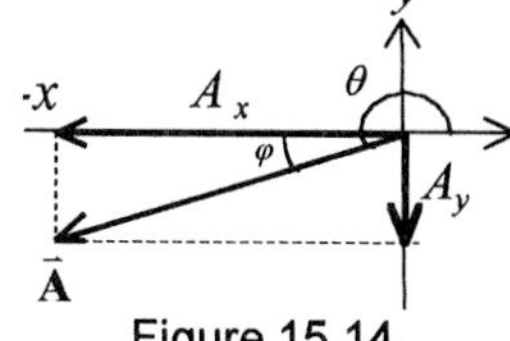

Figure 15.14

SOLUTION:

This figure does *not* have the same orientation as Figure 15.11 above, so we need to develop a new set of formulas just for this situation. We start by defining an angle φ, as sketched in Figure 15.14, where $\theta = \pi + \varphi$. With this new angle, we now have a right triangle made up of components A_x and A_y and the vector $\vec{\mathbf{A}}$:

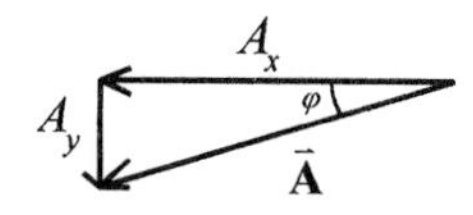

Using trigonometry on this triangle, we get $A_x = A\cos\varphi$ and $A_y = A\sin\varphi$:

$$A_x = A\cos\varphi = (10\text{ m})\cos\tfrac{\pi}{6} \approx 8.7\text{ m}$$
$$A_y = A\sin\varphi = (10\text{ m})\sin\tfrac{\pi}{6} = 5\text{ m}$$

Trigonometry is covered in Section 13.6.

More on quadrants in Section 13.2.2.

However, before our answer is complete, we must take into account the quadrant in which the vector lies. From the above figure, we see that the vector is in the third quadrant. According to Figure 15.12, our answers are then:

$$A_x = -8.7\text{ m} \text{ and } A_y = -5\text{ m}$$

EXERCISE 15.10: Vector Components, Part III

A vector has components $A_x = -3$ m and $A_y = 7$ m. Find its magnitude and direction.

SOLUTION:

Using the above equation $A = \sqrt{A_x^2 + A_y^2}$ for the vector magnitude in terms of its components, we can calculate the magnitude as:

$$A = \sqrt{(-3\text{ m})^2 + (7\text{ m})^2} = \sqrt{58\text{ m}^2} \approx 7.6\text{ m}$$

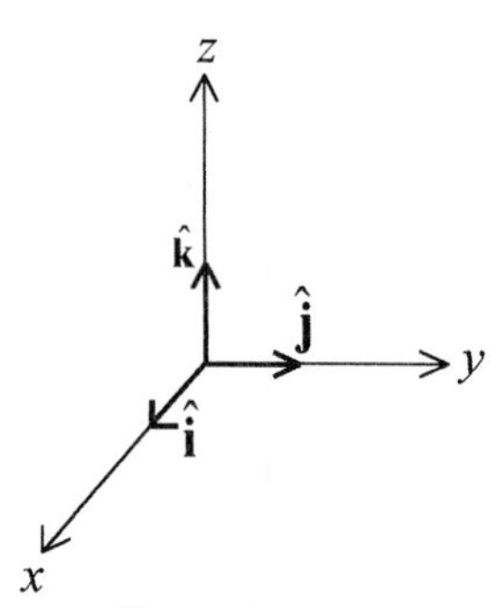

Figure 15.15 (a)

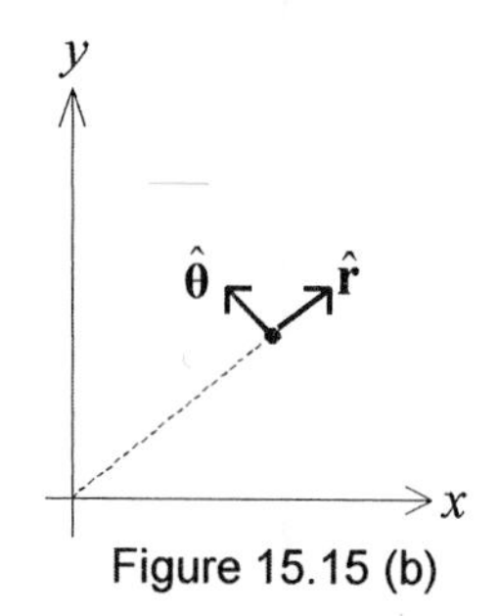

Figure 15.15 (b)

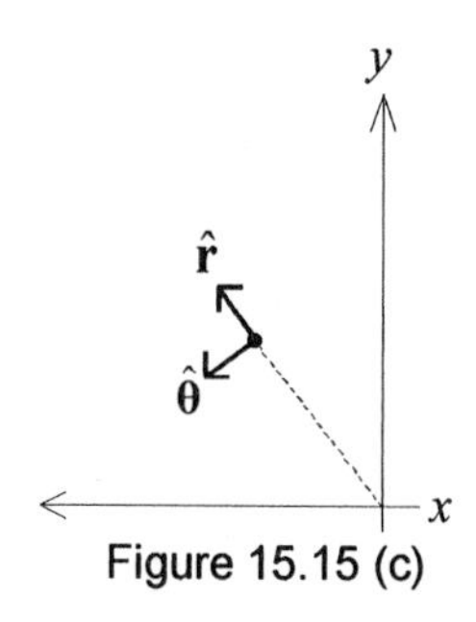

Figure 15.15 (c)

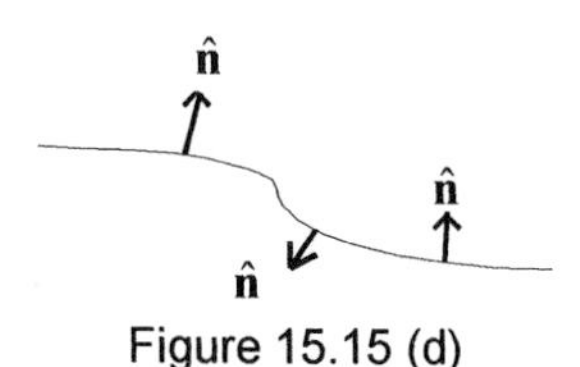

Figure 15.15 (d)

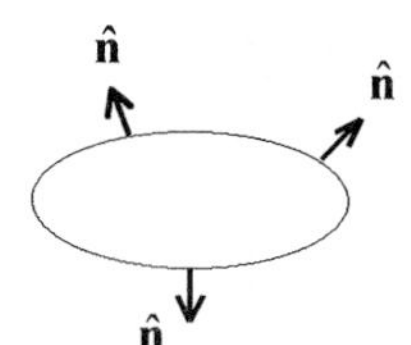

Figure 15.15 (e)

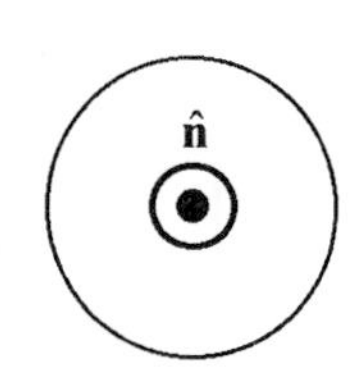

Figure 15.15 (f)

Exercise 15.10 Continued...

Next, to find the direction of the vector, we use our equation:

$$\theta = \tan^{-1}\left(\frac{A_y}{A_x}\right) = \tan^{-1}\left(\frac{7\text{ m}}{-3\text{ m}}\right) \approx -1.16\text{ rads}$$

Now, our vector $(-3\text{ m}, 7\text{ m})$ lies in the II quadrant (check it!). But the value given by the equation above lies in the IV quadrant (check this, too!). Therefore, the inverse tangent gave us the *smallest* angle, not the actual angle we desire. So, we have to adjust our answer by adding π to it:

$$\Rightarrow \theta = \pi + -1.16 = 1.98\text{ rads} \approx 2.0\text{ rads}$$

15.3.3 UNIT VECTORS

A **unit vector** is a vector with a magnitude of one, and a direction along a coordinate axis. As such, unit vectors are objects that define *directions*. Unit vectors are usually written in bold face (**i**) or as symbols topped by a caret ($\hat{\mathbf{i}}$). The latter symbol, used in this book and strongly suggested for your own use, is read, "i hat."

TYPES OF UNIT VECTORS

The following types of unit vectors are important for introductory physics:

- **Cartesian unit vectors**, written $\hat{\mathbf{i}}$, $\hat{\mathbf{j}}$, and $\hat{\mathbf{k}}$, all intersect at the origin, and are all mutually perpendicular (Fig. 15.15 (a)).* These unit vectors never change position.
- **Polar unit vectors**, labeled $\hat{\mathbf{r}}$ and $\hat{\boldsymbol{\theta}}$, intersect perpendicularly at the location being described (Fig. 15.15 (b)). $\hat{\mathbf{r}}$ points directly outwards from the point, along a line connecting the origin to the point, while $\hat{\boldsymbol{\theta}}$ points $\frac{\pi}{2}$ rads counter-clockwise (in the direction of increasing θ) from $\hat{\mathbf{r}}$. Unlike the definitions of Cartesian unit vectors, therefore, the directions of polar unit vectors $\hat{\mathbf{r}}$ and $\hat{\boldsymbol{\theta}}$ *change* depending upon the location being described (Fig. 15.15 (c))!
- **Normal unit vectors**, labeled $\hat{\mathbf{n}}$, point in the direction *perpendicular*, or **normal**, to a surface *at the point of contact*. If the surface is *open*, like the surface sketched in Figure 15.15 (d), normal vectors point in either of the two perpendicular directions. If the surface is *closed*, like the surface of Figure 15.15 (e), then normal vectors point outwards from the surface. In the special case of planar figures (Section 13.5.3), the normal vector is directed *perpendicular* to the plane of the figure (Figure 15.15 (f)).

---

* Sometimes these unit vectors are written (perhaps more logically) as $\hat{\mathbf{x}}$, $\hat{\mathbf{y}}$, and $\hat{\mathbf{z}}$.

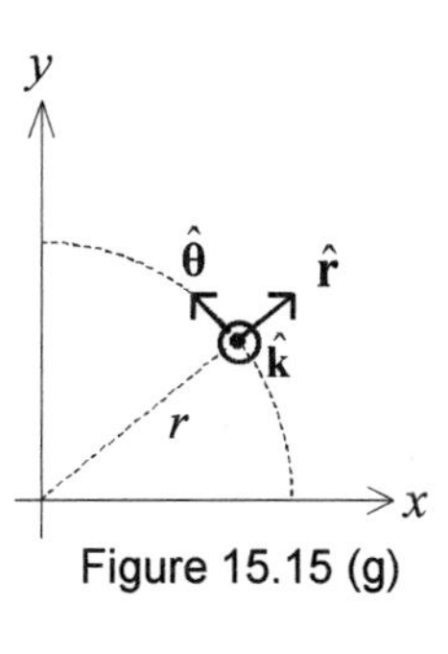

Figure 15.15 (g)

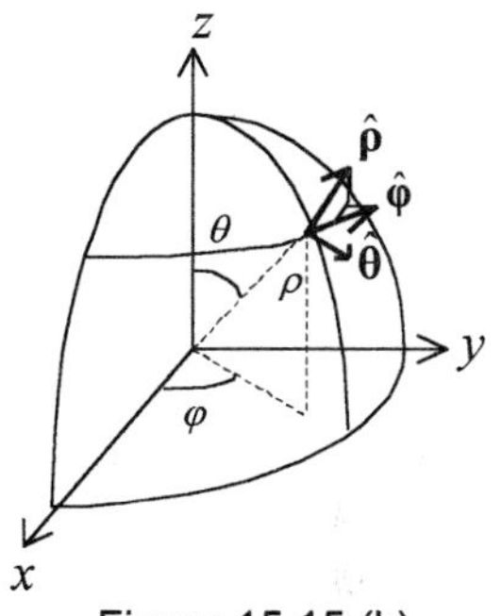

Figure 15.15 (h)

- **Cylindrical unit vectors**, notated $\hat{\mathbf{r}}$, $\hat{\boldsymbol{\theta}}$, and $\hat{\mathbf{k}}$, are made up of the polar unit vectors $\hat{\mathbf{r}}$ and $\hat{\boldsymbol{\theta}}$, and the Cartesian unit vector $\hat{\mathbf{k}}$, such that all three vectors intersect perpendicularly at the location being described (Fig. 15.15 (g)). These unit vectors, like the polar unit vectors above, vary with location.
- **Spherical unit vectors**, labeled $\hat{\boldsymbol{\rho}}$, $\hat{\boldsymbol{\varphi}}$, and $\hat{\boldsymbol{\theta}}$, all intersect perpendicularly at the location being described (Fig. 15.15 (h)). $\hat{\boldsymbol{\rho}}$ points directly outwards from the point, as if along a line drawn from the origin to the point, while $\hat{\boldsymbol{\theta}}$ points $\frac{\pi}{2}$ rads downwards from $\hat{\boldsymbol{\rho}}$, along an imaginary line tangent to a circle with radius ρ. $\hat{\boldsymbol{\varphi}}$ points $\frac{\pi}{2}$ rads counter-clockwise from $\hat{\boldsymbol{\rho}}$, parallel to the x-y plane. Spherical unit vectors vary with location.

VECTOR NOTATION

When a unit vector is multiplied by a scalar, the result is a regular vector. We can use this fact to write *any* vector in terms of scalar components and unit vectors, in what is called the **vector notation** of $\vec{\mathbf{A}}$. The Cartesian vector notation of $\vec{\mathbf{A}}$ is:

$$\vec{\mathbf{A}} = A_x\hat{\mathbf{i}} + A_y\hat{\mathbf{j}} + A_z\hat{\mathbf{k}}$$

Table 15.1 lists vector notations for other kinds of unit vectors. Note that the vector $\vec{\mathbf{A}}$ must begin at the origin for the component equations in the right hand column of the table to apply.

Table 15.1: Vector Notation

UNIT VECTOR	VECTOR NOTATION	COMPONENT EQUATIONS
POLAR	$\vec{\mathbf{A}} = A_r\hat{\mathbf{r}} + A_\theta\hat{\boldsymbol{\theta}}$	$A_r = \frac{xA_x + yA_y}{\sqrt{x^2+y^2}}$ and $A_\theta = \frac{-yA_x + xA_y}{\sqrt{x^2+y^2}}$
NORMAL	$\vec{\mathbf{A}} = A\hat{\mathbf{n}}$	A is the magnitude of the vector.
CYLINDRICAL	$\vec{\mathbf{A}} = A_r\hat{\mathbf{r}} + A_\theta\hat{\boldsymbol{\theta}} + A_z\hat{\mathbf{k}}$	$A_r = \frac{xA_x + yA_y}{\sqrt{x^2+y^2}}$, $A_\theta = \frac{-yA_x + xA_y}{\sqrt{x^2+y^2}}$, and $A_z = A_z$
SPHERICAL	$\vec{\mathbf{A}} = A_\rho\hat{\boldsymbol{\rho}} + A_\varphi\hat{\boldsymbol{\varphi}} + A_\theta\hat{\boldsymbol{\theta}}$	$A_\rho = \frac{xA_x + yA_y + zA_z}{\sqrt{x^2+y^2+z^2}}$, $A_\theta = \frac{-yA_x + xA_y}{\sqrt{x^2+y^2}}$, and $A_\varphi = \frac{xzA_x + yzA_y - (x^2+y^2)A_z}{\sqrt{x^2+y^2}\sqrt{x^2+y^2+z^2}}$

EXERCISE 15.11: Unit Vectors

Write the vector $\vec{\mathbf{A}}$ of Example 15.9 above in terms of (a) Cartesian and (b) polar unit vectors.

Exercise 15.11 Continued...

SOLUTION:

(a) Since we don't have a z term in this equation, we can simply write $\vec{\mathbf{A}} = A_x\hat{\mathbf{i}} + A_y\hat{\mathbf{j}} = (-8.7\text{ m})\hat{\mathbf{i}} + (-5.0\text{ m})\hat{\mathbf{j}}$.

(b) First, we have to figure the values of A_r and A_θ, using the equations in Table 15.1:

$$A_r = \frac{xA_x + yA_y}{\sqrt{x^2+y^2}} \text{ and } A_\theta = \frac{-yA_x + xA_y}{\sqrt{x^2+y^2}}$$

$$\Rightarrow A_r = \frac{-8.7x - 5y}{\sqrt{x^2+y^2}} \text{ and } A_\theta = \frac{8.7y - 5x}{\sqrt{x^2+y^2}}$$

Notice that these magnitudes will *vary* depending upon the position of the tip of the vector (x, y). Now, we can write the vector as:

$$\vec{\mathbf{A}} = A_r\hat{\mathbf{r}} + A_\theta\hat{\boldsymbol{\theta}} = \frac{1}{\sqrt{x^2+y^2}}\left[(-8.7x - 5y)\hat{\mathbf{r}} + (-5x + 8.7y)\hat{\boldsymbol{\theta}}\right]$$

It is legitimate to choose one coordinate system to describe the location of a point, and a different system for the unit vectors there.

EXAMPLE 15.12: Mixing Systems

For example, we can mix a polar function with Cartesian unit vectors:

$$\vec{\mathbf{A}} = f(r,\theta)\hat{\mathbf{i}} + g(r,\theta)\hat{\mathbf{j}}$$

To convert unit vectors between coordinate systems, use Table 15.2.

Table 15.2: Coordinate System Conversions

	POLAR	CYLINDRICAL	SPHERICAL
Convert FROM Cartesian TO...	$\hat{\mathbf{r}} = \cos\theta\hat{\mathbf{i}} + \sin\theta\hat{\mathbf{j}}$ $\hat{\boldsymbol{\theta}} = -\sin\theta\hat{\mathbf{i}} + \cos\theta\hat{\mathbf{j}}$	$\hat{\mathbf{r}} = \cos\theta\hat{\mathbf{i}} + \sin\theta\hat{\mathbf{j}}$ $\hat{\boldsymbol{\theta}} = -\sin\theta\hat{\mathbf{i}} + \cos\theta\hat{\mathbf{j}}$ $\hat{\mathbf{z}} = \hat{\mathbf{z}}$	$\hat{\boldsymbol{\rho}} = \cos\theta\sin\varphi\hat{\mathbf{i}} + \sin\theta\sin\varphi\hat{\mathbf{j}} + \cos\varphi\hat{\mathbf{k}}$ $\hat{\boldsymbol{\theta}} = -\sin\theta\hat{\mathbf{i}} + \cos\theta\hat{\mathbf{j}}$ $\hat{\boldsymbol{\varphi}} = \cos\theta\cos\varphi\hat{\mathbf{i}} + \cos\theta\sin\varphi\hat{\mathbf{j}} + \sin\varphi\hat{\mathbf{k}}$
Convert FROM ... TO Cartesian	$\hat{\mathbf{i}} = \cos\theta\hat{\mathbf{r}} - \sin\theta\hat{\boldsymbol{\theta}}$ $\hat{\mathbf{j}} = \sin\theta\hat{\mathbf{r}} + \cos\theta\hat{\boldsymbol{\theta}}$	$\hat{\mathbf{i}} = \cos\theta\hat{\mathbf{r}} - \sin\theta\hat{\boldsymbol{\theta}}$ $\hat{\mathbf{j}} = \sin\theta\hat{\mathbf{r}} + \cos\theta\hat{\boldsymbol{\theta}}$ $\hat{\mathbf{z}} = \hat{\mathbf{z}}$	$\hat{\mathbf{i}} = \cos\theta\sin\varphi\hat{\boldsymbol{\rho}} - \sin\theta\hat{\boldsymbol{\theta}} + \cos\theta\cos\varphi\hat{\boldsymbol{\varphi}}$ $\hat{\mathbf{j}} = \sin\theta\sin\varphi\hat{\boldsymbol{\rho}} + \cos\theta\hat{\boldsymbol{\theta}} + \cos\theta\sin\varphi\hat{\boldsymbol{\varphi}}$ $\hat{\mathbf{k}} = \cos\varphi\hat{\boldsymbol{\rho}} - \sin\varphi\hat{\boldsymbol{\varphi}}$

15.3.4 ADDING & SUBTRACTING VECTORS COMPONENT-WISE

While writing vectors component-wise takes some extra effort (luckily, this process becomes less painful the more you do it), the ease by which we can do vector mathematics—addition, subtraction, multiplication, and division—is well worth the extra trouble.

We used this technique for adding vectors in relative motion, Section 5.3.3.

For example, adding and subtracting vectors component-wise is very straightforward. The result of adding two vectors $\vec{\mathbf{A}} = A_x\hat{\mathbf{i}} + A_y\hat{\mathbf{j}}$ and $\vec{\mathbf{B}} = B_x\hat{\mathbf{i}} + B_y\hat{\mathbf{j}}$ is $\vec{\mathbf{R}} = R_x\hat{\mathbf{i}} + R_y\hat{\mathbf{j}}$, such that:

$$R_x = A_x + B_x$$
$$R_y = A_y + B_y$$

That is, the x-component of $\vec{\mathbf{R}}$ is simply the *sum* of the x-components of the original vectors $A_x + B_x$; and y-component is just the sum of the y-components of the original vectors $A_y + B_y$.

Subtraction works in just the same way, but with negative vectors.

EXERCISE 15.13: Adding Vectors Component-Wise

Velocity is defined in Section 5.3.1.

Find the sum of the two velocity vectors $\vec{\mathbf{A}} = (3\ \text{cm/s})\hat{\mathbf{i}} + (5\ \text{cm/s})\hat{\mathbf{j}}$ and $\vec{\mathbf{B}} = (2\ \text{cm/s})\hat{\mathbf{i}} + (-3\ \text{cm/s})\hat{\mathbf{j}}$.

SOLUTION:

From these expressions, we see that $A_x = 3$ cm/s, $A_y = 5$ cm/s, $B_x = 2$ cm/s, and $B_y = -3$ cm/s. Thus,

$$R_x = A_x + B_x = 3\ \text{cm/s} + 2\ \text{cm/s} = 5\ \text{cm/s}$$
$$R_y = A_y + B_y = 5\ \text{cm/s} - 3\ \text{cm/s} = 2\ \text{cm/s}$$

The resultant vector is therefore:

$$\vec{\mathbf{R}} = (5\ \text{cm/s})\hat{\mathbf{i}} + (2\ \text{cm/s})\hat{\mathbf{j}}$$

15.3.5 MULTIPLYING & DIVIDING VECTORS COMPONENT-WISE

There are *three* kinds of vector multiplication, compared with only one kind of scalar multiplication.

Newton's Second Law, $\vec{\mathbf{F}}_{net} = m\vec{\mathbf{a}}$, is this kind of product. See Section 6.3 for more.

MULTIPLICATION BY A SCALAR

The multiplication of a vector $\vec{\mathbf{A}}$ and a scalar s results in a vector $s\vec{\mathbf{A}}$:

$$s\vec{\mathbf{A}} = s(A_x\hat{\mathbf{i}} + A_y\hat{\mathbf{j}} + A_z\hat{\mathbf{k}}) = (sA_x)\hat{\mathbf{i}} + (sA_y)\hat{\mathbf{j}} + (sA_z)\hat{\mathbf{k}}$$

where we simply multiplied every component in the vector by s.

If $s > 0$, $s\vec{\mathbf{A}}$ will point in the *same* direction as $\vec{\mathbf{A}}$; if $s < 0$, $s\vec{\mathbf{A}}$ will point in the *opposite* direction of $\vec{\mathbf{A}}$. If $|s| > 1$, the new vector will be *longer* than the original vector; if $|s| < 1$, the new vector will be *shorter* than the original vector.

EXERCISE 15.14: Multiplication by a Scalar

Find the new vector obtained when we multiply $s = 2$ kg by the vector $\vec{\mathbf{A}} = (7\ \text{m/s}^2)\hat{\mathbf{i}} + (-2\ \text{m/s}^2)\hat{\mathbf{j}}$.

SOLUTION:

The new vector is $s\vec{\mathbf{A}} = (2\ \text{kg})[(7\ \text{m/s}^2)\hat{\mathbf{i}} + (-2\ \text{m/s}^2)\hat{\mathbf{j}}] = (14\ \text{kg}\cdot\text{m/s}^2)\hat{\mathbf{i}} + (-4\ \text{kg}\cdot\text{m/s}^2)\hat{\mathbf{j}}$. This vector has a magnitude twice as long as the original, but points in the same direction:

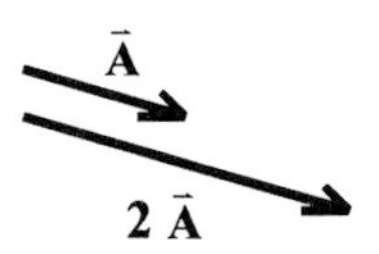

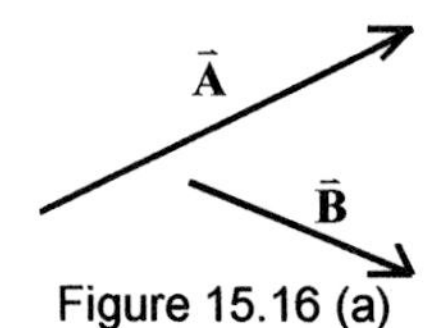

Figure 15.16 (a)

DOT PRODUCT

The **dot product** $\vec{\mathbf{A}}\cdot\vec{\mathbf{B}}$ between two vectors $\vec{\mathbf{A}}$ and $\vec{\mathbf{B}}$ (Figure 15.16 (a)) results in a scalar. There are two ways to calculate the dot product.

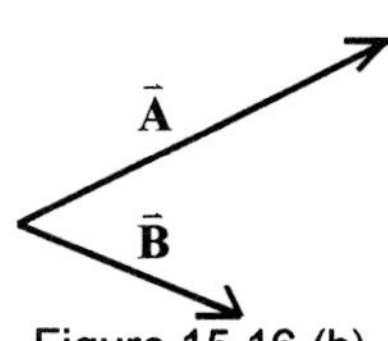
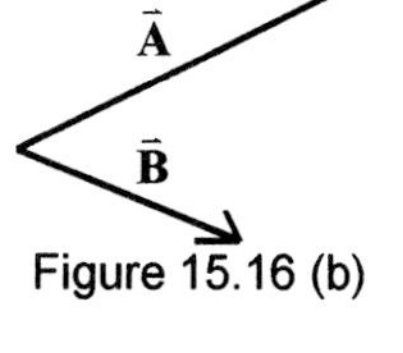

Figure 15.16 (b)

Method 1: *The Angle Way*

1. Place $\vec{\mathbf{A}}$ and $\vec{\mathbf{B}}$ tail to tail (Fig. 15.16 (b)).
2. Define the *smallest* angle between $\vec{\mathbf{A}}$ and $\vec{\mathbf{B}}$, measured *from* $\vec{\mathbf{A}}$ to $\vec{\mathbf{B}}$, as ϕ (in Fig. 15.16 (c); note that θ is *not* the correct angle).
3. The dot product is then:

$$\vec{\mathbf{A}}\cdot\vec{\mathbf{B}} \equiv AB\cos\phi$$

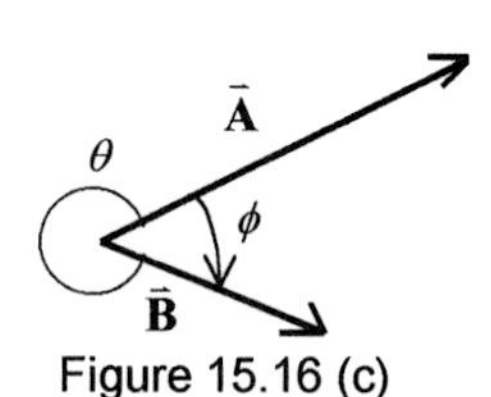

Figure 15.16 (c)

DEFINITION 15.15: Dot Products

With this definition, it is easy to see that the dot product can be positive, negative, or zero, depending upon the value of cos ϕ. When $\vec{\mathbf{A}}$ and $\vec{\mathbf{B}}$ are perpendicular ($\phi = \frac{\pi}{2}$ rads), the dot product is zero. When they are parallel ($\phi = 0$), or anti-parallel ($\phi = \pi$ rads), the dot product is equal to the product of the vector magnitudes AB.

Work (Section 8.5.1) is calculated via a dot product.

EXERCISE 15.16: Dot Product, Part I

What is the dot product between vectors $\vec{\mathbf{A}}$ and $\vec{\mathbf{B}}$, if $A = 4$ N, $B =$ 6 m, and $\phi = 0.45$ rads?

SOLUTION:

We can solve this problem by using the first method above. We already have the magnitudes of the vector (Step 1), and the angle between them (Step 2), so we merely need to apply the formula in Step 3:

$$\vec{\mathbf{A}} \cdot \vec{\mathbf{B}} \equiv AB\cos\phi = (4\text{ N})(6\text{ m})\cos(0.45\text{ rads}) \approx 21.6\text{ N}\cdot\text{m}$$

This first method of calculating dot products leads to a natural interpretation of dot products as the product of the *magnitude* of $\vec{\mathbf{A}}$ times the *component of the magnitude* of $\vec{\mathbf{B}}$ *in the direction* of $\vec{\mathbf{A}}$ $[A(B\cos\phi)]$, or visa versa $[B(A\cos\phi)]$.

Method 2: *The Component Way*

1. Calculate the components of $\vec{\mathbf{A}}$ and $\vec{\mathbf{B}}$.
2. The dot product is then the sum of the component products:

$$\vec{\mathbf{A}} \cdot \vec{\mathbf{B}} \equiv A_x B_x + A_y B_y + A_z B_z$$

EXERCISE 15.17: Dot Product, Part II

What is the dot product between vectors $\vec{\mathbf{A}} = 1\hat{\mathbf{i}} - 8\hat{\mathbf{j}}$ and $\vec{\mathbf{B}} = 5\hat{\mathbf{i}} + 2\hat{\mathbf{j}}$?

SOLUTION:

We can solve this problem by using the second method for calculating the dot product above. We already have the components of this vector (Step 1), so we merely need to sum their products (Step 2):

$$\vec{\mathbf{A}} \cdot \vec{\mathbf{B}} = A_x B_x + A_y B_y$$

$$\Rightarrow \vec{\mathbf{A}} \cdot \vec{\mathbf{B}} = (1)(5) + (-8)(2) = -11$$

Note that the negative sign on this dot product does *not* indicate a direction; the dot product is a scalar.

The order in which you take the dot product is irrelevant: $\vec{\mathbf{A}} \cdot \vec{\mathbf{B}} = \vec{\mathbf{B}} \cdot \vec{\mathbf{A}}$ (**commutative law of vector multiplication**).

Torque is calculated via a cross-product. See Section 6.5.2 for more information.

CROSS PRODUCT

The **cross product** $\vec{\mathbf{A}} \times \vec{\mathbf{B}}$ between two vectors $\vec{\mathbf{A}}$ and $\vec{\mathbf{B}}$ results in a vector. There are two ways to calculate the cross product.

Method 1: *The Angle Way*

1. Calculate ϕ in the same way as in Steps 1 and 2 of the dot product Method 1 above.
2. The *magnitude* of the cross product is then:

$$\left|\vec{\mathbf{A}} \times \vec{\mathbf{B}}\right| = AB \sin \phi$$

3. The *direction* of the vector $\vec{\mathbf{A}} \times \vec{\mathbf{B}}$ is obtained via a **cross-product right hand rule**. There are *four* versions of this rule, as listed below. *Pick the one way that is most comfortable for you, and use it for all cross products you encounter.* Note, however, that in all cases, the direction of $\vec{\mathbf{A}} \times \vec{\mathbf{B}}$ is *always* perpendicular to the plane made by $\vec{\mathbf{A}}$ and $\vec{\mathbf{B}}$.

Version I (Fig. 15.17 (a)): Palm Way

a. Point the thumb in the direction of the first vector $\vec{\mathbf{A}}$.
b. Point the other fingers in the direction of the second vector $\vec{\mathbf{B}}$.
c. The palm will indicate the direction of $\vec{\mathbf{A}} \times \vec{\mathbf{B}}$.

Version II (Fig. 15.17 (b)): Three Finger Way

a. Point your index finger in the direction of the first vector $\vec{\mathbf{A}}$.
b. Point the middle finger in the direction of the second vector $\vec{\mathbf{B}}$, making sure that your fingers are separated by $90°$.
c. Your thumb will indicate the direction of $\vec{\mathbf{A}} \times \vec{\mathbf{B}}$.

Version III (Fig. 15.17 (c)): Thumb Way

a. Point the fingers in the direction of the first vector $\vec{\mathbf{A}}$.
b. Curl your fingers towards the direction of the second vector $\vec{\mathbf{B}}$.
c. Your thumb will indicate the direction of $\vec{\mathbf{A}} \times \vec{\mathbf{B}}$.

Version IV (Fig. 15.17 (d)): Screwdriver Way

a. Imagine a screw located at the intersection of $\vec{\mathbf{A}}$ and $\vec{\mathbf{B}}$.
b. Turn the screw with your hand *from* $\vec{\mathbf{A}}$ *to* $\vec{\mathbf{B}}$, keeping in mind the screwdriver rule: "right to tighten."
c. Your thumb will indicate the direction of $\vec{\mathbf{A}} \times \vec{\mathbf{B}}$.

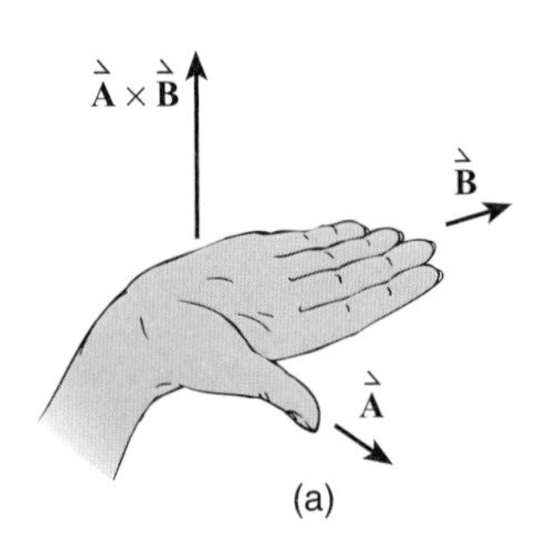

(a)

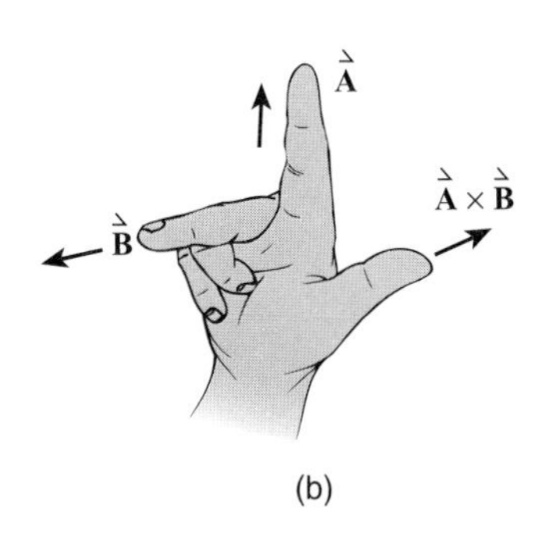

(b)

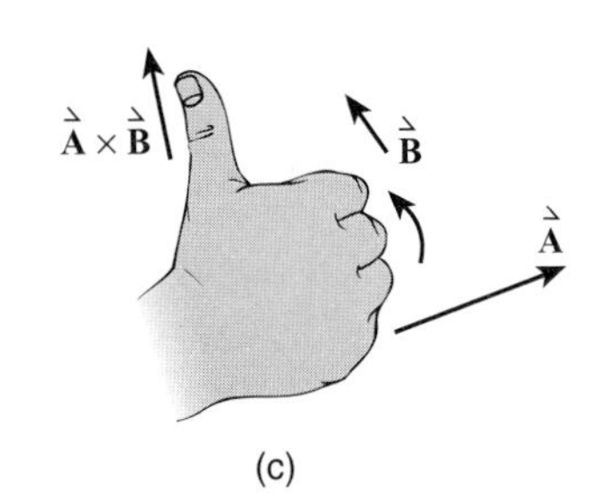

(c)

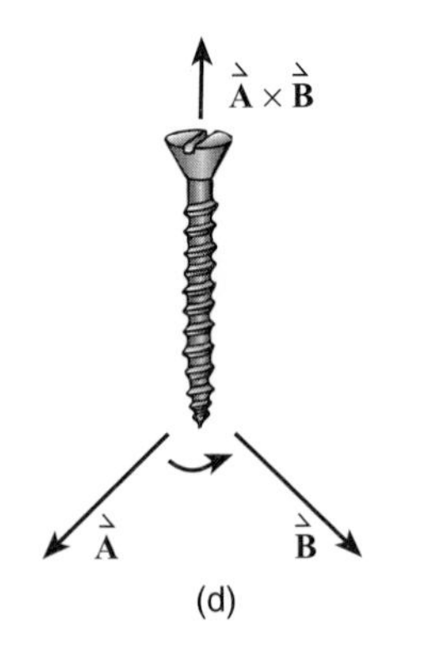

(d)

Figure 15.17

WARNING!

Be *very* careful with these rules! *Sloppiness results in wrong answers.* If you use your left hand to do a right hand rule (for example, if you write with your right hand and forget to put down your pencil before you do the rule), you will get the direction WRONG! Or, if you do not take extra care to line up your hand with the direction of motion *exactly*, you may get the wrong result as well.

DEFINITION 15.18: Cross Products

With this first definition, it is easy to see that the cross product can be positive, negative, or zero, depending upon the value of cos ϕ. When $\vec{\mathbf{A}}$ and $\vec{\mathbf{B}}$ are parallel ($\phi = 0$), or anti-parallel ($\phi = \pi$ rads), the cross product is zero. When $\vec{\mathbf{A}}$ and $\vec{\mathbf{B}}$ are perpendicular ($\phi = \frac{\pi}{2}$ rads), the cross product is equal to AB.

EXERCISE 15.19: Cross Product, Part I

Calculate the cross product for the vectors in Exercise 15.17. The angle between the vectors in that case is $\phi = 1.33$ rads.

SOLUTION:

First, we have to figure out the magnitudes of $\vec{\mathbf{A}}$ and $\vec{\mathbf{B}}$:

$$A = \sqrt{(1)^2 + (-8)^2} \approx 8.1$$

$$B = \sqrt{(5)^2 + (2)^2} \approx 5.4$$

Next, we can apply the cross product formula to find the magnitude:

$$\left|\vec{\mathbf{A}} \times \vec{\mathbf{B}}\right| = AB \sin\phi = (8.1)(5.4)\sin 1.33 \approx 42$$

Now, we must use one of the cross-product right hand rules to find the direction of the vector. To do this, we first sketch the situation:

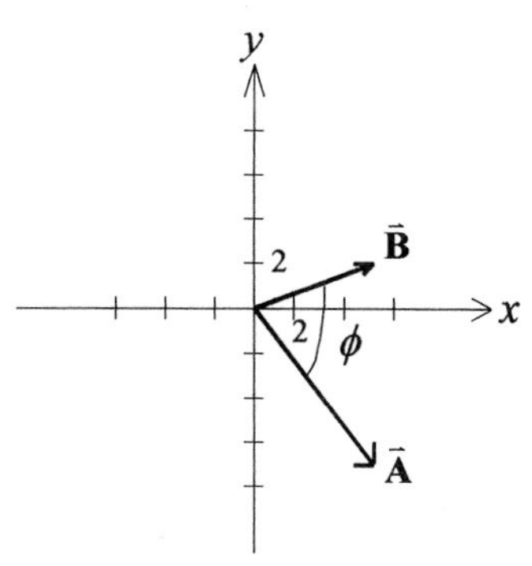

Finally, we apply a right hand rule of our choice, sweeping from $\vec{\mathbf{A}}$ to $\vec{\mathbf{B}}$ as necessary. This yields our direction as upwards ($+\hat{\mathbf{k}}$).

The first method of cross-products lends itself to two nice interpretations: (1) the cross-product is the *area* of the parallelogram created by vectors $\vec{\mathbf{A}}$ and $\vec{\mathbf{B}}$; and (2) cross-product is also the product of the *magnitude* of one vector and the *magnitude of the perpendicular component* of the other vector [$A(B \cos \phi)$ or $B(A \cos \phi)$].

The order of cross-multiplication is important; if you reverse the order, you obtain the negative result: $\vec{\mathbf{A}} \times \vec{\mathbf{B}} = -\vec{\mathbf{B}} \times \vec{\mathbf{A}}$ (**anti-commutative law of vector multiplication**).

Don't believe it? Check out the parallelogram area formula in Section 13.5.2, and then figure out what h is in terms of $\vec{\mathbf{A}}$ and $\vec{\mathbf{B}}$.

Method 2: *Determinant Way*

1. Calculate the components of $\vec{\mathbf{A}}$ and $\vec{\mathbf{B}}$.
2. The cross-product—magnitude *and* direction together—is:

$$\vec{\mathbf{A}} \times \vec{\mathbf{B}} = (A_y B_z - A_z B_y)\hat{\mathbf{i}} - (A_x B_z - A_z B_x)\hat{\mathbf{j}} + (A_x B_y - A_y B_x)\hat{\mathbf{k}} \equiv \begin{vmatrix} \mathbf{i} & \mathbf{j} & \mathbf{k} \\ A_x & A_y & A_z \\ B_x & B_y & B_z \end{vmatrix}$$

The box in the last term of this expression is called a **determinant**. Determinants make the complicated cross-product equation above easier to remember. To solve any determinant, follow these steps:

$$\begin{vmatrix} \hat{\mathbf{i}} & \hat{\mathbf{j}} & \hat{\mathbf{k}} \\ A_x & A_y & A_z \\ B_x & B_y & B_z \end{vmatrix}$$

Figure 15.18 (a)

a. Draw a vertical line through the first column of the determinant. Draw a horizontal line through the first row of the determinant (Fig. 15.18 (a)).
b. Four terms will remain unlined. Cross-multiply these, obtaining:

$$A_y B_z - A_z B_y$$

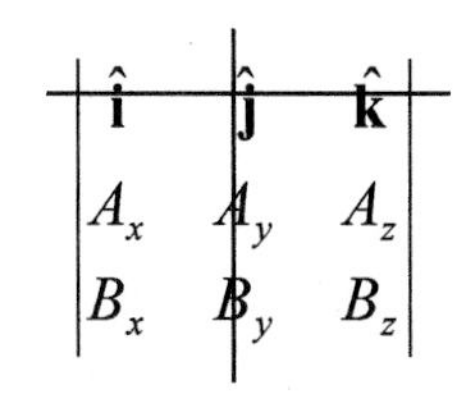

Figure 15.18 (b)

c. Multiply this entire expression by $\hat{\mathbf{i}}$, the unit vector that is crossed by both lines.
d. Erase the vertical line of Step 1. Draw a vertical line through the second column (Fig. 15.18 (b)).
e. Cross-multiply the remaining terms as:

$$A_x B_z - A_z B_x$$

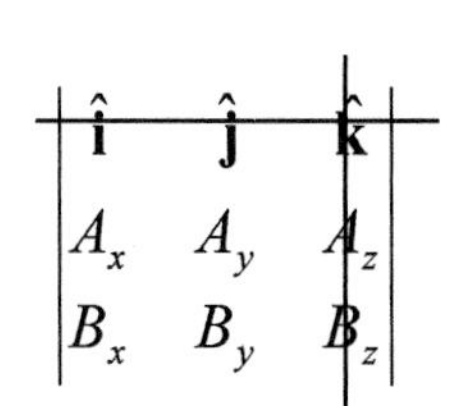

Figure 15.18 (c)

f. Multiply this entire expression by $-\hat{\mathbf{j}}$, the negative of the unit vector that is crossed by both lines.
g. Erase the vertical line of Step 4. Draw a vertical line through the third column (Fig. 15.18 (c)).
h. Cross-multiply the remaining terms as:

$$A_x B_y - A_z B_y$$

i. Multiply this entire expression by $\hat{\mathbf{k}}$, the unit vector that is crossed by both lines.
j. The final answer is the sum of Steps c, f, and i.

EXERCISE 15.20: Cross Product, Part II

Calculate the cross product for the vectors in Exercise 15.17 using the second method.

SOLUTION:

$$\vec{\mathbf{A}} \times \vec{\mathbf{B}} = \begin{vmatrix} \hat{\mathbf{i}} & \hat{\mathbf{j}} & \hat{\mathbf{k}} \\ 1 & -8 & 0 \\ 5 & 2 & 0 \end{vmatrix}$$

$$= [(-8)(0) - (2)(0)]\hat{\mathbf{i}} - [(1)(0) - (5)(0)]\hat{\mathbf{j}} + [(1)(2) - (5)(-8)]\hat{\mathbf{k}}$$

$$= 42\hat{\mathbf{k}}$$

VECTOR DIVISION

Unlike the three types of vector multiplication, there is only one type of vector division: division of a vector by a scalar. We *cannot* divide vectors by other vectors.

To divide a vector by a scalar, follow the same procedure as for scalar multiplication (*i.e.*, multiply the vector by s) EXCEPT that in this case s is a fraction (less than one).

VECTOR THEOREMS

The following theorems combine the various vector multiplication methods. These theorems can sometimes be helpful when rewriting knotty algebraic expressions:

$$s(\vec{\mathbf{A}}+\vec{\mathbf{B}}) = s\vec{\mathbf{A}} + s\vec{\mathbf{B}}$$
$$(r+s)\vec{\mathbf{A}} = r\vec{\mathbf{A}} + s\vec{\mathbf{A}}$$
$$(rs)\vec{\mathbf{A}} = r(s\vec{\mathbf{A}}) = s(r\vec{\mathbf{A}})$$
$$(s\vec{\mathbf{A}})\cdot\vec{\mathbf{B}} = \vec{\mathbf{A}}\cdot(s\vec{\mathbf{B}}) = s(\vec{\mathbf{A}}\cdot\vec{\mathbf{B}})$$
$$(s\vec{\mathbf{A}})\times\vec{\mathbf{B}} = \vec{\mathbf{A}}\times s\vec{\mathbf{B}} = s(\vec{\mathbf{A}}\times\vec{\mathbf{B}})$$
$$\vec{\mathbf{A}}\cdot(\vec{\mathbf{B}}+\vec{\mathbf{C}}) = \vec{\mathbf{A}}\cdot\vec{\mathbf{B}} + \vec{\mathbf{A}}\cdot\vec{\mathbf{C}}$$
$$\vec{\mathbf{A}}\cdot(\vec{\mathbf{B}}\times\vec{\mathbf{C}}) = \vec{\mathbf{B}}\cdot(\vec{\mathbf{A}}\times\vec{\mathbf{C}}) = \vec{\mathbf{C}}\cdot(\vec{\mathbf{A}}\times\vec{\mathbf{B}})$$
$$\vec{\mathbf{A}}\times(\vec{\mathbf{B}}+\vec{\mathbf{C}}) = (\vec{\mathbf{A}}\times\vec{\mathbf{B}}) + (\vec{\mathbf{A}}\times\vec{\mathbf{C}})$$
$$\vec{\mathbf{A}}\times(\vec{\mathbf{B}}\times\vec{\mathbf{C}}) = (\vec{\mathbf{A}}\cdot\vec{\mathbf{C}})\vec{\mathbf{B}} - (\vec{\mathbf{A}}\cdot\vec{\mathbf{B}})\vec{\mathbf{C}}$$

15.4 VECTOR CALCULUS

Like addition and multiplication, calculus with vectors is not the same as calculus using scalars. This section covers some basic concepts of vector calculus.

Refer to Chapter 14 for more on calculus.

15.4.1 DERIVATION

Derivatives with vectors take on several forms, depending upon whether the function to be derived has scalar or vector terms. Functions consisting of scalars are called **scalar functions**, while functions using vector notation are **vector** or **vector-valued functions**. Tables 15.3 (a) and (b) below summarize these derivatives.

See Section 13.4.12 for more on scalar functions.

Table 15.3 (a): Important Vector Derivatives with Respect to Scalar Functions

DERIVATIVE ON A...	WITH RESPECT TO A...	NAME	MEANING	DERIVATIVE	RESULT IS A...
SCALAR FUNCTION $f(x, y, z)$	SCALAR	GRADIENT	Points in the direction in which the function increases most rapidly	$\nabla f = \frac{\partial f}{\partial x}\hat{\mathbf{i}} + \frac{\partial f}{\partial y}\hat{\mathbf{j}} + \frac{\partial f}{\partial z}\hat{\mathbf{k}}$	VECTOR
	SCALAR	LAPLACIAN		$\nabla^2 f = \frac{\partial^2 f}{\partial x^2} + \frac{\partial^2 f}{\partial y^2} + \frac{\partial^2 f}{\partial z^2}$	SCALAR

Table 15.3 (b): Important Vector Derivatives with Respect to Vector Functions

DERIVATIVE ON A...	WITH RESPECT TO A...	NAME	MEANING	DERIVATIVE	RESULT IS A...
VECTOR FUNCTION $\vec{\mathbf{A}} = A_x\hat{\mathbf{i}} + A_y\hat{\mathbf{j}} + A_z\hat{\mathbf{k}}$	SCALAR			$\frac{d\vec{\mathbf{A}}}{ds} = \frac{dA_x}{ds}\hat{\mathbf{i}} + \frac{dA_y}{ds}\hat{\mathbf{j}} + \frac{dA_z}{ds}\hat{\mathbf{k}}$	VECTOR
	SCALAR	LAPLACIAN		$\nabla^2\vec{\mathbf{A}} = \nabla^2 A_x\hat{\mathbf{i}} + \nabla^2 A_y\hat{\mathbf{j}} + \nabla^2 A_z\hat{\mathbf{k}}$ *	VECTOR
	SCALAR	DIVERGENCE	The net rate of flow away from point (x, y, z)	$\text{div}\vec{\mathbf{A}} = \nabla \cdot \vec{\mathbf{A}} = \frac{\partial A_x}{\partial x} + \frac{\partial A_y}{\partial y} + \frac{\partial A_z}{\partial z}$	SCALAR
	VECTOR	CURL	The net rate of flow around an axis through the point (x, y, z)	$\text{curl}\vec{\mathbf{A}} = \nabla \times \vec{\mathbf{A}} = \left(\frac{\partial A_z}{\partial y} - \frac{\partial A_y}{\partial z}\right)\hat{\mathbf{i}} + \left(\frac{\partial A_x}{\partial z} - \frac{\partial A_z}{\partial x}\right)\hat{\mathbf{j}} + \left(\frac{\partial A_y}{\partial x} - \frac{\partial A_x}{\partial y}\right)\hat{\mathbf{k}}$	VECTOR

EXERCISE 15.21: Derivatives of Scalar Functions

Calculate the (a) gradient and the (b) Laplacian of the scalar function $f(x,y,z) = x^2 + 2y - z$.

SOLUTION:

(a) According to the gradient formula in Table 15.3 (a) above, we see that we must calculate partial derivatives of $f(x, y, z)$. So, we calculate these as:

Partial derivatives are in Section 14.2.3.

$$\frac{\partial f}{\partial x} = 2x, \ \frac{\partial f}{\partial y} = 2, \text{ and } \frac{\partial f}{\partial z} = -1$$

Then, we plug these into the gradient formula:

$$\nabla f = 2x\hat{\mathbf{i}} + 2\hat{\mathbf{j}} - \hat{\mathbf{k}}$$

(b) From the Laplacian formula above, we know that we need to find second derivatives of the scalar function from part (a). But because only the x term in the first derivatives above contained a variable, it is the only term left in the Laplacian:

$$\nabla^2 f = \frac{\partial^2 f}{\partial x^2} = 2$$

* where $\nabla^2 A_\alpha = \frac{\partial^2 A_\alpha}{\partial x^2} + \frac{\partial^2 A_\alpha}{\partial y^2} + \frac{\partial^2 A_\alpha}{\partial z^2}$; α is x, y, and z, respectively.

EXERCISE 15.22: Derivatives of Vector Functions

Calculate (a) the scalar derivative (with respect to x), (b) the Laplacian, (c) the divergence, and (d) the curl of the vector function $\vec{\mathbf{A}} = 5x^2\hat{\mathbf{i}} - 7x\hat{\mathbf{j}}$.

SOLUTION:

(a) To take the scalar derivative of a vector function, according to Table 15.3 (b) above, we simply take the ordinary derivative of each term, separately, (without regard for the unit vectors):

$$\frac{d\vec{\mathbf{A}}}{dx} = \frac{d}{dx}(5x^2)\hat{\mathbf{i}} - \frac{d}{dx}(7x)\hat{\mathbf{j}} = 10x\hat{\mathbf{i}} - 7\hat{\mathbf{j}}$$

(b) To take the Laplacian of a vector function, according to Table 15.3 (b), we must calculate three terms of the form $\nabla^2 A_\alpha$. Now, from the problem statement, we know that $A_x = 5x^2$, $A_y = -7x$, and $A_z = 0$. Because the only variable in any of these terms is x, all of the first partial derivatives with respect to y and z *must* be zero:

$$\frac{\partial A_x}{\partial y} = 0 = \frac{\partial A_x}{\partial z},\ \frac{\partial A_y}{\partial y} = 0 = \frac{\partial A_y}{\partial z},\ \text{and}\ \frac{\partial A_z}{\partial y} = 0 = \frac{\partial A_z}{\partial z}$$

Therefore, the second derivatives with respect to y and z are zero, too:

$$\frac{\partial^2 A_x}{\partial y^2} = 0 = \frac{\partial^2 A_x}{\partial z^2},\ \frac{\partial^2 A_y}{\partial y^2} = 0 = \frac{\partial^2 A_y}{\partial z^2},\ \text{and}\ \frac{\partial^2 A_z}{\partial y^2} = 0 = \frac{\partial^2 A_z}{\partial z^2}$$

Next, because $A_z = 0$, its partial derivative with respect to x is zero as well. However, A_x and A_y are not zero, so we must take their first and second partial derivatives with respect to x:

$$\frac{\partial A_x}{\partial x} = 10x \Rightarrow \frac{\partial^2 A_x}{\partial x^2} = 10\ \text{and}\ \frac{\partial A_y}{\partial x} = -7 \Rightarrow \frac{\partial^2 A_y}{\partial x^2} = 0$$

So, only one term out of the entire original mess survived! Thus,

$$\nabla^2\vec{\mathbf{A}} = 10\hat{\mathbf{i}}$$

(c) To find the divergence, we need to take some partial derivatives. We know from part (b) above the value of the following derivatives:

$$\frac{\partial A_y}{\partial y} = 0,\ \frac{\partial A_z}{\partial z} = 0\ \text{and}\ \frac{\partial A_x}{\partial x} = 10x$$

So, the divergence of the vector is:

$$\mathrm{div}\vec{\mathbf{A}} = 10x$$

(d) Like part (b) above, to find the curl, we must take many partial derivatives. Luckily, as we saw in part (b) above, all of the partial derivatives with respect to y and z must be zero, as well as the partial derivative of A_z. Furthermore, because there is no derivative of A_x with respect to x in the equation for the curl, the only term left is the derivative of A_y with respect to x:

$$\frac{\partial A_y}{\partial x} = -7$$

$$\Rightarrow \mathrm{curl}\vec{\mathbf{A}} = -7\hat{\mathbf{k}}$$

Vector derivatives obey the following distributive laws:

$$\frac{d}{ds}(r\vec{\mathbf{A}}) = \frac{dr}{ds}\vec{\mathbf{A}} + r\frac{d\vec{\mathbf{A}}}{ds}$$

$$\frac{d}{ds}(\vec{\mathbf{A}}\cdot\vec{\mathbf{B}}) = \frac{d\vec{\mathbf{A}}}{ds}\cdot\vec{\mathbf{B}} + \vec{\mathbf{A}}\cdot\frac{d\vec{\mathbf{B}}}{ds}$$

$$\frac{d}{ds}(\vec{\mathbf{A}}\times\vec{\mathbf{B}}) = \frac{d\vec{\mathbf{A}}}{ds}\times\vec{\mathbf{B}} + \vec{\mathbf{A}}\times\frac{d\vec{\mathbf{B}}}{ds}$$

$$\text{curl}(f\vec{\mathbf{A}}) = (\nabla f)\times\vec{\mathbf{A}} + f(\text{curl}\vec{\mathbf{A}})$$

$$\text{div}(\vec{\mathbf{A}}\times\vec{\mathbf{B}}) = \vec{\mathbf{B}}\cdot(\text{curl}\vec{\mathbf{A}}) - \vec{\mathbf{A}}\cdot(\text{curl}\mathbf{B})$$

$$\nabla(\vec{\mathbf{A}}\cdot\vec{\mathbf{B}}) = \vec{\mathbf{B}}\times(\text{curl}\vec{\mathbf{A}}) + \vec{\mathbf{A}}\times(\text{curl}\vec{\mathbf{B}}) + (\vec{\mathbf{B}}\cdot\nabla)\vec{\mathbf{A}} + (\vec{\mathbf{A}}\cdot\nabla)\vec{\mathbf{B}}$$

$$\text{curl}(\vec{\mathbf{A}}\times\vec{\mathbf{B}}) = (\text{div}\vec{\mathbf{B}})\vec{\mathbf{A}} - (\text{div}\vec{\mathbf{A}})\vec{\mathbf{B}} + (\vec{\mathbf{B}}\cdot\nabla)\vec{\mathbf{A}} - (\vec{\mathbf{A}}\cdot\nabla)\vec{\mathbf{B}}$$

$$\text{div}(\vec{\mathbf{A}}+\vec{\mathbf{B}}) = \text{div}\vec{\mathbf{A}} + \text{div}\vec{\mathbf{B}}$$

$$\text{div}(f\vec{\mathbf{A}}) = \vec{\mathbf{A}}\cdot(\nabla f) + f(\text{div}\vec{\mathbf{A}})$$

$$\nabla(f+g) = \nabla f + \nabla g$$

$$\nabla(fg) = g\nabla f + f\nabla g$$

$$\text{curl}(f\vec{\mathbf{A}}) = (\nabla f)\times\vec{\mathbf{A}} + f(\text{curl}\vec{\mathbf{A}})$$

$$\text{curl}(\vec{\mathbf{A}}+\vec{\mathbf{B}}) = \text{curl}\vec{\mathbf{A}} + \text{curl}\vec{\mathbf{B}}$$

$$\text{curl}(\text{curl}\vec{\mathbf{A}}) = \nabla(\text{div}\vec{\mathbf{A}}) - \nabla^2\vec{\mathbf{A}}$$

where $(\vec{\mathbf{A}}\cdot\nabla)\vec{\mathbf{B}} = \left(A_x\frac{\partial B_x}{\partial x} + A_y\frac{\partial B_x}{\partial y} + A_z\frac{\partial B_x}{\partial z}\right)\hat{\mathbf{i}} + \cdots$.

15.4.2 INTEGRATION

There are two types of calculus integration important for introductory physics: line and surface.

Line Segment

Figure 15.19 (a)

LINE INTEGRATION

A **line integral** integrates an arbitrary vector $\vec{\mathbf{A}}$ over a line segment (Fig. 15.19 (a)), $\int_{line}\vec{\mathbf{A}}\cdot d\vec{\mathbf{s}}$, or a closed loop (Fig. 15.19 (b)), $\oint_{loop}\vec{\mathbf{A}}\cdot d\vec{\mathbf{s}}$.*

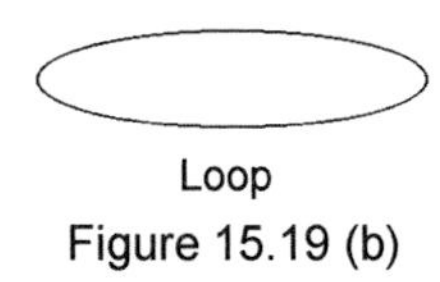

Loop

Figure 15.19 (b)

The integrand in these equations contains a dot product between $\vec{\mathbf{A}}$ and a differential $d\vec{\mathbf{s}}$, where $d\vec{\mathbf{s}}$ is a infinitesimally small vector that points in the direction of motion along the line or loop (Fig 15.19 (c)):

$$d\vec{\mathbf{s}} = dx\hat{\mathbf{i}} + dy\hat{\mathbf{j}} + dz\hat{\mathbf{k}}$$

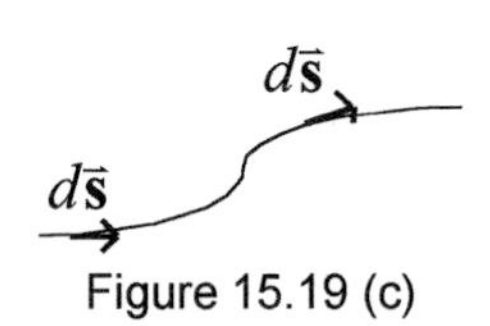

Figure 15.19 (c)

So, from the definition of dot products, we can write the line integral as:

$$\int_{line}\vec{\mathbf{A}}\cdot d\vec{\mathbf{s}} = \int_{line} A_x dx + A_y dy + A_z dz$$

* A line integral is said to be **independent of path** if it reduces to an ordinary integral with endpoints equal to the endpoints of the line: $\int_{line}\vec{\mathbf{A}}\cdot d\vec{\mathbf{s}} = \int_a^b \vec{\mathbf{A}}\cdot d\vec{\mathbf{s}}$. Any vector for which the line integral is independent of path is called **conservative**.

EXERCISE 15.23: Line Integral

Calculate the line integral of the vector from Exercise 15.22 over the line segment created by a parabola $y = x^2$ from $(0, 0)$ to $(2, \sqrt{2})$.

The line integral is important for calculations of the magnetic field. See Section 7.3.2 for more.

SOLUTION:

To calculate this integral, we first write the dot product in the integrand as:

$$A_x dx + A_y dy = 5x^2 dx - 7xdy$$

We want to write this totally in terms of the x variable (*i.e.*, we do not want the dy term), so we implicitly derive $y = x^2 \Rightarrow dy = 2xdx$. Therefore, we can rewrite the integrand as:

$$5x^2 dx - 7x(2xdx) = (5 - 14)x^2 dx = -9x^2 dx\,.$$

Since the integrand is now totally in terms of x, we can integrate normally. The endpoints of the curve are 0 and 2:

$$\int_0^2 -9x^2 dx = (-3x^3)\Big|_0^2 = -3(8) = -24$$

SURFACE INTEGRATION

A **surface integral** integrates an arbitrary vector $\vec{\mathbf{A}}$ over an open surface, $\int_{surface} \vec{\mathbf{A}} \cdot d\vec{\mathbf{a}}$, or a closed surface, $\oint_{surface} \vec{\mathbf{A}} \cdot d\vec{\mathbf{a}}$.

The surface integral is important for calculating Gauss' Law. See Section 7.3.2.

The integrand in these equations contains a dot product between $\vec{\mathbf{A}}$ and a vector differential $d\vec{\mathbf{a}}$, where $d\vec{\mathbf{a}}$ is a infinitesimally small area vector that points normal to the surface (Fig 15.20):

$$d\vec{\mathbf{a}} = da_x\hat{\mathbf{i}} + da_y\hat{\mathbf{j}} + da_z\hat{\mathbf{k}}$$

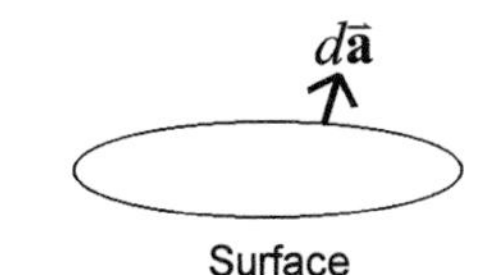

Figure 15.20

So, we can write the surface integral as:

$$\int_{surface} \vec{\mathbf{A}} \cdot d\vec{\mathbf{a}} = \int_{surface} A_x da_x + A_y da_y + A_z da_z$$

EXERCISE 15.24: Surface Integral

Calculate the surface integral of the vector from Exercise 15.22 over the open surface created by the circle $x^2 + z^2 = 25$.

SOLUTION:

Because the formula for the circle has only the variables x and z, we know that the circle lies in the x-z plane (in Section 13.5.3). The direction of the circle is perpendicular, or normal, to the circle. Therefore, the direction of the circle is in the positive y-direction ($+\hat{\mathbf{j}}$).

Exercise 15.24 Continued...

Using the equation $\int_{surface} \vec{\mathbf{A}} \cdot d\vec{\mathbf{a}} = \int_{surface} A_x da_x + A_y da_y + A_z da_z$ from above, we can rewrite the surface integral as:

$$A_x da_x + A_y da_y + A_z da_z = 0 - 7xdxdz + 0$$

We know the radius of the circle is 5. Thus, the x variable of the integral will range from 0 to 5. The range of the z can be determined by solving the circle equation above:

$$z = \sqrt{25 - x^2}$$

Using this information, we can now integrate normally:

$$\int_0^5 \int_0^{\sqrt{25-x^2}} -7xdzdx = \int_0^5 -7x(\sqrt{25-x^2})dx$$

We leave this ugly integral to integral tables to solve.

Sometimes a line or a surface integral has an easier solution when rewritten. The following theorems help us to rewrite these:

$$\oint_{loop} \vec{\mathbf{A}} \cdot d\vec{\mathbf{s}} = \int_{surface} (\nabla \times \vec{\mathbf{A}}) \cdot d\vec{\mathbf{a}}$$ **(Stoke's Theorem)**

$$\oint_{surface} \vec{\mathbf{A}} \cdot d\vec{\mathbf{a}} = \int \nabla \cdot \vec{\mathbf{A}} dV$$ **(Gauss' Divergence Theorem)**

$$\oint_{surface} \vec{\mathbf{A}} \times d\vec{\mathbf{a}} = -\int \mathrm{curl} \vec{\mathbf{A}} dV$$

$$\oint_{surface} f d\vec{\mathbf{a}} = \int \nabla f dV$$

$$\oint_{loop} f d\vec{\mathbf{s}} = -\int_{surface} \nabla f \times d\vec{\mathbf{a}}$$

15.5 CONCLUSION

Intuitively we would expect that physical behavior should not depend on the arbitrary location and orientation of the coordinate system we choose for describing the behavior.... Vectors were invented to take advantage of this experimental fact.
DONALD G. IVEY
American Engineer, 1935 -

This chapter introduced and discussed scalars, vectors, and vector mathematics. Although it may seem new and difficult now, with enough practice, vector math—like the scalar math of algebra, geometry, and trigonometry—will eventually become just another familiar tool in your physics toolbox. And as we know by now, the more tools we have in our toolbox, the easier it is to learn physics!

16
MATHEMATICS FOR DATA SETS

16.1 INTRODUCTION

The result of any experiment is a set of measurements called the **data set**, or the **data**. Experimental data sets can be grouped into two rough categories:

- **Analytical**: A *measured* quantity is compared against one or more *variable* quantities, in order to discover a relationship, if any, between them.
- **Statistical**: The *frequency* of a particular experimental outcome is *counted* and compared to the frequency of all other possible outcomes of the same experiment, in order to determine the relative frequencies of all events.

Science is an interconnected series of concepts and conceptual schemes that have developed as the result of experimentation and observation, and are fruitful of further experimentation and observation.
JAMES CONANT
President of Harvard, 1893 - 1978

EXAMPLE 16.1: Dueling Data Sets

As an example of the first kind of data set, think of an experiment in which a car speeds up. The task of the experimenter is to measure the car's *speed* at different *times*. Speed is thus the *measured* quantity, and time is the *variable* quantity. The purpose of this experiment is to determine if and how speed is *related* to time for the car.

Example 16.1 Continued...

As an example of statistical data sets, think of an experiment in which cards are dealt from a normal deck. The task of the experimenter, in this case, is to *count* the number of times a face card is played, as compared to the number of times a numbered card is dealt. The purpose of the experiment is to *compare* the frequency of face cards to the frequency of numbered cards.

In order for measurements to have any value to the experimenter, the data needs to be analyzed mathematically. The mathematics of data sets, and the related topic of probability, are the subjects of this chapter.

16.2 ANALYTICAL DATA SETS

In an analytical experiment, there are (at least) two kinds of data: the measured quantity, called the **dependent variable**, and the varied quantity, known as the **independent variable**. In order to correctly interpret the experiment, these quantities must be analyzed *together*. The analysis often includes organizing the data into tables and graphs, fitting the data to a known function, and analyzing errors in the measurements.

16.2.1 TABLES & GRAPHS

For the sake of persons of...different types, scientific truth should be presented in different forms, and should be regarded as equally scientific, whether it appears in the robust form and vivid coloring of a physical illustration, or in the tenuity and paleness of a symbolic expression.
JAMES CLERK MAXWELL
Scottish Physicist, 1831-1879

A **table** is a columnar listing of the independent and dependent variables, such that each variable is listed in a separate column (with the dependent variable(s) to the right). A good table has a title, column headings labeling the variables and their units, and a caption explaining the table's purpose (optional).

A **graph** is plot of the variables, such that each variable has a different axis (with the dependent variable on the vertical axis). A good graph has a title, axis tags that label the variables and their units, and a caption (optional). The plot is scaled so that the graph entirely fills the paper. Points on the graph are connected by a smooth line.*

EXAMPLE 16.2: Speed through Time

A ball falls from a tall building. Its speed is measured every second. The results of this experiment are tabulated and plotted below.

* Errors, if any, are indicated on the graph via proportionately sized symbols.

Example 16.2 Continued...

SPEED AS A FUNCTION OF TIME

Time (s)	Speed (m/s)	Time (s)	Speed (m/s)
0	0	5	52
1	7.2	6	60
2	22	7	71
3	32	8	77
4	39	9	89

This table records the speed of a falling ball, as measured every second.

SPEED AS A FUNCTION OF TIME

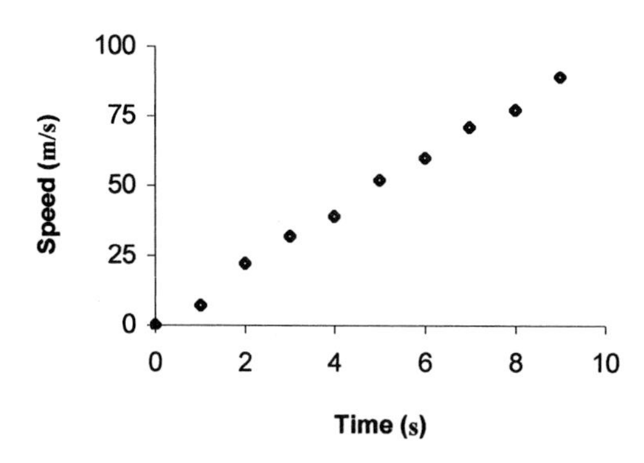

This plot records the speed of a falling ball, as measured every second.

16.2.2 CURVE-FITTING

Once it has been neatly presented in a table and a graph, the data set must be analyzed mathematically. We do this by *matching* the curve of the graph to a known function in a procedure called **curve fitting**.

Curve fitting can be a highly intricate process involving advanced mathematics: in other words, it is not the domain of introductory students. However, in many cases, we can skip the difficult math because only an *approximate* fit is necessary.

To make a *quick* curve fit, examine the plot and choose a function that (1) best matches the curve of your graph *or* that (2) is dictated by an appropriate theory. Note that the chosen function need not go through every point in the plot; it only needs to be suggestive of the general shape of the entire graph.

See Section 13.4.12 for some simple functions.

EXERCISE 16.3: A Simple Curve Fit

What function approximately fits the graph in Example 16.2?

SOLUTION:

A straight line will best fit this graph. We know this by the obvious shape of the plot, but we also know that, in a gravitational field, there

For more on this equation, see Section 5.4.2.

Exercise 16.3 Continued...

is a linear relationship between the speed of a falling object and time: *i.e.*, $v_y = v_{i,y} + gt$. Therefore, a line is the best choice for the curve fit.

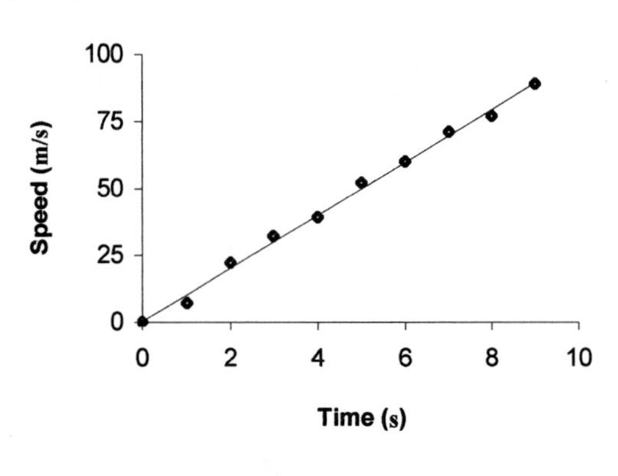

Man is the interpreter of nature; science, the right interpretation.
WILLIAM WHEWELL
English Philosopher, 1794 - 1866

In some cases, experimental data sets may have missing data or be otherwise incomplete. In these cases, you must *guess* the trend of the data in the missing regions in order to fit a curve to the plot. **Interpolation** is the process by which we guess the trend *between* real data points. **Extrapolation** is the process by which we guess the trend *outside* of known values.

EXERCISE 16.4: Interpolation and Extrapolation

In figure (a) below, *interpolate* the curve in the area of the missing data. In figure (b), *extrapolate* the curve into the region beyond the given data.

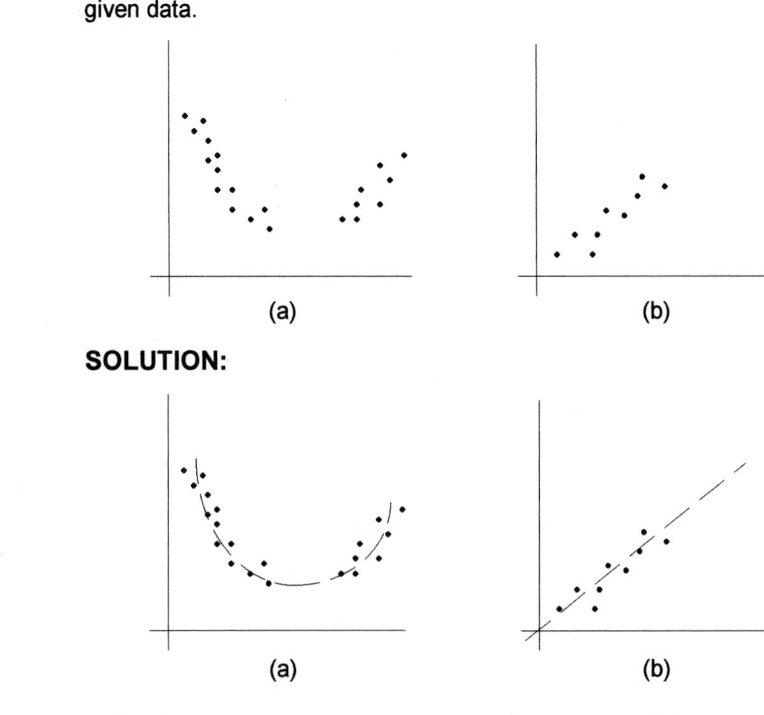

The purpose of curve fitting is to reduce the many measurements of an experiment to a few, convenient numerical **parameters** that (1) describe the data succinctly, and (2) relate the data to theoretical predictions.

EXAMPLE 16.5: Curve Fitting Explained!

For example, by fitting a line to the curve of Example 16.2, we can calculate a slope m and a y-intercept b in the point-slope equation $y = mx + b$. These two parameters (1) describe the line completely, in lieu of the entire set of data, and, more importantly, (2) have *physical meaning* (they correspond to the gravitational field g and the initial velocity v_i, respectively, in the equation $v_y = v_{i,y} + gt$).

For more on the point-slope equation, see Section 13.4.4.

In the special case that the curve fit is a line, you can roughly approximate the values of m and b without any difficult mathematics:

1. Sketch the best-fit line on the plot.
2. m is "rise over run," or $\Delta y / \Delta x$, where (x_1, y_1) and (x_2, y_2) are points at the top and bottom of the line (but are *not* data points).
3. b is the point where the line crosses the y-axis; you may have to extrapolate the best fit line to find it.

EXAMPLE 16.6: Quick Slope and y-Intercept

To quickly estimate the slope of the line in Example 16.2, we sketch a line on the plot (as has already been done for us in that example). Then, we pick two points on this line, one at the top and one at the bottom. We choose the points (8.5 s, 84 m/s) and (0.5 s, 6.0 m/s). So, the slope is:

$$m = \frac{y_2 - y_1}{x_2 - x_1} = \frac{(84\text{ m/s} - 6.0\text{ m/s})}{(8.5\text{ s} - 0.5\text{ s})} \approx 9.8\text{ m/s}^2$$

This is very close to the accepted value for g, as it should be, according to Example 16.5.

Next, we find the y-intercept, or the point where the graph crosses the y-axis. According to the graph in Exercise 16.2, the y-intercept is 0. Thus, the ball started its fall with zero initial velocity, $v_{i,y} = 0$ m/s. Again, this is an expected value, since the ball dropped from rest.

To find a function that *exactly* fits a graph—that is, to find the *exact* relationship between the independent and dependent variables—some fairly complicated mathematics is required.

The kind of mathematics used depends upon the apparent shape of the curve. If a graph appears to be linear—that is, if the data seems to fit the function $y = mx + b$—we use the technique of **least squares fitting**

or **linear regression** to find the best-fit line. Given N pairs of x (independent variable) and y (dependent variable) measurements, we can calculate the value of m and b using these linear regression formulas:

$$m = \frac{N(\sum xy) - (\sum x)(\sum y)}{N\sum x^2 - (\sum x)^2}$$

$$b = \frac{(\sum x^2)(\sum y) - (\sum x)(\sum xy)}{N\sum x^2 - (\sum x)^2}$$

EXERCISE 16.7: Linear Regression

What is the slope and y-intercept of the best-fit line through the graph of Example 16.2?

SOLUTION:

This graph has a linear shape, so we can find the best-fit line using linear regression. To find the slope, we use the equation:

$$m = \frac{N(\sum xy) - (\sum x)(\sum y)}{N\sum x^2 - (\sum x)^2}$$

To use this formula, we must separately calculate each of the sums:

$$\sum x = 0 + 1 + 2 + 3 + 4 + 5 + 6 + 7 + 8 + 9 = 45\ \text{s}$$
$$\sum x^2 = 0^2 + 1^2 + 2^2 + 3^2 + 4^2 + 5^2 + 6^2 + 7^2 + 8^2 + 9^2 = 285\ \text{s}^2$$
$$\sum y = 0 + 7.2 + 22 + 32 + 39 + 52 + 60 + 71 + 77 + 89 = 449.2\ \text{m/s}$$
$$\sum xy = (0)(0) + (1)(7.2) + (2)(22) + (3)(32) + (4)(39) + (5)(52) + (6)(60) + (7)(71) + (8)(77) + (9)(89) = 2837.2\ \text{m}$$

Further, since we have ten measurements in the experiment, $N = 10$. Now, we plug each of these values into the equation above:

$$\Rightarrow m = \frac{(10)(2837.2\ \text{m}) - (45\ \text{s})(449.2\ \text{m/s})}{(10)(285\ \text{s}^2) - (45\ \text{s})^2} = \frac{8158\ \text{m}}{825\ \text{s}^2} \approx 9.9\ \text{m/s}^2$$

Is this number correct? How do we know? Also, how good was the approximation in Example 16.6?

Now, we want to find the y-intercept of this line. To do so, we must use the equation:

$$b = \frac{(\sum x^2)(\sum y) - (\sum x)(\sum xy)}{N\sum x^2 - (\sum x)^2}$$

We have already figured the value of each of the sums in this equation, so we can plug these in without any changes:

$$b = \frac{(285\ \text{s}\ \ )(449.2\ \text{m/s}) - (45\ \text{s})(2837.2\ \text{m})}{(10)(285\ \text{s}^2) - (45\ \text{s})^2} = \frac{348\ \text{m}\cdot\text{s}}{825\ \text{s}^2} \approx 0.42\ \text{m/s}$$

Is *this* number correct? And how does *it* compare to our estimate?

Techniques similar to linear regression can help us fit graphs that are *not* linear. However, these fitting techniques are even more complex than linear regression.* Sometimes we can avoid these techniques by "forcing" non-linear graphs into a linear shape. Once this is done, we can then use linear regression to find the slope and y-intercept of the "forced" line.

The belief that we can start with pure observations alone, without anything in the nature of a theory, is absurd.
KARL POPPER
English Philosopher, 1902 -

There are three similar techniques for forcing non-linear graphs into a linear shape. Each technique involves plotting the curve in a non-standard fashion. Pick the method that best suits the form (or theoretical expectation of the form) of your curve:

- If the function is a polynomial ($y = ax^n$), you have two options, depending on the expected (or theoretical) value of n:
  1. If n is an integer, plot y versus x^n. The resultant graph will be a straight line with slope a and y-intercept equal to 0.
  2. For any value of n—integer or no—plot log y versus log x. The resultant graph will be a straight line with slope n and y-intercept log a.
- If the function is an exponential ($y = ae^{bx}$), plot log y versus x. The resulting graph will be a straight line with slope b and y-intercept log a.

For more on plotting logarithms, see Section 13.4.14.

EXERCISE 16.8: Forcing Linear Graphs

The experiment of Example 16.2 is repeated. The ball's position (rather than its speed) is measured every second:

POSITION AS A FUNCTION OF TIME

Time (s)	Position (m)	Time (s)	Position (m)
0	0	5	–130
1	–4.5	6	–185
2	–15	7	–230
3	–38	8	–305
4	–80	9	–405

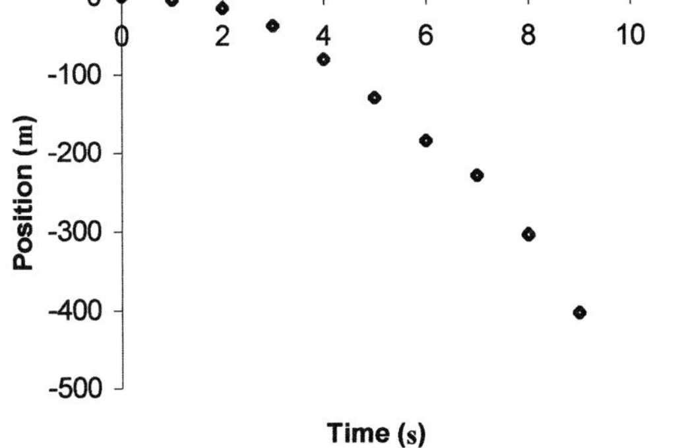

* For more information on these techniques, see, for example, Chapter 8 in the book *An Introduction to Error Analysis: The Study of Uncertainties in Physical Measurements* by John R. Taylor.

Exercise 16.8 Continued...

SOLUTION:

We know, from our understanding of the equations of motion in a gravitational field, that this graph is a polynomial; in particular, $y = -\frac{1}{2}gt^2$. Therefore, we choose the first method above to force a straight line out of this data. According to that method, we must match our theoretical formula $y = -\frac{1}{2}gt^2$ with the polynomial $y = ax^n$. By doing so, we see that $a = -\frac{1}{2}g$, $x = t$, and $n = 2$. So, we want to plot y versus t^2:

POSITION AS A FUNCTION OF TIME SQUARED

Time^2 (s^2)	Position (m)	Time^2 (s^2)	Position (m)
0	0	25	–130
1	–4.5	36	–185
4	–15	49	–230
9	–38	64	–305
16	–80	81	–405

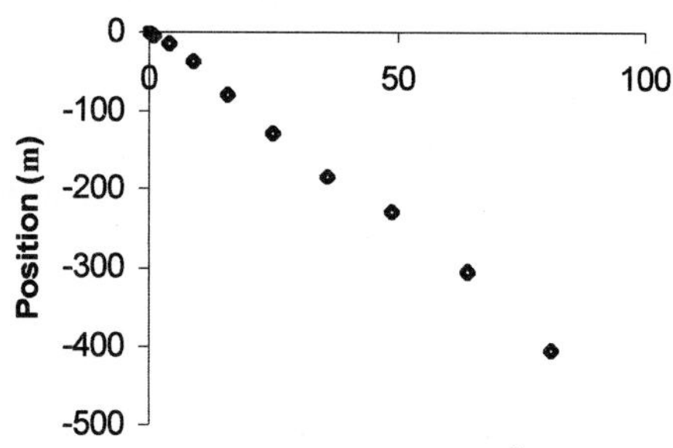

We see that this graph is clearly linear. Thus, we can apply the linear regression equations to calculate m and b. First, we do the sums, taking care to use the right definition of x and y:

$$\sum x = 0+1+4+9+16+25+36+49+64+81 = 285\ \text{s}^2$$

$$\sum x^2 = 0^2+1^2+4^2+9^2+16^2+25^2+36^2+49^2+64^2+81^2 = 15{,}333\ \text{s}^4$$

$$\sum y = 0-4.5-15-38-80-130-185-230-305-405 = -1392.5\ \text{m}$$

$$\sum xy = (0)(0)+(1)(-4.5)+(4)(-15)+(9)(-38)+(16)(-80)+(25)(-130)$$
$$+(36)(-185)+(49)(-230)+(64)(-305)+(81)(-405) = -75{,}191.5\ \text{m}\cdot\text{s}^2$$

Now, we find the slope...

$$m = \frac{(10)(-75{,}191.5\ \text{m}\cdot\text{s}^2)-(285\ \text{s}^2)(-1392.5\ \text{m})}{(10)(15{,}333\ \text{s}^4)-(285\ \text{s}^2)^2} = \frac{-355{,}052.5\ \text{m}\cdot\text{s}^2}{72{,}105\ \text{s}^4} \approx -4.9\ \text{m/s}^2$$

...and the y-intercept:

$$b = \frac{(15{,}333\ \text{s}^4)(-1392.5\ \text{m})-(285\ \text{s}^2)(-75{,}191.5\ \text{m}\cdot\text{s}^2)}{(10)(15{,}333\ \text{s}^4)-(285\ \text{s}^2)^2} \approx \frac{78{,}375\ \text{m}\cdot\text{s}^4}{72{,}105\ \text{s}^4} \approx 1.1\ \text{m}$$

You can do regression analyses easily using your calculator, a standard computer spreadsheet program, or (if any), the software in your physics lab.

16.2.3 ERROR ANALYSIS

Despite considerable effort to do experiments as carefully as possible, there is no such thing as a perfect measurement. *Every measurement must contain some* **error** *or* **uncertainty**.

These kinds of errors are *not* mistakes: mistakes, at least in principle, are avoidable; *errors are not.* A good experimenter works tirelessly to identify and eliminate all mistakes in each experiment, and then to identify, minimize, and catalog all errors.

All measurements, however careful and scientific, are subject to some uncertainties.
JOHN R. TAYLOR
American Physicist/Author, 1939 -

The following sections characterize different kinds of mistakes and errors seen in a typical introductory physics lab, so that you can take these into account when it comes time to do experiments on your own.

SOURCES OF MISTAKES

Experimental mistakes can come from many sources, such as:

- *Inappropriate choice of measuring device:*
  - The measuring device could be inappropriately-scaled, so that it may not be able to measure amounts small or large enough necessary for the experiment;
  - The measuring device could be broken.
- *Incorrect use of measuring device*: such as,
  - Inappropriate training on how to use the device;
  - Insufficient time spent making the measurement;
  - **Parallax**: looking at the measuring device from different angles, so that scale tick marks appear to move.
- *Inappropriate experiment design:* such as,
  - Scaling the experiment so that errors are introduced;
  - Failing to eliminate all possible independent variables, thereby introducing too much environment into the experiment;
  - A lack of appropriate planning.

In each of these cases, appropriate levels of training and care in the design and implementation of the experiment should clear up mistakes.

RANDOM ERROR

There are two types of error in a measurement. The first type is **random error**. The hallmark of random error is that every measurement is uncertain by a *different* amount: some measurements are slightly bigger, and some are slightly smaller, than the actual value.

EXAMPLE 16.9: Random Errors, Part I

As an example of a random error, think of an experiment in which a rod is sliced in half. Before the slice is made, the rod is carefully measured so that the location of the cut will be as precise as possible. In an ideal world, we could make the measurement perfectly, so that we could cut the rod *exactly* in half. In the real world, however, this kind of precision is impossible: one half of the rod will *always* be longer than the other half, introducing random error into the experiment.

With random errors, *the more measurements there are, the more the errors cancel each other out.* Thus, we minimize random errors by taking many measurements of the same type, and averaging the results.

EXAMPLE 16.10: Random Errors, Part II

In previous example, we could improve our experimental accuracy by measuring and cutting *several* rods. In this way, the *average* length of the half-rods would approach the ideal length.

In Section 16.3.2. 

The value of random error for each experiment is usually found in terms of the *standard deviation*, a quantity we will discuss later in the chapter.

SYSTEMATIC ERROR

The second type of error is **systematic error**. Systematic errors alter the measurement in a *single* direction, so these types of errors *don't* cancel out.

EXAMPLE 16.11: Systematic Errors, Part I

For more on friction and drag, see Sections 7.2.6 - 7.2.7. 

As an example of a systematic error, consider the experiment of Example 16.2. In an ideal situation, only the time of fall and the speed of throw determines the final speed of the dropping ball: $v_y = v_{i,y} + gt$. Thus, an experiment performed under ideal conditions will result in speed measurements *exactly* given by this equation.

However, in real life, we must deal with friction, air drag, and other effects that alter the time of fall. Our experimental result will therefore *not* equal the above, ideal value. The difference between our measured value and the calculated value is the systematic error.

Unlike random errors, we cannot reduce systematic errors by repeating the experiment many times. Therefore, we must keep anticipating and excluding all possible sources of systematic error in each experiment. With enough care, we should be able to reduce our systematic errors to such a degree that they become much smaller than any random errors. At that point, we can simply ignore them.

EXAMPLE 16.12: Systematic Errors, Part II

We can minimize the negative effects of friction and drag in the example above by making the ball smooth and very round. And, we could do the experiment in a vacuum. By taking the time and effort to design the experiment well, we should be able to reduce our systematic errors enough to ignore them.

If systematic errors in an experiment can't be ignored, the actual value for the error must be *estimated*. This is a rather tricky skill to master. So, in the typical introductory physics class, it is usually enough to (1) determine the direction in which the systematic error skews the experimental results, and (2) understand that very large systematic errors are *mistakes* in experimental design or execution, and therefore, if you have them, you usually need to go back and redo the experiment.

Aristotle could have avoided the mistake of thinking that women have fewer teeth than men by the simple device of asking Mrs. Aristotle to open her mouth.
BERTRAND RUSSELL
English Philosopher, 1922 - 1970

EXERCISE 16.13: Estimating Systematic Errors

How will the systematic errors in Example 16.11 affect the speed of the falling ball?

SOLUTION:

Friction and air drag *counteract* gravity. Therefore, the ball will drop at a *slower* speed than we would expect from the equation $v_y = v_{i,y} + gt$. So, your calculated value for g will be *smaller* than the accepted value.

PRESENTATION OF ERRORS

In order to fully understand experimental results, we must calibrate, track, and report all errors in every experiment. The general format for reporting error is:

$$(x \pm \delta x) \text{ units}$$

where x is the (average) measurement and δx is the error. This format defines a specific range, $x - \delta x < x < x + \delta x$, in which we have a high degree of confidence that the actual value for the measurement lies.* Note that we are *not absolutely sure* that the real value for the measurement is in this range; we are only *enough* sure to move on.

In this format, the error δx is always reported to *one* significant figure. The last significant figure in the measurement x has the same decimal position as the error.

Significant figures are covered in Section 11.2.

EXAMPLE 16.14: Error Presentation

Imagine an experiment in which the average mass of a rod is measured to be 16.345 g. The error in this measurement is found to be 0.573 g. We want to present this result in correct error format. First, we know that the error itself can only have one significant figure. Thus, we round the error to 0.6 g. Then, we know that the measurement can only be written out to the same decimal position as the error. Since the error is in the tenth decimal place, the measurement must also end in the first decimal position. Therefore, we round the result to 16.3 g. Finally, we want to put these together into the correct error format. Thus, our answer is written (16.3 ± 0.6) g.

Rounding is discussed in Section 11.4.

* In fact, we have a 68% confidence level. See Section 16.3.2 for more.

Occasionally, the error of a measurement is presented in terms of a ratio, called the **percent** or **fractional error**:

$$\text{percent error} = \frac{\delta x}{x}$$

EXAMPLE 16.15: Percent Error

In order to write our previous error in terms of percent error, we must divide the error by the measurement itself. Thus, the percent error in that example is $0.6/16.3 = 0.0368$ or 3.7%.

ESTIMATING ERRORS ON A SCALE

The error on reading a scale is found by one of the following two methods:

- When the scale divisions are close together, the error is $\pm$ half of the lowest scale division.
- When the scale divisions are far enough apart that you can make a good guess on the value of the measurement and the error, do.

EXAMPLE 16.16: Reading Scale Divisions

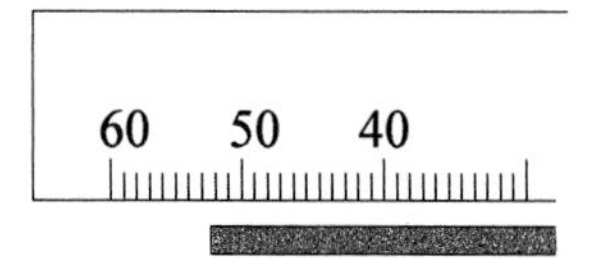

Figure 16.1 (a)

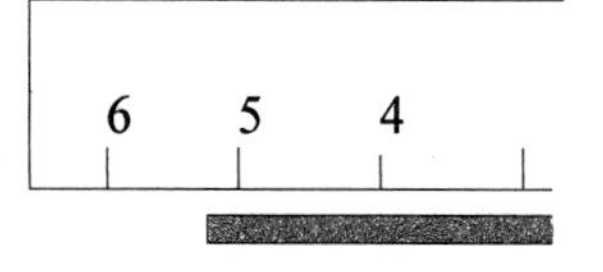

Figure 16.1 (b)

As an example of the first rule for finding the error on a scale, think about measuring the length of a rod (Fig. 16.1 (a)). The rod is lined up with a ruler marked in millimeters. In this measurement, we can see that the rod's length lies somewhere in between 53 and 54 mm. So, we can write:

$$(53.5 \pm 0.5)\ \text{mm},$$

where 53 mm represents our best estimate of the rod's length, and 0.5 mm is half of the lowest scale division on the ruler.

As an example of the second rule, think about measuring the same rod with a ruler that is marked in centimeters rather than millimeters (Fig. 16.2 (b)). Now, the demarcations on the ruler are wide enough so that we must *estimate* the length of the rod and its error. The rod's length is closer to 5 cm than 6 cm, so we guesstimate the length is about 5.3 cm, and error is give or take about 0.2 cm:

$$(5.3 \pm 0.2)\ \text{cm}$$

Notice that the error in this case is larger than in the previous one, since the measurement here is necessarily more uncertain.

PROPAGATING ERRORS

We must take special care when we mathematically combine different measurements x_1, x_2, ..., and their errors δx_1, δx_2, ..., because errors get worse, or **propagate**, when combined with other errors. We use the equations in Table 16.1 below to manipulate random errors.

Table 16.1: Equations for Manipulating Errors

PROCEDURE	GENERAL FORMULA	ERROR FORMULA
ADDING/ SUBTRACTING	$(x_1 \pm \delta x_1)+(x_2 \pm \delta x_2)=(x_1+x_2) \pm \sqrt{\delta x_1^2+\delta x_2^2}$	$\delta S=\sqrt{\delta x_1^2+\delta x_2^2}$
MULTIPLYING	$(x_1 \pm \delta x_1)(x_2 \pm \delta x_2) \approx (x_1 x_2)\left(1 \pm \left[\left(\frac{\delta x_1}{x_1}\right)^2+\left(\frac{\delta x_2}{x_2}\right)^2\right]^{1/2}\right)$	$\delta P=P\sqrt{\left(\frac{\delta x_1}{x_1}\right)^2+\left(\frac{\delta x_2}{x_2}\right)^2}$
DIVIDING	$\frac{(x_1 \pm \delta x_1)}{(x_2 \pm \delta x_2)} \approx \frac{x_1}{x_2}\left(1 \pm \left[\left(\frac{\delta x_1}{x_1}\right)^2+\left(\frac{\delta x_2}{x_2}\right)^2\right]^{1/2}\right)$	$\delta Q=Q\sqrt{\left(\frac{\delta x_1}{x_1}\right)^2+\left(\frac{\delta x_2}{x_2}\right)^2}$
MULTIPLYING EXACT NUMBERS	If $y=Bx$,	Then $\delta y=B\delta x$
POLYNOMIALS	If $y=x^n$,	Then $\frac{\delta y}{y}=n\frac{\delta x}{x}$
FUNCTIONS	If f is a function, $f(x)$,	Then $\delta y=\left\lvert\frac{dy}{dx}\right\rvert\delta x$

EXERCISE 16.17: Error Propagation

A rod is measured and found to have length of 74.1 ± 0.4 cm. Use this result, and the rules of error propagation above, to find:

(a) The length of a new rod made by gluing two rods together.

(b) The speed of an ant crawling along one rod, if it took the ant a time of 30.5 ± 0.8 s to make the trip.

(c) The circumference of a circle with the rod as its radius.

(d) The result of the function $f(x)=x^2+1$, where x is the length of the rod.

SOLUTION:

(a) To find the length of the new rod, we must add the lengths and errors of the original rods:

$$
\begin{aligned}
(L_1+\delta L_1)+(L_2+\delta L_2) &= (L_1+L_2)+\sqrt{(\delta L_1)^2+(\delta L_2)^2} \\
&= (74.1+74.1) \pm \sqrt{(0.4)^2+(0.4)^2} \\
&= (148.2 \pm 0.6)\ \text{cm}
\end{aligned}
$$

Exercise 16.17 Continued...

(b) To find the speed of the ant, we must divide the length of the rod by the time taken to traverse it:

$$\frac{(L+\delta L)}{(t+\delta t)} = \frac{L}{t}\left\{1 \pm \left[\left(\frac{\delta L}{L}\right)^2 + \left(\frac{\delta t}{t}\right)^2 + \cdots\right]^{1/2}\right\}$$

$$= \frac{74.1}{30.5}\left\{1 \pm \left[\left(\frac{0.4}{74.1}\right)^2 + \left(\frac{0.8}{30.5}\right)^2\right]^{1/2}\right\}$$

$$= (2.43 \pm 0.06)\ \text{cm/s}$$

(c) The circumference of a circle is found via the equation $C = 2\pi r$, where 2π is an exact number, and r, in this case, is given by the measured length of the rod. Thus, the circumference is $C = 2\pi$ $(74.1\ \text{cm}) = 465.6\ \text{cm}$.

Now, according to the rule above, the error on C is found from the equation $\delta C = 2\pi\delta r$. Since we know the error on r is 0.4 cm, we find that $\delta C = 2\pi(0.4\ \text{cm}) = 2\ \text{cm}$.

Finally, we must write our answer in correct error notation (including rounding!). We therefore combine our results as $C = (466 \pm 2)$ cm.

(d) When we input our measurement into the function $f(x)$, we get $f(74.1\ \text{cm}) = (74.1\ \text{cm})^2 + 1 = 5491$ cm. Then, when we input the error into the derivative of the function $f'(x) = 2x$, according to the rule above, we get $\delta f = 2(74.1\ \text{cm})(0.4\ \text{cm}) = 59\ \text{cm}^2$. Thus, our final answer (with appropriate rounding) is $(5500 \pm 60)\ \text{cm}^2$.

16.3 STATISTICAL DATA SETS

There are three kinds of lies: lies, damned lies, and statistics.
BENJAMIN DISRAELI
English Statesman, 1804 - 1881

Statistical measurements are essentially *counts* of the frequency of certain events, relative to counts of other events from the same experiment. Statistical analysis involves organizing the count data into tables and graphs, characterizing the shape of the graph, and using the results to predict future experiments.

16.3.1 GRAPHING & HISTOGRAMS

Statistical analysis begins by organizing data into tables and graphs, both of which are called **frequency distributions** f. While statistical tables are similar to those we saw in the last section, statistical graphs—called **histograms**—look completely different from those seen above.

A histogram is a *bar graph*: the height of each bar in the graph is proportional to the frequency of a certain number of events; while the width of each bar (called the **bin**) is proportional to the number of events counted in each bar. Usually (but not always) each bar in a given graph has the same bin so that the bars can be fairly compared with each other.

EXAMPLE 16.18: A Histogram

As an example of a histogram, consider an experiment in which the heights of 3000 men are measured. Our data set thus consists of 3000 different measurements.

Note that a plot of 3000 single heights would exhibit no trend. Therefore, we must sort the measurements into *bins*. For example, all men with heights between 65.5 and 66.5 inches will be binned into *one* category of 66 in. Thus, the *bin size*, in this example, is one inch wide.

When the binning is complete, our frequency distributions are:

HEIGHTS OF 3000 MEN (in Inches)

x (in)	f	x (in)	f	x (in)	f
55	0	62	175	69	177
56	1	63	317	70	97
57	1	64	393	71	46
58	6	65	462	72	17
59	23	66	458	73	7
60	48	67	413	74	4
61	90	68	264	75	0

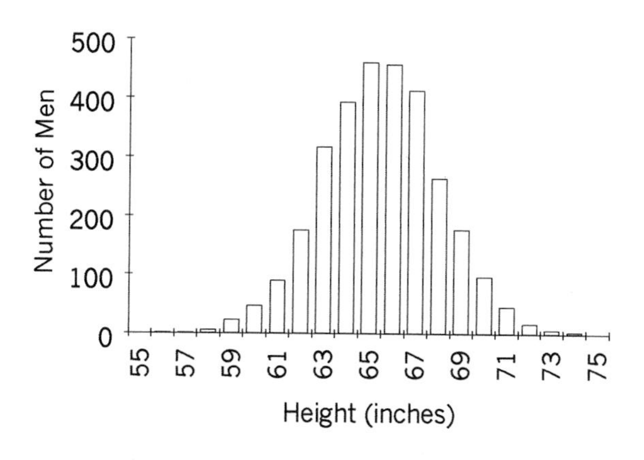

Statistical data sets are sometimes graphed in the form of a **probability distribution** p rather than a frequency distribution f. In such a distribution, the frequency of *each* measurement is divided by the total number of *all* measurements N, creating a fraction $p = f / N$. This fraction represents the *probability* that a measurement of a certain type will be made.

EXERCISE 16.19: A Probability Distribution, Part I

Recast the frequency distributions of the previous example in terms of probability distributions. What is the probability for an adult male of being 63 in tall?

SOLUTION:

To determine the probability distribution, we must divide each measurement in the frequency distribution by the total number of measurements made. In Example 16.18, there are 3000 men, so, for example, the probability of an adult male being 63 inches high is $317/3000 = 0.10$. Our complete probability distributions become:

PROBABILITY OF HEIGHTS IN ADULT MEN

x (in)	p	x (in)	p	x (in)	p
55	0	62	0.058	69	0.059
56	0.00033	63	0.10	70	0.032
57	0.00033	64	0.13	71	0.015
58	0.0020	65	0.15	72	0.0057
59	0.0077	66	0.15	73	0.0023
60	0.016	67	0.14	74	0.0013
61	0.030	68	0.088	75	0

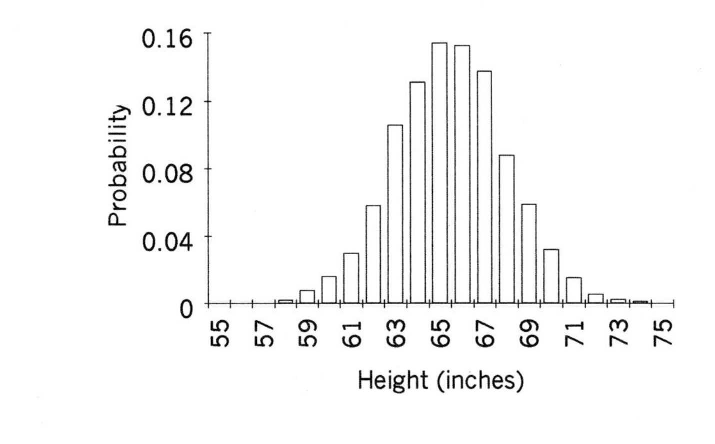

Science has, as its whole purpose, the rendering of the physical world understandable and beautiful. Without this, you have only tables and statistics.
J. ROBERT OPPENHEIMER
American Physicist, 1904 - 1967

As the number of measurements N in a statistical data set gets very high, the distribution stops looking like a bar graph and starts to look like a smooth curve or function, called the **limiting distribution** $f(x)$ or $p(x)$. Whereas frequency distributions are approximate representations of the data, a true limiting distribution is *exact.*

EXAMPLE 16.20: A Limiting Distribution

For example, if we expanded our experiment of the last two exercises to include 3,000,000 men (rather than 3000), the bins of the histogram would have to get smaller to accommodate the increased "volume" of measurements. As a result, the curve would appear smoother, more like a function:

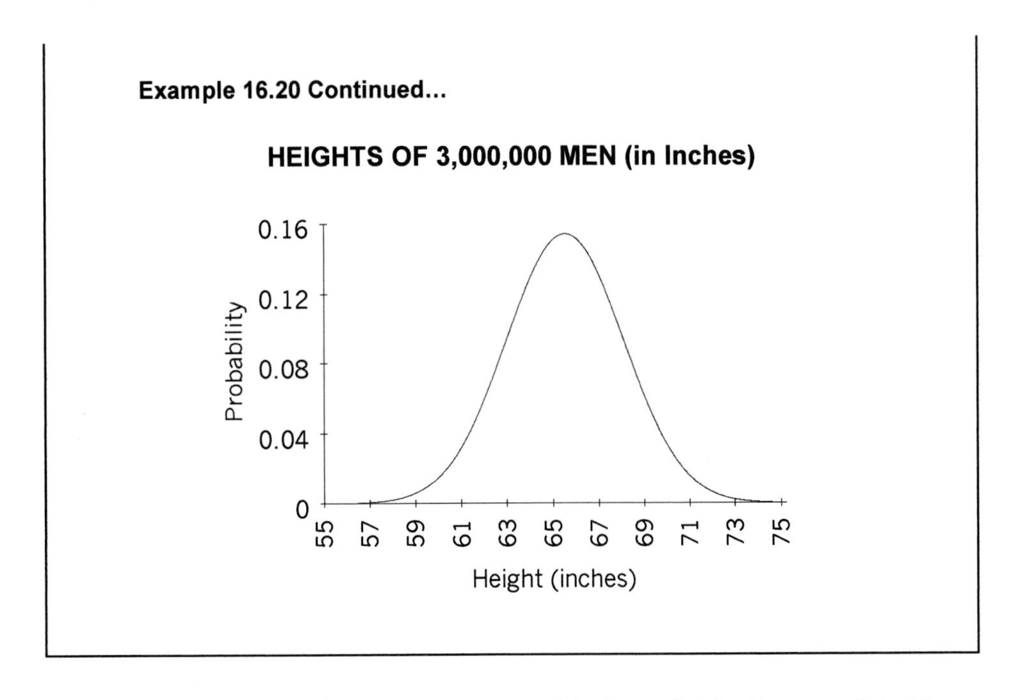

There are several important types of limiting distributions, as listed in Table 16.2 below.

Table 16.2: Important Limiting Distributions

NAME	WHERE IT ARISES	FORMULA	PARAMETERS
GAUSSIAN	Experiments measuring a large number of random, independent variables	$f(x)=\frac{1}{\sigma\sqrt{2\pi}}e^{-(\bar{x}-x)^2/2\sigma^2}$	σ = Width Parameter $\bar{x}$ = Center of Curve
BINOMIAL	Experiments in which there are only two possible outcomes; *i.e.*, heads *or* tails.	$f(x)=\frac{N!}{x!(N-x)!}p^x q^{N-x}$	p = Probability of First Outcome q = Probability of Second Outcome = $1-p$ N = Number of Trials x = Number of Times First Outcome Recorded
RANDOM WALK	Special case of the binomial distribution when $p=q=\frac{1}{2}$.	$f(x)=\frac{1}{2^N}\frac{N!}{x!(N-x)!}$	N = Number of Trials x = Number of Times First Outcome Recorded
POISSON	Experiments measuring events that occur randomly, but at a low, constant rate μ.	$f(x)=e^{-\mu}\frac{\mu^N}{N!}$	μ = Average Rate of Events
EXPONENTIAL	Experiments in which waiting times occur randomly, but at a constant average rate λ.	$f(x)=\lambda e^{-\lambda x}$	λ = Average Rate of Waiting Times

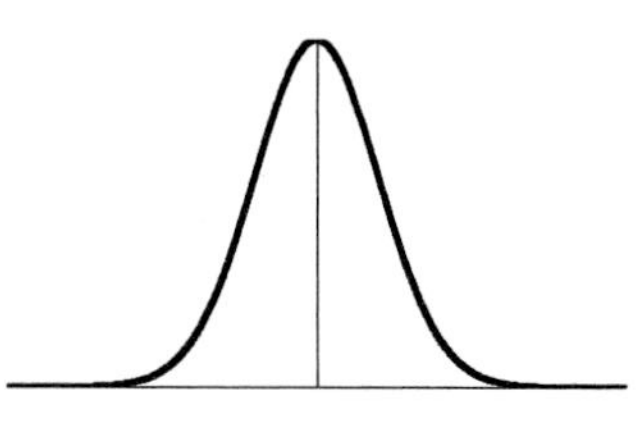

Figure 16.2

Of these five limiting distributions, by far the most important is the **Gaussian distribution**. The importance of this distribution is underscored by its many different names: the bell-shaped distribution, the normal distribution, and the error distribution (Fig. 16.2).

EXAMPLE 16.21: The Gaussian Distribution

For example, the heights of 3000 men in Example 16.18 has a Gaussian limiting distribution, as sketched in Example 16.20.

16.3.2 MEASURES OF THE CURVE

In addition to graphing frequency distributions, we wish to mathematically characterize their shape in terms of **parameters**.

There are three main types of distribution parameters: those that describe the location of the "center" of the curve; those that describe how "wide" the distribution is around the center; and those that describe the general shape of the curve.

MEASURES OF LOCATION

Research is formalized curiosity. It is poking and prying with a purpose.
ZORA NEALE HURSTON
American Author, 1903 - 1960

There are three commonly used measures for the location of the "center" of a distribution: the mean, the median, and the mode.

The **mean** or **average** of the distribution, $\overline{x}$, is by far the most important measure of the three. It is simply the sum of all measurements x in the data set, divided by the total number of measurements N:

$$\overline{x} = \frac{1}{N}\sum x$$

EXERCISE 16.22: Finding the Mean, Part I

Consider an experiment in which a box of thirty quarters is flipped ten times. After each flip, the number of heads is recorded, as tabulated:

NUMBER OF HEADS PER FLIP				
11	16	17	15	17
16	19	18	15	13

What is the mean number of heads flipped in this experiment?

SOLUTION:

To solve this problem, we use the formula given above, $\overline{x} = \frac{1}{N}\sum x$:

$$\Rightarrow \overline{x} = \frac{1}{10}(11+16+17+15+17+16+19+18+15+13) = 15.7$$

Thus, the mean (or average) value of this data set is 15.7.

We can also calculate the mean via the frequency distribution f...

$$\overline{x} = \frac{1}{N}\sum fx$$

...or the probability distribution p:

$$\overline{x} = \sum px$$

These methods of finding the mean are sometimes called **weighted averages**, because each measurement is *weighted* by its frequency or probability.

EXERCISE 16.23: Finding the Mean, Part II

Calculate the mean number of heads for the coin-flip experiment in Exercise 16.22, using weighted averages.

SOLUTION:

To use the above formulas, we first must develop frequency and probability distributions for the experiment. We do this by simply counting up the incidence of each event:

x	11	12	13	14	15	16	17	18	19
f	1	0	1	0	2	2	2	1	1
p	0.1	0.0	0.1	0.0	0.2	0.2	0.2	0.1	0.1

Our formula for the frequency weighted average is $\overline{x} = \frac{1}{N}\sum fx$, so:

$$\overline{x} = \frac{1}{10}[(1)(11) + (1)(13) + (2)(15) + (2)(16) + (2)(17) + (1)(18) + (1)(19)] = 15.7$$

Likewise, the formula using the probability distribution is $\overline{x} = \sum px$, so:

$$\begin{aligned}\overline{x} &= [(0.1)(11) + (0.0)(12) + (0.1)(13) + (0.0)(14) + (0.2)(15) \\ &\quad +(0.2)(16) + (0.2)(17) + (0.1)(18) + (0.1)(19)] \\ &= 15.7\end{aligned}$$

Both times, we get the same answer as in the previous example.

The **median** of a distribution is the *middle* measurement of the data set—when all measurements are sorted from the highest to lowest. The median is sometimes used when there are many "outlying" measurements that make the mean unreliable.

The **mode** is the measurement that occurs *most often* in the data. A mnemonic to keep these two parameters separate is "the *mo*de occurs *mo*st frequently, while the *med*ian is in the *mid*dle."

EXERCISE 16.24: The Median and the Mode

Find the median and the mode for the height experiment of Example 16.18.

SOLUTION:

To find the median, we must first sort the frequency distribution into descending list. This has already been done for us (in that example), so we just examine this list and choose the number that is in the middle, counting equally from both ends of the list. This number is 66 in. Thus, the median for the data set is 66 in.

To find the mode, we choose the entry that occurs the most frequently; *i.e.*, has the largest value for f. This number is 65 in, which occurs 458 times in the data set. Thus, the mode is 65 in.

From this discussion, we see that the definition of the location of the "center" of the distribution depends on which parameter—mean, median, or mode—is chosen. Always label *your* choice.

MEASURES OF DISPERSION

Once we locate the center of the distribution, we next need to determine the "width" of the distribution. A distribution *has* width because measurements cluster around the central value, but rarely hit it exactly. The less the measurements cluster around the central value, the wider the curve, and the more random the measurements.

The general term for curve width is **dispersion**. There are many measures of dispersion, some more important than others. The first step in calculating any of these involves finding the **deviation** of each measurement, or the amount that each measurement differs from the mean:

$$d = \overline{x} - x$$

Deviations can be positive or negative.

EXERCISE 16.25: Deviation

What is the deviation for the first measurement in our coin-flip experiment?

SOLUTION:

The deviation of each measurement is just the difference between it and the mean, so, for the first measurement:

$$d = \overline{x} - x = 15.7 - 11 = +4.7$$

Different measures of dispersion are summarized in Table 16.3 below:

Table 16.3: Measures of Dispersion

NAME	FORMULA	NAME	FORMULA
MAXIMUM DEVIATION	Largest value of the deviation, without regard to sign	STANDARD DEVIATION II	$\sigma_{imp} = \sqrt{\frac{1}{N-1}\sum d^2}$
AVERAGE DEVIATION	$\bar{d} = \frac{1}{N}\sum \lvert d \rvert$	STANDARD DEVIATION FROM THE MEAN	$\sigma_{\bar{x}} = \frac{\sigma}{\sqrt{N}}$
STANDARD DEVIATION I	$\sigma = \sqrt{\frac{1}{N}\sum d^2}$	VARIANCE	$\sigma^2 = \frac{1}{N}\sum d^2$

It is sometimes easier to calculate formula I using this equation:

$$\sigma = \sqrt{\frac{1}{N}\left(\sum x^2 - N\bar{x}^2\right)}$$

The most important formula in the table above is the **standard deviation**, also known as the **root-mean-square value.*** It is a measure of the average uncertainty in *each* measurement of the data set. We use formula I for the standard deviation in the usual case when N is moderate to large (greater than 10); otherwise, we use formula II.

The standard deviation from the mean is a measure of the average uncertainty of the *entire* data set.

EXERCISE 16.26: Dispersions

Find (a) the maximum deviation, (b) the average deviation, (c) the standard deviation, (d) the standard deviation from the mean, and (e) the variance for the coin flip experiment. (f) Which would we use, typically, in an introductory lab to indicate the random error?

SOLUTION:

(a) The maximum deviation is simply the largest deviation of the data set, without regard to sign. Therefore, the maximum deviation of this experiment is 4.7.

(b) To find the average deviation, we sum the deviations of the data set, without regard to sign, and divide by the total number of measurements:

$$\bar{d} = \frac{1}{N}\sum \lvert d \rvert = \frac{1}{10}(4.7+2.7+0.7+0.7+0.3+0.3+1.3+1.3+2.3+3.3) = 1.7$$

(c) To find the standard deviation, we first have to pick the right equation. The number of measurements, ten, in this experiment, is not too low, so we can safely use formula I. To do so, we sum the squares of the deviations, divide by N, and take the square root:

* Sometimes the standard deviation is symbolized by x_{rms} rather than σ.

Exercise 16.26 Continued...

$$\sigma = \sqrt{\frac{1}{10}\left[(4.7)^2 + (2.7)^2 + (0.7)^2 + (0.7)^2 + (0.3)^2 + (0.3)^2 + (1.3)^2 + (1.3)^2 + (2.3)^2 + (3.3)^2\right]} = 2.2$$

(d) The standard deviation from the mean is just the standard deviation calculated above divided by the square root of N:

$$\sigma_{\bar{x}} = \frac{2.2}{\sqrt{10}} = 0.7$$

(e) The variance is just the square of the standard deviation:

$$\sigma^2 = (2.2)^2 = 5.0$$

(f) The standard deviation in (c) above is usually chosen to represent the value for random error in most introductory labs.

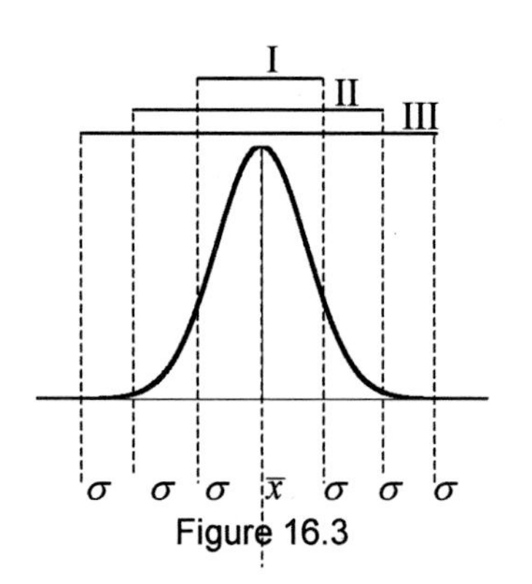

Figure 16.3

The standard deviation has a special meaning in regards to the Gaussian distribution. It demarcates the curve into well-defined regions (Fig. 16.3). In Region I (the area delineated by one standard deviation either side of the mean), the region marked out is 68.4% of the total area und\er the curve. In Region II (the area marked out by two standard deviations), 95.4% of the area is covered, and in Region III (the area defined by three standard deviations), 99.7% of the area is covered.

EXAMPLE 16.27: IQ

The intelligence quotient, or IQ, is described by a Gaussian, with a mean of 100 and a standard deviation of 15. Thus, a genius, with an IQ of 145, has an IQ three standard deviations above the average. Only 0.3% of the population, therefore, are considered geniuses.

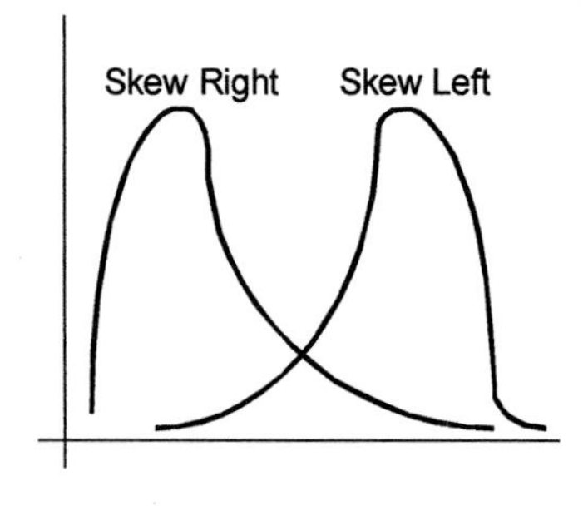

Figure 16.4 (a)

High Kurtosis

Low Kurtosis

Figure 16.4 (b)

A group of measurements is considered **accurate** when the average value is very close to the accepted value for the measurement. A group of measurements is considered **precise** when the standard deviation is low.

MEASURES OF CURVE SHAPE

There are two ways to characterize the general shape of a distribution: **skew**, which describes how much the curve is "weighted" to the left or right (Fig. 16.4 (a)), and **kurtosis**, a description of how "pointy" the peak is (Fig. 16.4 (b)).

16.3.3 PROBABILITY & STATISTICAL DATA SETS

A wonderful aspect of statistical data sets is the ability to *predict* the frequency of future events using the results of previous experiments.

Given a data set with very many measurements, we can use probability distributions to calculate the **probability** that similar events will occur. The more measurements that we have in the set, the better our predictions will be.

When it is not in our power to determine what is true, we ought to act in accordance with what is most probable.
RENE DESCARTES
French Mathematician, 1596 - 1650

EXERCISE 16.28: Probability Distribution, Part I

In the coin-flip experiment of Exercise 16.22, what is the probability that we would obtain 15 heads on a next throw? 14 throws? Anything peculiar about these results?

SOLUTION:

According to the probability distribution for this experiment (Exercise 16.23), the probability we would throw 15 heads in the next trial is $2/10 = 20\%$. On the other hand, according to this same set, the probability that we would obtain 14 heads is $0/10 = 0\%$.

These results are peculiar: we would expect that the probabilities of throwing 14 and 15 heads in an experiment with thirty coins would be *about the same*. Therefore, the results above do not "ring true."

The reason these results are so different from our expectations is that they are based upon an experiment with too few data points.

EXERCISE 16.29: Probability Distribution, Part II

Imagine that we repeat the coin-flip experiment, flipping the coins 100 times, rather than ten, as tabulated below:

NUMBER OF HEADS PER FLIP

11	16	17	15	17	16	19	18	15	13
11	17	17	12	20	18	11	16	17	14
16	12	15	10	18	17	13	15	14	15
16	12	11	22	12	20	12	15	16	12
16	10	15	13	14	18	15	16	13	18
14	14	13	16	15	19	21	14	12	15
16	19	16	14	17	14	11	18	17	16
19	15	14	12	18	15	14	21	11	16
17	17	12	13	14	17	9	13	16	13
17	18	12	16	14	14	15	19	13	14

Now what is the probability that we would obtain 15 heads on the next throw of the die? 14 throws? Do these results make more sense than the previous set?

Exercise 16.29 Continued...

SOLUTION:

To find the probabilities of 14 and 15 heads, we must first count the frequency of each event. 15 heads occurs eleven times in the data; 14 heads occurs fourteen times. Thus, the probability of throwing 15 heads on the next trial is $11/100 = 11\%$. On the other hand, the probability of throwing 14 heads is $14/100 = 14\%$.

These probabilities are approximately equal to each other, as we would expect. Furthermore, they are in the ballpark of the probability we calculated for 15 heads in the last example, showing that even small data sets can sometimes obtain rough predictions for future experiments.

I shall never believe that God played dice with the universe.
ALBERT EINSTEIN
American Physicist, 1879 - 1955

Stop telling God what to do.
NEILS BOHR, to Albert Einstein
Danish Physicist, 1885 - 1962

To calculate probabilities in the case of limiting distributions, plug into the function both the necessary parameters (found in the statement of the problem), and the number of events you are predicting. Then do the indicated math.

EXAMPLE 16.30: More Probability Distributions

Let's calculate the probability of throwing 9, 15, and 17 heads in the coin-flip experiment, assuming a Gaussian distribution applies.

First, we need the Gaussian distribution, $f(x) = \frac{1}{\sigma\sqrt{2\pi}} e^{-(\bar{x}-x)^2 / 2\sigma^2}$.

Then, we need the location of the center and the width parameters of this distribution. A Gaussian that fits the curve of Exercise 16.19 has a center at 15 and a width parameter of 2.5. We plug these numbers into the Gaussian, getting:

$$f(x) = \frac{1}{2.5\sqrt{2\pi}} e^{-(15-x)^2 / 2(2.5)^2} \approx 0.160 e^{-(15-x)^2 / 12.5}$$

Now, we use this "template" function to calculate the probabilities requested in the problem statement by simply plugging in 9, 15, and 17:

$$f(9) \approx 0.160 e^{-(15-9)^2 / 12.5} = 0.160 e^{-2.88} = 0.0090$$

$$f(15) \approx 0.160 e^{-(15-15)^2 / 12.5} = 0.160 e^{0} = 0.16$$

$$f(15) \approx 0.160 e^{-(15-17)^2 / 12.5} = 0.160 e^{-0.32} = 0.12$$

16.4 ELEMENTARY PROBABILITY THEORY

The true logic of this world
is in the language of probabilities.
JAMES CLERK MAXWELL
Scottish Physicist, 1831 - 1879

Probability theory is the study of the frequency at which certain experimental events occur, in order to *predict* the outcomes of future experiments. In general, probability is a very difficult subject to master.

However, in the special case of an experiment with N different, but *equally likely* outcomes, the probability of the occurrence of any one event M is easily found via the equation:

$$p = \frac{M}{N}$$

EXAMPLE 16.31: Probability and Cards, Part I

As an example of elementary probability theory, think of an experiment on a normal deck of cards. What is the probability that you will draw the queen of clubs out of the deck?

In this experiment, all events are *equally likely*: the probability that you will draw the queen of clubs is the same as the probability that you will draw the ace of clubs or the probability that you will draw the five of diamonds. Thus, once we make the calculation for the queen of clubs, we will really have *two* results: the probability that we draw the queen of clubs *and* the probability that we will draw *any* single card.

Let's do the calculation. Our formula is $p = M / N$. There are 52 cards in a deck, so there are $N = 52$ possible experimental results. However, there is only *one* queen of clubs, so $M = 1$ in this case. Therefore, the probability that you will draw the queen of clubs out of a regular deck of cards is $p = 1/52 = 0.0192$, or about 2%.

EXERCISE 16.32: Probability and Cards, Part II

What is the probability that you will draw the following cards out of a regular deck?

(a) One of the 4 aces (b) A diamond

SOLUTION:

(a) $N = 52$, as before. However, there are *four* possibilities for drawing an ace, so $M = 4$. Therefore, $p = 4/52 = 0.077 \approx 8\%$.

(b) There are 13 diamonds in a regular deck of cards, so the probability of drawing one of them is $p = 13/52 = 0.25 = 25\%$.

Sometimes we wish to find the probability of event A *or* event B. We do this by simply adding the probability of each event together:

$$p_{A\ or\ B} = p_A + p_B$$

To find the total probability of event A *and* event B, multiply each of the individual probabilities together:

$$p_{A\ and\ B} = (p_A)(p_B).$$

Be careful, though, when calculating p_A and p_B in this case! The occurrence of event A may *change* the subsequent probability of event B.

EXERCISE 16.33: Probability and Cards, Part III

What is the probability that you will successively draw the following cards out of a regular deck?

(a) The queen of clubs *or* the queen of diamonds.

(b) All four aces (and nothing else).

(c) All of the diamonds (and nothing else).

SOLUTION:

(a) We saw in Example 16.32 that the probability of drawing the queen of clubs was $1/52$. This is the same probability as drawing the queen of diamonds. To find the total probability that we *either* draw the queen of clubs, *or* we draw the queen of diamonds, we must *add* these two probabilities together:

$$\tfrac{1}{52} + \tfrac{1}{52} = \tfrac{2}{52} = 0.038$$

(b) We know that the probability of drawing any one card out of the deck is $1/52$. However, once this card is drawn, the probability of drawing *another* card is *not* $1/52$. There are no longer 52 cards in the deck; there's only 51! Thus, the probability of drawing another card is $1/51$; the probability of drawing the next card is $1/50$, and so on. We would like to draw *four* cards, the aces, from a deck. So, we multiply the probability of each successive event together and get:

$$\left(\tfrac{1}{52}\right)\left(\tfrac{1}{51}\right)\left(\tfrac{1}{50}\right)\left(\tfrac{1}{49}\right) = (\,0.0192)(\,0.0196)(\,0.02)(\,0.0204) = 1.53 \times 10^{-7}$$

This is a very small probability indeed.

(c) There are 13 diamonds in a regular deck of cards, so the probability of successively drawing *all* of them (and nothing else) is:

$$\left(\tfrac{1}{52}\right)\left(\tfrac{1}{51}\right)\left(\tfrac{1}{50}\right)\left(\tfrac{1}{49}\right)\left(\tfrac{1}{48}\right)\left(\tfrac{1}{46}\right)\left(\tfrac{1}{45}\right)\left(\tfrac{1}{44}\right)\left(\tfrac{1}{43}\right)\left(\tfrac{1}{42}\right)\left(\tfrac{1}{41}\right)\left(\tfrac{1}{40}\right)\left(\tfrac{1}{39}\right) = 6.48 \times 10^{-24}$$

16.5 CONCLUSION

This chapter discussed the details of analyzing experimental data sets, including elementary probability theory. Experiments are critical aspects of the scientific method, enabling you to prove that a theory is or is not correct. So without the methods of analysis discussed in this chapter, we would not be able to use the scientific method. You will especially find these techniques useful in the laboratory portion of your physics class.

All science requires mathematics.
ROGER BACON
English Philosopher, 1220 - 1292

APPENDIX: STUDYING FOR SUCCESS

A.1 INTRODUCTION

Research shows that many college-level students do not know how to study properly. You may wonder, How can this be? After all, most college-level students have been attending school for their entire lives!

There are many reasons for this situation. Some students were able to get through their early school years without ever opening a book; therefore, they never learned how to study. Other students, over time, fell into the habit of studying obsessively—but ineffectively—for hours on end; therefore, these students never learned how to study efficiently. Unfortunately, such students will find out the hard way that their past study experiences did not prepare them for college level coursework.

At the college-level, proper studying technique is *essential* for success. For one thing, course material is just too difficult to absorb in a single sitting (say, at a lecture or during a reading assignment). For another, there is usually not enough time in most students' schedules (between classes, social events, extracurricular activities, and so on) to waste on inefficient study patterns.

Therefore, at the college level, it is critical to know how to study properly. After all, studying is a *learned skill*, not an inborn talent: anyone can learn how to study appropriately, given the right instruction and a little dedicated practice. Yes, learning to study does take a bit of extra time at the beginning, but it is well worth it: efficient study patterns *save* you time over the long run.

When I really understand something, it is as if I had discovered it myself.
RICHARD FEYNMAN
American Physicist, 1918 - 1988

The purpose of this Appendix, then, is to provide you with a full range of study skills required for success at the college level. Everything from general study skills advice, to test taking skills, to relaxation techniques, are discussed here, each with its own section.

Like the rest of the book, however, the useful material in this Appendix is presented in a series of quick tips, rather than as a detailed discussion: there is enough here to get you started, but perhaps less than you need to fully grasp the underlying concepts. If, after reading this Appendix, you still need more information, please see the excellent book *How to Study in College* by Walter Pauk (Houghton Mifflin College, 6th Edition, 1996, ISBN 0395830621).

A.2 HOW TO STUDY

Good studiers have two main attributes. First, they possess a assortment of efficient study skills. Second, they have many positive personal and inter-personal skills.

This section provides you with some introductory principles in both areas, in order to start improving your overall study skills right away. More detailed information on specific study topics follows in subsequent sections.

The following sections, which provide more detailed study advice, can be roughly divided into two parts. The first half, comprising sections A.3 - A.16, discuss specific study skills, such as how to read a textbook and how to take lecture notes. The remaining sections, A.17 – A.23, are devoted to study advice of a more personal nature, such as how to improve your concentration and learning to relax.

To study well, then, follow these general tips:

Personal Attributes:

- *Check your attitude.* Your attitude towards studying is *the* most important part of your study success: you must continually strive to stay positive. The key is *positive self-talk*; praise yourself when you complete a goal; and refrain from berating yourself when you don't.

More on time management in Section 4.5 and Section A.3 below.

- *Plan your time wisely.* Half the battle in being a good student is keeping track of all of your commitments! Create a timetable, in which you schedule all of your study time as well as other things you want and need to do (including break time), in such a way that you can stick to the plan that you create. Then, stick to it!

Goal setting is in Section A.22.

- *Set reasonable goals for yourself.* Make progress in your study life by setting goals, and working towards them. The trick is to set goals that have the right degree of challenge: they should not be so easy that you swipe through them without effort, but they also shouldn't be so hard that you get discouraged, either.

- *Use lists.* Implement your plans and your goals with lists of things to do. Cross off items as you complete them, and rewrite your lists as your priorities change through time.
- *Recognize your strengths and weaknesses.* Not everyone can get straight-A's. Strive to do the best that *you* can, and let that be reward enough.
- *Manage distractions.* Don't waste too much time with external distractions such as music, computers, phones, friends, etc. Be strict: when you are working, these things are off limits! Also, watch out for internal distractions like daydreaming, personal problems, and anxiety. If difficulties like these are taking up too much of your time, seek help from a counselor.
- *Take appropriate breaks.* Don't feel guilty about taking needed breaks; as long as you've worked your level best up until then, you've *earned* it!
- *Reward your hard work.* Make learning a game: every time you make the most out of a study situation, give yourself a point. Then, after you accumulate a certain number of points, compensate yourself with a small reward like extra sleep time or a hot fudge sundae.

Relaxation techniques are discussed in detail in Section A.19.

- *Keep your body healthy.* Exercise regularly and eat properly. Healthy bodies think quicker and work longer.
- *Learn to relax.* Stress is cumulative, and can negatively affect your health: the longer you go without relaxation, the more stress has a chance to produce unpleasant symptoms, such as headaches, anxiety, and, in severe cases, depression.

General Study Skills:

- *Ask for help when you need it.* Always try to figure out any difficulties on your own first, but when you are truly stuck, find someone to help you! Don't waste time feeling frustrated or stupid; instead, use your time *productively* by getting help from your instructor, a friend, the librarian, a tutor, an advisor, a counselor, your parents, etc.
- *Join a study group.* Learning is easier and more effective in a group, not to mention more fun. Just make sure that you are carrying your own weight, and not just copying other people's efforts (and that everyone else is doing the same!); conversely, make sure that you don't have *so* much fun that you don't get any work done.

See Section 4.2 for more on physics teams.

- *Be a better reader.* Because so much material is covered in a college-level class, not all of it can be presented in lecture. Therefore, reading your textbook is a critical part of the learning process. If you are not a good reader, take the extra time to beef up your skills now, before you get overwhelmed with work.

Section A.4 below can help you improve your reading skills.

More on libraries in Section A.8. 

- *Check out the library.* There is a wealth of information available at your fingertips in the library; don't let this invaluable opportunity pass you by!

Refer to Section A.16 for more on lecture. 

- *Use lecture time wisely.* Don't waste class-time by not paying attention to the instructor, sleeping, or worse. Consider the lecture as structured study time with the instructor, and you will reap benefits almost immediately.

Sections A.10 - 15 below discuss test taking techniques. 

- *Get ready for tests.* Test preparation is a continuous duty, not a night-before-the-exam task. NEVER CRAM. Instead, spend time *every day* reviewing old material and relating it to new information. This way, your test reviews will be less painful and more productive when they do arrive. Then, when it is time to get ready for a test, follow these tips:
  - *Start studying at least a week before the exam.*
  - *Rewrite your notes,* and/or *create flashcards* with vocabulary or important facts on them (Fig A.1). Review the notes/flashcards over and over until you can remember their content without looking. Keep the flashcards handy, so if you have time on your hands, you have something to do.
  - *Keep re-reading of texts to a minimum.* The goal is to *review* material you already know, not learn new material. So don't waste time rereading the book; instead, scan your notes from class and the book.
  - *Avoid marathon study sessions.* Break up study tasks into manageable pieces of about an hour or less. Then take short breaks (10-15 minutes each) between chunks so that you remain refreshed.

Important Fact

FLASH CARD FRONT

Quiz Question

FLASH CARD BACK

Figure A.1

A.3 MANAGING YOUR TIME

College life is exciting in part because there are so many things that need to be done, new experiences to try, and fascinating people to meet. It is easy to get lost in all of this excitement, however, so that you fall behind in your coursework.

As a result, one of the most important skills that you must master is *scheduling* all of your conflicting demands. This section will give you tips on how to best manage your time.

- *Create a realistic weekly schedule of all your regular activities, including relaxation and rest.* At the beginning of each semester, write out a timetable that blocks out time every week for those activities that must be completed on a regular basis: such as classes, study, team practice and/or workout time, eating, sleeping, etc. Use the following rules of thumb to make the most out of your schedule:

1. You will need about two hours of study time for each hour of class.
2. Try to study during the day, if you can, perhaps during your free periods: research says that one hour of daytime study is worth one-and-half hours of study at night.
3. Schedule your study hours during your "peak times," those times when *you* work at your best.
4. Try to study at the same time every day, in the same place every day, to help establish a study routine.
5. Be sure to include relaxation and sleeping in your timetable.
6. Keep your schedule flexible: don't schedule every second of every day; instead, be open to serendipity. Things happen in life; you need to be able to respond to them.
7. Then, stick with your schedule. Be strict with yourself!

- *Keep a calendar to schedule longer-term commitments, like exams or term papers.* Use this calendar to *plan* extra study time before exams, papers, etc., are due, by scheduling these in at the beginning of the term (or as soon as you find out about them).
- *Plan your days.* Each morning (or night before), take some time to plan the upcoming day. Referring to both your weekly schedule and your long-term calendar, list all of your day's activities, prioritizing them in order of necessity. Then, during the day, try your best to complete the list. If you don't succeed, however, refuse to get stressed: instead, develop the willpower to accept that you have accomplished the *essential* tasks, and defer the rest to tomorrow without guilt.

Some other time management tips include:

- *Learn to say no.* You can't do *everything.* Part of college life is knowing how to pick and choose the things that really mean something to you, and then saying no to everything else.
- *Manage freedom.* Many college classes have no deadlines at all, except for a final exam or paper. It is very easy to get in the trap of not doing anything for these classes—since nothing is due—until it is far too late. For these kinds of classes, you must set your *own* study schedule, including reading and library research times. Ask your instructor for help setting up a study plan, if necessary.
- *Avoid procrastination.* The best cure for wasting time is to admit you have a problem and then try to solve it. Some tips for getting past procrastination include:

- Do psych yourself up for tasks ahead with positive self-talk. Don't berate yourself for not having worked enough yet.
- Start with an easy task just to get going.
- Promise yourself a break or other reward when you complete a task, or after a certain amount time.
- Then, try to eke out just one more task before you take a break.
- Ask yourself: Would I pay myself for the work I have done so far? If the answer is no, get going!
- Finally, ask yourself if the task at hand is truly important. If it isn't, defer the task to another list and move on.

- *Use waiting time.* Think of all the hours that you spend waiting in lines, waiting for classes to start, etc. Always bring a little work with you—for example, short readings or flashcards—and use your waiting time to study: these short times will add up!
- *Don't worry about one problem when dealing with another.* When you are working on one project, don't worry about other projects in your life; they just distract you from your present duties.

A.4 EFFECTIVE READING: SQ3R

At one time, you may have been able to earn good grades without reading your textbook. However, at the college level, this approach (unfortunately!) does not work. This is because the college-level textbook often covers additional, required material that is not discussed in class. It also provides context, background, and other specifics that help to clarify your understanding. Therefore, you *must* be able to read textbooks effectively in order to succeed at the college level.

The SQ3R (*S*urvey, *Q*uestion, *R*ead, *R*ecite, and *R*eview) method is an excellent technique for efficient and effective textbook reading. This technique may seem time-consuming at first, but give it a try! It really works. To use SQ3R, *every* time you need to read a passage or a chapter, follow these steps:

Before You Begin: *Prepare yourself for reading effectively.*

- Look over the *entire* reading assignment. Make an honest assessment of how much you can complete in one sitting: Do you need an hour? a day? a week? Schedule your reading accordingly.
- Get comfortable in a good study area. The library is the best place, but if you study in your room, make sure you turn off the stereo, log off of the computer, put away other reading material, and shut your door. Have a good source of lighting, pens/pencils and paper, and a dictionary within your reach. Finally, find a comfortable seat, one

that is away from windows or doors which might distract you. Just take care not to get *too* comfortable: you don't want to fall asleep!

- Don't allow yourself a "warm up period." Get started right away, and then later, after you've read for a while, take a 15-minute break.

<u>Survey</u>: *Look over the reading assignment to discover main ideas.*

- Using the book's Table of Contents, examine the headings and sub-headings for the reading assignment. Try to determine the *one or two* main topics that will be discussed. From these, sketch out your own rough outline for the chapter.
- Now turn to the reading assignment itself, and familiarize yourself with the material to come:
  - Read the introduction.
  - Read the first sentence or two under each heading and sub-heading.
  - Read the last paragraph or summary page.
  - Look at the illustrations, maps, charts, and graphs. Read their captions.
  - Note all occurrences of typographical aides, such as words in italics, boldface, etc.
- Then, skim through some of the questions and/or problems at the end of the chapter, to get an idea of what the *author* thinks is most important to take away from the reading.

<u>Question:</u> *Ask yourself questions regarding the content.*

- Return to the beginning of the chapter. Turn each heading and sub-heading in your outline into a question to be answered by the text. Use words like "how," "what," or "why" to start each question. For example, if a section title is "The Four Types of Reactions," your question might be, "What are the four types of reactions?" Jot your questions in the margin of the text or on another piece of paper.

<u>Read:</u> *Read the text.*

- Working one section at a time, read the text so that you are able to answer the question posed for that section.
- Underline or take notes on information that helps you answer your question.
- Work any example problems in the section along with the book.

<u>Recite:</u> *Answer your questions from information in the reading.*

- After *each section*, recite and answer that section's question. If you cannot answer your question, you may have to re-read or skim through the section or notes again. This is perfectly okay, just don't move on until you can do this step.

- Write your answer down, in outline form. Use your own words, and name an applicable example or problem that illustrates your point. Then, read what you have written aloud.

Review: *Review the material without looking at the book.*

- After completing the entire reading, see if you can recall the main points of the chapter without looking at the book.
- Then, cover up the answer to each question on your list, and see if you can answer it without peeking. Each time you can, place a small check mark next to the question; each time you can't, write a small ×.
- Repeat the previous step several times.When you have three checks in a row next to one point, you have learned the material. So, circle your checks so that you don't review this material again until test time.
- Look over all of your markings or notes one last time. Try to draw an overall picture of the entire story of the reading assignment.

Afterwards: *Think about what you have read.*

- Reflect on the purpose of the material, and consider how it relates to what you knew before.
- When you are done, take a well-deserved break! Get up and move around. Put the reading away for another day.

Once you have mastered the SQ3R reading method, the following tips will help you fine tune your reading skills even further.

- *Read a lot.* The more you read, the better you get at it. So practice, practice, practice!
- *Improve your vocabulary.* This advice will increase your word power:
    - Keep a dictionary close by, and *use it*.
    - Use the context of the sentence to help you determine word meaning.
    - Learn word roots, so that you can translate your knowledge of one word to another. For example, the word root *dermis* means *skin*; therefore, the words *epidermis* and *dermatitis* both refer to skin in some way.
    - Learn about the origin of words.
- *Concentrate on learning material the first time you read it.* You should never have to re-read material to *learn* something, only to review what you already have learned.

- *Retain what you read.* What's the point of reading if you can't remember what was said? The following tips will help you maximize you recall of reading material.
  - Before you begin to read a new passage, review previous or related material, so that you have a good idea of what is about to come.
  - Then, after you complete each passage, stop to review what you have read.
  - To organize the reading material, use the reading's paragraph structure as an outline.
  - Look for words that indicate changes in topic or point of view.
- *Learn how to read faster, so that you can cover more material in less time.* Slow readers read one word at a time. Fast readers move their eyes much more efficiently. The following advice will help you read faster:
  - Train your eyes to focus on word *units* (*e.g.*: phrases) rather than individual words or letters.
  - Keep your eyes focused on the middle of the page, rather than swinging them from edge to edge.
  - Read slightly above the line.
  - Keep your eyes moving forward and down the page.
  - Don't move your lips or read aloud. However, it *is* okay to keep a running "dialogue" going in your mind.
- *Change your reading pace depending upon the type of material being read.* The average person reads easy material at 250 words per minute; medium material at 200 wpm; and difficult material at 100 wpm. Don't make the mistake of reading too fast or too slow for the material at hand.

A.5 EFFECTIVE TEXTBOOK MARKING

Marking your textbook can be an efficient way to retain reading material, since a well-marked chapter can be reviewed in less than half the time it would take to re-read it from scratch. This section shows you how to mark books most effectively.

However, you should not mark all textbooks! In general, books used for humanities courses are good candidates for marking; technical textbooks are *not*. Because of their natural outline structure, technical texts are better suited for note-taking than for marking.

Section A.6 below discusses note-taking from books.

- *Choose Your Weapon.* There are two ways to mark books: with a pen/pencil (underlining passages), or a with a highlighter (drawing lines over the passage). The best way to mark your book, however, is to use *both* at the same time: highlight important passages with the highlighter, and use the pencil to make special marks or notes in the margin.

- *Mark Neatly but Swiftly.* Spend a little time making your marks look neat. For example, you might want to use a ruler to make your lines straight. However, don't spend *too* much time on this task; you are trying to learn new material, not create art!
- *Understand First, Mark Later.* Don't mark your book until you have understood a significant portion of the reading. Then, and *only then*, go back and mark the main points; they should be easier to make out now that you understand the author's argument. General Rule of Thumb: Always make sure you have a good reason for every mark you make *before* you make it.
- *Have a System.* The initial time spent in creating a consistent marking system will be more than recovered later when you go to review the material. The following tips will help.
  - *Make Your Marks Simple.* Simple marks are faster to make and easier to understand later.
  - *Indicate Relationships Between Marks With Brief Notes.* Use the margins or spaces between lines for very brief notes linking main ideas together. For example, your marks might:
    - Specify examples;
    - Identify cause and effect;
    - Indicate steps in a process;
    - Compare and contrast certain points;
    - Cross-reference the material from one page to material on another page.
  - *Other markings you might consider include:*
    - Marking the main points with "1," "2," and "3," and subpoints with "a," "b," "c," etc.
    - Double underlining main ideas, and single underlining supporting evidence.
    - Drawing a bracket in the margin around long passages, rather than underlining the whole thing.
    - Using stars to indicate important material, and question marks to indicate material you don't understand.
    - Circling key words and terms, and then jotting a quick definition in the margin.

- Boxing words of enumeration and transition, like *first* or *further.*
- Noting places where you disagree with the author.

A.6 EFFECTIVE NOTE-TAKING WHILE READING

Taking notes while you read is an important way to learn material the first time, rather than having to reread the entire passage again—especially during technical reading like that in physics textbooks. This section gives you some guidelines on effective note-taking:

- *Develop an outline.* Organize your notes according to the organization of the text. Use one of the two following formats. [Note: Technical textbooks are usually already arranged in outline form (chapter, heading, subheading, etc.), making note-taking in outline form a natural choice. But humanities texts often do not have such obvious organizational structures, so in that case, you will have to pick the note-taking style that is best for you.]
    - *Outline style.* Make an indented list of ideas, as illustrated below. Don't get carried away with elaborate indentation structures or long sentences, however: keep it simple and short!

Title

I. Main Idea
    A. Supporting Detail
        1. Sub-detail
        2. Sub-detail
    B. Supporting Detail
II. Main Idea
III. Main Idea....

- *Visual map.* Draw main ideas in the middle of the page, and circle them. Then, draw lines out from the circle that indicate supporting evidence, as shown. Again, don't get overly elaborate with your map: keep it simple!

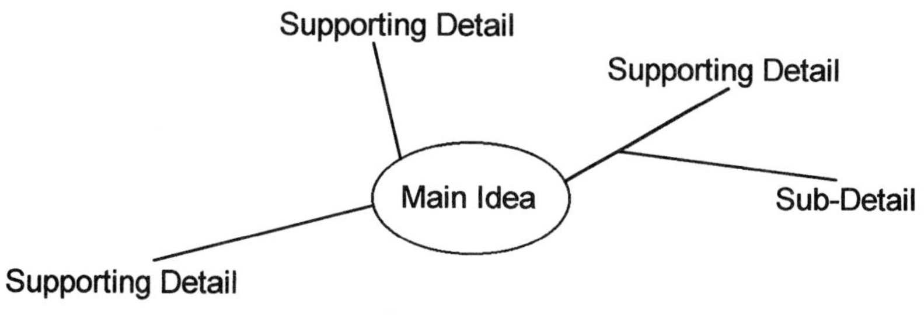

Figure A.2

- *In your outline, record only important information...* It is very easy to get into the trap of recording too much information, effectively re-writing the text into your notes. Don't do it! Keep your notes brief, use your own words, including only the main points and any supporting details *required* to prove them. Some important information to get into your notes consist of:
    - Names
    - Key terms and their definitions
    - Summarizing statements
    - Rules or laws
    - Important quotes
- *...But, record enough information to easily understand your notes without referring to the reading again.* It is also easy to get in the habit of *not* writing down enough information. Make sure that you have all of the information needed to reconstruct the author's argument without having to refer to the book again.

A.7 EFFECTIVE CLASSROOM NOTE-TAKING

Taking good notes during class time is an indispensable part of your study regimen. The Cornell Method of note-taking is a proven method for taking the best possible set of class notes. This technique may initially seem more time-consuming than your current method, but stick with it! In the long run, this method will actually *save* you study time, as well as help you learn class material more effectively.

Before the First Class Meeting:

- Buy one large, loose-leaf notebook (with dividers) for each class. These notebooks will organize your notes, handouts, worksheets, syllabus, assignment sheets, etc., all in one convenient place. Store the notebooks at home.
- Carry blank loose-leaf paper to each class for note taking.* This way, if you lose your notes on the way home from class, you will only lose one day's worth! After class, bring the notes and any class handouts home, and place them in the appropriate notebook right away.

Before Each Lecture:

- Choose a seat near the front and center of the classroom. Don't sit near friends or other distractions: this is study time, not social hour.
- Take a few minutes to look over your previous lecture notes so you can connect them with the lecture you are about to hear.

*Note: You can usually purchase bound tablets of paper, perforated at the top or sides, so that you can detach individual sheets of "loose leaf" paper. In this case, you can have the convenience of loose leaf paper for your class notebooks, without having the inconvenience of carrying around a sheaf of unbound paper.

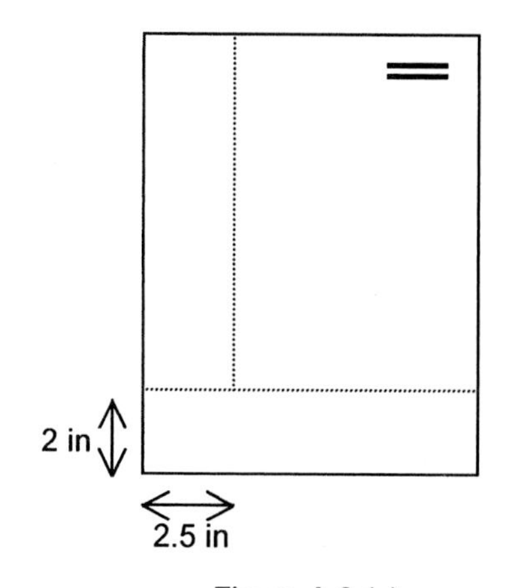

Figure A.3 (a)

- Prepare your note paper (Fig. A.3 (a)):
  - Write the date, subject, and source of information in the upper right hand corner of each page.
  - Draw a horizontal line two inches from the bottom of each page.
  - Draw a vertical line about 2.5 inches from the left edge of page, stretching from the top of the page to the horizontal line.
  - You will write your notes in the large space in the upper right hand corner.
- Sit up straight and get ready to listen.

During Each Lecture:

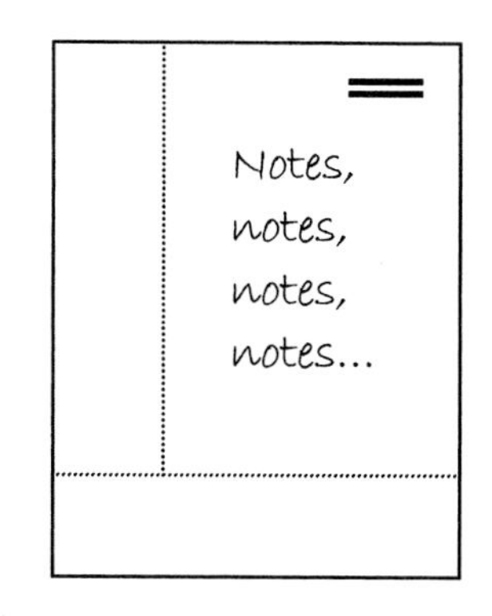

Figure A.3 (b)

- In your note-taking space, write down general facts, ideas, and examples covered in the lecture (Fig. A.3 (b)).
  - Write legibly and clearly, and leave plenty of white space between notes.
  - Use sentence fragments, not entire sentences. You will fill in specifics later. Try omitting words such as:
    - Unimportant verbs.
    - Words like *a, an,* and *the.*
    - BUT don't leave out words like *in*, *at*, *to*, *but*, *for*, and *key*...these add necessary structure to your notes.
  - Try to get some of the instructor's own quotes down, if possible.
  - <u>General Rule of Thumb</u>: Make your notes clear enough that you will be able to understand them years later, but not so clear that you are recording every spoken or written word.

- As you are writing, organize your notes. There are two methods:
  - *Outline style*. Make an indented list of ideas, as illustrated below. Don't get carried away with elaborate indentation structures or long sentences, however: keep it simple and short!

    Title
    I. Main Idea
        A. Supporting Detail
            1. Sub-detail
            2. Sub-detail
        B. Supporting Detail
    II. Main Idea
    III. Main Idea....

  - *Visual map*. Draw main ideas in the middle of the page, and circle them. Then, draw lines out from the circle that indicate supporting evidence, according to the following format. Again, don't get overly elaborate with your map: keep it simple!

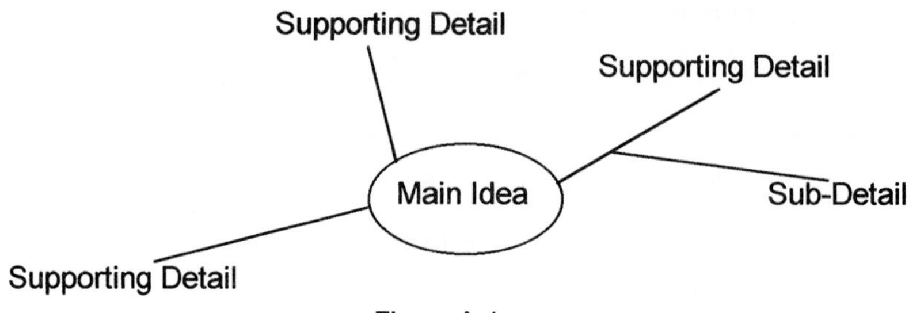

Figure A.4

- Be on the lookout for so-called "structure words" that may be mentioned in the lecture, like those in Table A.1 below:

Table A.1: Structure Words

CATEGORY	STRUCTURE WORDS
Example	To illustrate, For example, For instance
Addition	Furthermore, In addition, Moreover, Also
Chronology	Before, After, Formerly, Subsequently, Prior, Meanwhile
Cause & Effect	Therefore, As a result, If...then, Accordingly, Thus
Contrast	On the other hand, In contrast, Conversely, Pros and cons
Enumeration	The four steps are..., First/second/third..., Next, Finally
Emphasis	More importantly, Above all, Remember this
Repeat Words	In other words, In the vernacular, It simply means, That is, Briefly, In essence

- Use shorthand when you can. The following tips will help:
    - Use abbreviations, such as:
        - Write only the first syllable of a word (politics → pol), or the first syllable plus the next letter (subject → subj)
        - Drop the middle letters of a word (continued → con'd)
        - Drop the end letters of a word (government → gov)
        - Indicate the "ing" suffix with a "g" or "." (decreasing → decrg or decr.)
        - Indicate the "ed" suffix with a slash mark "/" (painted → paint/)
        - Use an "s" to indicate plurals (pols)
        - Use abbreviations for common or repeated words/phrases (United States of America → USA).

- Try stenographic abbreviations, such as:

c	about	#	number	=	equal
cf	compare	vs	versus/against	≠	not equal
fg	following	w/	with	<	less than
i.e.	that is	w/o	without	>	greater than
e.g.	for example	b/f	before	~	approximately
viz	namely	b/c	because	&	and
s/b	should be	*	important	$	cost, money
dept	department	**	very important	f	frequency
mx	maximum	∴	therefore	@	at
mn	minimum	%	percent		

- Here are some additional hints for fast, efficient note-taking:
    - Take care to listen carefully, and hear what is *really* being said. Judge the content of the lecture, not its delivery.
    - Write down any conclusions reached during class discussions. The purpose of discussion, in the instructor's mind, is to cover a certain set of material. The discussion will end when the material is covered. So write down any conclusions obtained as a class.
    - If the lecturer moves too fast, leave blanks to fill in later, when you have more time. Strive to capture idea fragments, not entire paragraphs word-for-word.
    - Don't stop and think too much. Keep up with the line of reasoning, but use lecture time to get information onto the page. Save your deep thinking for later.
    - Finally, don't wait for something "important" to be covered. It's all important! So take notes from class beginning to end.

As Soon as Possible After Each Lecture:

- Read through your notes, making any changes required so that the notes are clear, readable, and complete. Fill in missing information, if any, from your textbook.
- Now, in the blank left-hand column, jot down key words and phrases that *briefly* summarize ideas in the right-hand column. Use the TIPS format (Fig. A.5 (a)):

    Topic:
    Main Idea:
    Supporting Points: 1.
    2.
    3.
    Summary:

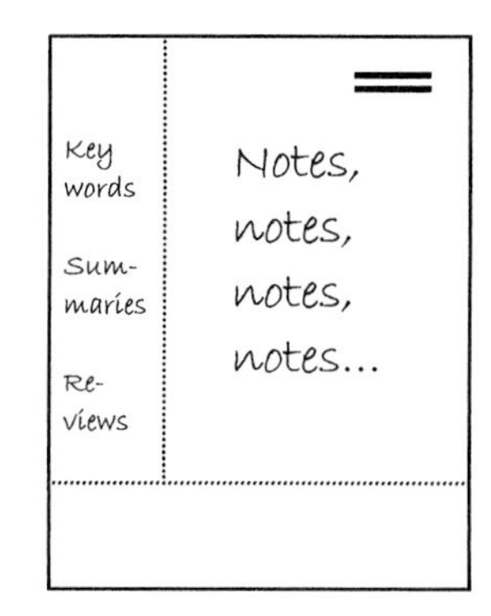

Figure A.5 (a)

- Then, cover up the right-hand column, and using only the brief summaries from the left-hand column, completely recite the content

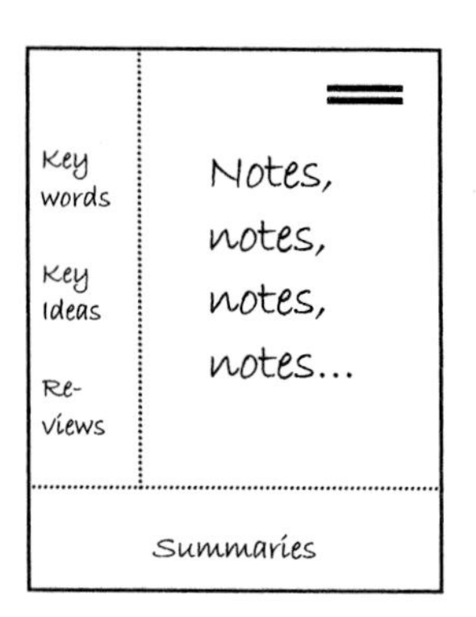

Figure A.5 (b)

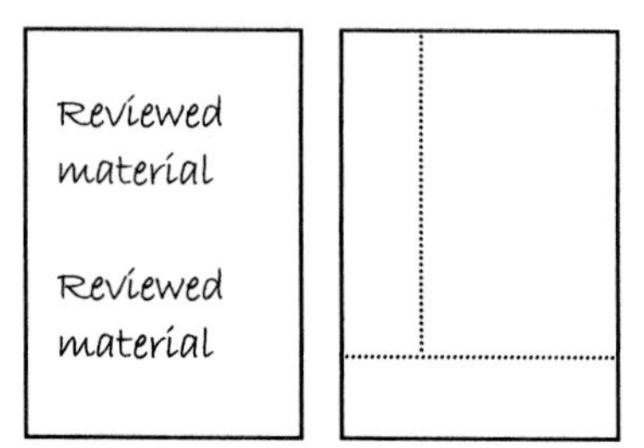

Back Side Front Side

Figure A.5 (c)

of the lecture *in your own words*. Be sure to uncover the right-hand column to verify your accuracy.

Each Week:

- For at least 10 minutes every week, review past lectures. Try to reorganize, rearrange, and reflect on how new material affects your understanding of the old.
- Use the blank at the bottom of each page to record your summaries (Fig. A.5 (b)).

When it Comes Time for Test Review:

- Use the blank page to the left of each note page (*i.e.*, the blank back side of the previous page of notes) to write concise summaries of class material, working to integrate *all* of the information you have learned between the time you took the notes and now (Fig. A.5 (c)).

A.8 USING THE LIBRARY

The library is the center of academic life: in it, a wealth of information is literally at your fingertips.

Unfortunately, many students feel nervous using this most important resource. Nonsense! The best way to get used to libraries is simply to wander around inside one for an hour or so; in fact, one of the first things you will notice is how many other people are just wandering around, too.

During your browse, you should see the following common library fixtures:

- *The Librarian:* If you can't find what you are looking for, just ask one of the friendly, knowledgeable librarians. That's what they're there for!
- *Computer Database:* The computer database is an index to all of the library holdings, sorted by author, title, subject, etc. Each record in the database has, among other things, a classification number which tells you where in the library to find each reference.
- *The Circulation Desk:* Check out (take home!) your reference here. You will usually need a library card to do this, but most libraries allow you to check out material the same as day you apply for a card.
- *Periodicals:* Most libraries have recent and back issues for a variety of regularly published periodicals. You can usually find your periodical reference in the computer database, but you can also use the "Reader's Guide to Periodical Literature." Ask the librarian how.
- *References:* Often a trip to the reference section will be all you need to find information you are seeking. There are many types of publications seen in a typical reference area:

- Dictionaries: Alphabetic listings of words and their definitions.
- Encyclopedias: Alphabetic summaries on various topics.
- Yearbooks and Almanacs: Single-volume, annual compilations of facts in a specific field.
- Directories: Special-interest alphabetic listings of people, places, services, and events.
- Atlases: Collections of maps and geographical data.
- Indexes and Bibliographies: Lists of books, pamphlets, and periodicals, arranged by subject and/or author.
- Catalogs: Complete listings of items on particular subjects (e.g.: *Books in Print*).
- Other References: *Current Biography, Congressional Directory, Who's Who In...., Bibliography of American Literature, Atlas of American History, Dictionary of American Biography,* etc.

- *Reserve:* This section is where instructors leave reading materials for their course, so students don't have to buy them (yeah!). However, items in this section are for use by many people at the same time, and so often materials cannot be checked out, or, can only be checked out for short time periods. Be sure and find out the policy of your reserve room.
- *Audio/Visual:* All audio/visual materials, including film-strips, audio cassettes, video tapes and discs, are located, and can be viewed, here.

A.9 WRITING EFFECTIVE REPORTS

Many courses in college require you to write reports. This section will give you some ideas for handling this basic requirement.

Before you begin:

- *Learn to type!* Typed papers are easier to read; and the easier a paper is to read, the better the grade the instructor will hand out.
- *Keep a dictionary or thesaurus close* (or their computer equivalents), so that your paper will be correctly spelled and interesting to read.
- *Set aside enough time to complete your research and writing.* Your first draft needs time to be re-written, edited, and re-arranged. Allot enough time to do this properly.

Quote or Fact

FRONT

Author
Publication
Publisher
Date
Page Number

BACK

Figure A.6

When you are ready to begin:

1. *Choose a topic.* Pick a subject you are interested in. Write down everything you can think of about this topic. Then, take some time to narrow these topics down to fit your assigned paper length.
2. *Go to the library and research your topic.* Take notes on note cards as you read (Fig. A.6). Write one quote, fact, or idea on each card, using quotation marks for direct quotes so that you don't accidentally plagiarize someone else's work. On the back of each card, record the author, publication, publisher, date of publication, and page numbers of the reference. Yes, these things take time to do—but it is much better to do them now than have to reconstruct your references after the paper is written!
3. *Think of your main points.* After you have thoroughly researched your topic, sort through your cards and identify your main points. What is the premise of the paper? Note: You may not need all of the information you have collected!
4. *Use the main points to develop a rough outline.* Write your outline down. The idea here is to "take a stand:" how do your main points *prove* your position? If they don't, throw them out! (Or, change your position!)
5. *Flesh out your outline.* What is the logical flow of your argument?
6. *Write a first draft straight from the outline.* That is, simply re-write your outline using complete sentences, adding transition sentences to make the draft smoother to read. Don't forget your footnotes!
7. *Put the paper away for awhile.* Don't look at it! The longer you have for this step, the better your final paper will be.
8. *Review and revise your paper.* Rethink your thesis. Fill in holes in your argument. Look for sentences that aren't completely clear. Throw away whole parts of the paper, if necessary.
9. *Repeat Steps 7 and 8 once more, if you have time.*
10. *Check your final draft* for spelling and punctuation mistakes, careful paragraphing, sentence structure, style, vocabulary, and sentence rhythm. Make a nice cover page, and be sure the paper as a whole looks neat and professional.

A.10 EFFECTIVE TEST TAKING

Tests are necessary parts of the learning process, for no one has (yet!) thought of a better way to make sure students have mastered necessary material.

To do your best on these inescapable aspects of your class, then, view each test as a valuable "self-portrait" that gives you insight into how well you understand the material. Just as you would prepare ahead to make sure a photographic portrait shows you at your best, you must prepare ahead to make sure a test will reflect your best abilities. This section will help.

At the Beginning of the Term:

- Look over the instructor's syllabus to see when tests are planned. Write these into your calendar and block out extra study time for them right away.

One Week Before the Test:

- Ask your instructor about the scope of the test. What is the format? What material will be covered? What are the instructor's goals for the test? Are there copies of old tests that can be reviewed?
- Begin your test review.

Twenty-Four Hours Before the Test:

- Try to get a good night's sleep. Do *not* cram until the wee hours of the morning: if you don't know the material, it is a far better use of your time to get some rest, so you can think flexibly in the morning, than to frantically cram half-understood material.
- Make sure you have all necessary supplies for the test, such as pencils, erasers, calculators, etc.
- Eat a good breakfast/lunch before the exam.
- If you exercise, try to go to the gym prior to the test to release some extra energy.
- Arrive early so that you can find your favorite seat and relax a bit. Choose a seat far away from any distractions, including your friends; there will be time enough *after* the test to talk to them!

At the Beginning of the Test:

- Listen carefully to and/or read *all* of the directions carefully. Do you have to answer all of the questions, or can you *choose* which ones to answer? Is there a time limit? Will some questions count more than others? The only way that you will know these things is if you read the instructions!
- Skim through the entire test before you begin. Identify questions you can tackle quickly, and those requiring more effort. Budget your time based on your skim-through. Don't spend too much time on any one question. Always read each question at least twice to be sure you completely understand its content and extent before you answer it.

During the Test:

- Go through the entire test *three* times: the first time through, answer all of the questions you know right away; on the second pass, answer all of the questions you can logically figure out; and on the final run-through, guess at the remaining answers.

There's more on test anxiety in Section A.15, and relaxation in Section A.19.

- Manage your anxiety. Some nervousness during exams is normal and useful, but too much will blow your careful preparation.
- If faced with a confusing question, circle or underline key words to help you focus on the central point. Then, rephrase the question in your own words. Or ask the instructor to explain the question. If all else fails, write what you think the answer *should* be in the margin of the test, along with a short sentence on why you are confused; you may at least get partial credit.
- Change answers only if you have a good reason to do so. Research shows that the first answer is right more often than not.
- Take advantage of the *full* time allotted for the test, using any extra time to check your work and look for answers you may have skipped. *Never* leave an exam early.

After the Test:

- Take a short, well-earned break.
- Then, start your next test review cycle. *Don't* put your books away until the next test!

Test review is in Section A.14.

- When you get your test back, review it carefully. Check your textbook or notes to find the correct answers to questions you missed. If you didn't do as well as you had hoped, make an appointment with the instructor to ask how you might do better next time.

The following five sections, A.11 through A.15, will help you cope with specific test issues, from problem-solving tests to test anxiety.

A.11 TAKING PROBLEM-SOLVING TESTS

Problem-solving exams—like those commonly seen in introductory physics classes—test your ability to *solve problems*, especially homework-style problems. The best way to study for these types of exams is to solve as many problems as you can, prior to the test. But the following hints will also help.

- Usually there are only a few problems—from three to five—on the entire problem-solving test.
    - Thus, each problem counts for a large portion of your total exam score. So, it *never* pays to skip a problem; instead, if you are

stuck, do your best to write *something* down to at least get partial credit. Try drawing a sketch, writing down applicable formulas, and so on.

- For the same reason, the instructor must extract as much meaning out of each problem as possible. This usually translates into each problem having multiple parts. Again, do your best not to skip any of the parts: one part may require an answer from a previous part. In the case that you need an answer from one, unsolvable part to work another part, make up a reasonable value and use that (writing a quick note to explain your actions).
- On such an exam, the instructor can usually only test *main* points. Therefore, your study regimen should concentrate (at the very least) on thoroughly understanding all of the main points discussed since the last exam.

- Since the problems on problem-solving exams are similar to homework problems, it is imperative that you study assigned problems prior to the exam. Many times an instructor will print a homework problem *verbatim* on a test, just to see if anyone is paying attention. If you have been, these will be free points!
- The best way to solve any problem—including those on problem-solving tests—is to use a consistent problem solving style. Chapter 4 of this book offers one version. The more you practice a problem-solving method, the faster and more effective at all problem solving you will be.
- Allot approximately the same amount of time for each problem. If you start to run over your allotted time on one problem, move on to the next one. Then, if you have extra time at the end of the exam, you can return to any unfinished work.
- If you are running out of time and can't complete a problem, write a short note to the instructor about what you would have done if there had been time. You might at least get some partial credit.
- Do not forget your calculator!
- Check your work at the end of the exam: Is it readable? Does it make sense?

A.12 TAKING OBJECTIVE TESTS

Objective tests—such as true-false, multiple choice, matching, or fill-in-the-blank tests—show your knowledge of *facts* and *details*. Thus, for these kinds of tests, there is no substitute for drilling facts before the exam. But the following hints can help too.

True-False Tests:

- Be on the lookout for key words such as "always," "never," "are," "guarantees," and "insures." Often these words alone determine the truth or falseness of the entire statement.
- Check each part of the statement for truth; if any part of it is not true, the *whole* statement must be false.
- Beware of negative word phrasings, as they make it harder to determine the truth of the statement.
- When something is given as the "reason" or "cause," or "because" of something, the statement will tend to be false.
- If you must guess, guess true: it occurs more often than predicted by chance.

Matching Tests:

- Read the longer column, if there is one, first, to save time.
- Mark only the matches you are sure of first. Cross these out.
- Then, go back to the skipped items and try to match them.
- Guess only when you are completely stumped.

Fill-In-The-Blank Tests:

- Let the context of the sentence guide your answer: Do you need a verb? Past tense or future? A noun?
- Be on the lookout for grammatical constructions. For example, the correct answer to a question ending with "an" would have to start with a vowel. Other clues to watch out for are subject and verb agreements.

Multiple-Choice Tests:

- Formulate your answer *before* you even look at the choices. Think of everything you know that relates to the question, and then think of your own answer—just as you would with any other kind of test.
- When have an answer of your own, read *all* of the choices, *especially* if the first choice seems correct. Sometimes the correct answer is "all of the above," which is at the bottom of the list.
- Always select the *most* correct answer.
- Look for differences in answers that all appear to be true; it is these differences that are probably being tested!
- Never be afraid to use common sense in determining your answer; sometimes you may attempt to recall the "right" answer, and skip the answer that makes good sense.
- Use answers from previous questions to help you answer other questions.

- Answer *all* questions, even if you must guess. The only exception is when there is a penalty for wrong answers. In that case, leave questions you do not know blank. If you must guess, use these hints:
    - Cross out all choices you know to be wrong, so that you can improve your chances of getting the question correct.
    - Choices that appear strange or foolish are usually wrong.
    - Look for key words such as "always," "never," "are," "guarantees," and "insures," which require that the condition for proof be very stringent. Thus, these answers are usually not the right choice.
    - On the other hand, answers with key words like "may sometimes be" and "can occasionally result in" tend to be correct more often than chance.
    - Often the longest answer, or the one with lots of jargon, is correct.
    - If there is more than one answer that looks the same, chances are that one of this group will be the correct answer.
    - If you are guessing from a list of numbers, choose a middle range value.
    - Inclusive statements like "all of the above" or "none of the above" tend to be correct more often than predicted by chance.
    - When all else fails, guess "B" or "C," as these answers have been shown to occur slightly more often than predicted by chance.

A.13 TAKING ESSAY TESTS

Essay tests—whether they are short or long answer—require you to recall *general* facts and ideas, as well as make and support valid generalizations, apply broad principles, and organize your thoughts into a coherent structure. Thus, for these tests, it is important to synthesize your thoughts before the exam into a "storyline" that both includes facts and states your opinions. This section will give you some hints on writing the best possible essays.

Before the Test:

- If your instructor has provided you with the essay questions prior to the test, make an outline of your answers for each. The outlines should consist of a few main ideas and a few supporting details for each. Make sure you allot enough time to memorize your answers before the exam.

- If your instructor did not provide you with the essay questions before the exam, try to predict possible test questions. Concentrate on material that your instructor has emphasized, but have a good grasp of *everything* that was covered. Then, practice answering your questions.
- Make sure that you understand, and can use comfortably, all appropriate terminology covered in the class.

At the Beginning of the Test:

- Read *all* of the questions before you begin writing, so that you will not make the mistake of answering questions with information best used in another question.
- If you have a choice of questions, answer the easiest ones, or those for which you are best prepared, first. Be on the lookout for questions with unequal weight: spend more time on questions worth more points.
- Read each question *twice* to be sure you completely understand it.
- Underline or circle key words in each question. These "code words" give specific instructions on how the instructor wants the question answered. Use Table A.2 to guide your answer:

Table A.2: Essay Exam Code Words

CATEGORY	SPECIFIC TERMS	ANSWER NEEDED
Demonstration	demonstrate, explain why, justify, prove, show, support	*Show*, do not state, why something is true or false, using factual evidence and logical arguments.
Description	describe, develop, discuss, diagram, illustrate, outline, review, sketch, summarize, trace	Talk about a particular topic with a certain amount of detail.
Evaluation	assess, comment, criticize, evaluate, interpret, propose	Give your opinion or judgment on a subject, but be sure to justify and support your thoughts with facts.
Identification	cite, define, enumerate, give, identify, indicate, list, mention, name, state	Provide a concise answer that presents only the bare facts, such as names, dates or phrases.
Relation	analyze, compare, contrast, differentiate, distinguish, relate	Describe the similarities and differences between two or more subjects.

- Using the key words you have circled, quickly outline your answer on the back of the test paper. Make sure you cover *all* parts of the question, and are answering the questions asked!
- Be sure that your answer is organized logically. *Now* is the time to plan your answer, not later.

When You Are Ready to Write:

- Make your opening sentence a *short* summary of what you are about to say. Be direct and forceful. Do not waste time with flowery language or a paragraph structure: your instructor will not appreciate it!
- In the body of your answer, use *facts* and *logic* to support the statement made in the first sentence. Do not waste time with vague impressions, unsupported opinions, or feelings.
- Write in careful and complete, but short, sentences. Be natural and sincere.
- *Don't* save your best idea for last. If it is not included right away, your instructor may miss it, or you may run out of time and never get to it.
- Include in your essay as much necessary information as possible, but, on the other hand, be concise! Say what you want to say, then *get out*. Padding answers makes you look less knowledgeable, not more.
- Use bullets to save time when writing lists. However, don't use them so much that your paper is reduced to an outline.
- If you make a mistake, cross it out with a single line (not a messy blackout), and write the correction in the line above it using a caret symbol '^'.
- At the end of your essay, include a summary sentence that states your main points and your conclusion in a straightforward manner.
- Always leave a few minutes at the end of the exam to re-read your answers for clarity. Check for spelling and grammar errors, awkward phrases, misplaced punctuation, transitions, etc. If you want to add a few sentences at this time, place an asterisk at the insert location, and write the addition in the margin.
- If you run out of time before you can write all you wish, include your remaining outline for partial credit.

General Hints for In-Class Essay Writing

- *Never* leave the page blank. Always write *something.*
- Neatness counts! Bottom line: neater essays earn better grades. So:
    - Write carefully and neatly.
    - Use ink, never pencil. Pencil smears as you write!
    - Leave margins and spaces between lines.
    - Write on one side of the paper only.
    - Start each new question on a fresh sheet.

- Qualify answers when in doubt. It is better to say "around the end of the nineteenth century" than "1897" if you can't remember the exact date.
- Be sure to include the instructor's pet ideas somewhere in the essay.

A.14 TEST REVIEW: LEARNING FROM MISTAKES

One way to improve your performance on future tests is to take a hard look at tests you have taken in the past and analyze your errors. Below are some common errors students make on their tests. Do you make any of these mistakes? If so, consider the associated advice.

- *You didn't know the material.* There's no substitute for learning what you need to know *before* you go in to a test. Regularly review class material as the class moves on, then begin test review a week before the exam.
- *Your answer was sketchy.* Sketchy answers are due to sketchy knowledge. Get over-prepared next time, rather than cramming at the last minute.
- *Your answer was overly long.* Plan your answer *first*, using a brief outline. Then stick to your plan. Write *only* what is needed to answer the question completely, and nothing more. Resist the impulse to recite everything covered in class: your instructor will *not* be impressed.
- *You answered with the wrong concept(s).* This is usually a result of not integrating the material together as a whole. To solve this problem, make sure that your study schedule includes time for connecting concepts to earlier and later material.
- *You answered the question the wrong way.* You probably read the question too fast. Always read each question *twice* before you answer it, to make sure that you fully understand the instructor's intent.
- *You forgot to answer all parts of the question.* When you answer a question, always re-read the question *and* your response to verify that you have completed the entire task.
- *You didn't understand key terms in the question, or use them in your answer.* The problem here is not enough time spent drilling vocabulary before the test. Next time, use lists or flashcards to review terms *before* the test.

Flashcards are in Section A.2. 

- *You didn't know how to apply material to new situations.* The only way you can do this is if you know the necessary material backwards and forwards before the test. When you study, make sure to set aside time to think about the *meaning* of each concept as you learn it. Then, invent new examples on your own, look for fresh relationships, and apply the material to new situations.

- *You didn't proof-read your answers.* Always leave time to check your work!

A.15 BEATING TEST ANXIETY

Feeling nervous before a test is perfectly normal. But, if you feel *so* anxious that your test performance is negatively affected, you have **test anxiety**. Symptoms of test anxiety include:

- An inability to concentrate or think about the task at hand;
- Excessive distraction by noise, temperature, and passersby;
- Fidgeting;
- Bodily symptoms that interfere with your performance, including "butterflies" in your stomach, quickened heart rate and/or breathing, nausea, sweaty palms, headache, etc.

Test anxiety is usually caused by feelings of pressure and fear of failure that build up prior to the exam. These pessimistic emotions create negative thought patterns in your mind. If you obsess over these thoughts too much, you will get anxiety.

The key to stopping this cycle is to *combat* your negative self-talk with positive counter-thoughts. Below are some common fears and sources of exam-related pressure at the college level, along with some positive self-talk ideas for combating them:

- *I can never study enough.* It's true, you probably do have a lot to study. But, if you stick to your study schedule, you *can* get through it all—and succeed!
- *My poor performance on past tests proves I can't do well on this test.* Past results *never* guarantee future performance. You can *always* do better than you did before. But, you have to have a plan. So get organized and get ready next time.
- *My performance on this test determines my entire future.* Don't put your whole future on the line with a single test! It is very unlikely that one test will "make or break it" for you forever.
- *I'll never be as good as my dad/sister/friend...* You are only competing with yourself on exams, not anyone else. Just do *your* best.
- *My performance on this test is going to let my parents down.* They will still love you, no matter how you do on this one test.
- *My mind wanders during tests.* Learn to improve your concentration *before* you get to the test so that your concentration *during* the test won't be affected by your nerves.

Section A.21 below will help you improve your concentration.

- *I get confused during tests.* Make sure you are *overprepared* for every test, so that questions can't shake you.

Relaxation techniques are in Section A.19. 

- *I remember the answers after I leave the test.* Use relaxation techniques to keep your mind sharp during the test.
- *I can't sleep the night before tests.* Try relaxation techniques right before you go to bed, to help you sleep better.

The best way to deal with test anxiety, however, is to deal with exam pressures *before* they have a chance to build up. Talk to your friends and parents about your feelings. See your instructor for help after class. Work out a plan for success. But whatever you do, do it head on!

A.16 DEALING WITH YOUR INSTRUCTOR

Your instructor, obviously, is a major part of your classroom experience (and grade). Therefore, it is in your best interest to treat your instructor with appropriate respect. The following tips can help:

- *Avoid excuses.* They've heard 'em all, and they can see 'em coming a mile away. Don't even bother.
- *Submit professional work.* Think of your instructor as an employer: would you hand in the same material to someone that was paying you?
- *Be attentive and participate.* Gets you noticed, in a *positive* way.
- *Accept criticism.* It's their *job* to criticize you; take it in stride.
- *Go to office hours.* There is usually lots of juicy information passed on here.
- *Get to know the person behind the instructor.* There may be a friend!

As for lecture, take it seriously. Try these tips:

- Sit in the front row.
- Focus your attention on the instructor, and really *listen* to what is being said.
- Ask questions and participate in discussions.

Learn about note-taking in Section A.6 - 7. 

- Take notes. You can't possibly remember everything that was said in the lecture, especially a week or two later, without them.

A.17 MATH ANXIETY

Many people feel slightly nervous when it comes to math. But if you feel an intrusive sense of *panic* at the thought of doing math, then you have **math anxiety**.

Math anxiety has many causes, such as learning math in a negative environment—where "wrong" answers earned a low grade or a sharp rebuke from the teacher—and learning math alone—so that you were never able to see other people solve problems or make mistakes. Furthermore, math anxiety can be *exacerbated* by the following misperceptions about math:

- Not feeling comfortable with mathematical language or symbolism.
- Never making the connection between mathematical concepts and "real life."
- Never taking mathematical risks, just to "see what happens if...."
- Learning *how* to do math, but never asking *why* or *what for.*
- Thinking that mathematical success depends on "smartness" or a "mathematical mind," rather than "effort" or "hard work."
- Giving up more easily in math than in other subjects.
- Distrusting one's own mathematical intuition.

Math anxiety *can* be remedied, but it takes effort on your part. You must take charge of your own mathematical success. The following tips will help:

Check Your Attitude:

- Admit that math is often *not obvious.* Realize that you—and everyone else—has to work hard to understand it. Then do it.
- Examine your feelings about mathematics: *why* does it make you so crazy? Are these reasons rational? Even if they are, should they be holding you back?
- Watch out for feelings of frustration. If you feel stress building up, stop and take a break. Then try relaxation exercises to release your stress.

More on relaxation exercises in Section A.19 below.

- When the going gets tough, don't give up...get help!
- Psych yourself up with positive self-talk: you *can* do it, so tell yourself so!
- Work in groups. This makes the learning go faster, with more fun. And, you have a natural arena to "vent" when you start to feel anxious.

Section 4.2 talks about physics teams.

Work on Your Math Skills:

- If you need mathematical remediation, get it. There is no shame in getting the help you truly need.
- When it comes to doing math, *just do it*. If you must reorganize the mathematical information to make it meaningful to you, do it! If you have to suspend your disbelief for a while, do it! Whatever you need to do, just get to work.
- Read mathematical material slowly, looking for details. Yet also try to see a "big picture." Do the same when working problems.
- If you make a mistake, don't worry or fret; instead, use it as an opportunity to probe your understanding of mathematics a little deeper.
- Try the *divided page* technique (Fig. A.7). Each time you begin a problem, divide your paper in half. On the left side, work the problem step by step. On the right side, after *each* step in the problem, write down *all* of your feelings and thoughts about the problem—no matter how ridiculous they may seem. This technique should slow you down to the proper problem-solving speed, even as it gives you an opportunity to examine your anxious feelings more closely at a later date.

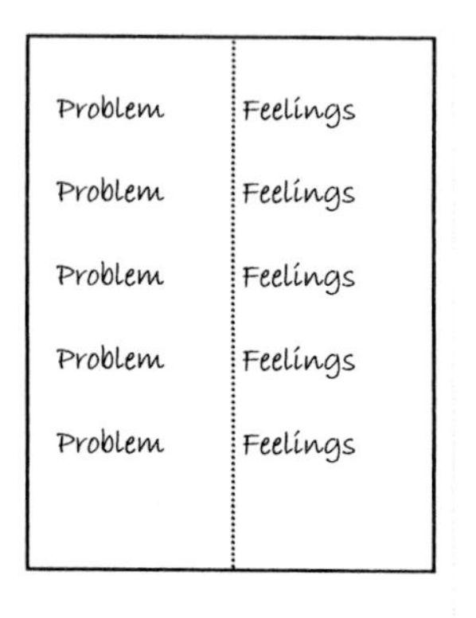

Figure A.7

A.18 BEATING STRESS

Stress is your body's natural reaction to any new, threatening, or exciting situation. In small doses, stress can be good for you, giving you an extra burst of energy during important events.

However, too much stress, or stress occurring over long periods of time, can be harmful to your health, causing headaches, backaches, loss of appetite, fatigue, depression, and other negative symptoms. Below are some common sources of stress at the college level, and how to deal with each of them.

- *Homesickness:* Keep in close touch with your family and friends with frequent phone calls, emails, letters, or visits. But also try to form new friendships with people around you. Getting involved with extracurricular activities such as clubs or sports will help you "get back in the saddle" and meet new people.
- *Personal Finances:* Create a budget that balances your expenditures with your income. Then stick to it! You may (unfortunately) have to eliminate some of your personal spending to make your budget balance. You should also try to save some money each month for unexpected expenses. And stay away from those credit cards!
- *Interpersonal Relationships:* Always try to listen to the other person's point of view and see things the way they do. If you do not agree, state your point of view calmly and carefully. Keep an open mind,

though, and change your opinions if someone makes a good argument. Give in once in a while!

- *Overwork:* One of the great challenges of college life is working in all of your new responsibilities. Try to create a schedule that has equal measure work andfun.
- *Roommate Situations:* When dealing with your roommate, it often helps to handle the situation more like a business relationship rather than a friendship. Follow the Golden Rule, and expect them to do so too.
- *Personal Problems:* Resolve conflicts when they arise. Seek help from a friend, your parents, a counselor, etc., if necessary, but before the problem gets out of hand.

To avoid stress in general, follow these tips:

- Do your very best on each task, within the appropriate time limitations and according to the relative importance of each task. To the best of your ability, concentrate on just one thing at a time.
- When the task is complete, evaluate its quality. What could you do differently next time?
- When you have done your very best work, take a well-deserved break.
- Never punish yourself for past faults; instead, praise yourself for things you did well, and vow to work on your areas that need improvement.
- Overprepare for exams or public speaking, so that you are ready for anything that might come up.
- Avoid stressful situations, when possible.
- If you do feel stressed, try talking about it to a friend, or taking a long walk or bath.
- Volunteer, or otherwise do nice things for other people.
- Make yourself available for other people to do nice things for you.

A.19 RELAXATION TECHNIQUES

The following techniques are good for relaxing and beating stress.

The first technique is called *deep breathing*. This is a great technique to get relaxed in any situation, including during a test or lecture.

1. Get comfortable and loosen your clothing.
2. Close your eyes.
3. Relax your upper body and shoulders.

4. Take a deep breath, filling your lungs as much as possible. As you breathe in, count to three, or say a calming word to yourself, such as "Relax" or "Calm."
5. Exhale, again counting to three, or repeating the calming word.
6. Repeat Steps 4 and 5 four more times.
7. Breathe normally and open your eyes.

The second technique is called *progressive relaxation*. It relaxes you to a greater degree than deep breathing alone.

1. Follow Steps 1 through 3 above.
2. Take a deep breath. Let it out slowly.
3. Repeat step 2 several times. Concentrate on slow and regular breathing throughout the exercise.
4. Tighten the muscles in your toes and hold for a count of ten. Release, feeling the tension flow out of your toes.
5. Repeat step 4, moving slowly up through your body: feet, legs, abdomen, back, neck, and face. The entire process should take about 15 minutes.
6. When you are done, breathe normally and open your eyes.

The third technique is called *visualization*. This method provides the deepest level of relaxation.

1. Follow Steps 1 through 3 for progressive relaxation.
2. Think of a special place. This can be a real place you have been, or an imaginary place you would like to go. Make sure you visualize the place in great detail.
3. Move slowly around in the space, looking at everything, and experiencing the surroundings with all of your senses. For example, if you are visualizing a beach, in your visualization you might see the white caps on the waves, while you listen to the gulls screeching in the distance, and feel the sand under your bare feet as you walk.
4. Keep the visualization as real and as pleasant as possible.
5. After several minutes of detailed visualization, bring yourself back to the present. Breathe regularly and open your eyes.

The last technique is good for dealing with sharp, sudden anxiety.

1. Breathe in deeply, until you can hold no more air.
2. Gasp in an additional short breath of air through your mouth.
3. Hold your breath for a count of 10.
4. Breathe out slowly.
5. Repeat Steps 1 through 3 at least seven times.

A.20 IMPROVING YOUR MEMORY

Remembering is a skill, not a talent, so *everyone* can learn to remember better. This section gives you some suggestions on how to improve your memory.

1. *Plan to remember.* Think about remembering *before* you need to remember; prepare your mind for the important information about to follow.

2. *Learn material the first time.* Study actively. Make the material familiar to you, as if you have always known it. Try these tips:

   - Learn from the general to the specific.
   - Relate the material to things that have meaning for you, or things that you already know.
   - Organize and consolidate the information.
   - Look for underlying principles rather than a list of small details.
   - Turn the concepts around in your mind. Look for different facets and aspects of the material that are not immediately obvious.

3. *Discuss.* The more you talk about a concept with other people, the easier it will be to recall.

4. *Apply the learning.* You will forget what you do not use. Do the homework problems. Study for the exam. Discuss the topic in class. In short, apply the learning to your life!

5. *Review.* Any chance you get, you should try to freshen up your understanding of the material. Not only will you solidify your memorization of the main points, but you will be able to add more and more detail each time you review.

6. *Memorize last.* When you have many projects to complete in one sitting, save your memorizing for last so that any new learning does not interfere with your recall ability.

7. *Use Mnemonics.* Tie all the topics you need to remember together using mnemonics, short but cute memory aids. For example:

   - Take the first letter of each concept and form a word or sentence. For example, the *acronym* WOC stands for the three states on the west coast from north to south (Washington, Oregon and California). The *acrostic* "Went Out for Cookies" stands for same thing.
   - Make up a song or poem with the information as lyrics. A familiar jingle, for example, is "i before e except after c."

- Associate the memorized word with another topic. For example, you might remember that the *Tropic of Capricorn* is the one below the equator, because *corns* grown on your *feet*.

8. If you need to memorize a long quotation, break it up into pieces, and memorize each piece separately. Once you have each piece memorized, string them together and memorize the whole.
9. *Use guided imagery*. Use the visual part of your brain to help you remember strings of information. For example:

   - Create a strange and unusual mental picture using each item on your list; the stranger the picture, the better you will recall the information. For instance, if you have to remember to pick up eggs and paper towels from the store, think of a picture in which the eggs smash themselves into tall and imposing paper towel rolls.
   - To memorize strings of numbers, assign each number to a word. For example, try this list:

one	bun	four	door	eight	plate
two	shoe	five	fish	nine	sign
three	tree	six	saxophone	zero	Nero
		seven	heaven		

   Then, to memorize the number, make a strange mental picture using these words—in the appropriate order. For example, to memorize the number pi, $\pi = 3.14...$, start with a mental picture of a big tree (3). Then, imagine yourself climbing up the tree to get a bun (1). Finally, imagine that you look at the bun, and realize there is a door (4) in it!

   - Think of a familiar place, and "walk" around it in your mind, always in the same order. For example, if you choose your house to visualize, you would always start at the entryway, then go to the living room, then go to the kitchen, and so on. Now, place each item in your list in a different location in the mental space. For instance, to remember your shopping list, you would place the eggs in the entryway, and the paper towels in the living room. When you get to the grocery store, remember your list by walking through your mental house, and picking up the items you "left" there!

A.21 IMPROVING YOUR CONCENTRATION

Concentration is a skill that can be improved with practice. These techniques will help.

- *Work in short blocks.* Most people can only concentrate for about an hour at a stretch. Therefore, you should schedule your study time in short blocks, rather than in long marathon sessions.
- *Cultivate a positive attitude.* There is nothing that will end your concentration faster than feeling bitter or upset. Think positively!
- *Make the material relevant.* The more you are interested in material, the more your concentration level will stay high. So make some effort to interest yourself in everything you study.
- *Have everything you need at your fingertips.* Be prepared! Put your pencil, paper, ruler, calculator, dictionary, and books all within arms reach, so that you do not have to interrupt yourself to get them.
- *Ignore distractions.* This is most easily done in quiet study spaces, like the library. But you can also create a quiet space at home, if you take care to turn off the music, log off of the computer, put away other reading materials, let the answering machine pick up your phone, and shut your door.
- *Deal with interruptions.* When you are distracted, or you start to daydream, the important thing is to *recognize* that you have lost your concentration. Then, remind yourself to get back to work. Don't berate yourself for lost time; instead, just get started again.
- *Deal with boredom.* If you find your mind wandering excessively, it may be because you are bored. Invent challenges to keep your concentration level high. For example, try writing 3 - 6 review questions about the material. Or, create a list of short term goals—each of length about 20 minutes—which will allow you to finish the boring material in small, manageable chunks.
- *Deal with your body's needs.* If you are truly hungry, eat. If you are truly sleepy, sleep. Also, exercise! Don't feel guilty about taking care of your legitimate bodily needs. But, be careful that you don't use these things as excuses to procrastinate.
- *Try relaxation or meditation.* The less stressed you are, the more you are capable of concentrating.

A.22 SETTING GOALS

Learning how to negotiate through all of the things you have to do in life is one of the most important lessons of the college experience. The skill of goal-setting helps you win this battle by providing direction, assigning priority, and organizing your work. This section will assist you in setting goals.

At its most basic level, a *goal* is just a statement about what you want to accomplish in a certain amount of time. A well-chosen goal is honest, realistic but challenging, and measurable:

- *Honest:* Goals should reflect what *you* need and value; not what other people think you need and value.
- *Realistic but Challenging:* Goals should not be so difficult to attain that you will never reach them; but they should also not be so easy to attain that you quickly achieve them and then become bored.
- *Measurable:* Goals should be stated in such a way that you can periodically evaluate your progress. You can only do this if your goal has a specific outcome after a specific amount of time. For example, a goal to "do better in Calculus" is not measurable in any concrete sense; a better goal would be to "get a B on my next Calculus test."

Once you have decided upon your goal, *write it down.* This will help to crystallize your thoughts. Phrase your goal in positive language: the idea here is to motivate yourself! Then post it somewhere that you see daily, like on your bathroom mirror.

The next step in the goal-setting process is to make a *plan* for achieving your goal. Ideally, you want to break the plan down into manageably small portions, so that each day, you can tackle a small part of the project.

Finally, you must *take action* to get your plan completed, and your goal achieved. Strive every day to meet your goal until you do. When you have met your goal, then, set another one!

Sometimes goals become obsolete through time. If you set a goal that no longer applies, then discard it without regret. But be sure and replace it with a new one!

A.23 LEARNING STYLES

Every person has a unique *learning style,* or consistent pattern of learning behavior that guides that person's ability to learn and accomplish tasks. There are three types of learning styles:

- *Visual*: The person learns best when *looking* at something;
- *Auditory*: The person learns best when *hearing* something;
- *Kinesthetic*: The person learns best when *touching* something or *doing* an activity.

Each person's learning style is a *mix* of these types, with one type dominant. The quiz at the end of this section will help you find your dominant learning style.

To maximize your study efficiency, then, you should match your study habits to your learning type. The following hints will help.

VISUAL LEARNERS: (Most people)

- The best type of studying for visual learners is reading.
- When you are done reading, try to visualize what you have learned in your mind. Be sure and check that you have it right!
- Pay close attention to charts, graphs, pictures, etc., in your lecture notes or textbook. Try making up your own when these are lacking.
- Rewrite your notes in a format which is easy to visualize: use indentation, color coding, underlining, highlighting, framing, capitalization, etc.
- Take mental pictures of experiments, board work, written materials, etc.
- Use flash cards to test yourself on spelling, facts, formulae, etc.

AUDITORY LEARNERS: (Most of the rest of people)

- The best type of studying for auditory learners is lecture. Make sure that you always attend yours!
- Read your material out loud whenever possible. Then cover it up and try to hear it in your mind. Be sure to check that you have it right!
- Read your notes into a tape recorder, then replay the tapes.
- Explain the material to a friend.

KINESTHETIC LEARNERS: (A few people)

- The best type of studying for kinesthetic learners is note-taking. Take notes in class and from your reading.
- Use manipulatives whenever possible. For example, volunteer to handle materials in your science classes.
- Volunteer to make posters, drawings, charts, etc., for class use, or make these for yourself.
- Use skits or creative movement to act out ideas you learn in class.

To determine your learning style, read the word in the left-hand column in Table A.3 below. Then answer the questions in the successive three columns, circling the one that best fits you. Your answers will fall into all three columns, but the column with the most circles is your dominant learning style.

Table A.3: Learning Styles

When you….	VISUAL	AUDITORY	KINESTHETIC
Spell	Do you try to see the word?	Do you sound out the word or use a phonetic approach?	Do you write the word down to find if it "feels" right?
Talk	Do you dislike listening for too long? Do you favor words like *see*, *picture*, and *imagine*?	Do you enjoy listening but are impatient to talk? Do you use words like *hear*, *tune*, and *think*?	Do you gesture and use expressive movements? Do you use words like *feel*, *touch*, and *hold*?
Concentrate	Do you become distracted by untidiness or movement?	Do you become distracted by sounds or noises?	Do you become distracted by activity around you?
Meet someone again	Do you forget names but remember faces or where people you have met?	Do you forget faces but remember names or remember what you talked about?	Do you remember best what you did together?
Contact people on business	Do you prefer direct, face-to-face, personal meetings?	Do you prefer the telephone?	Do you talk with people while walking or participating in an activity?
Read	Do you like descriptive scenes or pause in your reading to imagine the action?	Do you enjoy dialog and conversation or hear the characters talk?	Do you prefer action stories, or are you not a keen reader?
Do something new at work	Do you like to see demonstrations, diagrams, slides, or posters?	Do you prefer verbal instructions or talking about it with someone else?	Do you prefer to jump right in and try it?
Put something together	Do you look at the directions and the picture?	———	Do you ignore the directions and figure it out as you go?
Need help with a computer	Do you seek out pictures or diagrams?	Do you call the help desk, ask a neighbor, or growl at the computer?	Do you keep trying to do it, or try it on another computer?

BIBLIOGRAPHY

BOOKS REFERRED TO IN THE TEXT

Dwight, Herbert B. *Tables of Integrals and Other Mathematical Data.* (Macmillan).

Lide, David R. Jr. (ed.). *CRC Handbook of Chemistry and Physics, 79th Edition.* (CRC Press: 1998).

Taylor, John R. *An Introduction to Error Analysis: The Study of Uncertainties in Physical Measurements.* (University Science Books, 1997).

Tobias, Sheila. *Overcoming Math Anxiety.* (W.W. Norton, 1978, Houghton Mifflin paperback, 1980).

Tobias, Sheila. *Succeed With Math: Every Student's Guide to Conquering Math Anxiety.* (New York: The College Board, 1987).

Tobias, Sheila. *They're Not Dumb, They're Different: Stalking the Second Tier.* (Tucson, Arizona: Research Corporation, 1990).

Tobias, Sheila & Carl T. Tomizuka. *Breaking the Science Barrier: How to Explore and Understand the Sciences.* (New York: The College Board, 1992).

Pauk, Walter. *How To Study in College.* (Houghton Mifflin College: 1996).

SOURCES FOR FURTHER READING ON PHYSICS EDUCATION

Arons, Arnold B. "Student Patterns of Thinking and Reasoning," Parts 1 - 3, *The Physics Teacher, 21,* (1983), pp. 576-581; *22,* (1984), pp. 21-26; *22,* (1984), pp. 88-93.

Chi, M.T.H., P.J. Feltovich, & R. Glaser. "Categorization and Representation of Physics Problems by Experts and Novices," *Cognitive Science, 5,* (1981) pp. 121-152.

Clement, John. "Students' Preconceptions in Introductory Mechanics," *American Journal of Physics, 50,* (1982) pp. 66-71.

Hake, Richard R. "Promoting Student Crossover to the Newtonian World," *American Journal of Physics, 55,* (1987) pp. 878-884.

Halloun, I. & D. Hestenes. "The Initial Knowledge State of College Physics Students," and "Common Sense Concepts About Motion," *American Journal of Physics*, *53*, (1985) pp. 1043-1055 and 1056-1065.

Hestenes, D., M. Wells, & G. Swackhamer. "Force Concept Inventory," *The Physics Teacher, 30,* (March 1992) pp. 141-158.

Hestenes, D. and M. Wells. "A Mechanics Baseline Test," *The Physics Teacher, 30,* (March 1992) pp. 141-158.

Larkin, Jill H., John McDermott, Dorothea Simon, & Herbert A. Simon. "Expert and Novice Performance in Solving Physics Problems," *Science, 208,* (1980) pp. 1135-1342.

Mazur, Eric. *Peer Instruction: A User's Manual.* (Prentice Hall College Division: 1996).

Reif, F. & Heller, J.F. "Knowledge Structure and Problem Solving in Physics," *Educational Psychology, 17,* (1982) pp. 102-127.

Reif, F. "Teaching Problem Solving: A Scientific Approach," *The Physics Teacher, 19,* (1981) pp.310-316.

Seymour, Elaine & Nancy M. Hewitt. *Talking About Leaving: Why Undergraduates Leave the Sciences,* (Westview Press: 1997).

INDEX